MATHEMATICAL STUDIES STANDARD LEVEL

COURSE COMPANION

Peter Blythe
Jim Fensom
Jane Forrest
Paula Waldman de Tokman

OXFORD
UNIVERSITY PRESS

OXFORD
UNIVERSITY PRESS

Great Clarendon Street, Oxford OX2 6DP

Oxford University Press is a department of the University of
Oxford. It furthers the University's objective of excellence in
research, scholarship, and education by publishing worldwide in

Oxford   New York

Auckland   Cape Town   Dar es Salaam   Hong Kong   Karachi
Kuala Lumpur   Madrid   Melbourne   Mexico City   Nairobi
New Delhi   Shanghai   Taipei   Toronto

With offices in

Argentina   Austria   Brazil   Chile   Czech Republic   France
Greece   Guatemala   Hungary   Italy   Japan   Poland   Portugal
Singapore   South Korea   Switzerland   Thailand   Turkey
Ukraine   Vietnam

British Library Cataloguing in Publication Data

Data available

ISBN: 978-0-19-839013-8
10 9 8 7 6 5 4 3 2

Printed in Singapore by KHL Printing Co. Pte Ltd

Acknowledgments

The publishers would like to thank the following for permission
to reproduce photographs:

P3: PEKKA AHO/Associated Press; P20: kirych/Shutterstock; P22:
allOver photography/Alamy; P25: Ronald Sumners/Shutterstock;
P41: Christopher King/Dreamstime.com; P41: XYZ/Shutterstock;
P41: Ionia/Shutterstock; P43: Paul Brown/Rex Features; P45:
Gravicapa/Shutterstock; P45: Sergej Razvodovskij/Shutterstock;
P63: Stéphane Bidouze/Shutterstock; P69: Liv Falvey/Shutterstock;
P84: Paul Walters Worldwide Photography Ltd/Photo Library; P85:
David H.Seymour/Shutterstock; P85: SkillUp/Shutterstock; P85:
Nlshop/Shutterstock; P85: marina ljubanovic/Shutterstock; P87:
David Parker/Alamy; P130: Dietmar Höpfl/Shutterstock; P130:
pagadesign/istockphoto; P131: Professor Peter Goddardd/Science
Photo Library; P131: Dreamstime; P133: A777thunder; P165:
James Steidl/Shutterstock; P166: Tatiana53/Shutterstock; P166:
Hemera Technologies/Getty Images; P171: Smileus/Shutterstock;
P173: Dirk Ercken/Shutterstock; P173: Bradcalkin.../Dreamstime.
com; P174: Draghicich/Dreamstime.com; P175: sherpa/Shut-
terstock; P181: Yegor Korzh/Shutterstock; P183: dragon_fang/
Shutterstock; P201: NASA Archive; P203: Dmitrijs Dmitrijevs/
Shutterstock; P204: Zimmytws/Shutterstock; P214: Volosina/Shut-
terstock; P215: Elena Elisseeva/Shutterstock; P223: pandapaw/
Shutterstock; P224: Science Photo Library; P227: Lakhesis/shutter-
stock; P230: paul prescott /Shutterstock; P239: Erik Lam/Shutter-
stock; P241: Rakov Studio/Shutterstock; P252: Magalí Izaguirre/
Istock; P252: Maxx-Studio/Shutterstock; P225: italianestro/shut-
terstock; P278: ruzanna/Shutterstock; P293: Dmitry Rukhlenko/
Dreamstime.com; P293: Paul Wootton/Science Photo Library;
P292: Eugene Sim/Shutterstock; P293: PixAchi/Shutterstock;

P292: Jessmine/Shutterstock; P295: Annabelle496/Dreamstime.
com; P303: Rui Matos/Dreamstime.com; P304: Slidepix/Dream-
stime.com; P306: negative/Shutterstock; P308: Oleksandr Pekur/
Dreamstime.com; P310: Tupungato/Dreamstime.com; P312: Anna
Dudek/Dreamstime.com; P320: Stuart Key/Dreamstime.com;
P327: Seymour/Science Photo Library; P326: MoonBloom/Shut-
terstock; P327: Christian Delbert/Shutterstock; P327: GoodMood
Photo/Shutterstock; P329: Badzmanaoi.../Dreamstime.com; P350:
negative/Shutterstock; P352: Tatiana Popova/Shutterstock; P352:
Sinelyov/Shutterstock; P355: Roman Sigaev/Shutterstock; P361:
Sinelyov/Shutterstock; P365: grum_l/Shutterstock; P378: M&N/
Alamy; P379: Peter E Noyce/Alamy; P379: Tele52/Dreamstime.
com; P378: Oleksiy Mark/Shutterstock; P381: Comstock/Think-
stock; P403: Olga Utlyakova/Shutterstock; P419: FromOldBooks.
org/Alamy; P418: Briangoff/Dreamstime.com; P418: TerryM/
Shutterstock; P418: Bomshtein/Shutterstock; P419: Zack Clothier/
Shutterstock; P419: Anton Brand/Shutterstock; P421: Ahmet
Ihsan Ariturk/Dreamstime.com; P423: Sunnyi/Dreamstime.
com; P429: Sunnyi/Dreamstime.com; P452: Simon Colmer and
Abby Rex/Alamy; P452: Photo Researchers/Alamy; P452: Carlos
Caetano/Shutterstock; P452: Picsfive/Shutterstock; P520: Karin
Hildebrand Lau/Shutterstock; P524: Reeed/Shutterstock; P518:
De Agostini/Getty Images; P533: Science Source/Science Photo
Library; P539: Georgios Kollidas/Shutterstock.

Cover Image: JS. Sira/Photolibrary.

Every effort has been made to contact copyright holders of
material reproduced in this book. If notified, the publishers
will be pleased to rectify any errors or omissions at the earliest
opportunity.

Course Companion definition

The IB Diploma Programme Course Companions are resource materials designed to provide students with support through their two-year course of study. These books will help students gain an understanding of what is expected from the study of an IB Diploma Programme subject.

The Course Companions reflect the philosophy and approach of the IB Diploma Programme and present content in a way that illustrates the purpose and aims of the IB. They encourage a deep understanding of each subject by making connections to wider issues and providing opportunities for critical thinking.

The books mirror the IB philosophy of viewing the curriculum in terms of a whole-course approach; the use of a wide range of resources; international-mindedness; the IB learner profile and the IB Diploma Programme core requirements; theory of knowledge, the extended essay, and creativity, action, service (CAS).

Each book can be used in conjunction with other materials and indeed, students of the IB are required and encouraged to draw conclusions from a variety of resources. Suggestions for additional and further reading are given in each book and suggestions for how to extend research are provided.

In addition, the Course Companions provide advice and guidance on the specific course assessment requirements and also on academic honesty protocol.

IB mission statement

The International Baccalaureate aims to develop inquiring, knowledgable and caring young people who help to create a better and more peaceful world through intercultural understanding and respect.

To this end the IB works with schools, governments and international organizations to develop challenging programmes of international education and rigorous assessment.

These programmes encourage students across the world to become active, compassionate, and lifelong learners who understand that other people, with their differences, can also be right.

The IB learner profile

The aim of all IB programmes is to develop internationally minded people who, recognizing their common humanity and shared guardianship of the planet, help to create a better and more peaceful world. IB learners strive to be:

Inquirers They develop their natural curiosity. They acquire the skills necessary to conduct inquiry and research and show independence in learning. They actively enjoy learning and this love of learning will be sustained throughout their lives.

Knowledgable They explore concepts, ideas, and issues that have local and global significance. In so doing, they acquire in-depth knowledge and develop understanding across a broad and balanced range of disciplines.

Thinkers They exercise initiative in applying thinking skills critically and creatively to recognize and approach complex problems, and make reasoned, ethical decisions.

Communicators They understand and express ideas and information confidently and creatively in more than one language and in a variety of modes of communication. They work effectively and willingly in collaboration with others.

Principled They act with integrity and honesty, with a strong sense of fairness, justice, and respect for the dignity of the individual, groups, and communities. They take responsibility for their own actions and the consequences that accompany them.

Open-minded They understand and appreciate their own cultures and personal histories, and are open to the perspectives, values, and traditions of other individuals and communities. They are accustomed to seeking and evaluating a range of points of view, and are willing to grow from the experience.

Caring They show empathy, compassion, and respect towards the needs and feelings of others. They have a personal commitment to service, and act to make a positive difference to the lives of others and to the environment.

Risk-takers They approach unfamiliar situations and uncertainty with courage and forethought, and have the independence of spirit to explore new roles, ideas, and strategies. They are brave and articulate in defending their beliefs.

Balanced They understand the importance of intellectual, physical, and emotional balance to achieve personal well-being for themselves and others.

Reflective They give thoughtful consideration to their own learning and experience. They are able to assess and understand their strengths and limitations in order to support their learning and personal development.

A note on academic honesty

It is of vital importance to acknowledge and appropriately credit the owners of information when that information is used in your work. After all, owners of ideas (intellectual property) have property rights. To have an authentic piece of work, it must be based on your individual and original ideas with the work of others fully acknowledged. Therefore, all assignments, written or oral, completed for assessment must use your own language and expression. Where sources are used or referred to, whether in the form of direct quotation or paraphrase, such sources must be appropriately acknowledged.

How do I acknowledge the work of others?

The way that you acknowledge that you have used the ideas of other people is through the use of footnotes and bibliographies.

Footnotes (placed at the bottom of a page) or endnotes (placed at the end of a document) are to be provided when you quote or paraphrase from another document, or closely summarize the information provided in another document. You do not need to provide a footnote for information that is part of a "body of knowledge". That is, definitions do not need to be footnoted as they are part of the assumed knowledge.

Bibliographies should include a formal list of the resources that you used in your work. "Formal" means that you should use one of the several accepted forms of presentation. This usually involves separating the resources that you use into different categories (e.g. books, magazines, newspaper articles, Internet-based resources, CDs and works of art) and providing full information as to how a reader or viewer of your work can find the same information. A bibliography is compulsory in the extended essay.

What constitutes malpractice?

Malpractice is behavior that results in, or may result in, you or any student gaining an unfair advantage in one or more assessment component. Malpractice includes plagiarism and collusion.

Plagiarism is defined as the representation of the ideas or work of another person as your own. The following are some of the ways to avoid plagiarism:

- Words and ideas of another person used to support one's arguments must be acknowledged.
- Passages that are quoted verbatim must be enclosed within quotation marks and acknowledged.
- CD-ROMs, email messages, web sites on the Internet, and any other electronic media must be treated in the same way as books and journals.
- The sources of all photographs, maps, illustrations, computer programs, data, graphs, audio-visual, and similar material must be acknowledged if they are not your own work.
- Works of art, whether music, film, dance, theatre arts, or visual arts, and where the creative use of a part of a work takes place, must be acknowledged.

Collusion is defined as supporting malpractice by another student. This includes:

- allowing your work to be copied or submitted for assessment by another student
- duplicating work for different assessment components and/or diploma requirements.

Other forms of malpractice include any action that gives you an unfair advantage or affects the results of another student. Examples include, taking unauthorized material into an examination room, misconduct during an examination, and falsifying a CAS record.

Contents

About the book

The new syllabus for Mathematical Studies SL is thoroughly covered in this book. It is written by educators who were involved in the latest curriculum review. Each chapter is divided into lesson size chunks with the following features:

Investigations   Exploration suggestions

Examiner's tip   **Theory of Knowledge**

Did you know?   Historical exploration

It is intended for you to be able to navigate through the book in whatever order you choose. Before each chapter there is a short exercise on what a student should know before starting the chapter, and there is a chapter on prior knowledge. There are exam-style questions throughout and full solutions to these can be found on the website. There are final solutions to all the exercises at the end of the book.

The GDC chapter and the GDC screen shots throughout the book are from the TI-Nspire – there are also screen shots from the TI-84 Plus and Casio FX-9860GII GDCs on the CD-ROM. Questions that require a GDC have a calculator icon included beside them.

It is important to differentiate in the classroom. To help teachers with this, the authors have written the questions in each exercise to range from easy to difficult. There is extra extension work included on the CD. Some of the extension work will also be useful to the students when writing their projects. In order to gain full marks for the mathematical processes criterion, the calculations need to be done by hand. In the extension material this is clearly laid out.

There is also a chapter on the new assessment criteria for the project, along with hints for writing a good project. On the CD there are also some projects that the students can moderate to reinforce their understanding of the new criteria.

At the end of each chapter there is a summary of the most relevant skills that the student has learned in the chapter. This is followed by some interesting TOK pages to make students stop and think.

The language throughout is simple, concise and clear with international contexts that are interesting and relevant.

Note: US spelling has been used, with IB style for mathematical terms.

About the authors

Peter Blythe has been teaching the four IB Diploma Programme Mathematics courses for 25 years. He currently teaches at the United World College of S. E. Asia and is a deputy chief examiner for Mathematical Studies SL.

Jim Fensom has been teaching IB Mathematics courses for nearly 35 years. He is currently Mathematics Coordinator at Nexus International School in Singapore. He is an assistant examiner for Mathematics HL.

Jane Forrest has been teaching Mathematics for over 30 years. She is currently Head of School at Rotterdam International School in the

Netherlands. She was deputy chief examiner for Mathematical Studies SL for 5 years and is principal moderator for projects.

Paula Waldman de Tokman has been teaching Mathematics for over 20 years. She was a deputy chief examiner for Mathematical Studies for 6 years. She currently teaches the IB Mathematics courses at St. Andrews Scots School in Buenos Aires, Argentina.

Additional contributions for TOK sections from Paul La Rondie and all authors of *Mathematics Standard Level Course Companion*.

What's on the CD?

The material on your CD-ROM includes the entire student book as an eBook, as well as a wealth of other resources specifically written to support your learning. On these two pages you can see what you will find and how it will help you to succeed in your Mathematical Studies course.

The whole print text is presented as a user-friendly eBook for use in class and at home.

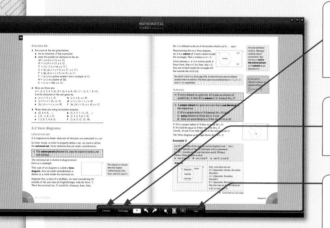

Extra content can be found in the Contents menu or attached to specific pages. This icon appears in the book wherever there is extra content.

Navigation is straightforward either through the Contents Menu, or through the Search and Go to page tools.

A range of tools enables you to zoom in and out and to annotate pages with your own notes.

The glossary provides comprehensive coverage of the language of the subject and explains tricky terminology. It is fully editable making it a powerful revision tool.

Extension material is included for each chapter containing a variety of extra exercises and activities. Full worked solutions to this material are also provided.

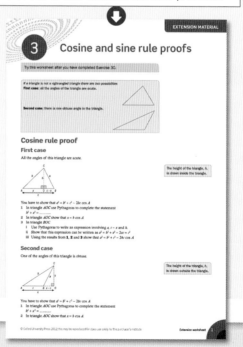

Glossary

Term	Definition	Notes
°C	Degrees Celsius, a unit of measurement for temperature. To convert to degrees fahrenheit, $°F = °C \times \frac{9}{5} + 32$	
absolute or global extrema	The highest or lowest value that a function can take	
acceleration	Rate of change of velocity	
accepted value	A value used when the exact value is not known	
acute	An acute angle has a measurement of less than 90 degrees	
adjacent (a)	The side in a right-angled triangle next to a given angle	
algebra	The study of operations and relations	
algebraic function	A function consisting of variables and rational coefficients	
alternate	Equal angles formed on opposite sides of a line that crosses two parallel lines, for example the inner angles of a Z	
alternative (H_1) hypothesis	This is what you accept if the observed value is a rare event when the null hypothesis is true	
altitude	Height. In a triangle, this is the perpendicular distance from the base to the apex	
ambiguous case (of triangles)	Given the size of one angle and the lengths of two sides a unique triangle can not be drawn	
amplitude	Half the distance between the minimum and maximum values of the range of a periodic function	
angle of depression	The angle formed below the horizontal to an object	
angle of elevation	The angle formed above the horizontal to the top of an object	
antecedent	The initial part , or cause, of an argument	

EXTENSION MATERIAL

3   Cosine and sine rule proofs

Try this worksheet after you have completed Exercise 3O.

If a triangle is not a right-angled triangle there are two possibilities
First case: all the angles of the triangle are acute.

Second case: there is one obtuse angle in the triangle.

Cosine rule proof
First case
All the angles of this triangle are acute.

The height of the triangle, h, is drawn inside the triangle.

You have to show that $a^2 = b^2 + c^2 - 2bc \cos A$
1   In triangle AOC use Pythagoras to complete the statement
$h^2 + x^2 = \_\_\_\_$
2   In triangle AOC show that $x = b \cos A$
3   In triangle BOC
  i   Use Pythagoras to write an expression involving a, $c - x$ and h.
  ii   Show that this expression can be written as $a^2 = b^2 + c^2 - 2xc + x^2$
  iii   Using the results from **1**, **2** and **3** show that $a^2 = b^2 + c^2 - 2bc \cos A$

Second case
One of the angles of this triangle is obtuse.

The height of the triangle, h, is drawn outside the triangle.

You have to show that $a^2 = b^2 + c^2 - 2bc \cos A$
1   In triangle AOC use Pythagoras to complete the statement
$h^2 + x^2 = \_\_\_\_$
2   In triangle AOC show that $x = b \cos A$

Extension worksheet

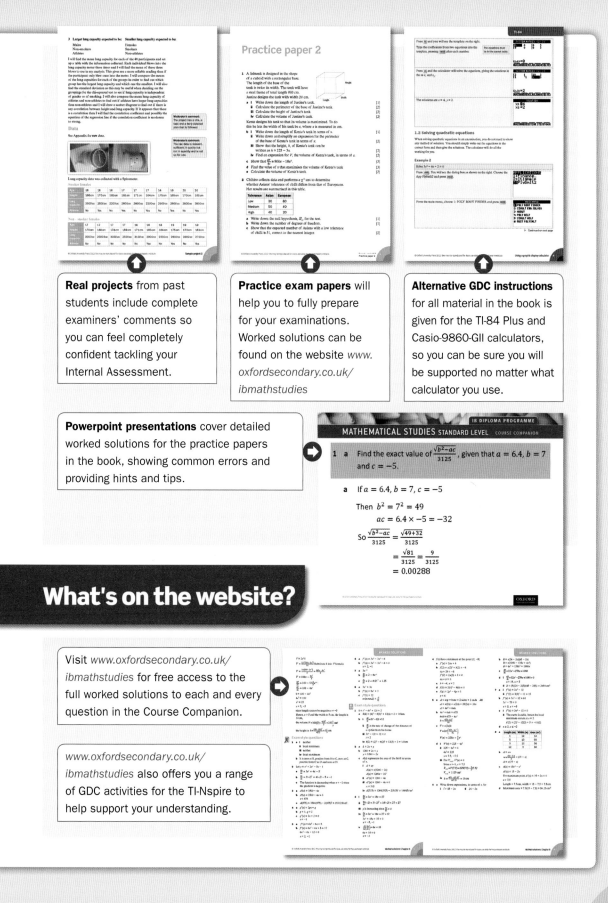

Real projects from past students include complete examiners' comments so you can feel completely confident tackling your Internal Assessment.

Practice exam papers will help you to fully prepare for your examinations. Worked solutions can be found on the website *www.oxfordsecondary.co.uk/ibmathstudies*

Alternative GDC instructions for all material in the book is given for the TI-84 Plus and Casio-9860-GII calculators, so you can be sure you will be supported no matter what calculator you use.

Powerpoint presentations cover detailed worked solutions for the practice papers in the book, showing common errors and providing hints and tips.

What's on the website?

Visit *www.oxfordsecondary.co.uk/ibmathstudies* for free access to the full worked solutions to each and every question in the Course Companion.

www.oxfordsecondary.co.uk/ibmathstudies also offers you a range of GDC activities for the TI-Nspire to help support your understanding.

1 Number and algebra 1

Before you start

You should know how to:

1 Substitute into formulae, e.g.
G and F are linked through the formula
$G = \dfrac{F-1}{\sqrt{F+2}}$. Find the value of G when
$F = 98$. $G = \dfrac{98-1}{\sqrt{98+2}} = 9.7$

2 Solve simple equations in one variable, e.g.

a $2x - 8 = 10$
$2x = 18$
$x = 9$

b $x^2 = 25$
$x = 5$ or $x = -5$

3 Calculate percentages, e.g.
Calculate 5% of 240. $\dfrac{5}{100} \times 240 = 12$

4 Solve inequalities and represent the solution on the number line, e.g.
$2x + 7 \le 10$
$2x \le 3$
$x \le 1.5$

5 Calculate the absolute value of a number,
e.g. $|2.5| = 2.5$, $|-1.3| = 1.3$,
$|0| = 0$, $|5 - 10| = 5$.

Skills check

1 Find the value of y when $x = -0.1$ if x and y are linked through the formula

a $y = 3x^2(x - 1)$ **b** $y = \dfrac{(x-1)^2}{x}$
c $y = (1 - x)(2x + 1)$.

2 Solve for x.

a $3x - 7 = 14$ **b** $2(x - 6) = 4$
c $\dfrac{1}{2}(1-x) = 0$ **d** $x^2 = 16$

3 Calculate

a 8% of 1200   **b** 0.1% of 234.

4 Solve the following inequalities. Represent their solutions on the number line.

a $10 - x \le 1$ **b** $3x - 6 > 12$
c $2x \le 0$

5 Calculate

a $|-5|$ **b** $\left|\dfrac{1}{2}\right|$

c $|5 - 7|$ **d** $\left|\dfrac{12-8}{8}\right| \times 100$

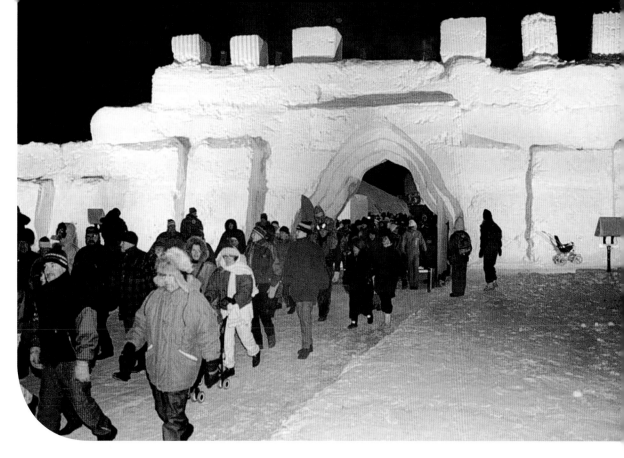

- The castle is 100 km south of the Arctic Circle.
- It takes approximately 6 weeks to build.
- The temperature has to be no more than −8 °C to prevent it melting.
- The castle's area varies each year. So far it has ranged from 13 000 to 20 000 m².
- Approximately 300 000 people from around the world visited the castle when it was first open.
- The castles have had towers taller than 20 m and walls longer than 1000 m.

▲ This is the biggest snow castle in the world, in northern Finland. First built in 1996, it has been rebuilt every winter when there has been enough snow.

These facts and figures about the snow castle use different types of number and different types of unit. Some are approximate values.

This chapter will help you to classify numbers, round numbers and make approximations, as well as showing you how to write very large or very small numbers in standard form, and convert between different units of measurement.

1.1 The number sets

These expressions use several different types of number.

- In Finland the lowest temperature in winter is around −45 °C.
- In 2010 unemployment in Ireland was more than 13%.
- Approximately $\frac{4}{5}$ of the world's population has a mobile or cell phone.
- Usain Bolt won the men's 100 metres at the 2008 Olympic Games with a world record time of 9.69 seconds.
- The area of a circle with a radius of 1 cm is π cm².

The numbers 60, -45, $\frac{1}{3}$, 9.69 and π belong to different **number sets**, which are described over the next few pages.

At the end of this section you will be able to classify them as elements of these sets.

Natural numbers, $\mathbb{N}$

> → The set of **natural numbers** $\mathbb{N}$ is $0, 1, 2, 3, 4, \ldots$

We use these numbers

- *to count*: for example '205 nations are expected to take part in the 2012 Olympic Games'
- *to order*: for example 'The Congo rainforest is the 2nd largest in the world'

You can represent the natural numbers on the number line by setting an **origin** and a **unit**.

We write $\mathbb{N} = \{0, 1, 2, 3, 4, 5, \ldots\}$
The curly brackets enclose the elements of a set.

Example 1

a Find the value of these expressions when $a = 5$ and $b = 7$.
  i $a + b$ **ii** $a \times b$ **iii** $a - b$ **iv** $b - a$
b State whether your answers to part **a** are natural numbers or not.

There are as many natural numbers as even numbers.

Answers
a **i** $5 + 7 = 12$ **ii** $5 \times 7 = 35$ **iii** $5 - 7 = -2$ **iv** $7 - 5 = 2$
b **i** natural **ii** natural **iii** not natural **iv** natural

Remember that the negative numbers are not in $\mathbb{N}$.

Exercise 1A

a Find the value of these expressions when $a = 2$ and $b = 4$.
  i $2a + b$ **ii** $2(a + b)$ **iii** $a^2 - b^2$ **iv** $(a - b)^2$
b State whether your answers to part **a** are natural numbers or not.

Investigation – natural numbers

State whether each statement is true or false. If it is false, give an example to show why.

a True or false? *Whenever you add two natural numbers the* **sum** *will be a natural number.*

If $a + b = c$, we say that c is the sum of a and b.

b True or false? *Whenever you multiply two natural numbers, the* **product** *will be a natural number.*

If $a \times b = c$, we say that c is the product of a and b.

c True or false? *Whenever you subtract two natural numbers the* **difference** *will be a natural number.*

If $a - b = c$, we say that c is the difference of a and b.

The set of integers, ℤ

In Example 1 you saw that the difference of two natural numbers is *not always* a natural number. So we need a new set as there are quantities that cannot be represented with natural numbers. The new set is ℤ, the set of integers.

ℤ is an extension of ℕ.

→ The set of **integers** ℤ is $\{\ldots, -4, -3, -2, -1, 0, 1, 2, 3, 4, \ldots\}$

Any natural number is also an integer but not all integers are natural numbers.

You can represent ℤ on the number line like this:

On this number line
● positive integers are placed to the right of zero
● negative integers are placed to the left of zero.
● Zero is neither positive nor negative.

Example 2

Solve each equation for x. State whether the solution to the equation is an integer or not.

a $x + 5 = 11$ **b** $-3x = 10$

Answers

a $x + 5 = 11$
$x = 6$ x is an integer.

b $-3x = 10$
$x = \dfrac{-10}{3}$ x is not an integer.

We use negative numbers to represent many everyday situations.
List at least three.

Example 3

a Find the value of the following expressions when $j = 4$ and $k = -2$.

i $\dfrac{5k - j}{k + j}$ **ii** $\dfrac{j^2 - k}{j^2 + 2k}$

b State whether your answers to part **a** are integers or not.

Brahmagupta lived from 589 to 669 CE in India. He is credited with writing the first book that included zero and negative numbers.

Answers

a i $\dfrac{5(-2) - 4}{-2 + 4} = \dfrac{-14}{2} = -7$

Write the expressions, substituting the numbers for the letters.

ii $\dfrac{4^2 - (-2)}{4^2 + 2(-2)} = 1.5$

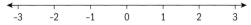

You can use your GDC to evaluate this.

When using your GDC to input fractional expressions, remember to use brackets to indicate clearly the numerator and the denominator, or use the fraction template.

b i integer **ii** not an integer

GDC help on CD: *Alternative demonstrations for the TI-84 Plus and Casio FX-9860GII GDCs are on the CD.*

Exercise 1B

1 a Solve the equation $4x + 2 = 0$.
  b State whether or not your solution to part **a** is an integer.

2 a Solve the equation $x^2 = 4$.
  b State whether or not your solutions to part **a** are integers.

3 a Find the value of these expressions when $a = -2$ and $b = 4$.

   i $\dfrac{a-b}{a+b}$ **ii** $3a^2 - \dfrac{9}{b}$

  b State whether or not your answers to part **a** are integers.

Investigation – integers

State whether each of these statements is true or false. If false, give an example to show why.

a The **sum** of two integers is always an integer.
b The **difference** of two integers is always an integer.
c The **quotient** of two integers is always an integer.
d The **product** of two integers is always an integer.

> If $\dfrac{a}{b} = c$ then we say that c is the **quotient** of a and b. Quotient means **ratio**.

The set of rational numbers, $\mathbb{Q}$

In the investigation you should have found that the quotient of two integers is not always an integer. So we need a new set as there are quantities that cannot be represented with integers. This set is $\mathbb{Q}$, the set of rational numbers.

> $\mathbb{Q}$ is an extension of $\mathbb{Z}$.

> → The set of **rational numbers** $\mathbb{Q}$ is
> $$\left\{ \frac{p}{q} \text{ where } p \text{ and } q \text{ are integers and } q \neq 0 \right\}$$

> Note that $q \neq 0$ as division by 0 is not defined.

This definition means that a number is rational if it can be written as the quotient of two integers. Here are examples of rational numbers.

> **The decimal expansion of a rational number** may have a finite number of decimal places (for example −1.5) or may recur (for example $0.\dot{6}$). A number with recurring digits has a **period**, which is the digit or group of digits that is repeated after the decimal point. For example, the period of 0.66666... is 6 and the period of 0.767676... is 76.

- 7 is a rational number as it can be written as $\dfrac{7}{1}$, and both 7 and 1 are integers.
- −3 is a rational number as it can be written as $\dfrac{-3}{1}$, and both −3 and 1 are integers.
- 0 is a rational number as it can be written as $\dfrac{0}{4}$, and both 0 and 4 are integers.
- −1.5 is a rational number as it can be written as $\dfrac{-3}{2}$, and both −3 and 2 are integers.
- $0.\dot{6} = 0.666...$ is a rational number as it can be written as $\dfrac{6}{9}$, and both 6 and 9 are integers.

From these examples we can see that any integer is also a rational number but not all rational numbers are integers. You can represent some of the rational numbers on the number line like this:

Find out more about the history of rational numbers on pages 40–41.

Example 4

a Express $1.\dot{3}$ as a fraction.

b Hence calculate $1.\dot{3} + \dfrac{4}{5}$

Give your answer as a fraction.

'**Hence**' is a command term that is frequently used in exams. If you read 'hence' then try to use the preceding work to find the required result.

Answers

a Let $a = 1.\dot{3}$ then

$$a = 1.3333\ldots$$
$$10a = 13.333\ldots$$

Multiply by 10 to obtain another number with the same period.

$$10a - a = 13.333\ldots - 1.3333\ldots$$

Subtract a from $10a$.

$$= 12$$
$$9a = 12$$

Divide both sides by 9.

$$a = \frac{12}{9} = \frac{4}{3}$$

Simplify to its simplest form.

b $1.\dot{3} + \dfrac{4}{5} = \dfrac{4}{3} + \dfrac{4}{5} = \dfrac{32}{15}$

Use a common denominator of 15 or your GDC.

Exercise 1C

1 a Find the decimal expansion of these fractions.

$$\frac{2}{3} \qquad -\frac{5}{4} \qquad \frac{2}{9} \qquad \frac{4}{7} \qquad \frac{-11}{5}$$

b For each fraction in **a**, state whether its decimal expansion
  i is finite **ii** recurs.

$\dfrac{2}{3} \longrightarrow 2 \div 3$, use your GDC.

2 a Express $0.\dot{5}$ as a fraction. **b** Express $1.\dot{8}$ as a fraction.
  c Hence calculate $0.\dot{5} + 1.\dot{8}$. Give your answer as a fraction.

3 a Write down a rational number whose decimal expansion is finite.
  b Write down a rational number whose decimal expansion recurs.
  c Write down a rational number whose decimal expansion has a period that starts in the fourth digit after the decimal point.

For any pair of rational numbers, you can always find a rational number that lies between them on the number line. For example, the **arithmetic mean** of two numbers is halfway between those numbers.

Express $1.\dot{9}$ as a fraction. What do you notice? Is it true that $1.\dot{9} = 2$?

Example 5

a Write down a rational number that lies on the number line between $\frac{2}{3}$ and 1.

b Write down a second rational number that lies on the number line between $\frac{2}{3}$ and 1.

c Write down a third rational number that lies on the number line between $\frac{2}{3}$ and 1.

'Write down' is a command term that means you don't need to show much or any working.

Answers

a $\dfrac{\frac{2}{3}+1}{2}=\dfrac{5}{6}$

Find the arithmetic mean of $\frac{2}{3}$ and 1. Use your GDC to simplify it.

b $\dfrac{\frac{2}{3}+\frac{5}{6}}{2}=\dfrac{3}{4}$

How many rational numbers are there between two rational numbers?

c $\dfrac{\frac{2}{3}+\frac{3}{4}}{2}=\dfrac{17}{24}$

→ A number is rational if
- it can be written as a quotient of two integers, or
- its decimal expansion is finite, or
- its decimal expansion is non-terminating but has a recurring digit or pattern of digits.

'Non-terminating' is the opposite of 'finite'.

Example 6

For each of the expressions **a** $(x+y)^2$ **b** $\sqrt{\dfrac{x+5}{y}}$

i Calculate the value when $x=-4$ and $y=\dfrac{1}{2}$.

ii State whether your answers to **i** are rational numbers or not. Justify your answer.

Answers

a i $\left(-4+\dfrac{1}{2}\right)^2=\left(-\dfrac{7}{2}\right)^2=\dfrac{49}{4}$

ii It is a rational number as it can be written as the quotient of two integers.

To justify your answer, explain how you know it is rational.

b i $\sqrt{\dfrac{-4+5}{\frac{1}{2}}}=\sqrt{\dfrac{1}{\frac{1}{2}}}=\sqrt{2}$

ii It is not a rational number. Its decimal expansion is 1.4142135... It does not have a finite number of decimal places and does not recur.

Exercise 1D

1 Write down three rational numbers that lie on the number line between 2 and $\frac{9}{4}$.

2 **a** Calculate the value of the expression $\sqrt{2(y - x)}$ when $y = 3$ and $x = -\frac{1}{8}$.

  b State whether your answer to part **a** is a rational number or not.

3 **a** Write down three rational numbers between $\frac{9}{5}$ and $\frac{11}{6}$.

  b i Write down three rational numbers between $-\frac{28}{13}$ and -2.

  ii How many rational numbers are there between $-\frac{28}{13}$ and -2?

Investigation – rational numbers

State whether each of these statements is true or false. If a statement is false, explain why by giving an example.

a The **difference** of two rational numbers is always a rational number.

b The **square** of a rational number is always a rational number.

c The **quotient** of two rational numbers is sometimes a rational number.

d The **square root** of a rational number is always a rational number.

The set of real numbers, $\mathbb{R}$

In the investigation you should have found that the square root of a rational number is not always a rational number. So we need a new set, as there are quantities that cannot be represented with rational numbers. For example, we could think of a circle with radius 1 cm.

What is the area, A, of this circle?

$A = \pi \times r^2$

$A = \pi \times (1 \, \text{cm})^2$

$A = \pi \, \text{cm}^2$

1 cm

Is π a rational number? The decimal expansion of π from the GDC is 3.141592654 – but these are just the first nine digits after the decimal point.

The decimal expansion of π has an infinite number of digits after the decimal point, and no **period** (no repeating pattern).

You can find the first ten thousand digits of π from this website: http://www.joyofpi.com/pi.html.

→ Any number that has a decimal expansion with an infinite number of digits after the decimal point and no period is an **irrational number**.

Irrational numbers include, for example, $\pi, \sqrt{2}, \sqrt{3}$.

> → The set of rational numbers together with the set of irrational numbers complete the number line and form the set of **real numbers**, $\mathbb{R}$.

How many real numbers are there? Can we count them?

Natural numbers $\mathbb{N}$

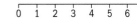

Integers $\mathbb{Z}$

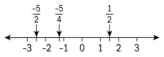

Rationals $\mathbb{Q}$

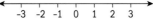

Real numbers $\mathbb{R}$ complete the number line

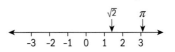

On March 14 (or, in month/day format, 3/14) a lot of people around the world celebrate **Pi Day**, as 3, 1 and 4 are the three most significant digits of π. Also March 14 is Albert Einstein's birthday so sometimes both events are celebrated together. **Pi Approximation Day** is July 22, or in the day/month format 22/7, which is an approximation to the value of π.

Example 7

Calculate each of these measurements and state whether it is rational or irrational.
a The length l of a diagonal of a square with side length of 1 cm.
b The area A of a circle with radius $\dfrac{1}{\sqrt{\pi}}$ cm.

1 cm

Answers

a $l^2 = 1^2 + 1^2$ *Use Pythagoras' theorem.*
$l^2 = 2$
$l = \sqrt{2}$ $\sqrt{2} = 1.4142...$
$\sqrt{2}$ is an irrational number   *It is not finite, not recurring.*

b $A = \pi r^2$ *Use the formula for the area of a circle.*
$A = \pi \times \left(\dfrac{1}{\sqrt{\pi}}\right)^2 = \pi \times \dfrac{1}{\pi}$

$A = 1 \text{ cm}^2$
1 is a rational number

Exercise 1E

1 a Calculate the length, h, of the hypotenuse of a right-angled triangle with sides 2 cm and 1.5 cm.
b State whether h is rational or irrational.

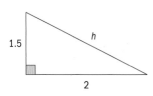

2 a Calculate the area, A, of a circle with diameter 10 cm.
b State whether A is rational or irrational.

Example 8

a Solve this inequality and represent the solution on the number line.
$8 + x > 5$

b State whether $p = -\pi$ is a solution to the inequality given in part **a**.

Answers

a $8 + x > 5$

$x > -3$

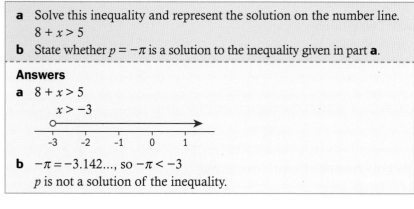

b $-\pi = -3.142...$, so $-\pi < -3$

p is not a solution of the inequality.

Do we all use the same notation in mathematics?

We are using an empty dot to indicate that $x = -3$ is *not* included. Different countries have different notations to represent the same thing. Furthermore, different teachers from the same country use different notations!

Exercise 1F

1 a Solve these inequalities.

    i $0.5 < \dfrac{x}{2} \leq 1.5$ **ii** $3 - x \geq 1$

b Represent the solution to part **a** on the number line.

c State whether the numbers $q = 1.5$ and $t = \sqrt{5}$ are solutions to the inequalities given in part **a**.

2 a Solve these inequalities.

    i $2x + 1 > -1$ **ii** $4 \leq x + 1 \leq 8$ **iii** $2 - x > -1$

b Represent the solutions to part **a** on the number line.

c Copy and complete this table. Put a ✓ if the number p is a solution to the inequality given.

Inequality / p	$2x + 1 > -1$	$4 \leq x + 1 \leq 8$	$2 - x > -1$
$\dfrac{-2}{3}$			
$\sqrt{10}$			
2π			

1.2 Approximations and error

It is important that you understand the difference between an **exact value** and an **approximate value**.

Sometimes, as in the following examples, we approximate a quantity because the exact values are not known (maybe because the instrument we use to take the measurements only reaches a certain accuracy).

- The approximate area of Ecuador is $283\,561$ km².
- The present height of the Great Pyramid of Giza is approximately 138.8 m.
- The weight of an apple is approximately 250 g.

Sometimes we approximate a quantity because we don't need the exact value, as in the following examples.

- India's population is about 1 800 000 000.
- I run for about 3 hours every Sunday.
- China's economy grew at an average rate of 10% per year during the period 1990–2004.

Rounding a number is the process of approximating this number to a given degree of accuracy.

Rounding numbers to the nearest unit, nearest 10, nearest 100, nearest 1000, etc.

> → Rounding a number to the **nearest 10** is the same as rounding it to the **nearest multiple of 10**.
>
> Rounding a number to the **nearest 100** is the same as rounding it to the **nearest multiple of 100**.

To round 3746 to the nearest hundred:

To round 81 650 to the nearest thousand:

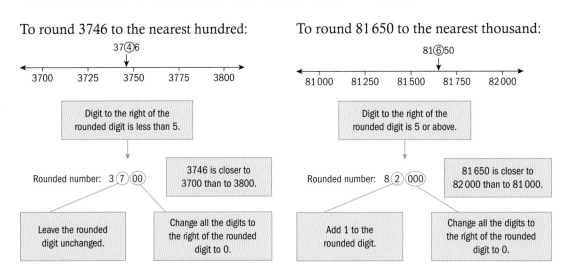

> → **Rules for rounding**
>
> If the digit after the one that is being rounded is less than 5 then keep the rounded digit unchanged and change all the remaining digits to the right of this to 0.
>
> If the digit after the one that is being rounded is 5 or more then add 1 to the rounded digit and change all remaining digits to the right of this to 0.

Example 9

> **a** Write down 247 correct to the nearest ten.
> **b** Write down 1050 correct to the nearest hundred.
>
> **Answers**
>
> | | |
> |---|---|
> | **a** 250 | *Both 240 and 250 are multiples of 10 but 250 is closer to 247.* |
> | **b** 1100 | *Both 1000 and 1100 are multiples of 100 and 1050 is exactly in the middle. Because the digit after the one being rounded is 5, round up.* |

Exercise 1G

1 Write these numbers correct to the nearest unit.
   a 358.4   **b** 24.5   **c** 108.9   **d** 10 016.01

2 Write these numbers correct to the nearest 10.
   a 246.25   **b** 109   **c** 1015.03   **d** 269

3 Write these numbers correct to the nearest 100.
   a 140   **b** 150   **c** 1240   **d** 3062

4 Write these numbers correct to the nearest 1000.
   a 105 607   **b** 1500   **c** 9640   **d** 952

5 Write down a number that correct to the nearest 100 is 200.

6 Write down a number that correct to the nearest 1000 is 3000.

7 Write down a number that correct to the nearest unit is 6.

Rounding numbers to a given number of decimal places (dp)

This is rounding numbers to the nearest tenth, to the nearest hundredth, etc.

> → Rounding a number **correct to one decimal place** is the same as rounding it to the **nearest tenth**.
>
> Rounding a number **correct to two decimal places** is the same as rounding it to the **nearest hundredth**.
>
> Rounding a number **correct to three decimal places** is the same as rounding it to the **nearest thousandth**.

To write 3.021 correct to 1 dp:

		Rounded digit	First digit to the right is less than 5	
NUMBER	3	• 0	2	1
ROUNDED NUMBER	3	• 0	......	......
		Rounded digit remains unchanged	Digits to the right of rounded digit are deleted	Digits to the right of rounded digit are deleted

$3.021 = 3.0 \ (1 \, dp)$

To write 10.583 correct to 2 dp:

NUMBER	1	0	• 5	8	3
ROUNDED NUMBER	1	0	• 5	8	......
				Rounded digit remains unchanged	Digits to the right of rounded digit are deleted

$10.583 = 10.58 \ (2 \, dp)$

To write 4.371 to 1 dp:

		Rounded digit	First digit to the right is more than 5	
NUMBER	4	• 3	7	1
ROUNDED NUMBER	4	• 4	......	......
		Rounded digit is changed to 1 more	Digits to the right of rounded digit are deleted	Digits to the right of rounded digit are deleted

$4.371 = 4.4 \ (1dp)$

> **→ Rounding rules for decimals**
> - If the digit after the one that is being rounded is less than 5 keep the rounded digit unchanged and delete all the following digits.
> - If the digit after the one that is being rounded is 5 or more then add 1 to the rounded digit and delete all the following digits.

Example 10

a Write down 10.045 correct to 2 dp.
b Write down 1.06 correct to 1 dp.

Answers

a $10.045 = 10.05 \ (2 \, dp)$	*10.045 Next digit is 5, so round up:* *10.05*
b $1.06 = 1.1 \ (1 \, dp)$	*1.06 Next digit is 6, so round up: 1.1*

Exercise 1H

1 Write these numbers correct to 1 dp.
  a 45.67   **b** 301.065   **c** 2.401   **d** 0.09

2 Write these numbers correct to 2 dp.
  a 0.0047   **b** 201.305   **c** 9.6201   **d** 28.0751

3 Write these numbers correct to the nearest thousandth.
  a 10.0485   **b** 3.9002   **c** 201.7805   **d** 0.008 41

4 Calculate $\dfrac{\sqrt{1.8}}{3.08 \times 0.012^2}$; use your GDC.
  Give your answer correct to

  a 1 dp   **b** 2 dp   **c** 3 dp   **d** nearest 100   **e** nearest 1000.

5 Given that $p = 3.15$ and $q = 0.8$, find the value of $\dfrac{(p+q)^3}{p+q}$ giving
  your answer correct to
  a 2 dp   **b** 3 dp   **c** nearest unit   **d** nearest ten.

6 Write down a number that correct to 2 dp is 2.37.

7 Write down a number that correct to 1 dp is 4.1.

Rounding numbers to a given number of significant figures (sf)

> → The number of **significant figures** in a result is the number of figures that are known with some degree of reliability.

This sometimes depends on the measurement that is being taken. For example, if the length of a pencil is measured with a ruler whose smallest division is 1 mm, then the measurement is only accurate to the nearest millimetre.

You can say: *The length of this pencil is 14.6 cm.*
But you cannot say: *The length of this pencil is 14.63 cm.*
The length of the pencil can be given correct to 3 sf but cannot be given correct to 4 sf.

Rules for significant figures:	
• All non-zero digits are significant.	2578 kg has 4 sf
• Zeros between non-zero digits are significant.	20 004 km has 5 sf
• Zeros to the left of the first non-zero digit are **not** significant.	0.023 g has 2 sf
• Zeros placed after other digits but to the right of the decimal point are significant.	0.100 ml has 3 sf

Make sure you understand when a digit is significant.

The rules for rounding to a given number of significant figures are similar to the ones for rounding to the nearest 10, 1000, etc. or to a number of decimal places.

This example shows you the method.

Example 11

a Write down 24.31 correct to 2 sf.
b Write down 1005 correct to 3 sf.
c Write down 0.2981 correct to 2 sf.

Answers

a 24.31 = 24 (2 sf)

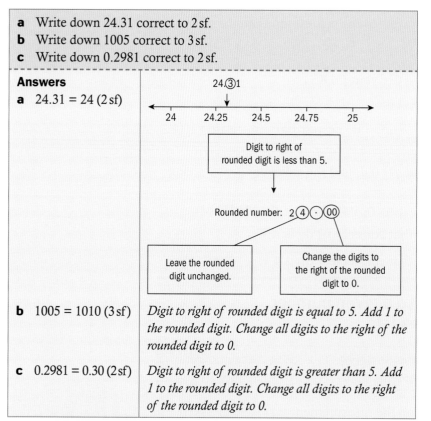

b 1005 = 1010 (3 sf) | *Digit to right of rounded digit is equal to 5. Add 1 to the rounded digit. Change all digits to the right of the rounded digit to 0.*

9 + 1 = 10 Replace the rounded digit with 0. Add 1 to the digit to the left of the rounded digit.

c 0.2981 = 0.30 (2 sf) | *Digit to right of rounded digit is greater than 5. Add 1 to the rounded digit. Change all digits to the right of the rounded digit to 0.*

→ **Rounding rules for significant figures**
- If the $(n+1)$th figure is less than 5 then keep the nth figure unchanged.
- If the $(n+1)$th figure is 5 or more then add 1 to this figure.
- In both cases all the figures to the right of figure n should be deleted if they are to the right of the decimal point and should be replaced by zeros if they are to the left of the decimal point.

Example 12

Let $t = \dfrac{12.4^3}{2.1 + \sqrt{3}}$.

a Write down the value of t giving the full calculator display.
b Write the answer to part **a** correct to
  i 3 significant figures   ii 2 significant figures.

▶ Continued on next page

Answers

a 497.5466391

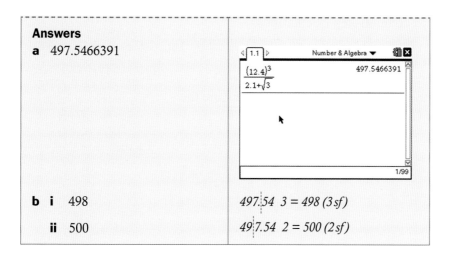

b i 498 $497.\overset{.}{5}4\ 3 = 498\ (3\,sf)$

  ii 500 $49\overset{.}{7}.54\ 2 = 500\ (2\,sf)$

Exercise 1I

1 Write the number of significant figures of each of these numbers.
  a 106   **b** 200   **c** 0.02   **d** 1290   **e** 1209

2 Write these numbers correct to 1 sf.
  a 280   **b** 0.072   **c** 390.8   **d** 0.001 32

3 Write these numbers correct to 2 sf.
  a 355   **b** 0.0801   **c** 1.075   **d** 1560.03

4 Write these numbers correct to 3 sf.
  a 2971   **b** 0.3259   **c** 10 410   **d** 0.5006

5 Calculate $\dfrac{\sqrt{8.7 + 2 \times 1.6}}{0.3^4}$.
  Give your answer correct to
  a 1 sf   **b** 3 sf   **c** 1 dp   **d** nearest hundredth.

6 Write the value of π correct to
  a nearest unit   **b** 2 dp   **c** 2 sf   **d** 3 dp.

7 Write down these numbers to the accuracy stated.
  a 238 (1 sf)   **b** 4609 (3 sf)   **c** 2.7002 (3 sf)

8 a Calculate $\dfrac{\sqrt[3]{3.375}}{1.5^2 + 1.8}$. Write down the full calculator display.
  b Give your answer to part **a** correct to
    i 2 sf   **ii** 3 sf   **iii** 4 sf.

Often in exams you need to do multi-step calculations.

In those situations, *keep at least one more significant digit in intermediate results* than needed in your final answer.

For instance, if the final answer needs to be given correct to 3 sf, then carry at least 4 sf in the intermediate calculations, or store the unrounded values in your GDC.

> The general rule in Mathematical Studies is *Unless otherwise stated in the question answers must be given exactly or correct to three significant figures.*

Example 13

The diagram represents a window grille made of wire, to keep pigeons out of the house. The small triangles are right-angled triangles and are all **congruent**. Their hypotenuse is 15 cm long. The other two sides are equal lengths. Find the total length of the wire, L. Give your answer correct to 3 significant figures.

> 'Congruent' means exactly the same shape and size.

Answers

Let x be the side length of the triangles.

$x^2 + x^2 = 15^2$

$\qquad 2x^2 = 225$

$\qquad x^2 = 112.5$

$\qquad x = \sqrt{112.5}$

$x = 10.6066\ldots$

First find the length of the shorter sides using Pythagoras.

Keep this value either exact or with more than three significant figures as this is just an intermediate value.

$L = 31 \times x + 12 \times 15$

$L = 31 \times 10.6066\ldots + 12 \times 15$

$L = 508.804\ldots$

$L = 509\,\text{cm}\ (3\,\text{sf})$

In the grille there are 31 sides of triangles with length x and 12 sides with length 15.

> Do not forget to write down the units in your answers.

Exercise 1J

EXAM-STYLE QUESTIONS

1 The area of a circle is $10.5\,\text{cm}^2$.

  a Find the length of its radius. Give your answer correct to four significant figures.

  b Find the length of its circumference. Give your answer correct to two significant figures.

2 Let the numbers $p = \sqrt{2}$ and $q = \sqrt{10}$.

  a Find the arithmetic mean of p and q. Give your answer correct to 4 sf.

  b Find the value of $(p + q)^2$. Give your answer correct to 3 sf.

  c Find the area of a rectangle whose sides are p cm and q cm long. Give your answer correct to 2 sf.

Estimation

An **estimate of a quantity** is an approximation that is usually used to check the reasonableness of an answer.

> → To estimate the answer to a calculation, round all the numbers involved to 1 sf.

Example 14

A theater has 98 rows; each row has 23 seats. Estimate the number of seats in the theater.

Answer

$100 \times 20 = 2000$ seats

Round 98 to 1 sf → 100
Round 23 to 1 sf → 20

Exact answer is $98 \times 23 = 2254$ seats.

Example 15

Estimate the average speed of a car that travels 527 km in 6 hours.

Answer

Average speed $= \dfrac{\text{distance traveled}}{\text{time taken}}$

$\dfrac{500}{5} = 100$ km h^{-1}

$527 \to 500$ (1 sf)
Round 6 down to 5 to make the division calculation easier.

Exact answer is
$\dfrac{527}{6} = 87.8$ km h^{-1} (3 sf)

Exercise 1K

1   Estimate the answers to these calculations.

  a 298×10.75 **b** 3.8^2 **c** $\dfrac{147}{11.02}$ **d** $\sqrt{103}$

2   A lorry is carrying 210 containers with pipes. There are 18 pipes in each container. Estimate the number of pipes that the lorry is carrying.

3   Japan covers an area of approximately 377 835 km² and in March 2009 Japan's population was 127 076 183. Estimate Japan's population density in 2009.

Population density $=$ $\dfrac{\text{total population}}{\text{land area}}$

4   A tree yields on average 9000 copy pages. Estimate the number of reams that can be made from one tree.

A ream has 500 pages.

5   Mizuki runs 33 km in 1.8 hours. Estimate Mizuki's average speed.

Average speed $=$ $\dfrac{\text{distance traveled}}{\text{time taken}}$

6 The Badaling Section and the Ming Mausoleums Scenic Area of the Great Wall are limited to 53 000 visitors per day. Estimate the number of visitors per year.

7 Peter calculates the area of this square as 1020.01 m². Use estimation to decide whether Peter is correct.

100.1 m

▲ The Great Wall of China

Percentage errors

Sometimes you need to know the difference between an estimated value and the exact value.

> → The difference between an estimated or **approximated value** and the **exact value** is called the **error**:
> Error = $v_A - v_E$
> where v_A is the approximated value and v_E is the exact value

Why do errors arise? What kind of errors do you know?
Do 'error' and 'mistake' have the same meaning?

Example 16

Olivia and Ramesh each went to a different concert.
In the concert that Olivia attended there were 1450 people and Olivia estimated that there were 1300.
In the concert that Ramesh attended there were 1950 people and Ramesh estimated that there were 1800.
Calculate the errors Olivia and Ramesh made in their estimations.

Answer

Olivia: Error = 1450 – 1300 Error = 150 people Ramesh: Error = 1950 – 1800 Error = 150 people	$v_A - v_E$ *is negative, so use* $v_E - v_A$ *instead.*

$|v_A - v_E|$ is the modulus, or positive value, of $v_A - v_E$.

In Example 16, Olivia and Ramesh both made the same error, 150. However, Ramesh's estimate is more accurate as 150 out of 1950 is a smaller proportion than 150 out of 1450.

Using **percentages**:

$\frac{150}{1450} \times 100\% = 10.3\%$ (3 sf)   and $\frac{150}{1950} \times 100\% = 7.69\%$ (3 sf)

Olivia's error represents 10.3% of the total.
Ramesh's error represents 7.69% of the total.
These percentages help us to have a better idea of the accuracy of the estimations. They are called **percentage errors**.

> → Percentage error = $\left| \frac{v_A - v_E}{v_E} \right| \times 100\%$
>
> where v_A represents **approximated value** or **estimated value** and v_E represents the **exact value**.

Sometimes we don't have the exact value. In these cases we replace v_E with the **accepted value**.

Example 17

> The size of angle M is 125.7°. Salomon measures M with a protractor as 126°. Find the percentage error he made in measuring angle M.

Answer

Percentage error	Percentage error
$= \left\| \dfrac{126 - 125.7}{125.7} \right\| \times 100\%$	$= \left\| \dfrac{v_A - v_E}{v_E} \right\| \times 100\%$
Percentage error	*with $v_A = 126$, $v_E = 125.7$*
$= 0.239\%$ (3 sf)	*Use your GDC. Round to 3 sf.*

Exercise 1L

EXAM-STYLE QUESTIONS

1 Let $a = 5.2$ and $b = 4.7$.

  a Find the exact value of $3a + b^3$.

Xena estimates that the answer to part **a** is 140.

  b Find the percentage error made by Xena in her estimation.

2 Ezequiel's Biology marks are 8.3, 6.8 and 9.4 out of 10. His final grade in Biology is the mean of these three marks.

  a Calculate Ezequiel's final grade in Biology.

Ezequiel wrote the three marks correct to the nearest unit to find his final grade in Biology.

  b Calculate the final grade that Ezequiel found.

  c Calculate the percentage error made by Ezequiel when finding his final grade in Biology.

3 The measurements of the length and width of a rectangular kitchen are 5.34 m and 3.48 m respectively.

  a Calculate in m² the exact area of the kitchen.

  b Write down both the length and the width correct to 1 dp.

  c Calculate the percentage error made if the area was calculated using the length and the width correct to 1 dp.

4 The area of a circular garden is 89 m².

  a Find the radius of the garden. Give your answer correct to three decimal places.

  b Find the perimeter of the garden.

José estimates that the perimeter of the garden is 30 m.

  c Use your answer to part **b** to find the percentage error made by José. Give your answer correct to two significant figures.

1.3 Standard form

- The number of internet users in the world up to June 2010 was 2×10^9.
- The mass of the Earth is about 5.97×10^{24} kg.
- An estimate for the average mass of a human cell is about 10^{-9} g.

These numbers are either very large or very small.

They are written in **standard form**: a way of writing very large or very small numbers without writing a lot of zeros.

> → A number is written in standard form if it is in the form $a \times 10^k$ where $1 \le a < 10$ and k is an integer.

> If we did not use standard form, we would write the mass of the Earth as
> 5 970 000 000 000 000 000 000 000 kg

> When numbers are written in standard form it is easier to
> - compare them
> - calculate with them.

> A **googol** is the number 1 followed by 100 zeros. In standard form it is 10^{100}. The name googol was invented by a 9-year-old, who was asked by his uncle, the American mathematician Edward Kasner, to think up a name for a very large number.
>
> The name of the company Google comes from a misspelling of the word googol and is related to the amount of information that the company handles.

Example 18

These numbers are written in standard form ($a \times 10^k$).
For each of them state the value of a and of k.

a 2×10^9 **b** 5.97×10^{24} **c** 10^{-9}

Answers

a $a = 2;\ k = 9$

b $a = 5.97;\ k = 24$

c $a = 1;\ k = -9$

Compare with $a \times 10^k$

> Abu Kamil Shuja (c. 850–c. 930), also know as al-Hasib al-Misri, meaning 'the calculator from Egypt', was one of the first to introduce symbols for indices, such as $x^m\ x^n = x^{m+n}$, in algebra.

Example 19

Decide which of these numbers are *not* written in the form $a \times 10^k$ where $1 \le a < 10$ and k is an integer. Justify your decisions.

a 2.06×10^{-5} **b** 13×10^{-1} **c** $6.13 \times 10^{\frac{1}{3}}$

d 7.05 **e** 0.12×10^6

Answers

b 13×10^{-1} is not written in standard form as 13 is greater than 10.

c $6.13 \times 10^{\frac{1}{3}}$ is not written in standard form as $\frac{1}{3}$ is not an integer.

e 0.12×10^6 is not written in standard form as 0.12 is smaller than 1.

Compare with $a \times 10^k$, where $1 \le a < 10$ and $k \in \mathbb{Z}$

Example 20

Write these numbers in standard form, showing your working.
a 257 000 000   **b**   0.00043

Answers

a 257 000 000

so $k = 8$

$257\,000\,000 = 2.57 \times 10^8$

First significant figure of 257 000 000 is 2. Place the decimal point immediately after 2.
Moving the decimal point 8 places to the right is the same as multiplying by 10^8.

b 0.000 43

so $k = -4$

$0.000\,43 = 4.3 \times 10^{-4}$

First significant figure of 0.000 43 is 4. Place the decimal point immediately after 4.
Moving the decimal point 4 places to the left is the same as multiplying by 10^{-4}.

> Tips to write a number in standard form:
> **1** Write down a: write down all the significant figures of the number and place the decimal point immediately after the first significant figure.
> **2** Find k.

Exercise 1M

1 Which of these numbers are written in standard form?

$$2.5 \times 10^{-3} \qquad 12 \times 10^5 \qquad 10^{10} \qquad 3.15 \times 10^{\frac{1}{2}} \qquad 0.81 \times 10^2$$

2 Write these numbers in standard form.
    a 135 600   **b** 0.002 45   **c** 16 000 000 000
    d 0.000 108   **e** 0.23×10^3

> Change them to decimal numbers, e.g 2.3×10^6
> $= 2\,300\,000$.
> A decimal number is a 'normal' number written to base 10. It doesn't necessarily have a decimal point or decimal places.

3 Write these numbers in ascending order.
    $2.3 \times 10^6 \qquad 3.4 \times 10^5 \qquad 0.21 \times 10^7 \qquad 215 \times 10^4$

4 Write these numbers in descending order.
    $3.621 \times 10^4 \qquad 31.62 \times 10^2 \qquad 0.3621 \times 10^4 \qquad 3.261 \times 10^3$

Example 21

Let $x = \dfrac{-5 + \sqrt{121}}{(7 - 1)^2}$.

a Calculate the value of x. Write down the full calculator display.
b Write your answer to part **a** correct to 3 sf.
c Write your answer to part **b** in the form $a \times 10^k$
    where $1 \le a < 10$ and $k \in \mathbb{Z}$.

▶ Continued on next page

Answers

a 0.1666666667

Use your GDC.

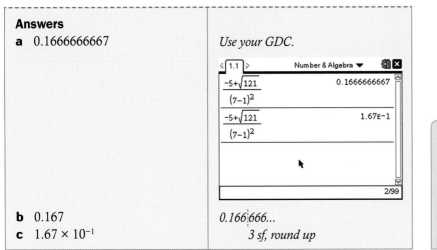

b 0.167

c 1.67×10^{-1}

0.166̇666...

3 sf, round up

Calculations with numbers expressed in standard form

You can use your GDC for calculations in standard form.

Example 22

Let $x = 2.4 \times 10^4$ and $y = 5.10 \times 10^5$.

a Find the value of $3x + y$.

b Write your answer to part **a** correct to 2 sf.

c Write your answer to part **b** in the form $a \times 10^k$ where $1 \le a < 10$ and k is an integer.

Answers

a $3 \times 2.4 \times 10^4 + 5.10 \times 10^5$
  $= 582\,000$

b $580\,000$

c 5.8×10^5

$3 \cdot 2.4 \cdot 10^4 + 5.1 \cdot 10^5$	582000.
$3 \cdot 2.4 \cdot 10^4 + 5.1 \cdot 10^5$	5.82E5

Always use a GDC in this type of question, but show the working as shown in **a**.

Exercise 1N

1 Given that $x = 6.3 \times 10^6$ and $y = 2.8 \times 10^{10}$, calculate the following. Give your answers in the form $a \times 10^k$ where $1 \le a < 10$ and $k \in \mathbb{Z}$.

  a $x \times y$ **b** $\dfrac{x}{y}$ **c** $\sqrt{\dfrac{x}{y}}$

2 Let $x = 2.5 \times 10^6$ and $y = 3.48 \times 10^6$.

  a Find the arithmetic mean of x and y. Give your answer in the form $a \times 10^k$ where $1 \le a < 10$ and $k \in \mathbb{Z}$.

  b Give your answer to part **a** correct to the nearest million.

3 Let $t = 22.05 \times 10^8$ and $q = 3.15 \times 10^6$

  a Write down t in the form $a \times 10^k$ where $1 \le a < 10$ and $k \in \mathbb{Z}$.

  b Calculate $\dfrac{t}{q}$.

  c Write your answer to part **b** in the form $a \times 10^k$ where $1 \le a < 10$ and $k \in \mathbb{Z}$.

4 Let $x = 225 \times 10^8$.

  a Write x in the form $a \times 10^k$ where $1 \le a < 10$ and $k \in \mathbb{Z}$.

  b State whether the following statement is true: $x^2 > 10^{20}$

    Justify your answer.

  c i Calculate $\dfrac{x}{\sqrt{x}}$.

    ii Give your answer to part **i** in the form $a \times 10^k$ where $1 \le a < 10$ and $k \in \mathbb{Z}$.

1.4 SI units of measurement

Ariel is baking a tuna pie.
He needs a tuna can whose net weight is 180 **g***.*
Another ingredient is 240 **ml** *of milk.*
He bakes the pie in a preheated oven to 200° **C** *for 20* **minutes***.*
Ariel recycles material. He has decided to use the metal from the can so he needs to take some measurements:
The height of the tuna can is 4 **cm***.*
The total area of metal used to make the tuna can was 219 **cm²***.*
The volume of the tuna can is 314 **cm³***.*

Here we have seen how in an everyday situation we deal with different kinds of units such as g, ml, °C, minutes, cm, cm², cm³. These units are internationally accepted and have the same meaning in any part of the world.

SI is the international abbreviation for the *International System of Units* (in French, *Système international d'unités*). There are seven **base units** (see table). Each unit is accurately defined and the definition is independent from the other six units.

> The 11th General Conference on Weights and Measures, CGPM, held in 1960, adopted the name Système International d'Unités. The CGPM is made up of representatives from 54 member states and 31 associate states and economies.

The seven base units and their respective quantities are given in the following table.

Base quantity	Base unit name	Base unit symbol
Length	metre	m
Mass	kilogram	kg
Time	second	s
Electric current	ampere	A
Temperature	kelvin	K
Amount of substance	mole	mol
Intensity of light	candela	cd

One metre is defined in the SI as the distance travelled by light in a vacuum in $\dfrac{1}{299\,792\,458}$ of a second.

In the **SI** there are other units, the **derived units**. These units are expressed in terms of the base units. Some of these units along with their quantities are listed below:

- The **square metre** (m^2) for area
- The **cubic metre** (m^3) for volume
- The **metre per second** ($m\,s^{-1}$) for speed or velocity
- The **kilogram per cubic metre** ($kg\,m^{-3}$) for density or mass density

Derived units are products of powers of **base units**.

→ In Mathematical Studies, the most common **SI base units** used are m, kg and s, and **derived units** are m^2 (area), m^3 (volume), $km\,h^{-1}$ (velocity), $kg\,m^{-3}$ (density).

Example 23

Write down the symbol used for the quantities in bold:
a The **velocity** of an object that travels 1000 km in 3 hours.
b The **density** of an object with a mass of 550 g and a volume of 400 cm³.

Answers

a $km\,h^{-1}$

b $g\,cm^{-3}$

Velocity is kilometres per hour.
Density is grams per cubic centimetre.

SI prefixes

To avoid writing very small or very large quantities we use prefix names and prefix symbols. Some of these are shown in this table.

Factor	Prefix	Symbol	Factor	Prefix	Symbol
10^3	kilo	k	10^{-3}	milli	m
10^2	hecto	h	10^{-2}	centi	c
10^1	deca	da	10^{-1}	deci	d

The kilogram is the only SI base unit with a prefix as part of its name.

Investigation – SI units

a How many prefix names and symbols are there nowadays?
b Six prefix names and their symbols are listed in the table. Find the others.
c Choose at least two of them and describe situations where they are used.

> Does the use of SI notation help us to think of mathematics as a 'universal language'?

Example 24

Convert each measurement to the stated unit.
a 1 dm to m   **b** 1 das to s   **c** 1 hg to g

Answers

a $1\,dm = 10^{-1}\,m$
b $1\,das = 10^{1}\,s$
c $1\,hg = 10^{2}\,g$

Use the information on prefixes given in the table on the previous page.
dm *reads decimetre*
das *reads decasecond*
hg *reads hectogram*

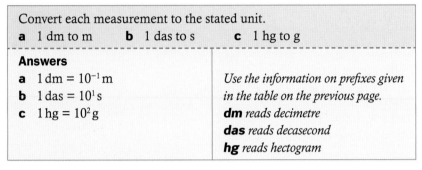

> This diagram will help you to convert between SI units.

Example 25

Convert each measurement to the stated unit. Give your answers in standard form.
a 2.8 m to hm   **b** 3200 s to ms   **c** 0.5 kg to dg

Answers

a $1\,m = 10^{-2}\,hm$
$2.8\,m = 2.8 \times 10^{-2}\,hm$

In this example replace 'SI unit' in the diagram with m.
To convert from m to hm divide by 10 twice therefore $1\,m = 10^{-2}\,hm$.

b $1\,s = 10^{3}\,ms$
$3200\,s = 3200 \times 10^{3}\,ms$
$\qquad = 3.2 \times 10^{6}\,ms$

In this example replace 'SI unit' in the diagram with s.
To convert from s to ms multiply by 10 three times therefore $1\,s = 10^{3}\,ms$.

c $1\,kg = 10^{4}\,dg$
$0.5\,kg = 0.5 \times 10^{4}\,dg$
$\qquad = 5 \times 10^{3}\,dg$

In this example replace 'SI unit' in the diagram with g.
To convert from kg to dg multiply by 10 four times therefore $1\,kg = 10^{4}\,dg$.

Exercise 10

1 Write down the symbol used for the quantities in bold.
   a The **acceleration** of an object that has units measured in kilometres per hour squared.
   b The **density** of an object with a mass of 23 kg and a volume of $1.5\,\text{m}^3$.
   c The average **speed** of an object that travels 500 m in 70 seconds.

2 Write down these units in words.
   a dag   **b** cs   **c** mm   **d** dm

3 Convert each of these to the stated unit.
   a 32 km to m   **b** 0.87 m to dam   **c** 128 cm to m

4 Convert each of these to the stated unit.
   a 500 g to kg   **b** 357 kg to dag   **c** 1080 dg to hg

5 Convert each of these to the stated unit.
   a 0.080 s to ms   **b** 1200 s to das   **c** 0.8 hs to ds

6 **a** Convert 67 800 000 mg to kg. Give your answer correct to the nearest kg.
   b Convert 35 802 m to km. Give your answer correct to the nearest km.
   c Convert 0.654 g to mg. Give your answer in the form $a \times 10^k$ where $1 \le a < 10$ and $k \in \mathbb{Z}$.

Area and volume SI units

Area

The diagrams show two different ways of representing $1\,\text{m}^2$.

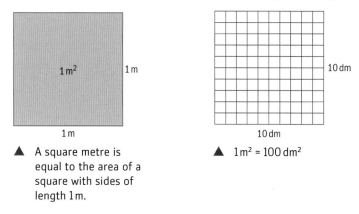

1 m²   1 m

10 dm

1 m

10 dm

▲ A square metre is equal to the area of a square with sides of length 1 m.

▲ $1\,\text{m}^2 = 100\,\text{dm}^2$

$1\,\text{m}^2 = 1\,\text{m} \times 1\,\text{m} = 10\,\text{dm} \times 10\,\text{dm} = 100\,\text{dm}^2$

To convert from m^2 to dm^2 we multiply by **100** or 10^2.
You can use the same method to convert from

- km^2 to hm^2
- hm^2 to dam^2
- dam^2 to m^2
- dm^2 to cm^2
- cm^2 to mm^2

$$\times 10^2 \ \times 10^2 \ \times 10^2 \ \times 10^2 \ \times 10^2 \ \times 10^2$$

km^2	hm^2	dam^2	m^2	dm^2	cm^2	mm^2

$$\div 10^2 \ \div 10^2 \ \div 10^2 \ \div 10^2 \ \div 10^2 \ \div 10^2$$

Example 26

Convert each measurement to the stated unit.
Give your answers in full.
a $1.5\,m^2$ to cm^2
b $3240\,m^2$ to km^2

Answers

a $1\,m^2 = 10^4\,cm^2$
Therefore
$1.5\,m^2 = 1.5 \times 10^4\,cm^2$
$\qquad\quad = 15\,000\,cm^2$

To convert from m^2 to cm^2 multiply by 10^2 twice; this is multiply by 10^4.
$\left(10^2\right)^2 = 10^4$

b $1\,m^2 = 10^{-6}\,km^2$
Therefore
$3240\,m^2 = 3240 \times 10^{-6}\,km^2$
$\qquad\qquad = 0.003\,240\,km^2$

To convert from m^2 to km^2 divide by 10^2 three times; this is divide by 10^6 or multiply by 10^{-6}.
$\left(10^2\right)^3 = 10^6$

Volume

The diagrams show two different ways of representing $1\,m^3$.

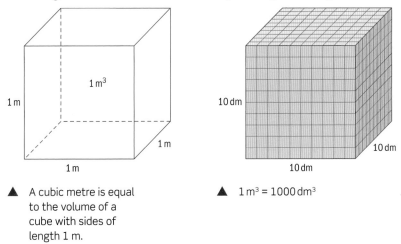

1 m³

1 m

1 m

1 m

10 dm

10 dm

10 dm

▲ A cubic metre is equal to the volume of a cube with sides of length 1 m.

▲ $1\,m^3 = 1000\,dm^3$

$1\,m^3 = 1\,m \times 1\,m \times 1\,m = 10\,dm \times 10\,dm \times 10\,dm = 1000\,dm^3$

To convert from m^3 to dm^3 we multiply by **1000** or 10^3.
You can use the same method to convert from

- km^3 to hm^3
- hm^3 to dam^3
- dam^3 to m^3
- dm^3 to cm^3
- cm^3 to mm^3

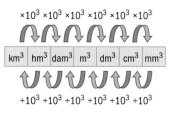

Example 27

Convert each measurement to the stated unit.
Give your answers in standard form.
a $0.8\,m^3$ to cm^3
b $15\,900\,cm^3$ to dam^3

Answers

a $1\,m^3 = 10^6\,cm^3$

Therefore

$0.8\,m^3 = 0.8 \times 10^6\,cm^3$

$= 8 \times 10^5\,cm^3$

To convert from m^3 to cm^3 multiply by 10^3 twice; this is multiply by 10^6.
$(10^3)^2 = 10^6$

b $1\,cm^3 = 10^{-9}\,dam^3$

Therefore

$15\,900\,cm^3$

$= 15\,900 \times 10^{-9}\,dam^3$

$= 1.59 \times 10^{-5}\,dam^3$

To convert from cm^3 to dam^3 divide by 10^3 three times; this is multiply by 10^{-9}.

Exercise 1P

1 Convert these measurements to the stated unit.
Give your answers in full.
a $2.36\,m^2$ to cm^2 **b** $1.5\,dm^2$ to dam^2
c $5400\,mm^2$ to cm^2 **d** $0.06\,m^2$ to mm^2
e $0.8\,km^2$ to hm^2 **f** $35\,000\,m^2$ to km^2

2 Convert these measurements to the stated unit.
Give your answers in the form $a \times 10^k$ where $1 \le a < 10$
and $k \in \mathbb{Z}$.
a $5\,m^3$ to cm^3 **b** $0.1\,dam^3$ to m^3
c $3\,500\,000\,mm^3$ to dm^3 **d** $255\,m^3$ to mm^3
e $12\,000\,m^3$ to dam^3 **f** $0.7802\,hm^3$ to dam^3

3 The side length of a square is $13\,cm$. Find its area in
a cm^2 **b** m^2

13 cm

4 The side length of a cube is $0.85\,m$. Find the volume of the cube in
a m^3 **b** cm^3

5 Write these measurements in order of size starting from the smallest.

$0.081\,dam^2$, $8\,000\,000\,mm^2$, $82\,dm^2$, $7560\,cm^2$, $0.8\,m^2$

Convert all to the same unit.

6 Write these measurements in order of size starting from the smallest.

$11.2\,m^3$, $1200\,dm^3$, $0.01\,dam^3$, $11\,020\,000\,000\,mm^3$, $10\,900\,000\,cm^3$

Convert all to the same unit.

Extension material on CD:
Worksheet 1 - Calculations with measures

Non–SI units accepted in the SI

→ There are some units that are **non-SI** units but are accepted for use with the SI because they are widely used in everyday life, for example, min, h, ℓ.

Each of these non-SI units has an exact definition in terms of an SI unit. The table below shows some of these units along with their equivalents in SI units.

Quantity	Name of unit	Symbol	Equivalents in SI units
time	minute	min	$1\,min = 60\,s$
	hour	h	$1\,h = 60\,min = 3600\,s$
	day	d	$1\,d = 24\,h = 86\,400\,s$
area	hectare	ha	$1\,ha = 1\,hm^2 = 10^4\,m^2$
volume	litre	L, ℓ	$1\,\ell = 1\,dm^3$
mass	tonne	t	$1\,t = 10^3\,kg$

The SI prefixes are used with ℓ, but *not* used with min, h and d.

Example 28

a Convert $3\,d\,15\,h\,6\,min$ to seconds.
b Convert the average speed of $12\,km\,h^{-1}$ to $m\,s^{-1}$.

Answers

a $1\,d = 86\,400\,s$
$\Rightarrow 3\,d = 259\,200\,s$
$1\,h = 3600\,s \Rightarrow 15\,h = 54\,000\,s$
$1\,min = 60\,s \Rightarrow 6\,min = 360\,s$
Therefore
$3\,d\,15\,h\,6\,min = 259\,200\,s$
$\qquad\qquad + 54\,000\,s + 360\,s$
$\qquad\qquad = 313\,560\,s$

1 day = 24 hours
$\quad = 24 \times 60\,min$
$\quad = 24 \times 60 \times 60\,s$

'$\Rightarrow$' means 'therefore' or 'implies'.

b Average speed = $12\,km\,h^{-1}$
$\Rightarrow$ in 1 h the object moved 12 km
$\Rightarrow$ in 3600 s it moved 12 000 m
Average speed = $\dfrac{12\,000\ m}{3600\ s}$
$\qquad\qquad = 3.33\,m\,s^{-1}\,(3\,sf)$

1 h = 60 min
$\quad = 60 \times 60\,s$
12 km = 12 000 m

Average speed
$= \dfrac{distance\ traveled}{time\ taken}$

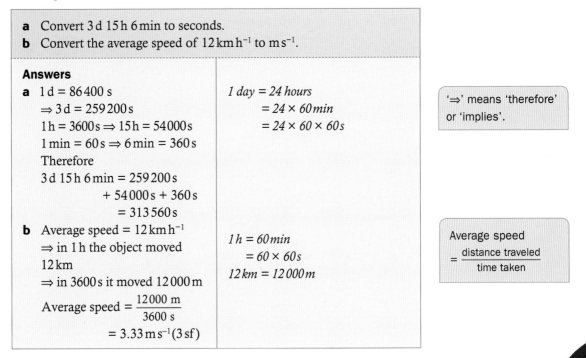

Example 29

> Convert
> **a** 120 hl to cl
> **b** 5400 ℓ to m³
>
> **Answers**
>
> **a** 120 hl = 120 × 10⁴ cl | *To convert from hl to cl, multiply by*
> = 1 200 000 cl | *10 four times, i.e. multiply*
> | *by 10⁴.*
>
> **b** 1 ℓ = 1 dm³
> ⇒ 5400 ℓ = 5400 dm³
> 5400 dm³ = 5400 × 10⁻³ m³ | *To convert from dm³ to m³ we divide*
>   = 5.4 m³ | *by 10³ once; this is multiply by 10⁻³.*

Exercise 1Q

1 a Convert 1 d 2 h 23 m to seconds.
  b Give your answer to part **a** correct to the nearest 100.

2 a Convert 2 d 5 m to seconds.
  b Give your answer to part **a** in the form $a \times 10^k$ where $1 \le a < 10$ and $k \in \mathbb{Z}$.

3 Convert these measurements to the stated unit.
   Give your answers in full.
   a 5 ℓ to ml   **b** 0.56 ml to hl   **c** 4500 dal to cl

4 Convert these measurements to the stated unit. Give your
   answers in the form $a \times 10^k$ where $1 \le a < 10$ and $k \in \mathbb{Z}$.
   a 500 ℓ to cm³   **b** 145.8 dl to dm³   **c** 8 hl to cm³

5 Convert these measurements to the stated unit.
   Give your answers to the nearest unit.
   a 12.5 dm³ to ℓ   **b** 0.368 m³ to hl   **c** 809 cm³ to cl

6 A particle travels 3000 m at an average speed of 40 m min⁻¹.
  a Find in minutes the time travelled by the particle.
  b Give your answer to part **a** in seconds.

EXAM-STYLE QUESTION
7 A cubic container has sides that are 1.5 m long.
  a Find the volume of the container. Give your answer in m³.
  b Give your answer to part **a** in dm³.
  c Decide whether 4000 ℓ of water can be poured in the
    container.
    Justify your answer.

8 The volume of a tea cup is $220\,cm^3$. Mercedes always serves a tea cup to $\dfrac{4}{5}$ of its capacity to avoid spilling any.

a Find, in ℓ, the amount of tea that Mercedes serves in a tea cup.

The volume of Mercedes' teapot is $1.5\,\ell$.

b Find the maximum number of tea cups that Mercedes can serve from one teapot.

9 The distance by air from Buenos Aires to Cape Town is $6900\,km$. An airplane flies at an average speed of $800\,km\,h^{-1}$.

a Find the time it takes for this airplane to fly from Buenos Aires to Cape Town.

Abouo takes this flight and then flies to Johannesburg, which is $1393\,km$ from Cape Town. The flight is 2 hours long.

b Find the average speed of this second airplane.

Abouo leaves Buenos Aires at 10:00 a.m. When he arrives at Cape Town he waits 1.5 hours until the next flight.

c Find the time at which he arrives at Johannesburg.

Temperature

→ There are three temperature scales:

- **kelvin** (K)
- **Celsius** (°C)
- **Fahrenheit** (°F)

The kelvin (K) is the only SI base unit of temperature and is mainly used by scientists. The °C is an SI derived unit. The Celsius scale is used in most countries but not in the United States, where the Fahrenheit scale is used. In the following table the freezing and boiling points of water for each of the three scales are shown.

Scale	Freezing point of water	Boiling point of water
Fahrenheit (°F)	32	212
Celsius (°C)	0	100
kelvin (K)	273.15	373.15

The formula used to convert from °C to °F is

$$t_F = \frac{9}{5} \times t_C + 32$$

The formula used to convert from K to °C is

$$t_C = t_K - 273.15$$

Fahrenheit 451 is the name of a book written by Ray Bradbury. The title refers to the temperature at which paper combusts. This temperature is also known as the flashpoint of paper.

In this formula t_C represents temperature in °C and t_F represents temperature in °F.

In this formula t_C represents temperature in °C and t_K represents temperature in K.

Example 30

Convert **a** 25°C to °F **b** 300 K to °C **c** 200°F to °C	

Answers

a $\dfrac{9}{5} \times 25 + 32 = 77\,°F$ *Use the formula* $t_F = \dfrac{9}{5} \times t_C + 32$

b $300 - 273.15 = 26.85\,°C$ *Use the formula* $t_C = t_K - 273.15$

c $200 = \dfrac{9}{5} \times t_C + 32$ *Rearrange to make* t_C *the subject of the formula.*

$\quad t_C = (200 - 32) \times \dfrac{5}{9}$

$\quad t_C = 93.3\,°C\ (3\,sf)$

> You will derive formulae like this to model real-life situtations in chapter 6.

Exercise 1R

1 Convert into °C. Give your answer correct to one tenth of a degree.
   a 280 K   **b** 80°F

2 Convert into °F. Give your answer correct to the nearest degree.
   a 21°C   **b** 2°C

3 **a** Convert 290 K to °C.
   b Hence convert 290 K to °F.

4 **a** The formula to convert from K to °C is $t_C = t_K - 273.15$.
     Find the formula used to convert from °C to K.

   b The formula to convert from °C to °F is $t_F = \dfrac{9}{5} \times t_C + 32$.
     Find the formula used to convert from °F to °C.

Review exercise

Paper 1 style questions

EXAM-STYLE QUESTION

1 Consider the numbers $5, \dfrac{\pi}{2}, -3, \dfrac{5}{4}, 2.\dot{3}$ and the number
sets $\mathbb{N}, \mathbb{Z}, \mathbb{Q}$ and $\mathbb{R}$.
Complete the following table by placing a tick ($\checkmark$) in the
appropriate box if the number is an element of the set.

	5	$\dfrac{\pi}{2}$	-3	$\dfrac{5}{4}$	$2.\dot{3}$
$\mathbb{N}$					
$\mathbb{Z}$					
$\mathbb{Q}$					
$\mathbb{R}$					

2 Given the numbers

14.1×10^{-1} 1.4×10^2 $\sqrt{2}$ $0.001\,39 \times 10^2$ 1414×10^{-2}

  a state which of these numbers is irrational

  b write down $\sqrt{2}$ correct to five significant figures

  c write down these numbers in ascending order.

3 The mass of a container is 2690 kg.

  a Write down this weight in the form $a \times 10^k$ where $1 \leq a < 10$ and $k \in \mathbb{Z}$.

  Nelson estimates that the mass of the container is 2.7×10^3 kg.

  b **i** Write down this mass in full.

  ii Find the percentage error made by Nelson in his estimation.

4 Light travels in empty space at a speed of $299\,792\,458\,\mathrm{m\,s^{-1}}$.

  a Write this value correct to three significant figures.

  b Use your answer to part **a** to find in km the distance that the light travels in 1 second.

  c Use your answer to part **b** to find in $\mathrm{km\,h^{-1}}$ the speed at which the light travels in empty space. Give your answer in the form $a \times 10^k$ where $1 \leq a < 10$ and $k \in \mathbb{Z}$.

5 The total mass of 90 identical books is 52 200 g.

  a Calculate the exact mass of one book in kg.

  b Write down your answer to part **a** correct to one significant figure.

  Matilda estimates that the mass of any of these books is 0.4 kg. She uses the answer to part **b** to find the percentage error made in her estimation.

  c Find this percentage error.

6 The volume, V, of a cubic jar is 1560 cm^3.

  a Write down V in dm^3.

  Sean works in the school cafeteria making juice. He pours the juice in these jars. He always fills the jars up to $\dfrac{3}{4}$ of their height.

  b Find in ℓ the amount of juice that Sean pours in each jar.

  He makes 25 ℓ of juice per day.

  c **i** Find the number of jars that Sean pours per day.

  ii Write down the amount of juice left.

7 Let $x = \dfrac{30y^2}{\sqrt{y+1}}$.

  a Find the exact value of x when $y = 1.25$.

  b Write down the value of x correct to three significant figures.

  c Write down your answer to part **b** in the form $a \times 10^k$ where $1 \leq a < 10$ and $k \in \mathbb{Z}$.

8 The side length of a square field is x m.

   a Write down in terms of x an expression for the area of the field.

   The area of the field is $2.56\,\text{km}^2$.

   b **i** Find the value of x.

      ii Find, in **metres,** the perimeter of the field.

9 The formula to convert from the kelvin scale to the Fahrenheit scale is

$$t_{\text{F}} = \frac{9}{5} \times t_{\text{K}} - 459.67$$

   where t_{K} represents temperature in K and t_{F} represents temperature in °F.

   a Find the temperature in °F for 300 K.

   b Find the temperature in K for 100 °F. Give your answer to the nearest unit.

10 Consider the inequality $2x + 5 > x + 6$.

   a Solve the inequality.

   b Represent the solution to part **a** on a copy of the number line.

   c Decide which of these numbers are solutions to the inequality given in part **a**.

$$1 \qquad \frac{\pi}{4} \qquad -5 \qquad \sqrt{3} \qquad 2.0\dot{6} \qquad \frac{101}{100} \qquad 1.2 \times 10^{-3}$$

11 The size of an A4 sheet is $210\,\text{mm} \times 297\,\text{mm}$.

   a Find the area of an A4 sheet. Give your answer in mm^2.

   b Give your answer to part **a** in m^2.

   One ream has 500 pages. It weighs $75\,\text{g m}^{-2}$.

   c Find the mass of one page.

   d Find the mass of one ream. Give your answer in kg.

Paper 2 style questions

1 The figure shows a rectangular field. The field is 1260 m wide and 2500 m long.

   a Calculate the perimeter of the field. Give your answer in km.

1260 m

2500 m

Figure not to scale

   The owner of the field, Enrico, wants to fence it. The cost of fencing is $327.64 per km.

   b Calculate the cost of fencing the field. Give your answer correct to two decimal places.

   Enrico estimates that the perimeter of the field is 7.6 km.
He uses this estimation to calculate the cost of fencing the field.

   c Calculate the percentage error made by Enrico when using his estimation of the perimeter of the field to calculate the cost of fencing.

   d Calculate the area of the field. Give your answer in square kilometres (km^2).

2 A running track is made of a rectangular shape 800 m by 400 m with semicircles at each end as shown in the figure below.

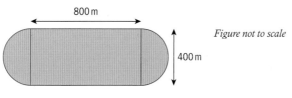

800 m

400 m

Figure not to scale

a Find the perimeter of the running track. Give your answer correct to the nearest metre.

Elger runs 14 200 m around the track.

b Find the number of complete laps that Elger runs around the running track.

Elger runs at an average speed of 19 km h^{-1}.

c Find the time it takes Elger to complete **one** lap. Give your answer in hours.

d Find the time **in minutes** it takes Elger to run 14 200 m. Give your answer correct to 5 sf.

Elger estimates that it takes him 44 minutes to run 14 200 m.

e Find the percentage error made by Elger in his estimation.

3 A chocolate shop makes spherical chocolates with a diameter of 2.5 cm.

a Calculate the volume of each of these chocolates in cm^3. Give your answer correct to two decimal places.

The chocolates are sold in cylindrical boxes which have a radius of 12.5 mm and a height of 15 cm.

b Calculate the volume of each of these cylindrical boxes in cm^3. Give your answer correct to two decimal places.

c Show that the maximum number of chocolates that fit in each of these boxes is 6.

The boxes are filled with 6 chocolates.

d Find the volume of the box that is **not** occupied by the chocolates.

e Give your answer to part **d** in mm^3.

f Give your answer to part **d** in the form $a \times 10^k$ where $1 \le a < 10$ and $k \in \mathbb{Z}$.

CHAPTER 1 SUMMARY

The number sets

- The set of **natural numbers** $\mathbb{N}$ is $\{0, 1, 2, 3, 4, 5, \ldots\}$
- The set of **integers** $\mathbb{Z}$ is $\{\ldots, -4, -3, -2, -1, 0, 1, 2, 3, 4, \ldots\}$.
- The set of **rational numbers** $\mathbb{Q}$ is $\left\{\dfrac{p}{q} \text{ where } p \text{ and } q \text{ are integers and } q \neq 0\right\}$.
  A number is rational if
  - it can be written as a quotient of two integers, or
  - its decimal expansion is finite, or
  - its decimal expansion has a recurring digit or pattern of digits.
- Any number that has a decimal expansion with an infinite number of digits after the decimal point and with no period is an **irrational number**.
- The set of rational numbers together with the set of irrational numbers complete the number line and form the set of **real numbers**, $\mathbb{R}$.

Approximations and error

- Rounding a number to the **nearest 10** is the same as rounding it to the **nearest multiple of 10**.
- Rounding a number to the **nearest 100** is the same as rounding it to the **nearest multiple of 100**.
- **Rules for rounding**
  - If the digit after the one that is being rounded is less than 5 then keep the rounded digit unchanged and change all the remaining digits to the right of this to 0.
  - If the digit after the one that is being rounded is 5 or more then add 1 to the rounded digit and change all remaining digits to the right of this to 0.
- Rounding a number **correct to one decimal place** is the same as rounding it to the **nearest tenth**.
- Rounding a number **correct to two decimal places** is the same as rounding it to the **nearest hundredth**.
- Rounding a number **correct to three decimal places** is the same as rounding it to the **nearest thousandth**.
- **Rounding rules for decimals**
  - If the digit after the one that is being rounded is less than 5 keep the rounded digit unchanged and delete all the following digits.
  - If the digit after the one that is being rounded is 5 or more then add 1 to the rounded digit and delete all the following digits.
- The number of **significant figures** in a result is the number of figures that are known with some degree of reliability.

Continued on next page

- **Rules for significant figures:**
  - All non-zero digits are significant.
  - Zeros between non-zero digits are significant.
  - Zeros to the left of the first non-zero digit are *not* significant.
  - Zeros placed after other digits but to the right of the decimal point are significant.
- **Rounding rules for significant figures**
  - If the $(n+1)$th figure is less than 5 then keep the nth figure unchanged.
  - If the $(n+1)$th figure is 5 or more then add 1 to this figure.
  - In both cases all the figures to the right of figure n should be deleted if they are to the right of the decimal point and should be replaced by zeros if they are to the left of the decimal point.
- To **estimate** the answer to a calculation, round all the numbers involved to 1 sf.
- The difference between an **estimated** or **approximated value** and the **exact value** is called the **error**:

$$\text{Error} = v_A - v_E$$

where v_A is the approximated value and v_E is the exact value.

- Percentage error $= \left| \dfrac{v_A - v_E}{v_E} \right| \times 100\%$

where v_A represents **approximated value** or **estimated value** and v_E represents the **exact value**.

Standard form

- A number is written in **standard form** if it is in the form $a \times 10^k$ where $1 \le a < 10$ and k is an integer.

SI units of measurement

- In Mathematical Studies the most common **SI base units** used are m, kg and s, and **derived units** are m² (area), m³ (volume), $km\,h^{-1}$ (velocity), $kg\,m^{-3}$ (density).
- To avoid writing very small or very large quantities we use **prefix** names and prefix symbols. Some of these are shown in this table.

Factor	Prefix	Symbol	Factor	Prefix	Symbol
10^3	kilo	k	10^{-3}	milli	m
10^2	hecto	h	10^{-2}	centi	c
10^1	deca	da	10^{-1}	deci	d

- There are some units that are **non-SI** units but are accepted for use with the SI because they are widely used in everyday life, for example, min, h, *l*.
- There are three temperature scales: **kelvin** (K), **Celsius** (°C) and **Fahrenheit** (°F).

A rational explanation

The Pythagorean School, around 2500 years ago, believed that all numbers were rational. This idea was expressed in terms of sticks of different lengths which could be measured exactly by a third, shorter stick.

For example these sticks:

can both be measured by this one: ▬

like this:

- What is the ratio of the shorter stick to the longer one?

- What fraction of the longer stick is the shorter stick?
- What fraction of the shorter stick is the longer stick?

Because a smaller stick exists which divides exactly into both larger sticks, the two sticks are said to be 'commensurable'. The early Pythagoreans believed that all numbers could be represented by a series of commensurable lines.

> The Pythagorean School had very strict rules and was a school of philosophy as well as of mathematics. Find out more about its principles and beliefs.

Brahmi		–	=	≡	+	Ⴑ	ℓ	૧	ৎ	?
Hindu		o	९	२	३	४	५	६	७	८	९
Arabic		.	١	٢	٣	٤	٥	٦	٧	٨	٩
Medieval		o	I	2	3	૪	୧	6	⅄	8	9
Modern		0	1	2	3	4	5	6	7	8	9

> ◀ Where do our numerals for zero to ten come from?
>
> ◀ When was zero discovered – or was it invented?

Hippasus demonstrates an irrational number

According to legend one of the Pythagoreans, Hippasus, first demonstrated that $\sqrt{2}$ was not rational. It is likely that Hippasus used the idea that sticks of length $\sqrt{2}$ and 1 could not both be measured by the same stick, no matter how small.

First, a few things Hippasus did know...

1 Pythagoras' theorem: this square with side length 1 has a diagonal of length $\sqrt{2}$.

Figure 1

2 If a stick could measure two larger sticks it could measure the difference between them. In the example above, the difference is 2 times the measuring stick.

So Hippasus reasoned that if there was a stick that could measure both the side and the diagonal of a square then it could measure the difference, shown in green in Figure 2.

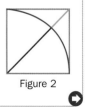

Figure 2

> William Dunham in his book *Journey through Genius* hints at how Hippasus might have done this.

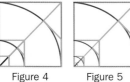

Figure 3

He knew enough circle theorems to deduce that all the green line segments (Figure 3) were the same length and so could be measured by the original stick. So could the part colored purple.

So he started again with the small square and diagonal and made the same picture within it... and again within that ...

Figure 4   Figure 5

He argued that because the square was getting smaller every time, the measuring stick must be even smaller, in the end vanishing, because the reduction could be repeated indefinitely. Because the stick got so small it would vanish, it must therefore not exist in the first place.

His colleagues were convinced, but they most certainly weren't happy, and they threw him off a ship leaving him to drown. No doubt the story has gained a few details over the years, but the discovery of irrational numbers did have a profound effect on the Greek mathematicians, who for several centuries left the study of Number and concentrated on the 'safe' topic of Geometry.

- Were irrational numbers created or discovered?
- Do irrational numbers exist?

Now we know that $\sqrt{2}$ was just the tip of the iceberg. Even though there are an infinite number of rationals, there are infinitely more irrationals.

Cantor's proof

Georg Cantor described infinitely big sets as either countably infinite or uncountably infinite. Countably infinite sets described a set where each member could be counted with the natural numbers: 1, 2, 3, 4, ... The process would go on forever, but because the members of the set had been put in some sort of order you could move forward counting them without leaving any out. He showed that rational numbers could be ordered in this way, but that it was impossible to do this with irrational numbers. Whatever order you devised, there would always be irrational numbers missing from the list.

Cantor's theories are today a standard (if slightly uncomfortable) part of mathematics, but at the time they caused even more controversy than Hippasus did in his day. Cantor was seen as undermining mathematics and his ideas were rejected by almost all contemporary mathematicians of

that time. He suffered from severe depression and finished his life in a sanatorium.

Cantor lived in Vienna during World War I, when the Austro-Hungarian Empire was collapsing. His fellow citizens were fearful of the change they saw around them. Perhaps Cantor 'changing' the concept of number was a step too far?

- Can mathematics develop 'in a bubble'?
- Can mathematicians ever be free from external influences?

Descriptive statistics

CHAPTER OBJECTIVES:

4.1–4.3 Discrete and continuous data: frequency tables; mid-interval values; upper and lower boundaries. Frequency histograms

4.4 Cumulative frequency tables; cumulative frequency curves; median and quartiles. Box and whisker diagrams

4.5 Measures of central tendency: mean; median; mode; estimate of a mean; modal class

4.6 Measures of dispersion: range, interquartile range, standard deviation

Before you start

You should know how to:

1 Collect and represent data using

  a a pictogram

Age 11	𝕏𝕏𝕏𝕏𝕏
Age 12	𝕏𝕏𝕏𝕏𝕏𝕏𝕏𝕏
Age 13	𝕏𝕏𝕏𝕏𝕏𝕏𝕏
Age 14	𝕏𝕏𝕏

  Key: 𝕏 = 1 Student

  b a bar chart

  c a pie chart

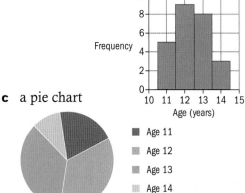

  - Age 11
  - Age 12
  - Age 13
  - Age 14

2 Set up axes on graphs using given scales.

Skills check

1 Maerwen wants to find out information about the numbers of men, women, boys and girls using a library. Design a suitable data-collection sheet to collect the information.

2 These data show the number of different colored sweets in a packet.

Color	blue	green	red	orange	yellow
Frequency	5	7	8	4	6

  a Draw a pictogram to represent these data.

  b Draw a bar chart to represent these data.

  c Draw a pie chart to represent these data.

3 On graph paper, draw a set of axes such that 1 cm represents 2 units on the x-axis and 1 cm represents 10 units on the y-axis.

Every country needs basic information on its population so that it can plan and develop the services it needs. For example, to plan a road network you need to know the size of the population so you can estimate the amount of traffic in an area.

To collect information on a population, governments often organize a census. A census is a survey of the **whole population** of a country.

The information collected includes data on age, gender, health, housing, employment and transport. The data are then analyzed and presented in tables, charts and spreadsheets. All data should be processed so that information on individuals is protected. The United Nations recommends that population censuses should be taken at least every 10 years.

In what other areas of society is mathematics used in a practical way?

What are the benefits of sharing and analyzing data from different countries?

When was the last census in your country? Is the census information published in the public domain? How has technology changed the way census data is collected and presented?

Investigation – population distribution

In the United Kingdom, there is a census every 10 years.

These population pyramids are based on information collected in the 2001 census. They show the distribution of age ranges in Tower Hamlets, London, and Christchurch, Dorset.

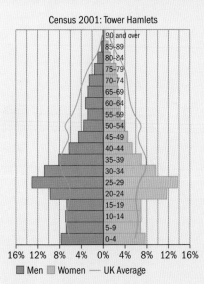

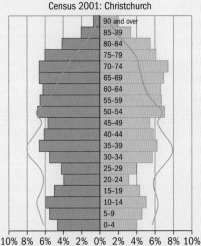

Compare the population pyramids for Tower Hamlets and Christchurch.

Simply based on these data, make a number of conjectures about these two areas.

Fully research the areas to test your conjectures. How accurate were you?

All information from the 2001 census can be found at www.ons.gov.uk by searching for '2001 census data'.

In this chapter you will organize data in frequency tables, graph data in a variety of diagrams, and analyze data using a range of measures.

2.1 Classification of data

There are two main types of data: **qualitative** and **quantitative**. Qualitative data are data that are not given numerically, for example, favorite color. Quantitative data are numerical.

Quantitative data can be further classified as **discrete** or **continuous**.

> → **Discrete data** are either data that can be **counted** or data that can only take **specific values**.

Examples of data that can be counted include the number of sweets in a packet, the number of people who prefer tea to coffee, and the number of pairs of shoes that a person owns.

How is education data used to investigate the link between the level of education and patterns of creating families and fertility?

Examples of data that can only take specific values include shoe size, hat size and dress size.

Is the number of grains of salt in a salt cellar discrete?

> → **Continuous data** can be **measured**. They can take any value within a range.

Examples of continuous data include weight, height and time.

Continuous data can be expressed to a required number of significant figures. The greater the accuracy required, the more significant figures the data will have.

The weighing scale was invented at a time when countries began trading materials and a standard measurement was required to ensure fair trading.

Time is a continuous measure because it can take any numerical value in a particular range. For example, the time taken for world-class sprinters to run 100 m can be recorded as any fraction of a second.

▲ The number of shoes and shoe size are examples of discrete data.

Population and sample

When conducting a statistical investigation, the whole of a group from which we may collect data is known as the **population**. It is not always possible, or even necessary, to access data for a whole population.

You can make conclusions about a population by collecting data from a sample. It is usually cheaper and quicker to collect data from a sample.

A **sample** is a small group chosen from the population.

A **random sample** is one where each element has the same chance of being included.

A **biased sample** is one that is not random.

It is important that a sample is random and not biased – it must be **representative** of the elements being investigated. To ensure that the different members of the population have an equal probability of being selected you could choose people by picking names out of a hat. Or you could assign a number to each member of the population and then choose numbers at random using the random number function on a GDC.

▲ A weighing scale gives us continuous data.

Can the wording of a survey question and the way the data are presented introduce bias?

Sampling will not be examined. However, if you use sampling when writing your Mathematical Studies project, you will need to discuss how you picked your sample and convince the moderator that it is indeed a random sample.

Are exit polls a good way of predicting the results of an election?

Example 1

Kiki wants to find out if there is any connection between eating breakfast and grades among students in her school. However, there are too many students in the school to ask everyone. She needs to pick a sample.

How can she make sure that the sample she picks is a random sample?

Answer

Kiki can use her GDC to generate random numbers and use the students who have those numbers on the school register.	*Does each student have the same chance of being included in her sample? If yes, it is a random sample.*

In market research, a sample of the population is interviewed in order to collect data about customers. Many research methods have been developed since companies began to carry out formal market research in the 1920s.

Example 2

Ayako is conducting a survey to find out how much money women who live in London spend on fashion in a month. She only interviews women coming out of Harrods (a very exclusive store). Is this a random sample?

Answer

No, because the sample will not come from the total population of women in London and some of the women she interviews may not even belong to the population.	*Is Ayako only asking 'women who live in London'?* *Do all women who live in London shop at Harrods?*

Exercise 2A

1 State whether these data are discrete or continuous.
   a The number of sweets in a packet
   b The heights of students in Grade 8
   c The dress sizes of a girls' pipe band
   d The number of red cars in a parking lot
   e The weights of kittens
   f The marks obtained by Grade 7 in a science test
   g The times taken for students to write their World Literature paper
   h The weights of apples in a 5 kg bag
   i The number of cm of rain each day during the month of April
   j The number of heads when a coin is tossed 60 times
   k The times taken for athletes to run a marathon
   l The number of visitors to the Blue Mosque each day.

2 State whether the following samples are random or biased.

   a When researching if people eat breakfast, only interview the people in the canteen.

   b When researching spending habits, interview every third person you meet.

   c When researching spending habits on cars, Josh interviews men exiting a garage.

   d When comparing GDP to child mortality, Eizo chooses the countries from a numbered list, by generating random numbers on his GDC.

   e When researching the sleeping habits of children, Adham distributes a questionnaire to the students in his school.

> GDP, gross domestic product, is the total value of goods produced and services provided in a country in a year.

2.2 Simple discrete data

When there is a large amount of data, it is easier to interpret if the data are organized in a **frequency table** or displayed as a graph.

Example 3

The numbers of sweets in 24 packets are shown below.

22 23 22 22 23 21 22 22 20 22 24 21
22 21 22 23 22 22 24 20 22 23 22 22

Organize this information in a frequency table.

Answer

Number of sweets	Tally	Frequency
20				2
21					3
22													13
23						4
24				2
	TOTAL	**24**

Draw a chart with three columns.

Write the possible data values in the 'Number of sweets' column.

Use tally marks to record each value in the 'Tally' column.

For each row, count up the tally marks and write the total in the 'Frequency' column.

Add up the values in the 'Frequency' column to work out the total frequency.

> Now you can see how many packets have each number of sweets.

Exercise 2B

1 The numbers of goals scored by Ajax football team during their last 25 games were:

   1 3 0 2 1 1 2 3 0 1 2 2 5 0 2 1 4 3 2 1 0 1 2 3 5

   Organize this information in a frequency table.

2 The numbers of heads obtained when twelve coins were tossed 50 times are recorded below.

8 3 5 7 1 9 2 10 5 12 7 6 6 8 12 4 10 2 6 6 8 4 5 11 3
4 6 8 6 7 5 3 11 2 10 5 6 7 5 8 9 2 10 11 0 12 3 6 6 5

Organize this information in a frequency table.

3 The ages of the girls in a hockey club are:

10 11 12 10 9 11 15 13 12 16 11 13 14 12 10 10 11 9 9 10
10 12 15 16 12 11 13 10 15 13 12 11 15 16 11 12 10 9 10 11

Organize this information in a frequency table.

4 It is stated that there are 90 crisps in a box.
Viktoras checked 30 boxes and the numbers of crisps in them are recorded below.

90 90 91 90 89 89 90 90 92 90 90 88 89 90 90
91 90 89 90 88 89 90 91 90 92 88 89 90 90 90

Organize this information in a frequency table.

5 Sean threw a dice 50 times. The numbers that appeared are shown below.

1 1 3 2 6 6 5 6 4 4 3 6 2 1 3 5 6 3 2 1 4 5 6 3 2
1 5 3 4 6 2 5 5 4 2 1 3 6 4 2 3 1 6 3 2 5 3 3 2 6

Organize this information in a frequency table.

6 The numbers of games played in matches at a badminton tournament are recorded below.

8 8 10 11 9 7 8 7 11 12 7 8 10 10 11 9 9 8 11 7 9 8

The raw data have been organized in the frequency table.

Games	Frequency
7	4
8	m
9	4
10	n
11	4
12	1

Write down the values of m and n.

2.3 Grouped discrete or continuous data

When there are a lot of data values spread over a wide range it is useful to **group** the data. Depending on the number of data values, there should be between 5 and 15 groups, or classes, of equal width.

The classes must cover the range of the values and they must not overlap – each data value must belong to only one class.

You can organize both discrete and continuous data in **grouped frequency tables**.

Example 4

Loni made 30 telephone calls one week. The times of the calls, in minutes, were recorded.

3.1	12.2	9.6	8.1	2.2	1.2	15.0	4.8	21.2	13.6
17.3	22.3	1.5	4.6	31.2	26.7	7.8	18.2	35.4	1.6
2.9	5.5	12.8	28.3	16.9	1.3	5.6	7.8	2.3	6.9

Organize this information in a grouped frequency table.

> The frequency table gives a much clearer picture of the data.

Answer

Time (t)	Frequency
$0 \leq t < 5$	10
$5 \leq t < 10$	7
$10 \leq t < 15$	3
$15 \leq t < 20$	4
$20 \leq t < 25$	2
$25 \leq t < 30$	2
$30 \leq t < 35$	1
$35 \leq t < 40$	1

First decide on the size and the number of classes:

Smallest number = 1.2 so classes start at 0.

Largest number = 35.4 so classes finish at 40.

Using a class width of 5, there will be (40 ÷ 5 =) 8 classes in total.

Exercise 2C

1 Organize each of these sets of data in a grouped frequency table.

a 2   5   12   21   7   9   25   31   17   19   22   23   15   24   5
    34   45   32   13   43   7   11   32   6   18   40   23   32   22   8

b 10   24   31   29   42   19   55   65   46   72   35   48   68   56   92
    12   33   77   56   45   82   76   56   34   12   78   89   45   59   32
    26   97   67   54   34   18   77   59   34   27   13   19   63   65   22

c 1   3   8   12   4   2   6   3   9   10   11   9   7   5   14   2   3   16
    9   5   13   14   4   8   17   3   15   19   5   3   9   10   11   14   15

Upper and lower boundaries

To find the **upper** and **lower boundaries** of a class, calculate the mean of the upper value from one class and the lower value from the following class.

Example 5

	Height (x cm)	Frequency
This table shows the heights of flowers in a garden. Write down	$0 \leq x < 10$	5
a the upper boundary of the first class	$10 \leq x < 20$	12
	$20 \leq x < 30$	21
b the lower boundary of the third class.	$30 \leq x < 40$	15
	$40 \leq x < 50$	6

Answers

a $\dfrac{10+10}{2} = 10$

Upper value of the first class is 10. Lower value of the second class is 10. The upper boundary of the first class is the mean of these two values.

b $\dfrac{20+20}{2} = 20$

Upper value of the second class is 20. Lower value of the third class is 20. The lower boundary of the third class is the mean of these two values.

Example 6

	Shoe size	Frequency
The table shows the numbers of pairs of shoes of each size sold in a shop one day. Write down	15–19	3
a the upper boundary of the first class and the last class	20–24	9
	25–29	12
	30–34	22
b the lower boundary of the first class and the fourth class.	35–39	45
	40–44	31

> These are European shoe sizes. What are the equivalent shoe sizes in your country?

Answers

a Upper boundary of the first class: $\dfrac{19+20}{2} = 19.5$

Upper boundary of the last class: $\dfrac{44+45}{2} = 44.5$

Upper value of the first class is 19. Lower value of the second class is 20. The upper boundary of the first class is the mean of these two numbers. Similarly for last class.

b Lower boundary of the first class: $\dfrac{14+15}{2} = 14.5$

Lower boundary of the fourth class: $\dfrac{29+30}{2} = 29.5$

Upper value of the previous class would be 14. Lower value of the first class is 15. The lower boundary of the first class is the mean of these two numbers. Similarly for fourth class.

> How could the shoe shop manager use this data?

1 Copy these tables and fill in the missing lower and upper boundary values.

a

Class	Lower boundary	Upper boundary
9–12		12.5
13–16		
17–20	16.5	
21–24		

b

Time (t seconds)	Lower boundary	Upper boundary
$2.0 \leq t < 2.2$		
$2.2 \leq t < 2.4$		
$2.4 \leq t < 2.6$		

Frequency histograms

A **frequency histogram** is a useful way to represent data visually.

> → To draw a frequency histogram, find the lower and upper boundaries of the classes and draw the bars between these boundaries. There should be no spaces between the bars.

In the Mathematical Studies course you will only deal with frequency histograms that have equal class intervals.

The class boundaries are plotted on the x-axis and the frequency values on the y-axis.

Here are the frequency histograms for Examples 5 and 6.

English statistician Karl Pearson (1857–1936) first used the term 'histogram' in 1895.

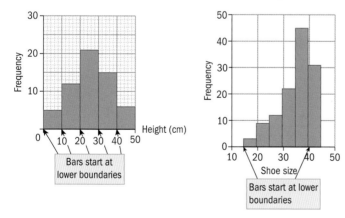

Bars start at lower boundaries

Bars start at lower boundaries

Exercise 2E

1 The costs, in euros, of 80 dinners are shown in the table.
Draw a histogram to display this information.

Cost of dinner in euros (c)	Frequency
$10 \leq c < 15$	2
$15 \leq c < 20$	8
$20 \leq c < 25$	11
$25 \leq c < 30$	25
$30 \leq c < 35$	14
$35 \leq c < 40$	11
$40 \leq c < 45$	6
$45 \leq c < 50$	3

2 The table shows the age distribution of teachers at Genius Academy.
  a Write down the lower and upper boundaries of each class.
  b Draw a histogram to represent the information.

Age (x)	Frequency
$20 \leq x < 30$	4
$30 \leq x < 40$	8
$40 \leq x < 50$	10
$50 \leq x < 60$	9
$60 \leq x < 70$	3

3 The masses of 150 melons are recorded in the table.
  a Write down the lower and upper boundaries of the third class.
  b Draw a histogram to represent the information.

Mass (x kg)	Frequency
$0.4 \leq x < 0.6$	21
$0.6 \leq x < 0.8$	36
$0.8 \leq x < 1.0$	34
$1.0 \leq x < 1.2$	29
$1.2 \leq x < 1.4$	18
$1.4 \leq x < 1.6$	12

4 The lengths of 100 worms (to the nearest cm) are given in the table.
  a Write down the lower and upper boundaries of each class.
  b Draw a histogram to represent the information.

Length (cm)	4	5	6	7	8	9	10
Frequency	18	20	26	15	8	6	7

5 50 people were asked how often they traveled by train each month. The results were:

8	7	10	5	23	4	16	9	62	28
14	53	29	11	34	33	68	75	12	79
22	54	67	55	13	32	41	58	36	2
26	80	65	38	52	71	2	16	36	40
18	24	52	64	76	16	6	18	28	40

  a Organize this information in a grouped frequency table.
  b Draw a histogram to represent the information graphically.

6 Yuri decided to count the number of weeds in one square metre of grass. He chose 80 plots of one square metre. The results for each square metre are:

22	24	21	12	8	14	34	62	54	6	28	42	35	22	14	18	9	24	12	18
31	47	17	9	35	24	41	52	38	19	5	23	31	65	32	46	15	13	74	22
9	13	22	55	47	52	14	13	21	19	52	33	71	12	22	17	58	42	31	16
2	15	31	73	45	31	12	8	4	33	42	57	61	48	43	27	14	5	14	26

a Organize this information in a grouped frequency table.

b Draw a histogram to represent the information graphically.

7 Simi recorded the numbers of vans per five minutes that drove down her street over a period of eight hours. Her results were:

Number of vans (x)	Frequency
$1 \leq x \leq 5$	12
$6 \leq x \leq 10$	23
$11 \leq x \leq 15$	31
$16 \leq x \leq 20$	13
$21 \leq x \leq 25$	9
$26 \leq x \leq 30$	5
$31 \leq x \leq 35$	2
$36 \leq x \leq 40$	1

a Write down the lower and upper boundaries of the fourth class.

b Draw a histogram to represent the information.

EXAM-STYLE QUESTION

8 The numbers of visitors per hour to the Taj Mahal are recorded in the table.

Time (t)	Number of visitors
$09{:}00 \leq t < 10{:}00$	324
$10{:}00 \leq t < 11{:}00$	356
$11{:}00 \leq t < 12{:}00$	388
$12{:}00 \leq t < 13{:}00$	435
$13{:}00 \leq t < 14{:}00$	498
$14{:}00 \leq t < 15{:}00$	563
$15{:}00 \leq t < 16{:}00$	436
$16{:}00 \leq t < 17{:}00$	250
$17{:}00 \leq t < 18{:}00$	232

Draw a histogram to represent this information.

2.4 Measures of central tendency

Data can be summarized by using a measure of central tendency such as the mode, median or mean.

> → The **mode** of a data set is the value that occurs most frequently.
>
> The **median** of a data set is the value that lies in the middle when the data are arranged in size order.
>
> The **mean** of a data set is the sum of all the values divided by the number of values.

When there are two 'middle' values, the median is the midpoint between the two middle values. To find the midpoint, add the two middle values and divide by two.

Example 7

Here is a set of data: 5 4 8 4 4 7 8 9 11 1 5
Find the mode, median and mean.

Answer

5 ④ 8 ④ ④ 7 8 9 11 1 5   *The value '4' occurs three times.*
Mode = 4

1 4 4 4 5 ⑤ 7 8 8 9 11   *First arrange the data in size order.*
Median = 5   *There are 11 entries, so the*

$$\text{median is the } \frac{11+1}{2} = 6\text{th value.}$$

Mean =   *The mean is the*

$$\frac{1+4+4+4+5+5+7+8+8+9+11}{11} \qquad \frac{\textit{sum of all the values}}{\textit{number of values}}$$

$$= \frac{66}{11}$$

Mean = 6

 You can also use a GDC to calculate the median and mean.
Enter the data values:

1.1	1.2 ▶	Desc Stats ▼		
A	B	C	D	
1	5			
2	4			
3	8			
4	4			
5	4			
A1	5			

How do you know which measure of central tendency is the best to use?

Can you mislead people by quoting statistics? For example, the numbers 1, 1, 100 have mode = 1, median = 1 and mean = 34.

You have to be aware that there may be outliers (isolated points outside the normal range of values) that skew the statistics.

What are the ethical implications of using statistics to mislead people?

GDC help on CD: *Alternative demonstrations for the TI-84 Plus and Casio FX-9860GII GDCs are on the CD.*

For help with entering data values, see Chapter 12, Section 2.1.

The GDC screen is too small to display all of the values in the list. Scroll down to see the remaining values.

The value of the mean is given by $\bar{x}$ (pronounced 'x-bar'):

1.1	1.2 ▷	Desc Stats ▼	⬚⬚
"Title"	"One−Variable Statistics"		
"x̄"	6.		
"Σx"	66.		
"Σx²"	478.		
"sx := Sn-1x"	2.86356		
"σx := σnx"	2.7303		
"n"	11.	▶	
"MinX"	1.		
"Q₁X"	4.	▶	
"Median X"	5		

2/99

The value of the median is shown as 'MedianX':

1.1	1.2 ▷	Desc Stats ▼	⬚⬚
"sx := Sn-1x"	2.86356		
"σx := σnx"	2.7303		
"n"	11.	▶	
"MinX"	1.		
"Q₁X"	4.		
"MedianX"	5.		
"Q₃X"	8.		
"MaxX"	11.		
"SSX := Σ(x−x̄)²"	82. I		

1/2

Exercise 2F

1 Calculate the mode, median and mean for each data set.
   a 7   3   8   9   1   10   1
   b 3   4   8   2   5   6   11   13   3   5   6   5

2 Calculate the values of **a**, **b**, **c**, **d** and **e** in this table.

Data	Median	Mode	Mean
Height (m): 1.52, 1.74, 1.83, 1.52, 1.67, 1.91	a	b	1.70
Age (years): 21, 34, 17, 22, 56, 38	28	none	c
Weight (kg): 54.7, 48.6, 63.2, 55.1, 77.9, 48.6	d	48.6	e

> The German psychologist Gustav Fechner (1801–1887) popularized the use of the median, although French mathematician and astronomer Pierre-Simon Laplace (1749–1827) had used it previously.

3 The weights of eight pumpkins are
   26.3 kg, 12.6 kg, 33.5 kg, 8.9 kg, 18.7 kg, 22.6 kg, 31.8 kg and 45.3 kg.
   a Find the median weight.
   b Calculate the mean weight.

EXAM-STYLE QUESTIONS

4 For these data the mode is 5, the median is 6 and the mean is 6.5.
   1   1   2   3 s 5   5   7   8   9   10 t 12   12

   Given that $s < t$, find the values of s and t.

5 Jin obtained marks of 76, 54 and 65 in his Physics, Biology and History examinations respectively.

   a Calculate his mean mark for the three examinations.
   b Find the mark that Jin must achieve in Mathematics so that the mean mark for the four examinations is exactly 68.

6 Zoe and Shun compared their test scores. Zoe had a mean of 81 after taking five tests and Shun had a mean of 78 after taking three tests. Each of them took one more test and ended up with the same mean score of 80.

  a Find the grade that Zoe gained on her sixth test.

  b Find the grade that Shun gained on his fourth test.

Mean, median and mode from a frequency table

> → For data in a frequency table, the **mode** is the entry that has the largest frequency.
>
> The **median** is the middle entry as the entries in the table are already in order. For n pieces of data, the median is the $\frac{n+1}{2}$th value.

The next example shows how to calculate the mean from a frequency table.

Example 8

Calculate the mode, median and mean of these data.

Number of sweets	Frequency
20	2
21	3
22	13
23	4
24	2
TOTAL	**24**

Sometimes a question asks for the 'modal value'. This means 'the mode'.

Answer

Mode = 22

Median = 22

'22' has the highest frequency (13).

Median is the $\frac{24+1}{2} = 12.5$th entry, so it is between the 12th and 13th entry. Both the 12th and 13th entries are 22, so the median = 22.

To calculate the mean: Label the first column x_i. Label the second column f_i. Add a third column and label it $f_i x_i$.

Number of sweets, x_i	Frequency, f_i	$f_i x_i$
20	2	40
21	3	63
22	13	286
23	4	92
24	2	48
TOTAL	**24**	**529**

$\text{Mean} = \dfrac{529}{24} = 22.0 \text{ (to 3 sf)}$

Work out $f_i \times x_i$ for each row:
$2 \times 20 = 40$
$3 \times 21 = 63$
$13 \times 22 = 286$
$4 \times 23 = 92$
$2 \times 24 = 48$
Work out the total of the f_i column and the total of the $f_i x_i$ column.

$\text{Mean} = \dfrac{\text{total of } f_i x_i}{\text{total of } f_i}$

→ The **mean** from a frequency table is:

$$\text{mean} = \frac{\text{total of } f_i \times x_i}{\text{total frequency}}$$

where f_i is the frequency of each data value x_i and $i = 1, ..., k$, where k is the number of data values.

 You can also use your GDC to calculate the mean and median from a frequency table.

Enter the data values: The value of the mean is given by $\bar{x}$:

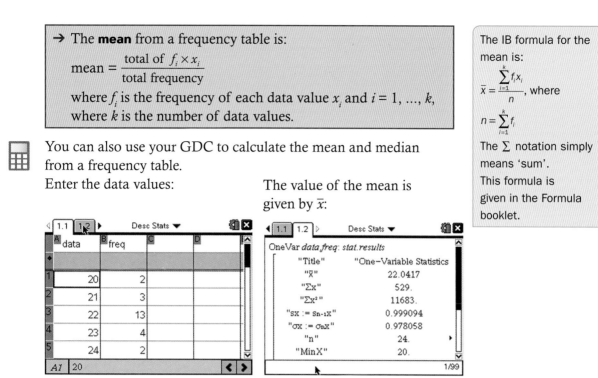

The IB formula for the mean is:

$$\bar{x} = \frac{\sum_{i=1}^{k} f_i x_i}{n}, \text{ where}$$

$$n = \sum_{i=1}^{k} f_i$$

The $\sum$ notation simply means 'sum'.

This formula is given in the Formula booklet.

The value of the median is shown as 'MedianX':

"σX := σnX"	0.978058
"n"	24.
"MinX"	20.
"Q₁X"	22.
"MedianX"	22.
"Q₂X"	22.5
"MaxX"	24.
"SSX := Σ(x−x̄)²"	22.9583

Exercise 2G

1 A dice is thrown 29 times and the score noted. The results are shown in the table.
  a Write down the modal score.
  b Write down the median score.
  c Calculate the mean score.

Score	Frequency
1	4
2	7
3	3
4	8
5	5
6	2

EXAM-STYLE QUESTION

2 The table shows the frequency of the number of visits to the doctor per year for a group of children.
  a How many children are in the group?
  b Write down the modal number of visits.
  c Calculate the mean number of visits.

Number of visits	0	1	2	3	4	5
Frequency	4	3	8	5	4	1

3 A bag contains six balls numbered 1 to 6. A ball is drawn at random and its number noted. The ball is then returned to the bag. The numbers for the first 30 draws are:

Number	1	2	3	4	5	6
Frequency	4	5	3	n	6	5

  a Write down the value of n.

  b Calculate the mean number.

  c Write down the modal number.

4 The table gives the frequency of grades achieved by students in an IB school.

  a Calculate the mean grade.

  b What percentage of students achieved a grade 4 or 5?

  c Write down the modal grade.

Grade	Frequency
1	1
2	6
3	19
4	34
5	32
6	18
7	10

Mean, median and mode for grouped data

For grouped data, you can find the modal class and an **estimate of the mean**.

> → For grouped data, the **modal class** is the group or class interval that has the largest frequency.

The next example shows how to calculate an estimate of the mean.

Example 9

The times, in seconds, taken to complete 200 bouts of sumo wrestling are shown in the table.

Time (t seconds)	Frequency
$0 \le t < 20$	37
$20 \le t < 40$	62
$40 \le t < 60$	46
$60 \le t < 80$	25
$80 \le t < 100$	11
$100 \le t < 120$	9
$120 \le t < 140$	6
$140 \le t < 160$	4
TOTAL	200

You do not know the exact data values for each group. Use the midpoint of each class interval as an estimate of the values in each group. You may also find the midpoint referred to as the 'mid-interval value'.

To find the midpoint of a class, find the mean of the class limits.

$$\text{midpoint} = \frac{\text{lower boundary} + \text{upper boundary}}{2}$$

Calculate **a** the modal class and **b** an estimate of the mean.

▶ Continued on next page

Answer

a Modal class = $20 \leq t < 40$

b

Time (t seconds)	Frequency, f_i	Midpoint, x_i	$f_i\, x_i$
$0 \leq t < 20$	37	10	370
$20 \leq t < 40$	62	30	1860
$40 \leq t < 60$	46	50	2300
$60 \leq t < 80$	25	70	1750
$80 \leq t < 100$	11	90	990
$100 \leq t < 120$	9	110	990
$120 \leq t < 140$	6	130	780
$140 \leq t < 160$	4	150	600
TOTAL	200		9640

Mean $= \dfrac{9640}{200} = 48.2$ (to 3 sf)

This class interval has the largest frequency (62).

*To work out an estimate of the mean, you must first work out the **midpoint** of each class interval. Add a third column and label it 'Midpoint, x_i'. Work out each midpoint:*

Midpoint of $0 \leq t < 20$: $\dfrac{0+20}{2} = 10$

Midpoint of $20 \leq t < 40$: $\dfrac{20+40}{2} = 30$

Midpoint of $40 \leq t < 60$: $\dfrac{40+60}{2} = 50$

Next add a fourth column and label it '$f_i x_i$'. Then work out $f_i \times x_i$ for each row:

$9 \times 110 = 990$
$6 \times 130 = 780$

Work out the total of the f_i column and the total of the $f_i\, x_i$ column.

Mean $= \dfrac{total\ of\ f_i x_i}{total\ f_i}$

→ To calculate an estimate of the **mean** from a grouped frequency table, use $\dfrac{total\ of\ f_i \times x_i}{total\ frequency}$

where f_i is the frequency and x_i is the corresponding midpoint of each class.

Why does this give an estimate of the mean and not an exact value?

You can also use a GDC to calculate an estimate of the mean from a grouped frequency table.

Enter the data values:

	A mid_pt	B freq	C	D
1	10	37		
2	30	62		
3	50	46		
4	70	25		
5	90	11		

1.1 1.2 ▶ Desc Stats ▼

A1 10

For help with entering data values, see Chapter 12, Section 2.2.

GDC help on CD: *Alternative demonstrations for the TI-84 Plus and Casio FX-9860GII GDCs are on the CD.*

The value of the mean is given by $\bar{x}$:

You were not asked for the median but the GDC works it out as part of the calculation screen (this too is only an estimate as we do not have all the individual values):

1.1	1.2 ▷	Desc Stats ▼	
OneVar *mid_pt,freq*: *stat.results*			
"Title"	"One–Variable Statistics"		
"$\bar{x}$"	48.2		
"Σx"	9640.		
"Σx^2"	686400.		
"$sx := s_{n-1}x$"	33.3816		
"$\sigma x := \sigma_n x$"	33.298		
"n"	200.	▶	
"MinX"	10.		

1/1

1.1	1.2 ▷	Desc Stats ▼	
"$\sigma x := \sigma_n x$"	33.298		
"n"	200.	▶	
"MinX"	10.		
"$Q_1 X$"	30.		
"MedianX"	50.		
"$Q_3 X$"	70.		
"MaxX"	150.		
"$SSX := \Sigma(x-\bar{x})^2$"	221752.		

1/99

Exercise 2H

1 The table shows the times taken for 25 cheetahs to cover a distance of 50 km.
   a Write down the modal class.
   b Calculate an estimate of the mean time taken.

Time taken (t minutes)	Frequency
$20 \le t < 22$	2
$22 \le t < 24$	5
$24 \le t < 26$	8
$26 \le t < 28$	4
$28 \le t < 30$	3
$30 \le t < 32$	2
$32 \le t < 34$	1

2 The speeds of vehicles passing under a bridge on a road are recorded in the table.
   a Write down the modal class.
   b Calculate an estimate of the mean speed of the vehicles.

Speed ($s\,km\,h^{-1}$)	Frequency
$60 \le s < 70$	8
$70 \le s < 80$	15
$80 \le s < 90$	12
$90 \le s < 100$	10
$100 \le s < 110$	8
$110 \le s < 120$	3
$120 \le s < 130$	4

3 The results of a Geography test for 25 students are given in the diagram.
   a Write down the modal class.
   b Calculate an estimate of the mean grade.

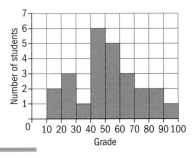

2.5 Cumulative frequency curves

> → The **cumulative frequency** is the sum of all of the frequencies up to and including the new value. To draw a **cumulative frequency curve** you need to construct a cumulative frequency table, with the upper boundary of each class interval in one column and the corresponding cumulative frequency in another. Then plot the upper class boundary on the *x*-axis and the cumulative frequency on the *y*-axis.

Example 10

A supermarket is open 24 hours a day and has a free car park. The number of parked cars each hour is monitored over a period of several days. The results are shown in the table.
Organize this information in a cumulative frequency table.
Draw a graph of the cumulative frequency.

Number of parked cars per hour	Frequency
0–49	6
50–99	23
100–149	41
150–199	42
200–249	30
250–299	24
300–349	9
350–399	5

Answer

Number of parked cars per hour	Frequency	Upper boundary	Cumulative frequency
0–49	6	49.5	6
50–99	23	99.5	29
100–149	41	149.5	70
150–199	42	199.5	112
200–249	30	249.5	142
250–299	24	299.5	166
300–349	9	349 5	175
350–399	5	399.5	180

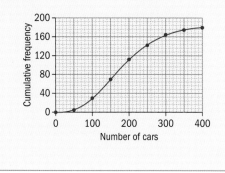

Add a third column and label it 'Upper boundary'.
Work out the upper boundary of each class:
$$Upper\ boundary = \frac{49 + 50}{2} = 49.5$$
$$Upper\ boundary = \frac{99 + 100}{2} = 99.5$$
$$Upper\ boundary = \frac{149 + 150}{2} = 149.5$$
Now add a fourth column and label it 'Cumulative frequency'.
Work out the cumulative frequency for each row:

$6 + 23 = 29$
$29 + 41 = 70$
$166 + 9 = 175$
$175 + 5 = 180$

> The final cumulative frequency value should equal the total frequency value. Cumulative frequency is always plotted on the **vertical** axis.

To draw the graph of the cumulative frequency, plot the value of the upper boundary against the cumulative frequency value.

Interpreting cumulative frequency graphs

We can use the cumulative frequency curve to find estimates of the **percentiles** and **quartiles**.

Percentiles separate large ordered sets of data into hundredths. Quartiles separate large ordered sets of data into quarters.

When the data are arranged in order, the lower quartile is the 25th percentile, the median is the 50th percentile (middle value) and the upper quartile is the 75th percentile.

> → To find the **lower quartile**, Q_1, read the value on the curve corresponding to $\frac{n+1}{4}$ on the cumulative frequency axis, where n is the total frequency.
>
> To find the median, read the value on the curve corresponding to $\frac{n+1}{2}$ on the cumulative frequency axis.
>
> To find the **upper quartile**, Q_3, read the value on the curve corresponding to $\frac{3(n+1)}{4}$ on the cumulative frequency axis.
>
> To find the **percentiles**, $p\%$, read the value on the curve corresponding to $\frac{p(n+1)}{100}$ on the cumulative frequency axis.
>
> To find the **interquartile range** subtract the lower quartile from the upper quartile: IQR $= Q_3 - Q_1$.

> The cumulative frequency curve is sometimes called an ogive.

> Per cent means out of 100.
> $\frac{1}{4} = 25\%$
> $\frac{1}{2} = 50\%$
> $\frac{3}{4} = 75\%$

> There are no universally agreed formulae for the quartiles. For large n and grouped data: n rather than $n + 1$ may be used.

> The IQR shows the spread of the middle 50% of the data

For any set of data:
- 25% or one-quarter of the values are between the smallest value and the lower quartile
- 25% are between the lower quartile and the median
- 25% are between the median and the upper quartile
- 25% are between the upper quartile and the largest value
- 50% of the data lie between the lower and upper quartiles.

In this cumulative frequency diagram (from the data in Example 10), $n = 180$.

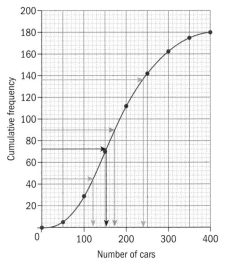

Lower quartile ≈ 120 (blue)

This is the value corresponding to $\frac{180+1}{4} = 45.25$.

Median ≈ 173 (green)

This is the value corresponding to $\frac{180+1}{2} = 90.5$.

Upper quartile ≈ 238 (orange)

This is the value corresponding to $\frac{3(180+1)}{4} = 135.75$.

40th percentile ≈ 153 (brown)

This is the value corresponding to $\frac{40(180+1)}{100} = 72.4$.

The interquartile range $\approx 238 - 120 = 118$

Example 11

50 contestants play the game of Oware. In total they have to play 49 games to arrive at a champion. The average times for the 49 games are given in the table.

Time (t minutes)	Frequency
$3 \leq t < 4$	4
$4 \leq t < 5$	12
$5 \leq t < 6$	18
$6 \leq t < 7$	9
$7 \leq t < 8$	3
$8 \leq t < 9$	2
$9 \leq t < 10$	1

The game of Oware is played all over the world and there is even an Oware Society.

Why must 50 contestants play 49 games to arrive at a champion? Can you prove this?

a Construct a cumulative frequency table for these data.

b Draw a cumulative frequency graph for these data.

c Use your graph to estimate
  i the lower quartile   **ii** the median
  iii the upper quartile   **iv** the interquartile range
  v the 30th percentile.

Answers

a

Time (t minutes)	Frequency	Upper boundary	Cumulative frequency
$3 \leq t < 4$	4	4	4
$4 \leq t < 5$	12	5	16
$5 \leq t < 6$	18	6	34
$6 \leq t < 7$	9	7	43
$7 \leq t < 8$	3	8	46
$8 \leq t < 9$	2	9	48
$9 \leq t < 10$	1	10	49

Check:
Total frequency: 4 + 12 + 18 + 9 + 3 + 2 + 1 = 49
Final cumulative frequency value = 49

b

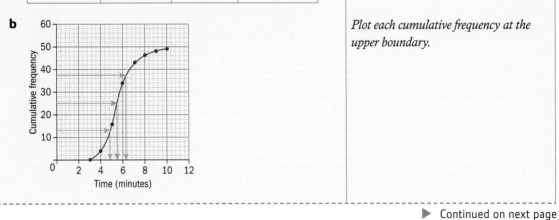

Plot each cumulative frequency at the upper boundary.

▶ Continued on next page

c i $n = 49$

$$\frac{n+1}{4} = \frac{50}{4} = 12.5$$

Lower quartile ≈ 4.7 minutes

> 25% of games last 4.7 minutes or less.

Read across from 12.5 on the vertical axis, then down to the horizontal axis.

ii $\dfrac{n+1}{2} = \dfrac{49+1}{2} = 25$

Median ≈ 5.5 minutes

> 50% of games last 5.5 minutes or less.

This is the value on the horizontal axis corresponding to 25 on the vertical axis.

iii $\dfrac{3(n+1)}{4} = \dfrac{3(49+1)}{4} = 37.5$

Upper quartile ≈ 6.4 minutes

> 75% of games last 6.4 minutes or less.

This is the value on the horizontal axis corresponding to 37.5 on the vertical axis.

iv Interquartile range $= 6.4 - 4.7$
$= 1.7$ minutes

> The 'middle' 50% of games last between 4.7 and 6.4 minutes.

v $\dfrac{30(n+1)}{100} = \dfrac{30(49+1)}{100} = 15$

30th percentile ≈ 4.9 minutes

> 30% of games last 4.9 minutes or less.

This is the value on the horizontal axis corresponding to 15 on the vertical axis.

Example 12

From this cumulative frequency graph find
i the median
ii the interquartile range
iii the 70th percentile.

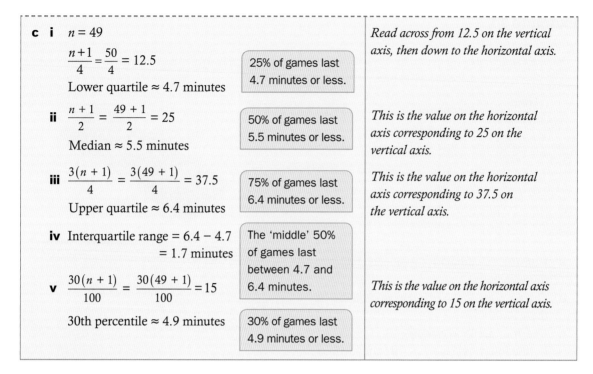

Answers

i $n = 120$

$$\frac{n+1}{2} = \frac{121}{2} = 60.5$$

Median ≈ 35

$n = 120$ from graph
Median is the value on the horizontal axis corresponding to
60.5 on the vertical axis.

ii Lower quartile $\dfrac{120+1}{4} = 30.25$th value

Lower quartile $= 26$

Upper quartile $\dfrac{3(120+1)}{4} = 90.75$th value

Upper quartile $= 46$

Interquartile range $\approx 46 - 26 = 20$

Interquartile range = upper quartile – lower quartile
Upper quartile is the value corresponding to 90.75 on the vertical axis.
Lower quartile is the value corresponding to 30.25 on the vertical axis.

iii $\dfrac{70(120+1)}{100} = 84.7$th value

70th percentile ≈ 43

This is the value corresponding to 84.7 on the vertical axis.

Exercise 2I

EXAM-STYLE QUESTIONS

1 A dice is tossed 50 times.
The number shown is
recorded each time and the
results are given in the table.

a Write down the value of N.

b Find the values of a, b and c.

Number	Frequency	Cumulative frequency
1	6	6
2	a	14
3	10	24
4	b	c
5	5	43
6	7	50
	N	

2 The table shows the percentages scored by candidates in a test.

Marks (%)	0–9	10–19	20–29	30–39	40–49	50–59	60–69	70–79	80–89	90–100
Frequency	1	5	7	11	19	43	36	15	2	1

Here is the cumulative
frequency table for the marks.

a Calculate the values of s and of t.

b Draw a cumulative frequency
graph for these data.

c Use your graph to estimate

   i the median mark

   ii the lower quartile

   iii the pass mark, if 40% of the
candidates passed.

Marks (%)	Cumulative frequency
< 9.5	1
< 19.5	6
< 29.5	s
< 39.5	24
< 49.5	43
< 59.5	86
< 69.5	t
< 79.5	137
< 89.5	139
≤ 100	140

3 A safari park is open to visitors
every day of the year. The numbers
of cars that pass through the park
each day for a whole year were
recorded and are shown in the table.

a Draw a cumulative frequency
graph to represent this
information.

b Find the median and the
interquartile range.

c On what percentage of days
were there more than 800 cars in the park?

Number of cars (n)	Frequency
$0 < n \le 150$	25
$150 < n \le 300$	36
$300 < n \le 450$	68
$450 < n \le 600$	102
$600 < n \le 750$	64
$750 < n \le 900$	41
$900 < n \le 1050$	19
$1050 < n \le 1200$	10

4 Sofia studied an article in the *Helsingborgs Dagblad*. She recorded the numbers of words in each sentence in the frequency table.

a Draw a cumulative frequency graph to represent this information.

b Work out the lower quartile, the median and the upper quartile of the data.

Number of words	Frequency
1–4	4
5–8	19
9–12	38
13–16	23
17–20	8
21–24	4
25–28	2
29–32	1
33–36	1

EXAM-STYLE QUESTIONS

5 A salmon farmer records the lengths of 100 salmon, measured to the nearest cm. The results are given in the table.

a Construct a cumulative frequency table for the data in the table.

b Draw a cumulative frequency curve.

c Use the cumulative frequency curve to find

  i the median length of salmon

  ii the interquartile range of salmon length.

Length of salmon (x cm)	Number of salmon
$25 < x \le 28$	3
$28 < x \le 31$	4
$31 < x \le 34$	11
$34 < x \le 37$	23
$37 < x \le 40$	28
$40 < x \le 43$	15
$43 < x \le 46$	12
$46 < x \le 49$	4
TOTAL	100

6 The table shows the times taken by 100 students to complete a puzzle.

Time (t minutes)	11–15	16–20	21–25	26–30	31–35	36–40
Number of students	6	13	27	31	15	8

a Construct a cumulative frequency table.

b Draw a cumulative frequency graph.

c Use your graph to estimate

  i the median time

  ii the interquartile range of the time

  iii the time within which 75% of the students completed the puzzle.

2.6 Box and whisker graphs

Another useful way to represent data is a **box and whisker graph** (or box and whisker plot).

A box and whisker graph looks something like this.

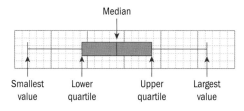

> → To draw a box and whisker graph, five pieces of information are needed: calculate the lower quartile, median and upper quartile for the data. Find the smallest and largest values.

Draw the box and whisker graph to scale on graph paper.

Note:

An **outlier** is a value that is much smaller or much larger than the other values.

Normally we consider an outlier to be a point with a value:

- less than 'the lower quartile − 1.5 × the interquartile range' or
- greater than 'the upper quartile + 1.5 × the interquartile range'.

> Outliers will not be examined but they may be useful for projects.

Example 13

A yacht club hosts an annual race. The numbers of people in each yacht are recorded in the table.

a Find the median number of people in a yacht.

b Find the upper and lower quartiles.

c Draw a box and whisker graph to represent the information.

Number of people	Frequency
4	1
5	8
6	16
7	25
8	28
9	16
10	5
TOTAL	99

▶ Continued on next page

Answers

a $n = 99$, so the median is the number of people in the $\dfrac{99 + 1}{2} = \dfrac{100}{2} = 50$th yacht.

The 50th yacht is in the group highlighted red.

Number of people	Frequency	Cumulative frequency
4	1	1
5	8	9
6	16	25
7	25	50
8	28	78
9	16	94
10	5	99

The median number of people is 7.

b The lower quartile is the number of people in the $\dfrac{99 + 1}{4} = 25$th yacht.

The 25th yacht is in the group highlighted green.

The lower quartile is 6.

The upper quartile is the number of people in the $\dfrac{3(99 + 1)}{4} = 75$th yacht.

The 75th yacht is in the group highlighted blue.

The upper quartile is 8.

c

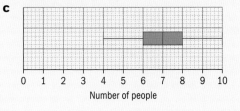

Number of people

Need five pieces of information to draw a box and whisker graph:
Smallest number of people = 4
*Lower quartile = 6 (from part **b**)*
*Median = 7 (from part **a**)*
*Upper quartile = 8 (from part **b**)*
Largest number = 10

You can also find all the data for the box and whisker graph using your GDC.

Enter the 'Number of people' and 'Frequency' into lists named 'Number' and 'Freq' in a Lists & Spreadsheets page. Add a Data & Statistics page and press MENU 2: Plot Properties | 5: Add X Variable with Frequency and select the two lists. To read the values use the touchpad to move the arrow over them.

These GDC screenshots show the median and the upper quartile (Q_3).

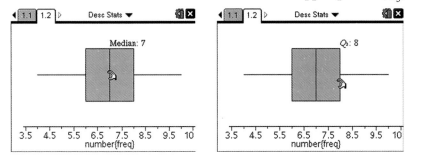

GDC help on CD: *Alternative demonstrations for the TI-84 Plus and Casio FX-9860GII GDCs are on the CD.*

Example 14

The weights, in kilograms, of 25 koala bears are:
4.3, 7.2, 5.6, 4.8, 10.7, 9.7, 5.6, 7.8, 8.2, 11.4, 7.9, 12.6, 13.1,
5.7, 9.9, 11.3, 13.4, 8.8, 7.5, 5.8, 9.2, 10.3, 12.1, 6.5, 8.6
Draw a box and whisker graph to represent the information.

Answer

First arrange the data in ascending order:
4.3, 4.8, 5.6, 5.6, 5.7, 5.8, 6.5,
7.2, 7.5, 7.8, 7.9, 8.2, 8.6, 8.8,
9.2, 9.7, 9.9, 10.3, 10.7, 11.3,
11.4, 12.1, 12.6, 13.1, 13.4
$n = 25$

Lowest value = 4.3

Lower quartile: $\dfrac{25+1}{4} = 6.5$,

so between 6th and 7th value
6th value = 5.8,
7th value = 6.5,

6.5th value = $\dfrac{5.8+6.5}{2} = 6.15$

Median = 8.6

$\left(\text{the } \dfrac{25+1}{2} = 13\text{th value}\right)$

Upper quartile: $\dfrac{3 \times 26}{4} = 19.5$,

so between 19th and 20th value
19th value = 10.7,
20th value = 11.3,

19.5th value = $\dfrac{10.7+11.3}{2} = 11$

Largest value = 13.4

To find the 6.5th value, calculate the mean of the 6th and 7th values.

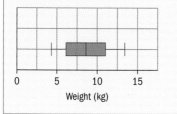

0   5   10   15
Weight (kg)

Need five pieces of information to plot a box and whisker graph.

Using a GDC:
Enter the data into a list. You do not need to put it in order.
These GDC screenshots show the median and the upper
quartile (Q₃).

GDC help on CD: *Alternative demonstrations for the TI-84 Plus and Casio FX-9860GII GDCs are on the CD.*

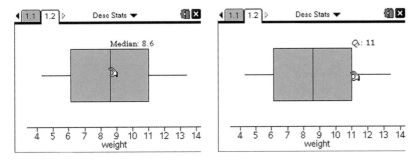

You cannot use a GDC to draw box and whisker graphs for grouped frequency tables.

Exercise 2J

1 The numbers of sweets in 45 bags are:

34 33 35 33 32 33 34 34 32 35 33 32 36 31 33 34
33 34 33 32 35 31 33 32 32 34 33 36 33 30 33 32
34 35 32 33 33 32 33 31 34 33 32 33 34

  a Construct a frequency table to represent the information.
  b Find the median, the lower quartile and the upper quartile.
  c Draw a box and whisker graph to represent this information.
Use a GDC to check your answer.

2 An experiment was performed 60 times. The scores from the
experiment were recorded in the table.
  a Find the median, the lower quartile and the upper quartile.
  b Draw a box and whisker graph to represent this information.
Use a GDC to check your answer.

Score	Frequency
1	6
2	12
3	13
4	15
5	8
6	6

EXAM-STYLE QUESTION

3 The cumulative frequency graph shows the weights,
in kg, of 200 sumo wrestlers.
  a Write down
    i the median
    ii the lower quartile
    iii the upper quartile.

The lightest wrestler weighs 125 kg and the heaviest
weighs 188 kg.
  b Draw a box and whisker graph to represent
the information.

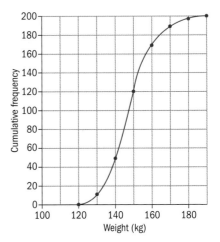

4 The heights, in cm, of 180 students are given in the
cumulative frequency table.
  a Draw a cumulative frequency diagram to represent
    this information.
  b Write down
    i the median
    ii the lower quartile and the upper quartile.
  c The smallest student is 146 cm and the tallest is 183 cm.
    Represent this information on a box and whisker graph.

Height (x cm)	Cumulative frequency
$x \le 145$	0
$x \le 150$	26
$x \le 155$	81
$x \le 160$	119
$x \le 165$	142
$x \le 170$	154
$x \le 175$	167
$x \le 180$	174
$x \le 185$	180

5 The table shows the heights, in cm, of 50 kangaroos.
  a Construct a cumulative frequency table and use it to
    draw the cumulative frequency curve.
  b Write down the median.
  c Find the lower quartile and the upper quartile.
  The smallest kangaroo is 205 cm and the tallest is 258 cm.
  d Draw a box and whisker graph to represent the
    information.

Height (x cm)	Frequency
$200 \le x < 210$	4
$210 \le x < 220$	6
$220 \le x < 230$	11
$230 \le x < 240$	22
$240 \le x < 250$	5
$250 \le x < 260$	2

Interpreting box and whisker graphs

For any set of data:
- 25% or one-quarter of the values are between the smallest
  value and the lower quartile
- 25% are between the lower quartile and the median
- 25% are between the median and the upper quartile
- 25% are between the upper quartile and the largest value
- 50% of the data lie between the lower and upper quartiles.

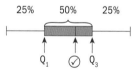

Example 15

The box and whisker graph shows the times, in hours, that it takes to build an igloo.
  a Write down the median time.
  b Find the interquartile range.
  c Write down the percentage of people who took less than 5.2
    hours to build an igloo.
  d x% of the people took more than 6.1 hours to build an igloo.
    Write down the value of x.

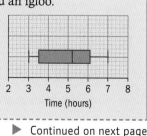

▶ Continued on next page

<table>
<tr><td colspan="2">Answers</td></tr>
<tr><td>a The median time is 5.2 hours.</td><td></td></tr>
<tr><td>b The interquartile range = 6.1 − 3.5
                = 2.6 hours.</td><td>From the graph, upper quartile = 6.1,
lower quartile = 3.5</td></tr>
<tr><td>c 50% of the people took less than
   5.2 hours to build an igloo.</td><td>5.2 hours = median (from part a)
50% of data are at or below this value</td></tr>
<tr><td>d 25% of the people took more than
   6.1 hours to build an igloo.</td><td>Upper quartile = 6.1
75% of data are at or below this value</td></tr>
</table>

Exercise 2K

1 The box and whisker graphs represent the scores on a Psychology test for 40 boys and 40 girls.

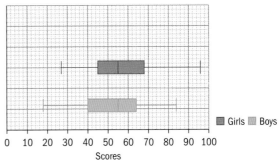

  a Find the median score for the boys and the girls.

  b Write down the interquartile range for the boys' scores and the girls' scores.

  c Write down the percentage of boys that scored more than 55.

  d Write down the percentage of girls that scored more than 68.

2 The box and whisker graph represents the number of faults made by horses in a jumping competition. Write down

  a the lowest number of faults

  b the median

  c the interquartile range

  d the largest number of faults

  e the percentage of horses that had fewer than six faults.

EXAM-STYLE QUESTION

3 The box and whisker graph represents the ages of the teachers at Myschool High.

  a Write down the age of the youngest teacher.

  b Write down the median age.

  c If 25% of the teachers are older than x, write down the value of x.

  d Find the interquartile range of the ages.

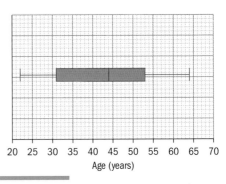

Extension material on CD:
Worksheet 2 - Standard deviation, standardization and outliers

2.7 Measures of dispersion

Measures of dispersion measure how spread out a set of data is. The simplest measure of dispersion is the **range**.

> → The **range** is found by subtracting the smallest value from the largest value.

Example 16

The numbers of piglets in the litters of 10 pigs are:

| 10 | 12 | 12 | 13 | 15 | 16 | 9 | 10 | 14 | 11 |

Find the range.

Answer

| Range = 16 − 9 = 7 | *Identify the largest value (16) and the smallest value (9).* |

The **interquartile range** is found by subtracting the lower quartile, Q_1, from the upper quartile, Q_3: IQR = $Q_3 - Q_1$.

To work out the lower and upper quartiles the values must be arranged in size order

Example 17

Find the interquartile range of this data set.

| 4 | 5 | 6 | 6 | 7 | 8 | 10 | 10 | 11 | 14 | 15 |

Answer

| Q_1 is the $\frac{11+1}{4}$ = 3rd number, so $Q_1 = 6$. Q_3 is the $\frac{3(11+1)}{4}$ = 9th number, so $Q_3 = 11$. IQR = 11 − 6 = 5 | *There are 11 numbers so n = 11.* |

Using a GDC:
Enter the data into a list. Then use One Variable Statistics. Scroll down to find the quartiles.
The value of Q_1 is given as 'Q_1X' and Q_3 as 'Q_3X'.

You can use a GDC for drawing graphs for frequency tables but not for grouped frequency tables.

1.1	1.2 ▷	Desc Stats ▼	
"σx := σnx"		3.48551	
"n"		11.	▶
"MinX"		4.	▶
"Q₁X"		6.	
"MedianX"		8.	
"Q₂X"		11.	
"MaxX"		15.	
"SSX := Σ(x−x̄)²"		133.636	

2/99

GDC help on CD: *Alternative demonstrations for the TI-84 Plus and Casio FX-9860GII GDCs are on the CD.*

For finding the interquartile range from a cumulative frequency graph see page 62. For finding the interquartile range from a box and whisker graph see page 71.

Exercise 2L

1 For each set of data calculate
  i the range **ii** the interquartile range.
  a 6 3 8 5 2 9 11 21 15 8
  b 5 3 6 8 9 12 10 9 8 13 16 12 9 11 8
  c

Price of main course in euros	Frequency
18	6
19	4
20	5
21	8
22	3
23	2
24	5
25	4

Standard deviation

The **standard deviation** is a measure of dispersion that gives an idea of how the data values are related to the mean.

Example 18

Find the mean and standard deviation of this data set.
4 5 6 8 12 13 2 5 6 9 10 9 8 3 5

Answer
Mean = 7
Standard deviation = 3.10 (to 3 sf)

Using a GDC:
Enter the data.
Mean is indicated by $\bar{x}$.
Standard deviation is indicated by σ_x.

1.1 1.2 ▷	*Desc Stats ▼	🔲🗙
OneVar *number*, 1: *stat.results*		
"Title"	"One–Variable Statistics"	
"$\bar{x}$"	7.	
"Σx"	105.	
"Σx^2"	879.	
"sx := Sn-1x"	3.20713	
"σx := σnx"	3.09839	
"n"	15.	▶
"MinX"	2.	
		1/99

When is the standard deviation of a set of data small?
Can the standard deviation equal zero?

Why do we take the square root to find the standard deviation?

Why is the standard deviation sometimes called the root-mean-square deviation?

You are expected to use a GDC to calculate standard deviations.

GDC help on CD: *Alternative demonstrations for the TI-84 Plus and Casio FX-9860GII GDCs are on the CD.*

Example 19

50 students were asked the total number of points that they received on their IB Diploma. The results are shown in the table.

Score on IB Diploma	Boys	Girls
31	0	3
32	2	4
33	6	3
34	11	5
35	4	3
36	1	2
37	0	1
38	1	2
39	0	2

Use your GDC to calculate the mean and standard deviation for the boys and girls separately and comment on your answer.

Answer

Boys' mean = 34

Boys' standard deviation = 1.23 (to 3 sf)

Girls' mean = 34.3 (to 3 sf)

Girls' standard deviation = 2.41 (to 3 sf)

Both the boys and the girls have a mean of about 34 points. The standard deviation for the boys is small, which implies that most boys achieved close to 34 points. However, the standard deviation for the girls is larger which implies that some girls will have much less than 34 points and some will have much more.

Using a GDC:

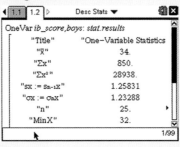

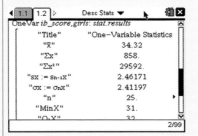

To make a comment, compare the mean to the corresponding standard deviation.

> Is standard deviation a mathematical discovery or an invention?

It is often impossible to find the mean and standard deviation for a whole population. This could be due to time restrictions, financial constraints or other reasons.

If we have, say, a random sample of 12 babies' heights from the UK, then the standard deviation of those 12 babies' heights is given as 'σ_x' on a GDC. This is the one we use for Mathematical Studies.

If we wanted to estimate the standard deviation of all the babies' heights in the UK, based on our random sample, then we would use 's_x' on the GDC.

The IB notation for standard deviation is s_n. When you use your GDC, choose σ_x

Exercise 2M

1 For each set of data calculate the standard deviation.

a 5   3   6   8   9   12   10   9   8   13   16   12   9   11   8

b

Price of main course in euros	Frequency
18	6
19	4
20	5
21	8
22	3
23	2
24	5
25	4

2 Calculate the mean and standard deviation for these data.

6   3   8   5   2   9   11   21   15   8

3 An experiment was performed 50 times. The scores from the experiment were recorded in the table.
  a Write down the range.
  b Find the interquartile range.
  c Find the mean and standard deviation.

Score	Frequency
1	4
2	12
3	11
4	15
5	6
6	2

4 A boat club hosts an annual race. The numbers of people in each boat are recorded in the table.
  a Write down the range.
  b Find the interquartile range.
  c Find the mean and standard deviation.

Number of people	Frequency
4	2
5	7
6	25
7	15
8	30
9	16
10	5

5 The numbers of telephone calls to a call center were monitored every hour for a month. The data collected are shown in the table.

Use your GDC to find
a the mean number of calls per hour
b the standard deviation
c the range
d the interquartile range.

Number of calls per hour	Frequency
60	18
62	45
64	40
66	55
68	31
70	32
72	15
74	13
76	14
78	16

6 The mean of these numbers is 33.

16   41   24 x 62   18   25

a Find the value of x.
b Calculate the standard deviation.
c Find the range.
d Find the interquartile range.

7 80 plants were measured and their heights (correct to the nearest cm) recorded in the table.
a Write down the value of m.
b Find the mean height.
c Find the standard deviation of the heights.
d Find the interquartile range of the heights.

Height (cm)	Frequency
10	7
11	m
12	21
13	22
14	11
15	7
16	3

8 The 60 IBDP students at Golden Globe Academy complete a questionnaire about the number of pairs of shoes that they own. The results are shown in the table.
a Find the range and interquartile range.
b Find the mean and standard deviation.

Pairs of shoes	Frequency
5	6
6	8
7	15
8	10
9	5
10	12
11	1
12	3

9 The times taken for 50 students to complete a crossword puzzle are shown in the table.

Time (*m* minutes)	Frequency
$15 \leq m < 20$	3
$20 \leq m < 25$	7
$25 \leq m < 30$	10
$30 \leq m < 35$	11
$35 \leq m < 40$	12
$40 \leq m < 45$	5
$45 \leq m < 50$	2

> Use the midpoint of each class to estimate the mean and the standard deviation of grouped data.

Find an approximation for the mean and standard deviation.

10 The percentage marks obtained for an ITGS (Information Technology for a Global Society) test by the 25 boys and 25 girls at Bright High are shown in the table.

  a Calculate an estimated value for the mean and standard deviation for the girls and the boys separately.

  b Comment on your findings.

Girls' frequency	Percentage mark	Boys' frequency
0	$0 \leq x < 10$	2
0	$10 \leq x < 20$	1
0	$20 \leq x < 30$	1
3	$30 \leq x < 40$	1
5	$40 \leq x < 50$	5
7	$50 \leq x < 60$	9
8	$60 \leq x < 70$	2
2	$70 \leq x < 80$	0
0	$80 \leq x < 90$	2
0	$90 \leq x < 100$	2

Review exercise

Paper 1 style questions

1 The mean of the twelve numbers listed is 6.

  3 4 *a* 8 3 5 9 5 8 6 7 5

  a Find the value of *a*.
  b Find the median of these numbers.

2 The mean of the ten numbers listed is 5.

  4 3 *a* 6 8 4 6 6 7 5

  a Find the value of *a*.
  b Find the median of these numbers.

3 For the set of numbers

3 4 1 7 6 2 9 11 13 6 8 10 6

   a calculate the mean

   b find the mode

   c find the median.

4 The lengths of nine snakes, in metres, are:

6.5 4.6 7.2 5.0 2.4 3.9 12.9 10.3 6.1

   a **i** Find the mean length of the snakes.

     ii Find the standard deviation of the length of the snakes.

   b Find the median length of the snakes.

5 A survey was conducted of the number of bathrooms in 150 randomly chosen houses. The results are shown in the table.

Number of bathrooms	1	2	3	4	5	6
Number of houses	79	31	22	10	5	13

   a State whether the data are discrete or continuous.

   b Write down the mean number of bathrooms per house.

   c Write down the standard deviation of the number of bathrooms per house.

6 The table shows the age distribution of members of a chess club.

Age (years)	Number of members
$20 \leq x < 30$	15
$30 \leq x < 40$	23
$40 \leq x < 50$	34
$50 \leq x < 60$	42
$60 \leq x < 70$	13

   a Calculate an estimate of the mean age.

   b Draw a histogram to represent these data.

7 Using the cumulative frequency graph, write down the value of

   a the median

   b the lower quartile

   c the upper quartile

   d the interquartile range.

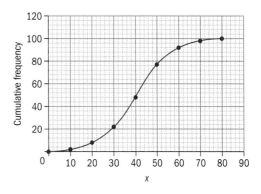

8 The numbers of horses counted in 35 fields are represented in the table.
Draw a box and whisker graph to represent this information.

Number of horses	Frequency
8	4
10	9
12	7
15	12
21	3

Paper 2 style questions

1 Nineteen students carried out an experiment to measure gravitational acceleration in cm s^{-2}.
The results are given to the nearest whole number.

96 97 101 99 100 98 99   94 96 100

97 98 101 98   99 96 96 100 97

a Use these results to find an estimate for
  i the mean value for the acceleration
  ii the modal value for the acceleration.
b **i** Construct a frequency table for the results.
  ii Use the table to find the median value and the interquartile range.

2 A gardener wanted to estimate the number of weeds on the sports field.
He selected at random 100 sample spots, each of area 100 cm^2, and counted the number of weeds in each spot.

The table shows the results of his survey.

Number of weeds	Frequency
0–4	18
5–9	25
10–14	32
15–19	14
20–24	7
25–29	4

a **i** Construct a cumulative frequency table and use it to draw the cumulative frequency curve.
  ii Write down the median number of weeds.
  iii Find the percentage of spots that have more than 19 weeds.
b **i** Estimate the mean number of weeds per spot.
  ii Estimate the standard deviation of the number of weeds per spot.

The area of the field is 8000 m^2.
  iii Estimate the total number of weeds on the field.

3 The marks for a test are given in the frequency table.
a Complete a cumulative frequency table and use it to draw the cumulative frequency curve.
b Find the median mark.
c Find the interquartile range.

60% of the candidates passed the examination.
d Find the pass mark.
e Given that the lowest mark was 9 and the highest was 98, draw a box and whisker graph to represent the information.

Mark, x	Frequency
$0 \le x < 10$	3
$10 \le x < 20$	14
$20 \le x < 30$	21
$30 \le x < 40$	35
$40 \le x < 50$	42
$50 \le x < 60$	55
$60 \le x < 70$	43
$70 \le x < 80$	32
$80 \le x < 90$	15
$90 \le x < 100$	10

4 The cumulative frequency graph shows the monthly incomes, in South African Rand, ZAR, of 150 people.

a Write down the median and find the interquartile range.

b Given that the lowest monthly income is 6000 ZAR and the highest is 23 500 ZAR, draw a box and whisker graph to represent this information.

c Draw a frequency table for the monthly incomes.

d Use your GDC to find an estimate of the mean and standard deviation of the monthly incomes.

5 The weights of 200 female athletes are recorded in the table.

a Write down the modal group.

b Calculate an estimate of the mean and the standard deviation.

c Construct a cumulative frequency table and use it to draw the cumulative frequency graph.

d Write down the median, the lower quartile and the upper quartile.

e The lowest weight is 47 kg and the heaviest is 76 kg. Use this information to draw a box and whisker graph.

Weight (w kg)	Frequency
$45 \leq w < 50$	4
$50 \leq w < 55$	16
$55 \leq w < 60$	45
$60 \leq w < 65$	58
$65 \leq w < 70$	43
$70 \leq w < 75$	28
$75 \leq w < 80$	6

6 A group of 60 women were asked at what age they had their first child. The information is shown in the histogram.

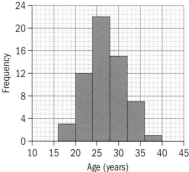

a Calculate an approximation for the mean and standard deviation.

b Write down the modal class.

c Construct a cumulative frequency table for the data and draw the cumulative frequency curve.

d Use your graph to find the median and interquartile range.

e Given that the youngest age was 16 and the oldest was 39, draw a box and whisker graph to represent the information.

7 The average times, to the nearest second, that 100 participants waited for an elevator are shown in the table.

a Write down the modal class.

b Calculate an estimate of the mean time and the standard deviation.

c Construct a cumulative frequency table and use it to draw the cumulative frequency graph.

d Write down the median and interquartile range.

Time (t seconds)	Frequency
$0 \leq t < 10$	5
$10 \leq t < 20$	19
$20 \leq t < 30$	18
$30 \leq t < 40$	22
$40 \leq t < 50$	16
$50 \leq t < 60$	12
$60 \leq t < 70$	8

8 The cumulative frequency graph shows the daily number of visitors to the Mausoleum on Tiananmen Square in the month of January.

a Write down the median, the lower quartile and the upper quartile.

b Given that the least number of visitors was 4000 and the most was 5700, draw a box and whisker graph to represent the information.

c Construct a frequency table for this information.

d Write down the modal class.

e Calculate an estimate of the mean and the standard deviation.

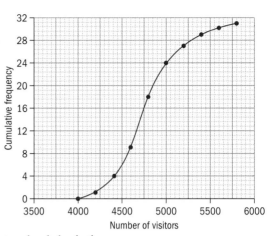

9 The cumulative frequency graph shows the weights, in kg, of 200 professional wrestlers.

a Construct a grouped frequency table for this information.

b Write down the modal class.

c Calculate an estimate of the mean weight.

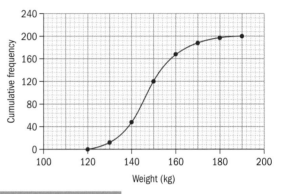

CHAPTER 2 SUMMARY

Classification of data

- **Discrete data** are either data that can be counted or data that can only take specific values.
- **Continuous data** can be measured. They can take any value within a range.

Grouped discrete or continuous data

- To draw a **frequency histogram**, find the lower and upper boundaries of the classes and draw the bar between these boundaries. There should be no spaces between the bars.

Measures of central tendency

- The **mode** of a data set is the value that occurs most frequently.
- The **median** of a data set is the value that lies in the middle when the data are arranged in size order.
- The **mean** of a data set is the sum of all the values divided by the number of values.
- For data in a frequency table, the **mode** is the entry that has the largest frequency.

Continued on next page

- The **median** is the middle entry as the entries in the table are already in order. For n pieces of data, the median is the $\frac{n+1}{2}$th value.
- The **mean** from a frequency table is:
$$\text{mean} = \frac{\text{total of } f_i \times x_i}{\text{total frequency}}$$
where f_i is the frequency of each data value x_i and $i = 1, ..., k$, where k is the number of data values.
- For grouped data, the **modal class** is the group or class interval that has the largest frequency.
- To calculate the **mean** from a grouped frequency table, an estimate of the mean is
$$\frac{\text{total of } f_i \times x_i}{\text{total frequency}}$$
where f_i is the frequency and x_i is the corresponding midpoint of each class.

Cumulative frequency curves

- The **cumulative frequency** is the sum of all of the frequencies up to and including the new value. To draw a **cumulative frequency curve** you need to construct a cumulative frequency table, with the upper boundary of each class interval in one column and the corresponding cumulative frequency in another. Then plot the upper class boundary on the x-axis and the cumulative frequency on the y-axis.
- To find the **lower quartile**, Q_1, read the value on the curve corresponding to $\frac{n+1}{4}$ on the cumulative frequency axis, where n is the total frequency.
- To find the median, read the value on the curve corresponding to $\frac{n+1}{2}$ on the cumulative frequency axis.
- To find the **upper quartile**, Q_3, read the value on the curve corresponding to $\frac{3(n+1)}{4}$ on the cumulative frequency axis.
- To find the **percentiles**, $p\%$, read the value on the curve corresponding to $\frac{p(n+1)}{100}$ on the cumulative frequency axis.
- To find the **interquartile range** subtract the lower quartile from the upper quartile: $\text{IQR} = Q_3 - Q_1$.

Box and whisker graphs

- To draw a box and whisker graph, five pieces of information are needed: calculate the lower quartile, median and upper quartile for the data. Find the smallest and largest values.

Measures of dispersion

- The **range** is found by subtracting the smallest value from the largest value.
- The **interquartile range** is found by subtracting the lower quartile, Q_1, from the upper quartile, Q_3: $\text{IQR} = Q_3 - Q_1$.
- The standard deviation is often referred to as the 'root-mean-square deviation' because we find the **deviation** of each entry from the mean, then we **square** these values and find the **mean** of the squared values, and, finally, we take the square **root** of this answer.

Statistically speaking

Descriptive statistics describe the basic features of a data set.

Descriptive statistics reduce lists of data into a simple summary such as a single average (a number) or a visual form such as a graph or diagram.

Morals and statistics

Case study 1

A company has 3 employees and a boss.

The employees earn 2500 euros a month and the boss earns 25 000 euros a month.

A report in the local newspaper states that the average salary in the company is 8125 euros a month.

- Which average has the newspaper used?

- Does this average give a fair representation of the average salary?

- Which would be the most appropriate average to use? Why?

Case study 2

Ten appliances were tested and the number of faults each one had recorded below.

0 0 0 0 0 0 15 19 25 31

The company advertises that the average number of faults is 0.

- Which average has the company used?

- Is the company misleading people?

- Is it morally acceptable for the company to advertise the 'facts' in this manner?

> *"Statistical thinking will one day be as necessary for efficient citizenship as the ability to read and write."*
>
> H. G. Wells (1866–1946)

- What do you think H. G. Wells meant?
- Do you agree with him?

- How accurate are these visual representations:
  - X-rays
  - Snapshots
  - Paintings?

> *"There are three kinds of lies: lies, damned lies, and statistics."*
>
> Benjamin Disraeli (1084–1881)
> Popularized by
> Mark Twain (1835–1910)

- Do statistics 'lie'?
- Are all statistics 'accurate'?

Misleading graphs

▼ What is wrong with this graph?

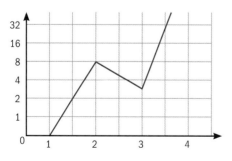

▼ What is wrong with this 3D histogram?

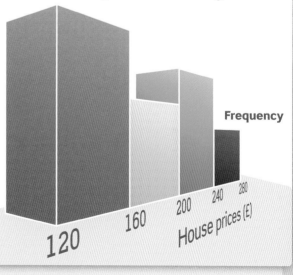

3 Geometry and trigonometry 1

CHAPTER OBJECTIVES:

5.1 Gradient; intercept; equation of a line in two dimensions; point of intersection of two lines; parallel lines; perpendicular lines

5.2 Use of sine, cosine and tangent ratios to find the sides and angles of a right-angled triangle; angles of depression and elevation

5.3 Use of the sine rule and the cosine rule; use of area of a triangle; construction of labeled diagrams from verbal statements

Before you start

You should know how to:

1 Use Pythagoras' theorem, e.g. Find the length of side AC if $AB = 2\,cm$ and $BC = 5\,cm$.

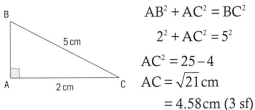

$$AB^2 + AC^2 = BC^2$$
$$2^2 + AC^2 = 5^2$$
$$AC^2 = 25 - 4$$
$$AC = \sqrt{21}\,cm$$
$$= 4.58\,cm \text{ (3 sf)}$$

2 Find the midpoint of a line and the distance between two given points, e.g. If A is $(-3, 4)$ and B is $(1, 2)$:

a Midpoint of AB is $\left(\dfrac{-3+1}{2}, \dfrac{4+2}{2}\right)$

$= (-1, 3)$

b Distance AB is

$$\sqrt{(1-(-3))^2 + (2-4)^2} = \sqrt{4^2 + 2^2}$$
$$= \sqrt{20}$$
$$= 4.47 \text{ (3 sf)}$$

Skills check

1 a Find the height h of triangle ABC.

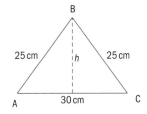

b Find the side length of a square if the length of its diagonal is $10\,cm$.

2 a A is the point $(-3, 5)$ and B is the point $(3, 7)$.
  i Find the midpoint of AB.
  ii Find the distance AB.

b The midpoint between C$(2, p)$ and D$(q, -4)$ is M$(2.5, 1)$. Find the values of p and q.

When a lighthouse is designed, distances and angles are involved. The lighthouse needs to be tall enough for the light to be seen from a distance. Also, if a boat comes close into shore, could it still see the light?

In a manned lighthouse, if the keeper lowers his eyes and looks down to a boat, he can use this angle and the height of the lighthouse to calculate how far out the boat is. Problems like this can be solved using **trigonometry** – the part of mathematics that links the angles and lengths of a triangle. Using trigonometry you can calculate lengths that cannot be measured directly, such as the distance from a boat to the base of the lighthouse, the height of a tree or a building, the width of a river, etc.

This chapter will show you how to draw diagrams to represent these types of problem, and use trigonometry to solve them.

▲ *Les Eclaireurs Light-house*, in Tierra del Fuego, Argentina, is near Ushuaia, the southernmost city in the world. It has been guiding sailors since 1920.

This lighthouse is sometimes called *the Lighthouse at the End of the World*, as in the Jules Verne novel. However, the writer was inspired by the lighthouse *San Juan de Salvamento*, on another island nearby.

Geometry came before trigonometry. In Egypt, after the flood seasons, nobody would know the borders of their lands so geo-metry, the art of 'earth-measuring', was invented. Geometry and trigonometry complement each other and are used extensively in a number of fields such as astronomy, physics, engineering, mechanics and navigation.

3.1 Gradient of a line

A bread factory has two bread-making machines, A and B.
Both machines make 400 kg of bread per day **at a constant rate**.
Machine A makes 400 kg in 10 hours.
Machine B makes the 400 kg in 8 hours.
For each machine, these graphs show the number of kilograms of
bread made, y, in x hours. For example, in 2 hours machine A makes
80 kilograms of bread and machine B makes 100 kilograms of bread.

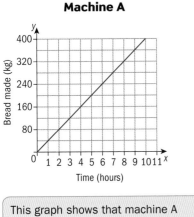

Machine A

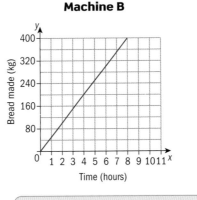

Machine B

> This graph shows that machine A makes **40** kg of bread per hour.

> This graph shows that machine B makes **50** kg of bread per hour.

The next graph shows the number of kilograms of bread
made by both machines.

The line for machine B is steeper than the line for machine A.
The **gradient** of a line tells you how steep it is. The
gradient of line B is greater than the gradient of line A.

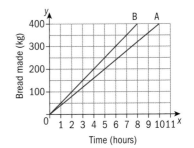

The gradient of a line $= \dfrac{\text{vertical step}}{\text{horizontal step}}$

Gradient of line A $= \dfrac{\text{vertical step}}{\text{horizontal step}}$

$= \dfrac{400}{10} = 40$

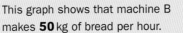

> The gradient tells you the rate at
> which the machine is working:
> A's rate = 40 kg per hour and
> B's rate = 50 kg per hour

Gradient of line B $= \dfrac{\text{vertical step}}{\text{horizontal step}}$

$= \dfrac{400}{8} = 50$

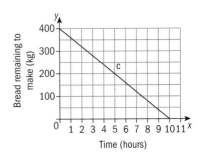

This graph shows the number of kilograms of bread still
to be made by machine A. At the beginning of the day the
machine has 400 kg to make, after 1 hour the machine has
360 kg to make, and so on.

Line C has a negative gradient; it slopes downwards from left to right.

$$\text{Gradient of line C} = \frac{\text{vertical step}}{\text{horizontal step}}$$

$$= \frac{-400}{10}$$

$$= -40$$

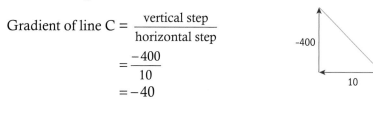

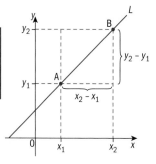

Each hour there is 40 kg less bread to be made.

→ If $A(x_1, y_1)$ and $B(x_2, y_2)$ are two points that lie on line L, the gradient of L is $m = \dfrac{y_2 - y_1}{x_2 - x_1}$

Note that the suffix order 2, then 1 in the gradient formula is the same in both the numerator and denominator.

Example 1

Find the gradient of the line L that passes through the points
a $A(1, 5)$ and $B(2, 8)$
b $A(0, 4)$ and $B(3, -2)$
c $A(2, 6)$ and $B(-1, 6)$
d $A(1, 5)$ and $B(1, -2)$

Answers

a $\left.\begin{array}{l} x_1 = 1 \\ y_1 = 5 \\ x_2 = 2 \\ y_2 = 8 \end{array}\right\} \Rightarrow m = \dfrac{y_2 - y_1}{x_2 - x_1}$

$$= \frac{8 - 5}{2 - 1} = 3$$

Substitute into the gradient formula.

Gradient = 3
For each 1 unit that x increases, y increases 3 units.

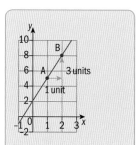

b $\left.\begin{array}{l} x_1 = 0 \\ y_2 = 4 \\ x_2 = 3 \\ y_2 = -2 \end{array}\right\} \Rightarrow m = \dfrac{y_2 - y_1}{x_2 - x_1}$

$$= \frac{-2 - 4}{3 - 0} = -2$$

Substitute into the gradient formula.

Gradient = −2
*For each 1 unit that x **increases**, y **decreases** by 2 units.*

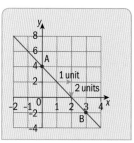

c $\left.\begin{array}{l} x_1 = 2 \\ y_1 = 6 \\ x_2 = -1 \\ y_2 = 6 \end{array}\right\} \Rightarrow m = \dfrac{y_2 - y_1}{x_2 - x_1}$

$$= \frac{6 - 6}{-1 - 2} = 0$$

Substitute into the gradient formula.
Gradient = 0
*For each 1 unit that x **increases**, y **remains constant**. The line is horizontal.*

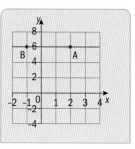

▶ Continued on next page

d $\left.\begin{array}{l} x_1 = 1 \\ y_1 = 5 \\ x_2 = 1 \\ y_2 = -2 \end{array}\right\} \Rightarrow m = \dfrac{y_2 - y_1}{x_2 - x_1}$ *Substitute into the gradient formula.*

$$= \frac{-2-5}{1-1} = \frac{-7}{0}$$

Remember *that division by zero is not defined therefore* **the gradient of this line is not defined**. *The line is vertical.*

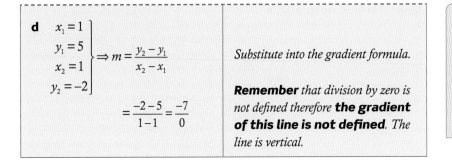

Exercise 3A

1 Plot the points A(2, 7), B(0, 9), C(0, −9) and D(2, −7) on a graph.
Find the gradients of the lines

  a AB   **b** AC

  c BD   **d** CD

2 For each of these lines

    i write down the coordinates of the points A and B

    ii find the gradient.

> The scales on the x-axis and y-axis are not always the same.

a

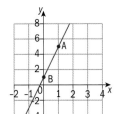

b

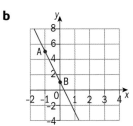

c

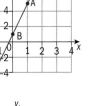

d

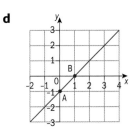

e

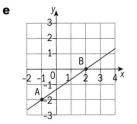

f

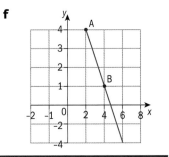

Example 2

a Draw a line that passes through the point A(1, 4) with gradient –1.

b Draw a line that passes through the point A(0, –2) with gradient $\frac{2}{3}$.

> Road gradients are often given as percentages or ratios. How do road signs show gradient in your country?

Answers

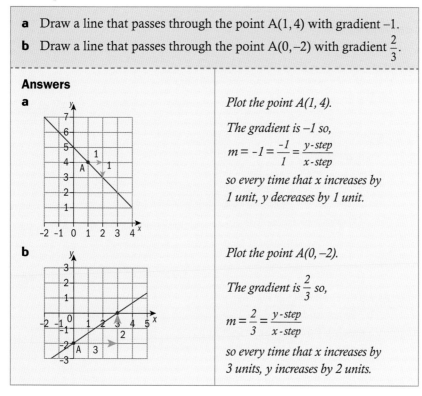

a Plot the point A(1, 4).

The gradient is –1 so,

$$m = -1 = \frac{-1}{1} = \frac{y\text{-}step}{x\text{-}step}$$

so every time that x increases by 1 unit, y decreases by 1 unit.

b Plot the point A(0, –2).

The gradient is $\frac{2}{3}$ so,

$$m = \frac{2}{3} = \frac{y\text{-}step}{x\text{-}step}$$

so every time that x increases by 3 units, y increases by 2 units.

Exercise 3B

1 a Draw a line with gradient $\frac{1}{2}$ that passes through the point A(0, 3).

 b Draw a line with gradient –3 that passes through the point B(1, 2).

 c Draw a line with gradient 2 that passes through the point C(3, –1).

2 For each of these lines, points A, B and C lie on the same line.

   i find the gradient of line AB.

   ii find the second coordinate of point C:

 a A(2, 5), B(3, 7) and C(4, p)

 b A(0, 2), B(1, 6) and C(2, t)

 c A(0, 0), B(1, –5) and C(2, q)

 d A(0, –1), B(1, 0) and C(4, s)

 e A(–5, 1), B(–6, 4) and C(–4, r)

> You may use a graph or the gradient formula, $m = \frac{y_2 - y_1}{x_2 - x_1}$.

3 The gradient of the line passing through points P(–1, 5) and Q(a, 10) is 4.

 a Write down an expression in terms of a for the gradient of PQ.

 b Find the value of a.

4 In line MN, every time that x increases by 1 unit, y increases by 0.5 units. Point M is (2, 6) and point N is (–3, t).

 a Write down the gradient of MN.

 b Write down an expression for the gradient of MN in terms of t.

 c Find the value of t.

Parallel lines

> → **Parallel lines** have the **same gradient**. This means that
> - if two lines are parallel then they have the same gradient
> - if two lines have the same gradient then they are parallel.

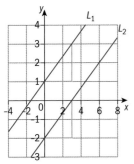

The symbols, $L_1 \parallel L_2$ mean 'L_1 is parallel to L_2'.

Note that, although the gradient of a vertical line is not defined, two vertical lines are parallel.

Example 3

Line L_1 passes through the points A$(0,-3)$ and B$(-7,4)$.
a Find the gradient of L_1. **b** Draw and label L_1.
c Draw and label a second line L_2 passing through the origin and parallel to L_1.

Answers

a $m = \dfrac{4-(-3)}{-7-0} = -1$

Substitute into the gradient formula.

b and **c**

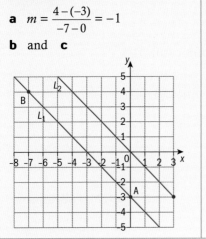

For L_1, plot A and B and join them. For L_2, draw a line through the origin parallel to L_1.

Remember that the **origin** is the point O$(0,0)$, the point where the x-axis and the y-axis meet.

Exercise 3C

1 Line L_1 passes through the points A(2, 5) and B(0, −4).
   a Find the gradient of L_1. **b** Draw and label L_1.
   c Draw and label a second line L_2 passing through the point C$(0,2)$ and parallel to L_1.

2 Decide whether each line is parallel to the y-axis, the x-axis or neither:
   a the line passing through the points P(1, 7) and Q(12, 7)
   b the line passing through the points P(1, 7) and T(1, −3)
   c the line passing through the points P(1, 7) and M(2, 5).

3 Complete these statements to make them true.

    a Any horizontal line is parallel to the _____-axis.

    b Any vertical line is parallel to the _____-axis.

    c Any horizontal line has gradient equal to _____.

4 PQ is parallel to the x-axis. The coordinates of P and Q are $(5, 3)$ and $(8, a)$ respectively. Write down the value of a.

5 MN is parallel to the y-axis. The coordinates of M and N are respectively $(m, 24)$ and $(-5, 2)$. Write down the value of m.

Perpendicular lines

→ Two lines are **perpendicular** if, and only if, they make an angle of 90°.

This means that
- if two lines are perpendicular then they make an angle of 90°
- if two lines make an angle of 90° then they are perpendicular.

> The x-axis and the y-axis are perpendicular.

The next example shows you the **numerical relationship** between the gradients of two perpendicular lines that are not horizontal and vertical.

> Any vertical line is perpendicular to any horizontal line.

Example 4

The diagram shows two perpendicular lines L_1 and L_2.

    a Find the gradients of L_1 and L_2.

    b Show that the product of their gradients is equal to -1.

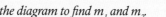

Answers

a Let m_1 be the gradient of L_1 and m_2 the gradient of L_2.

    $m_1 = 2$ and $m_2 = -\dfrac{1}{2}$

b $2 \times -\dfrac{1}{2} = -1$

Use the diagram to find m_1 and m_2.

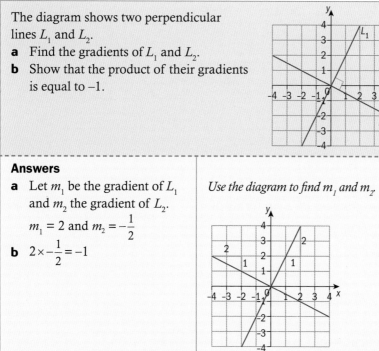

> Note that the gradient of L_1 is positive and the gradient of L_2 is negative.

> In general, if the gradient of a line is m, the gradient of a perpendicular line is $-\dfrac{1}{m}$

> a and b are reciprocal if $a \times b = 1$ or $a = \dfrac{1}{b}$
> For example:
> 2 and $\dfrac{1}{2}$, $\dfrac{4}{3}$ and $\dfrac{3}{4}$

→ Two lines are **perpendicular** if the product of their gradients is -1.

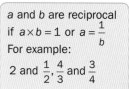

Exercise 3D

1 Which of these pairs of numbers are negative reciprocals?

a 2 and $-\dfrac{1}{2}$ **b** $-\dfrac{4}{3}$ and $\dfrac{3}{4}$ **c** 3 and $\dfrac{1}{3}$ **d** -1 and 1

2 Which of these pairs of gradients are of perpendicular lines?

a $\dfrac{2}{5}$ and $\dfrac{5}{2}$ **b** $\dfrac{4}{3}$ and $-\dfrac{3}{4}$ **c** -3 and $-\dfrac{1}{3}$ **d** 1 and -1

3 Find the gradient of lines that are perpendicular to a line with gradient

a -3 **b** $\dfrac{2}{3}$ **c** $-\dfrac{1}{4}$ **d** 1 **e** -1

4 Find the gradient of any line perpendicular to the line passing through the points

a $A(-2,6)$ and $B(1,-1)$ **b** $A(5,10)$ and $B(0,-2)$

5 Each diagram shows a line and a point A.

i Write down the gradient of the line.

ii Write down the gradient of any line that is perpendicular to this line.

iii Copy the diagram and draw a line perpendicular to the red line passing through the point A.

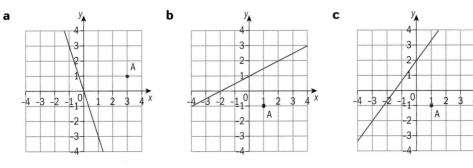

EXAM-STYLE QUESTIONS

6 Line L_1 passes through the points $P(0,3)$ and $Q(-2,a)$.

a Find an expression for the gradient of L_1 in terms of a.

L_1 is perpendicular to line L_2. The gradient of L_2 is 2.

b Write down the gradient of L_1.

c Find the value of a.

7 The points $A(3,5)$ and $B(5,-8)$ lie on the line L_1.

a Find the gradient of L_1.

A second line, L_2, is perpendicular to L_1.

b Write down the gradient of L_2.

L_2 passes through the points $P(5,0)$ and $Q(t,2)$.

c Find the value of t.

3.2 Equations of lines

The coordinates x and y of **any** point on a line L are linked by an equation, called the **equation of the line**.

This means that:

- If a point Q lies on a line L then the coordinates of Q satisfy the equation of L.
- If the coordinates of any point Q satisfy the equation of a line L, then the point Q lies on L.

> → The equation of a straight line can be written in the form $y = mx + c$, where
> - m is the **gradient**
> - c is the y-**intercept** (y-coordinate of the point where the line crosses the y-axis).

$y = mx + c$ is the **gradient-intercept** form of the straight line equation.

Example 5

The line L passes through the point A$(1, 7)$ and has gradient 5.
Find the equation of L.
Give your answer in the form $y = mx + c$.

Answer
Let P(x, y) be **any** point on L.
The gradient of L is 5

$$\left.\begin{array}{l} x_1 = 1 \\ y_1 = 7 \\ x_2 = x \\ y_2 = y \end{array}\right\} \Rightarrow \frac{y - 7}{x - 1} = 5$$

Use the gradient formula with A and P, and equate to 5.

$y - 7 = 5(x - 1)$ *Multiply both sides by $(x - 1)$*
$y - 7 = 5x - 5$ *Expand brackets.*
$y = 5x + 2$ *Add 7 to both sides.*
  $y = mx + c$ where $m = 5$ and
  $c = 2$

Use A$(7, 1)$ to check:
$7 = 5 \times 1 + 2$

Check that
- *the coordinates of the point A$(1, 7)$ satisfy the equation of the line.*

<aside>
Values for variables x and y are said to **satisfy** an equation if, when the variables are replaced by the respective values, the two sides of the equation are equal.
</aside>

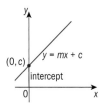

<aside>
The equation $y = mx + c$ is in the Formula booklet. You will revisit this equation again in Chapter 4.
</aside>

<aside>
As well as $y = mx + c$, some people express the equation of a line as $y = ax + b$ or $y = mx + b$
</aside>

<aside>
Note that in the equation $y = 5x + 2$
- 5 multiplies x, and the gradient of the line is $m = 5$
- Putting $x = 0$ in the equation of L,
$y = 5 \times 0 + 2 = 2$
Therefore the point $(0, 2)$ lies on L.
</aside>

Example 6

The line L has gradient $\frac{1}{3}$ and passes through A$(2, -1)$.

a Find the equation of L. Give your answer in the form $y = mx + c$.
b Write down the point of intersection of L with the y-axis.
c Find the point of intersection of L with the x-axis.
d Draw the line L showing clearly the information found in **b** and **c**.

Answers

a $y = \frac{1}{3}x + c$

$$-1 = \frac{1}{3} \times 2 + c$$

$$-1 = \frac{2}{3} + c$$

$$c = -\frac{5}{3}$$

$$y = \frac{1}{3}x - \frac{5}{3}$$

b $\left(0, -\frac{5}{3}\right)$

c $0 = \frac{1}{3}x - \frac{5}{3}$

$$\frac{1}{3}x = \frac{5}{3}$$

$$x = 5$$

Therefore L intersects the
x-axis at the point $(5, 0)$.

*Substitute $m = \frac{1}{3}$ in the equation
$y = mx + c$.*

*Substitute the coordinates of point
A(2,−1) in the equation of the line.*

Make c the subject of the equation.

*Substitute c in the equation of the
line.*
*The line crosses the y-axis at the
point (0, c).*

*Any point on the x-axis has the form
(k, 0).*
*Substitute y = 0 in the equation
of L.*

> Note that you could
> find the equation
> of L using the
> same method as in
> Example 5.

d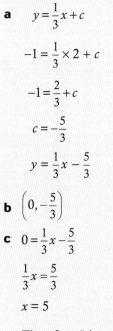

Exercise 3E

1 Find the equation of a line with
  a gradient 3 that passes through the point A$(1, 4)$
  b gradient $\frac{5}{3}$ that passes through the point A$(4, 8)$
  c gradient -2 that passes through the point A$(-3, 0)$
  Give your answers in the form $y = mx + c$.

2 For each of these lines write down

    i the gradient

    ii the point of intersection with the *y*-axis

    iii the point of intersection with the *x*-axis.

    a $y = 2x + 1$ **b** $y = -3x + 2$ **c** $y = -x + 3$ **d** $y = -\dfrac{2}{5}x - 1$

EXAM-STYLE QUESTIONS

3 A line has equation $y = \dfrac{3(x-6)}{2}$.

    a Write the equation in the form $y = mx + c$.

    b Write down the gradient of the line.

    c Write down the *y*-intercept.

    d Find the point of intersection of the line with the *x*-axis.

4 The line AB joins the points A$(2, -4)$ and B$(1, 1)$.

    a Find the gradient of AB.

    b Find the equation of AB in the form $y = mx + c$

5 The line PQ joins the points P$(1, 3)$ and Q$(2, 5)$.

    a Find the gradient of PQ.

    b Find the equation of PQ in the form $y = mx + c$

    c Find the gradient of all lines perpendicular to PQ.

    d Find the equation of a line perpendicular to PQ that passes through A$(0, 2)$.

6 Line L_1 has gradient 3 and is perpendicular to line L_2.

    a Write down the gradient of L_2.

    Line L_2 passes through the point P$(5, 1)$.

    b Find the equation of L_2. Give your answer in the form $y = mx + c$

    c Find the *x*-coordinate of the point where L_2 meets the *x*-axis.

7 Find the equations of these lines, in the form $y = mx + c$

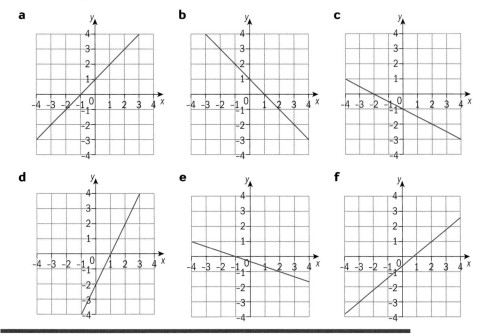

Example 7

a Line L joins the points A(–3, 5) and B(1, 2).
   Find the equation of line L.
   Give your answer in the form $ax + by + c = 0$ where $a, b, c \in \mathbb{Z}$

b The point $Q\left(\dfrac{5}{3}, t\right)$ lies on L. Find the value of t.

> The equation
> $ax + by + d = 0$ is
> called **the general
> form** and is also in
> the Formula booklet.

Answers

a The gradient of L is
   $$m = \frac{2-5}{1-(-3)} = -\frac{3}{4}$$

   Use the gradient formula with the coordinates of A and B.

   Let P(x, y) be **any** point on L.
   The gradient of L is also

   Use the gradient formula with A and P (or B and P).

   $$\left.\begin{array}{l} x_1 = -3 \\ y_1 = 5 \\ x_2 = x \\ y_2 = y \end{array}\right\} \Rightarrow m = \frac{y-5}{x-(-3)}$$

   $$\frac{y-5}{x-(-3)} = -\frac{3}{4}$$

   Equate gradients.

   $$4(y - 5) = -3(x + 3)$$

   Cross multiply.
   $$\frac{a}{b} = \frac{c}{d} \Leftrightarrow a \times d = b \times c$$

   $$4y - 20 = -3x - 9$$
   $$3x + 4y - 11 = 0$$

   Expand brackets.
   Rearrange equation to form
   $ax + by + d = 0$
   $a = 3, b = 4, d = -11$
   Check that both points A and B satisfy the equation of the line.

> Note that any multiple
> of this equation would
> also be correct as
> long as
> $a, b, d \in \mathbb{Z}$, e.g.
> $-3x - 4y + 11 = 0$ or
> $6x + 8y - 22 = 0$

b The point $Q\left(\dfrac{5}{3}, t\right)$ lies on L
   so its coordinates must satisfy
   the equation of L.
   $$3x + 4y - 11 = 0$$
   $$3 \times \frac{5}{3} + 4 \times t - 11 = 0$$
   $$5 + 4t - 11 = 0$$
   $$4t - 6 = 0$$
   $$4t = 6$$
   $$t = 1.5$$

   Substitute the coordinates of Q in the equation of L.

   Solve for t.

> **Discuss**: How many
> points do we need to
> determine a line?
> **Investigate:** the
> meaning of the word
> 'collinear'. When do
> we say that three
> or more points are
> collinear?

→ The equation of a straight line can be written in the form
   $$ax + by + c = 0$$

   where a, b and $c \in \mathbb{Z}$.

Exercise 3F

1 Find the equations of these lines. Give your answers in the form
$ax + by + c = 0$ where $a, b, c \in \mathbb{Z}$.
  a A line with gradient -4 that passes through the point $A(5,0)$.
  b A line with gradient $\frac{1}{2}$ that passes through the point $A(2,3)$.
  c The line joining the points $A(3,-2)$ and $B(-1,3)$.
  d The line joining the points $A(0,5)$ and $B(-5,0)$.

2 Rewrite each of these equations in the form $y = mx + c$.

> Make y the subject of the formula

  a $3x + y = 0$ b $x + y + 1 = 0$ c $2x + y - 1 = 0$
  d $2x - 4y = 0$ e $6x + 3y - 9 = 0$

3 The line L has equation $3x - 6y + 6 = 0$.
  a Write down the equation of L in the form $y = mx + c$.
  b Write down the x-intercept.
  c Write down the y-intercept.

4 The equation of a line is $y = 2x - 6$
  a Which of these points lie on this line?

  $A(3,0), B(0,3), C(1,-4), D(4,2), E(10,12), F(5,4)$

  b The point $(a, 7)$ lies on this line. Find the value of a.
  c The point $(7, t)$ lies on this line. Find the value of t.

5 The equation of a line is $-6x + 2y - 2 = 0$
  a Which of these points lie on this line?

  $A(1,4), B(0,1), C(1,0), D(2,6), E(-\frac{1}{3}, 0), F(-1, 2)$

  b The point $(a, 3)$ lies on this line. Find the value of a.
  c The point $(10, t)$ lies on this line. Find the value of t.

6 The table has four equations and four pairs of conditions. Match
each equation with the pair of conditions that satisfies that line.

Equation		Conditions	
A	$6x - 3y + 15 = 0$	E	The x-intercept is 2.5 and the y-intercept is 5
B	$y = 2x - 5$	F	The gradient is -2 and the line passes through the point $(1,-7)$
C	$10x + 5y + 25 = 0$	G	The line passes through the points $(0, -5)$ and $(2.5,0)$
D	$y = -2x + 5$	H	The y-intercept is $(0,5)$ and the gradient is 2

EXAM-STYLE QUESTION

7 The line L_1 has equation $2x - y + 6 = 0$
  a Write down the gradient of L_1.
  b Write down the y-intercept of L_1.
  c The point $A(c, 1.5)$ lies on L_1. Find the value of c.
  d The point $B(5, t)$ lies on L_1. Find the value of t.

  Line L_2 is parallel to L_1.
  e Write down the gradient of L_2.
  f Find the equation of L_2 if it passes through $C(0,4)$.

8 The line L_1 joins the points A $(1, 2)$ and B $(-1, 6)$.

   a Find the equation of L_1.

C is the point $(10, -16)$.

   b Decide whether A, B and C are collinear, giving a reason for your answer.

Vertical and horizontal lines

Vertical lines are parallel to the y-axis.
Horizontal lines are parallel to the x-axis.

Investigation – vertical and horizontal lines

The diagram shows two **vertical** lines, L_1 and L_2.

1 a Write down the coordinates of at least five points lying on L_1.
   b What do you notice about the coordinates of the points from **a**? What do their coordinates have in common?
   c What is the condition for a point to lie on L_1? Write down this condition in the form $x = k$ where k takes a particular value.

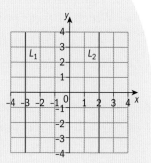

2 a Write down the coordinates of at least five points lying on L_2.
   b What do you notice about the coordinates of the points from **a**? What do their coordinates have in common?
   c What is the condition for a point to lie on L_2? Write down this condition in the form $x = k$ where k takes a particular value.

3 What is the equation of a vertical line passing through the point $(1, -3)$?

The diagram shows two **horizontal** lines, L_3 and L_4.

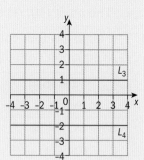

4 a Write down the coordinates of at least five points lying on L_3.
   b What do you notice about the coordinates of the points from **a**? What do their coordinates have in common?
   c What is the condition for a point to lie on L_3? Write down this condition in the form $y = k$ where k takes a particular value.

5 a Write down the coordinates of at least five points lying on L_4.
   b What do you notice about the coordinates of the points from **a**? What do their coordinates have in common?
   c What is the condition for a point to lie on L_4? Write down this condition in the form $y = k$ where k takes a particular value.

6 What is the equation of a horizontal line passing through the point $(1, -3)$?

> • The equation of any vertical line is of the form $x = k$ where k is a constant.
> • The equation of any horizontal line is of the form $y = k$ where k is a constant.

Intersection of lines in two dimensions

> → If two lines are parallel then they have the same gradient and do not intersect.

Parallel lines L_1 and L_2 can be:

• *Coincident lines (the same line)*

   e.g. $2x + y = 3$ and
      $6x + 3y = 9$

   $L_1 = L_2$ therefore they have the same gradient and the same y-intercept. There is an infinite number of points of intersection.

• *Different lines*

   e.g. $2x + y = 3$ and
      $2x + y = -1$

   L_1 and L_2 have the same gradient but different y-intercepts. There is no point of intersection.

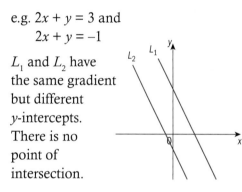

> → If two lines L_1 and L_2 are not parallel then they intersect at just one point.

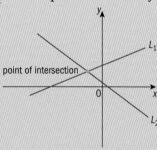

To find the point of intersection write $m_1 x_1 + c_1 = m_2 x_2 + c_2$ and solve for x.

Example 8

Find the point of intersection of the lines $y = 2x + 1$ and $-x - y + 4 = 0$.

Answer

Algebraically
$y = 2x + 1$ and $y = -x + 4$
$2x + 1 = -x + 4$
   $3x = 3$
    $x = 1$
 so $y = 2 \times 1 + 1$
     $= 3$
The point of intersection is $(1, 3)$.

Write both equations in the gradient–intercept form.
Equate expressions for y.
Solve for x.

Substitute for x in one of the equations to find y.

▶ Continued on next page

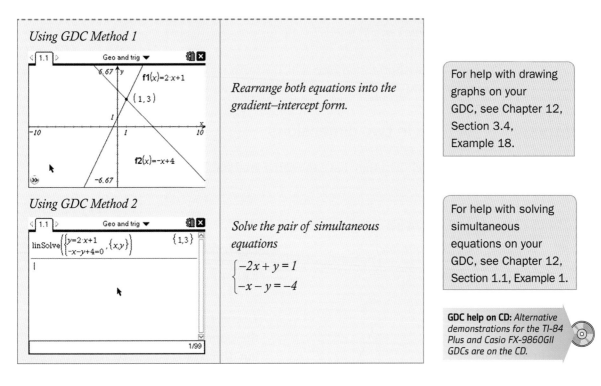

Using GDC Method 1

Rearrange both equations into the gradient–intercept form.

For help with drawing graphs on your GDC, see Chapter 12, Section 3.4, Example 18.

Using GDC Method 2

Solve the pair of simultaneous equations

$$\begin{cases} -2x + y = 1 \\ -x - y = -4 \end{cases}$$

For help with solving simultaneous equations on your GDC, see Chapter 12, Section 1.1, Example 1.

GDC help on CD: *Alternative demonstrations for the TI-84 Plus and Casio FX-9860GII GDCs are on the CD.*

Exercise 3G

1 Write down the equations of these lines.

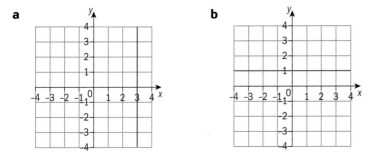

a

b

2 Find the point of intersection of each pair of lines.

a $y = 3x - 6$ and $y = -x + 2$

b $-x + 5y = 0$ and $\dfrac{1}{5}x + y - 2 = 0$

c $y = 3$ and $x = -7$

d $y = 1.5x + 4$ and $y = 1$

e $-x + 2y + 6 = 0$ and $x + y - 3 = 0$

f y-axis and $y = 4$

3 Show that the lines L_1 with equation $-5x + y + 1 = 0$ and L_2 with equation $10x - 2y + 4 = 0$ are parallel.

4 State, with reasons, whether each pair of lines meet at
    i only one point   **ii** an infinite number of points
    iii no point.

  a $y = 3(x - 5)$ and $x - \dfrac{1}{3}y + 6 = 0$

  b $\dfrac{y+1}{x-2} = -1$ and $y = -x + 1$

  c $y = 4x - 8$ and $4x - 2y = 0$

  d $x - y + 3 = 0$ and $3x - 3y + 9 = 0$

EXAM-STYLE QUESTION

5 Line L_1 has gradient 5 and intersects line L_2 at the point
A (1, 0).

  a Find the equation of L_1.

  Line L_2 is perpendicular to L_1.

  b Find the equation of L_2.

> Point A lies on both lines.

3.3 The sine, cosine and tangent ratios

> Some textbooks use 'right triangle' instead of right-angled triangle.

Trigonometry is the study of lengths and angles in triangles. This section looks at trigonometry in *right-angled* triangles.

In a **right-angled triangle** the side opposite the right angle is the **hypotenuse**, which is the **longest side**.

- AC is the hypotenuse
- AB is adjacent to angle A (Â)
- BC is opposite Â

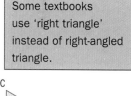

Investigation – right-angled triangles

Draw a diagram of two triangles like this.

1 Measure the angles at E and C. What do you notice?

2 Measure the lengths AB and AD. Calculate the ratio $\dfrac{AD}{AB}$

3 Measure the lengths AE and AC. Calculate the ratio $\dfrac{AE}{AC}$

4 Measure the lengths DE and BC. Calculate the ratio $\dfrac{DE}{BC}$

What do you notice about your answers to questions 2 to 4?

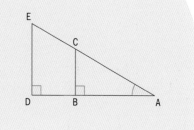

In the diagram the right-angled triangles ABC and ADE have the same angles, and corresponding sides are in the same ratio.

The ratios $\dfrac{AB}{AC}$, $\dfrac{BC}{AC}$ and $\dfrac{BC}{AB}$ in triangle ABC are respectively

equal to the ratios $\dfrac{AD}{AE}$, $\dfrac{DE}{AE}$ and $\dfrac{DE}{AD}$ in triangle ADE.

> Two triangles with the same angles and corresponding sides in the same ratio are called **similar triangles**.

Therefore

$$\frac{AB}{AC} = \frac{AD}{AE} = \frac{\text{Adjacent to } \hat{A}}{\text{Hypotenuse}}$$

Note that both AB and AD are adjacent to $\hat{A}$, and AC and AE are the hypotenuses.

Some textbooks call the two shorter sides of a right-angled triangle the 'legs' of the triangle.

$$\frac{BC}{AC} = \frac{DE}{AE} = \frac{\text{Opposite } \hat{A}}{\text{Hypotenuse}}$$

Note that both BC and DE are opposite $\hat{A}$, and both AC and AE are the hypotenuses.

$$\frac{BC}{AB} = \frac{DE}{AD} = \frac{\text{Opposite } \hat{A}}{\text{Adjacent to } \hat{A}}$$

Note that both BC and DE are opposite $\hat{A}$, and both AB and AD are adjacent to $\hat{A}$.

In any triangle **similar** to triangle ABC these ratios will remain the same.

→ Three trigonometric ratios in a right-angled triangle are defined as

$$\sin \alpha = \frac{\text{Opposite side}}{\text{Hypotenuse}}$$

$$\cos \alpha = \frac{\text{Adjacent side}}{\text{Hypotenuse}}$$

$$\tan \alpha = \frac{\text{Opposite side}}{\text{Adjacent side}}$$

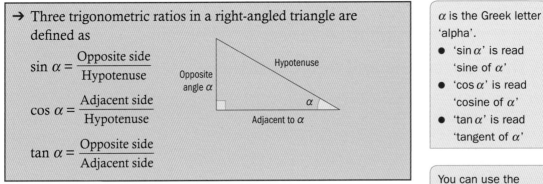

α is the Greek letter 'alpha'.
- 'sin α' is read 'sine of α'
- 'cos α' is read 'cosine of α'
- 'tan α' is read 'tangent of α'

You can use the acronym **SOHCAHTOA** to help you remember which ratio is which.

SOH as **S**in $\alpha = \dfrac{\mathbf{O}}{\mathbf{H}}$

CAH as **C**os $\alpha = \dfrac{\mathbf{A}}{\mathbf{H}}$

TOA as **T**an $\alpha = \dfrac{\mathbf{O}}{\mathbf{A}}$

Example 9

For each triangle, write down the three trigonometric ratios for the angle θ in terms of the sides of the triangle.

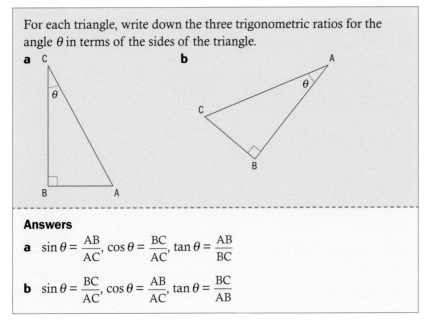

Answers

a $\sin \theta = \dfrac{AB}{AC}$, $\cos \theta = \dfrac{BC}{AC}$, $\tan \theta = \dfrac{AB}{BC}$

b $\sin \theta = \dfrac{BC}{AC}$, $\cos \theta = \dfrac{AB}{AC}$, $\tan \theta = \dfrac{BC}{AB}$

Example 10

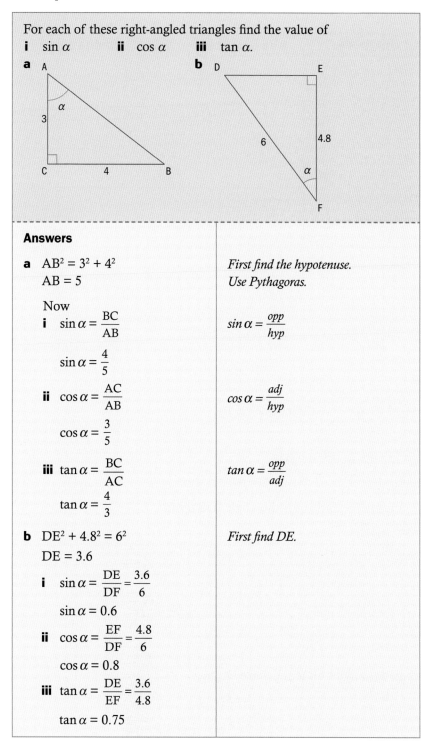

For each of these right-angled triangles find the value of
i $\sin \alpha$ **ii** $\cos \alpha$ **iii** $\tan \alpha$.

a

b

Answers

a $AB^2 = 3^2 + 4^2$ *First find the hypotenuse.*
     $AB = 5$ *Use Pythagoras.*

     Now

     i $\sin \alpha = \dfrac{BC}{AB}$ $sin\, \alpha = \dfrac{opp}{hyp}$

        $\sin \alpha = \dfrac{4}{5}$

     ii $\cos \alpha = \dfrac{AC}{AB}$ $cos\, \alpha = \dfrac{adj}{hyp}$

        $\cos \alpha = \dfrac{3}{5}$

     iii $\tan \alpha = \dfrac{BC}{AC}$ $tan\, \alpha = \dfrac{opp}{adj}$

        $\tan \alpha = \dfrac{4}{3}$

b $DE^2 + 4.8^2 = 6^2$ *First find DE.*
     $DE = 3.6$

     i $\sin \alpha = \dfrac{DE}{DF} = \dfrac{3.6}{6}$

        $\sin \alpha = 0.6$

     ii $\cos \alpha = \dfrac{EF}{DF} = \dfrac{4.8}{6}$

        $\cos \alpha = 0.8$

     iii $\tan \alpha = \dfrac{DE}{EF} = \dfrac{3.6}{4.8}$

        $\tan \alpha = 0.75$

Exercise 3H

1 Copy and complete this table.

Triangle	Hypotenuse	Side opposite α	Side adjacent to α
X, Z, Y triangle			
A, B, C triangle			
P, Q, R triangle			

2 Write down the three trigonometric ratios for the angle δ in terms of the sides of the triangle.

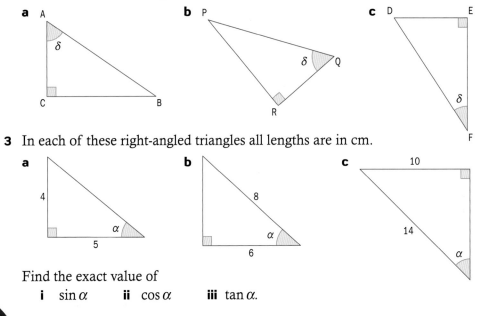

a A, C, B triangle

b P, Q, R triangle

c D, E, F triangle

3 In each of these right-angled triangles all lengths are in cm.

a triangle with sides 4 and 5, angle α

b triangle with sides 8 and 6, angle α

c triangle with sides 10 and 14, angle α

Find the exact value of

    i $\sin\alpha$ **ii** $\cos\alpha$ **iii** $\tan\alpha$.

4 For each triangle write down a trigonometric equation to link angle β and the side marked x.

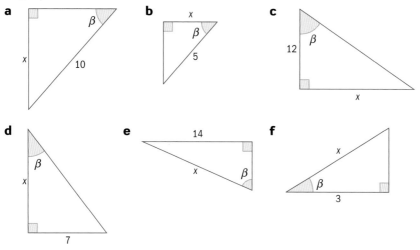

a

b

c

d

e

f

Finding the sides of a right-angled triangle

If you know the size of one of the acute angles and the length of one side in a right-angled triangle you can find

- the lengths of the other sides using trigonometric ratios
- the third angle using the sum of the interior angles of a triangle.

> Label the sides opposite, adjacent and hypotenuse so you can identify which ones you know.

Example 11

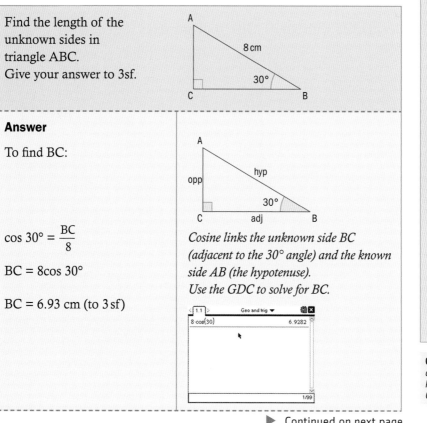

Find the length of the unknown sides in triangle ABC.
Give your answer to 3sf.

Answer

To find BC:

$$\cos 30° = \frac{BC}{8}$$

$$BC = 8\cos 30°$$

$$BC = 6.93 \text{ cm (to 3 sf)}$$

Cosine links the unknown side BC (adjacent to the 30° angle) and the known side AB (the hypotenuse).
Use the GDC to solve for BC.

Remember to set your GDC in **degrees**. To change to **degree mode** press ⌂ On and choose 5:Settings & Status | 2:Settings | 1:General

Use the tab key to move to Angle and select Degree. Press enter and then select 4:Current to return to the document.

GDC help on CD: *Alternative demonstrations for the TI-84 Plus and Casio FX-9860GII GDCs are on the CD.*

▶ Continued on next page

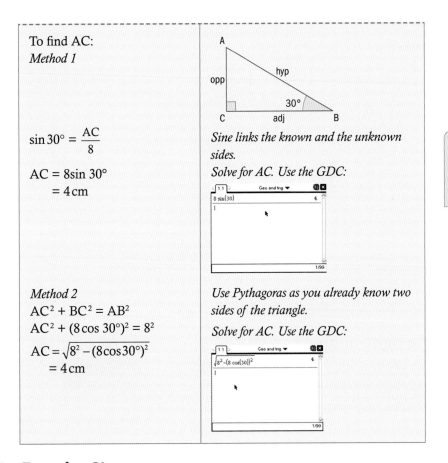

To find AC:

Method 1

$$\sin 30° = \frac{AC}{8}$$

$$AC = 8\sin 30°$$
$$= 4\,cm$$

Sine links the known and the unknown sides.

Solve for AC. Use the GDC:

Method 2

$$AC^2 + BC^2 = AB^2$$

$$AC^2 + (8\cos 30°)^2 = 8^2$$

$$AC = \sqrt{8^2 - (8\cos 30°)^2}$$
$$= 4\,cm$$

Use Pythagoras as you already know two sides of the triangle.

Solve for AC. Use the GDC:

You could also use tangent as you know the angle and the adjacent.

Exercise 3I

Find the lengths of the sides marked with letters. Give your answers correct to two decimal places.

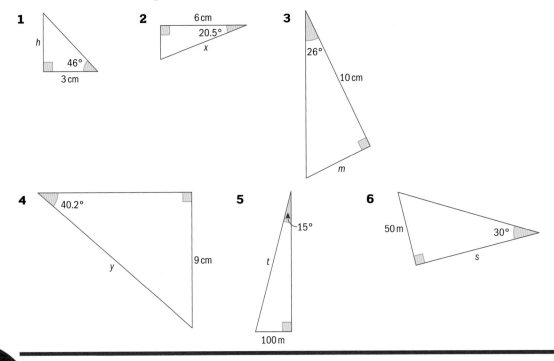

Example 12

The astronomer Aryabhata, born in India is about 476 CE, believed that the Sun, planets and stars circled the Earth in different orbits. He began to invent trigonometry in order to calculate the distances from planets to the Earth.

In triangle DEF, $\hat{E} = 90°$, $\hat{F} = 50°$ and DE = 7 m
  a Represent this information in a clear **labeled** diagram.
  b Find the size of $\hat{D}$.
  c Find EF.
  d Find DF.
Give your answers to 3 sf.

Answers

a

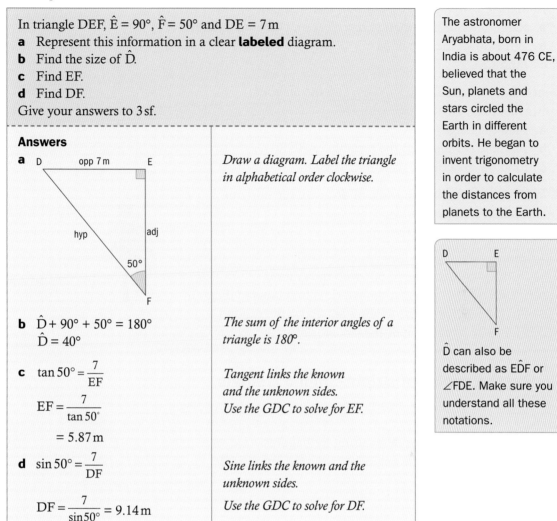

Draw a diagram. Label the triangle in alphabetical order clockwise.

b $\hat{D} + 90° + 50° = 180°$
$\hat{D} = 40°$

The sum of the interior angles of a triangle is 180°.

c $\tan 50° = \dfrac{7}{\text{EF}}$

$\text{EF} = \dfrac{7}{\tan 50°}$

$= 5.87\,\text{m}$

Tangent links the known and the unknown sides.
Use the GDC to solve for EF.

d $\sin 50° = \dfrac{7}{\text{DF}}$

$\text{DF} = \dfrac{7}{\sin 50°} = 9.14\,\text{m}$

Sine links the known and the unknown sides.

Use the GDC to solve for DF.

$\hat{D}$ can also be described as $\text{E}\hat{\text{D}}\text{F}$ or $\angle\text{FDE}$. Make sure you understand all these notations.

Exercise 3J

1 In triangle PQR, $\hat{R} = 90°$, $\hat{P} = 21°$, PR = 15 cm.
  a Represent this information in a clear and **labeled** diagram.
  b Write down the size of $\hat{Q}$.
  c Find QR.

Label the triangle in alphabetical order clockwise.

2 In triangle STU, $\hat{T} = 90°$, $\hat{U} = 55°$, SU = 35 cm.
  a Represent this information in a clear and **labeled** diagram.
  b Write down the size of $\hat{S}$.
  c Find TU.

3 In triangle ZWV, $\hat{V} = 90°$, $\hat{W} = 15°$, WV = 30 cm.
  a Represent this information in a clear and **labeled** diagram.
  b Write down the size of $\hat{Z}$.
  c Find VZ.

4 In triangle LMN, $\hat{N} = 90°$, $\hat{L} = 33°$, LN = 58 cm.

   a Represent this information in a clear and **labeled** diagram.

   b Write down the size of $\hat{M}$.

   c Find LM.

5 In rectangle ABCD, DC = 12 cm and the diagonal
   BD makes an angle of 30° with DC.

   a Find the length of BC.

   b Find the perimeter of the rectangle ABCD.

   c Find the area of the rectangle ABCD.

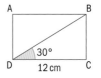

6 When the sun makes an angle of 46° with the horizon
   a tree casts a shadow of 7 m.
   Find the height of the tree.

7 A ladder 7 metres long leans against a wall, touching a window
   sill, and makes an angle of 50° with the ground.

   a Represent this information in a clear and **labeled** diagram.

   b Find the height of the window sill above the ground.

   c Find how far the foot of the ladder is from the foot of the wall.

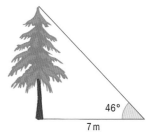

Finding the angles of a right-angled triangle

If you know the lengths of two sides in a right-angled triangle, you
can find

- the length of the other side by using Pythagoras
- the size of the two acute angles by using the appropriate
  trigonometric ratios.

Example 13

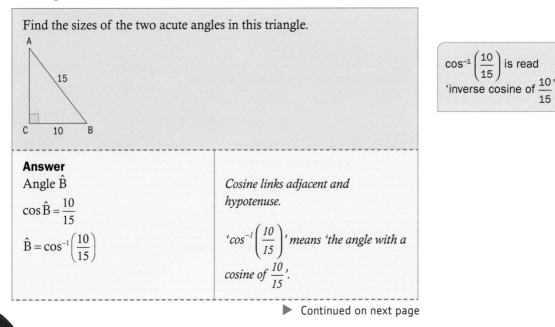

Find the sizes of the two acute angles in this triangle.

> $\cos^{-1}\left(\dfrac{10}{15}\right)$ is read
> 'inverse cosine of $\dfrac{10}{15}$'.

Answer

Angle $\hat{B}$

$\cos\hat{B} = \dfrac{10}{15}$

$\hat{B} = \cos^{-1}\left(\dfrac{10}{15}\right)$

*Cosine links adjacent and
hypotenuse.*

'$\cos^{-1}\left(\dfrac{10}{15}\right)$' means 'the angle with a

cosine of $\dfrac{10}{15}$'.

▶ Continued on next page

Therefore $\hat{B} = 48.2°$	*Use the GDC.* 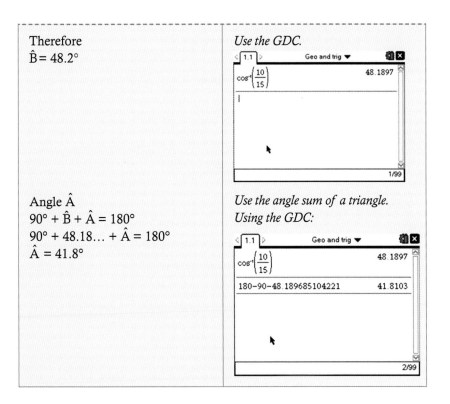
Angle $\hat{A}$ $90° + \hat{B} + \hat{A} = 180°$ $90° + 48.18... + \hat{A} = 180°$ $\hat{A} = 41.8°$	*Use the angle sum of a triangle.* *Using the GDC:*

Example 14

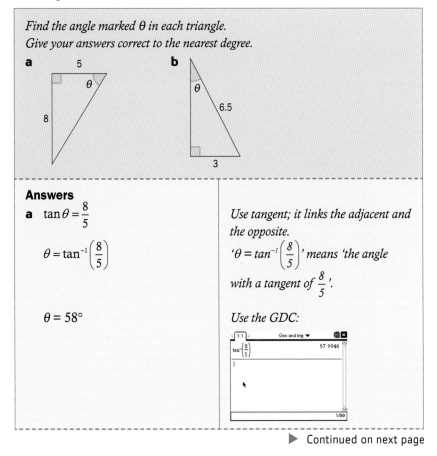

Find the angle marked θ in each triangle.
Give your answers correct to the nearest degree.

a 5, 8, θ

b θ, 6.5, 3

Answers

a $\tan\theta = \dfrac{8}{5}$

$\theta = \tan^{-1}\left(\dfrac{8}{5}\right)$

$\theta = 58°$

Use tangent; it links the adjacent and the opposite.

'$\theta = \tan^{-1}\left(\dfrac{8}{5}\right)$' means 'the angle with a tangent of $\dfrac{8}{5}$'.

Use the GDC:

▶ Continued on next page

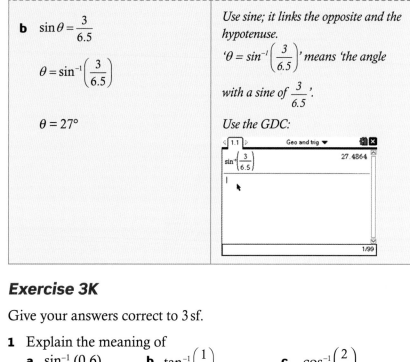

b $\sin\theta = \dfrac{3}{6.5}$

$\theta = \sin^{-1}\left(\dfrac{3}{6.5}\right)$

$\theta = 27°$

Use sine; it links the opposite and the hypotenuse.

'$\theta = \sin^{-1}\left(\dfrac{3}{6.5}\right)$' means 'the angle with a sine of $\dfrac{3}{6.5}$'.

Use the GDC:

Exercise 3K

Give your answers correct to 3 sf.

1 Explain the meaning of
   a $\sin^{-1}(0.6)$ **b** $\tan^{-1}\left(\dfrac{1}{2}\right)$ **c** $\cos^{-1}\left(\dfrac{2}{3}\right)$.

2 Calculate
   a $\sin^{-1}(0.6)$ **b** $\tan^{-1}\left(\dfrac{1}{2}\right)$ **c** $\cos^{-1}\left(\dfrac{2}{3}\right)$.

3 Find the **acute** angle α if
   a $\sin\alpha = 0.2$ **b** $\cos\alpha = \dfrac{2}{3}$ **c** $\tan\alpha = 1$.

4 Find the sizes of the two acute angles in these triangles.

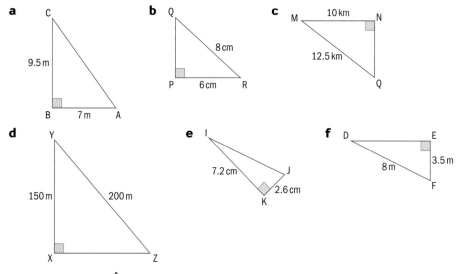

5 In triangle BCD, $\hat{D} = 90°$, BD = 54 cm, DC = 42 cm.
   a Represent this information in a clear and **labeled** diagram.
   b Find the size of $\hat{C}$.

6 In triangle EFG, $\hat{G} = 90°$, FG = 56 m, EF = 82 m.
   a Represent this information in a clear and **labeled** diagram.
   b Find the size of $\hat{F}$.

7 In triangle HIJ, $\hat{J} = 90°$, IJ = 18 m, HI = 25 m.
   a Represent this information in a clear and **labeled** diagram.
   b Find the size of $\hat{H}$.

8 In rectangle ABCD, BC = 5 cm and DC = 10 cm.

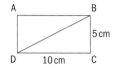

Find the size of the angle that the diagonal BD makes with the side DC.

9 The length and width of a rectangle are 20 cm and 13 cm respectively.
Find the angle between a diagonal and the shorter side of the rectangle.

10 A ladder 8 m long leans against a vertical wall.
The base of the ladder is 3 m away from the wall.
Calculate the angle between the wall and the ladder.

11 a On a pair of Cartesian axes plot the points A(3, 0) and B(0, 4). Use the same scale on both axes.
   b Draw the line AB.
   c Find the size of the **acute** angle that the line AB makes with the x-axis.

12 a On a pair of Cartesian axes plot the points A(−1, 0) and B(1, 4). Use the same scale on both axes.
   b Draw the line AB.
   c Find the size of the **acute** angle that the line AB makes with the x-axis.

Finding right-angled triangles in other shapes

So far you have found unknown sides and angles in right-angled triangles. Next you will learn how to find unknown sides and angles in triangles that are not right-angled and in shapes such as rectangles, rhombuses and trapeziums.

The technique is to break down the shapes into smaller ones that contain right-angled triangles.

Name of shape	Shape	Where are the right-angled triangles?
Isosceles or equilateral triangles		
Rectangles or squares		
Circle		

Investigation – 2-D shapes

How can you break these shapes into smaller shapes so that at least one of them is a right-angled triangle?

To do this you need to know the properties of 2-D shapes.

1 *Rhombus*

What is the property of the diagonals of a rhombus?
Make an accurate drawing of a rhombus on squared paper.
Draw the diagonals. How many right-angled triangles do you obtain? Are they congruent? Why?
Comment on your findings.

2 *Kite*

What is the property of the diagonals of a kite?
Make an accurate drawing of a kite on squared paper.
Draw the diagonals. How many right-angled triangles do you obtain? Are they congruent? Why? Comment on your findings.

3 *Parallelogram*

Draw a parallelogram like this one on squared paper.
There is a rectangle that has the same base and height as this parallelogram. Draw dotted lines where you would cut the parallelogram and rearrange it to make a rectangle.
How many shapes do you obtain? How many of them are right-angled triangles? Comment on your findings.

4 *Triangle*

Draw a triangle like this one.
Every triangle has three heights, one for each base (or side).
Draw the height relative to AC (this is the line segment drawn from B to AC and perpendicular to AC). You will get two right-angled triangles that make up the triangle ABC. Under what conditions would these triangles be congruent? Comment on your findings.

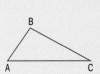

Continued on next page

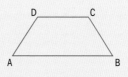

5 *Trapezium*
Draw a trapezium like this one.

Draw a line from D perpendicular to AB and a line from
C perpendicular to AB. You will get two right-angled triangles.
What is the condition for these triangles to be congruent?

6 *Regular polygon*
Here are a regular hexagon and a regular pentagon.

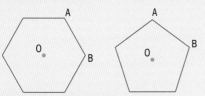

> A **regular polygon**
> has all sides equal
> lengths and all angles
> equal.

O is the center of each polygon.
For **each** polygon:
What type of triangle is ABO? Why? Draw a line from O perpendicular to the
side AB to form two right-angled triangles. These two triangles are congruent.

Explain why.

Example 15

Triangle ABC is isosceles. The two equal sides AB and BC are 10 cm
long and each makes an angle of 40° with AC.
a Represent this information in a clear and labeled diagram.
b Find the length of AC.
c Find the perimeter of triangle ABC.

Answers

a

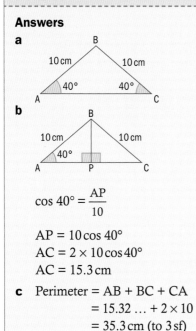

b

*In an isosceles triangle the
perpendicular from the apex to
the base **bisects** the base,
making two right-angled
triangles.*

> Bisect means 'cut in
> half'.

$$\cos 40° = \frac{AP}{10}$$

$$\cos = \frac{adj}{hyp}$$

$$AP = 10\cos 40°$$
$$AC = 2 \times 10\cos 40°$$
$$AC = 15.3\,\text{cm}$$

*Make AP the subject of the
equation.*
Use the fact that AC = 2 × AP.

c Perimeter = AB + BC + CA
$$= 15.32\ldots + 2 \times 10$$
$$= 35.3\,\text{cm (to 3 sf)}$$

Example 16

The diagonals of a rhombus are 10 cm and 5 cm. Find the size of the **larger** angle of the rhombus.

Answer

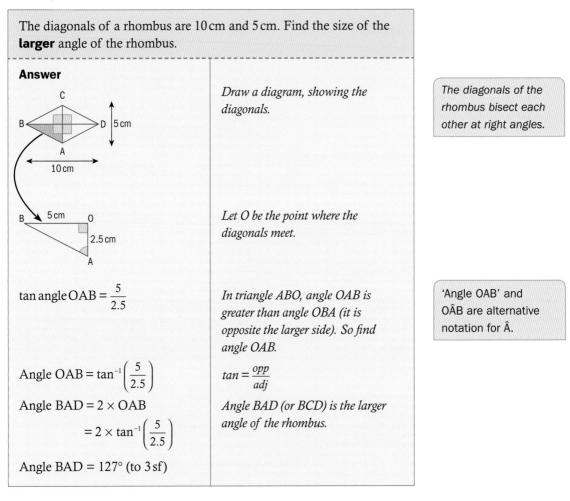

| | The diagonals of the rhombus bisect each other at right angles. |

Draw a diagram, showing the diagonals.

Let O be the point where the diagonals meet.

$$\tan \text{angle} \, OAB = \frac{5}{2.5}$$

In triangle ABO, angle OAB is greater than angle OBA (it is opposite the larger side). So find angle OAB.

'Angle OAB' and OÂB are alternative notation for Â.

$$\text{Angle} \, OAB = \tan^{-1}\left(\frac{5}{2.5}\right)$$

$$\text{Angle} \, BAD = 2 \times OAB$$

$$= 2 \times \tan^{-1}\left(\frac{5}{2.5}\right)$$

$$\text{Angle} \, BAD = 127° \, (\text{to 3 sf})$$

$$\tan = \frac{opp}{adj}$$

Angle BAD (or BCD) is the larger angle of the rhombus.

Investigation – rhombus

1 Use a ruler and a pair of compasses to construct a rhombus with a side length of 6 cm.
2 Construct another rhombus with a side length of 6 cm that is not congruent to the one you drew in **1**.
3 How many different rhombuses with a side length of 6 cm could you construct? In what ways do they differ?

Exercise 3L

1 Triangle ABC is isosceles. The two equal sides AC and BC are 7 cm long and they each make an angle of 65° with AB.
   a Represent this information in a clear and labeled diagram.
   b Find the length of AB.
   c Find the perimeter of triangle ABC (give your answer correct to the nearest centimetre).

2 The diagonals of a rhombus are 12 cm and 7 cm. Find the size of the **smaller** angle of the rhombus.

3 The size of the larger angle of a rhombus is 120° and the longer diagonal is 7 cm.

    a Represent this information in a clear and labeled diagram.

    b Find the length of the shorter diagonal.

EXAM-STYLE QUESTIONS

4 In the diagram ABCD is a trapezium where AD ∥ BC, CD = BA = 6 m, BC = 12 m and DA = 16 m

    a Show that DE = 2 m.

    b Find the size of $\hat{D}$.

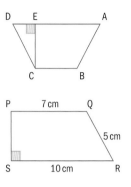

5 In the diagram PQRS is a trapezium, PQ ∥ SR, PQ = 7 cm, RS = 10 cm, QR = 5 cm and $\hat{S}$ = 90°

    a Find the height, PS, of the trapezium.

    b Find the area of the trapezium.

    c Find the size of angle SRQ.

6 The length of the shorter side of a rectangular park is 400 m. The park has a straight path 600 m long joining two opposite corners.

    a Represent this information in a clear and labeled diagram.

    b Find the size of the angle that the path makes with the longer side of the park.

7 **a** On a pair of Cartesian axes, plot the points A(3, 2), C(−1, −4), and D(−1, 2). Use the same scale on both axes.
B is a point such that ABCD is a rectangle.

    b **i** Plot B on your diagram.

       ii Write down the coordinates of the point B.

    c Write down the length of

       i AB   **ii** BC.

    d Hence find the size of the angle that a diagonal of the rectangle makes with one of the shorter sides.

Angles of elevation and depression

→ The **angle of elevation** is the angle you lift your eyes through to look at something above.

α is the angle of elevation.

→ The **angle of depression** is the angle you lower your eyes through to look at something below.

β is the angle of depression.

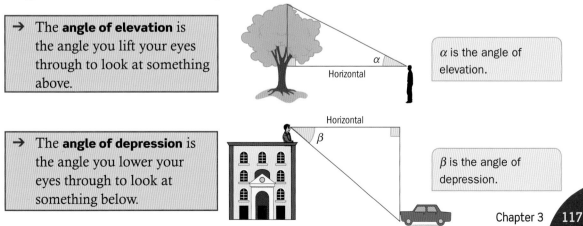

Notice that both the angle of elevation and the angle of depression are measured from the **horizontal**.

Example 17

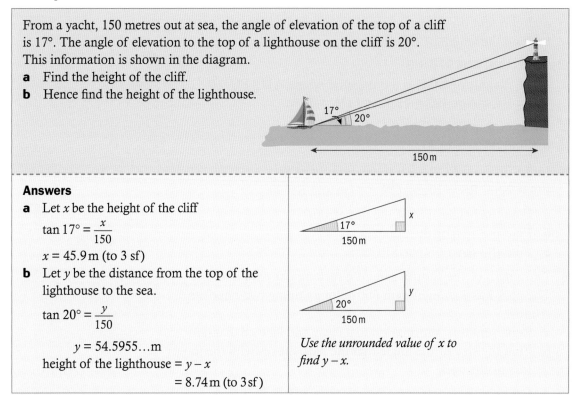

From a yacht, 150 metres out at sea, the angle of elevation of the top of a cliff is 17°. The angle of elevation to the top of a lighthouse on the cliff is 20°. This information is shown in the diagram.

a Find the height of the cliff.

b Hence find the height of the lighthouse.

Answers

a Let x be the height of the cliff

$$\tan 17° = \frac{x}{150}$$

$x = 45.9\,\text{m (to 3 sf)}$

b Let y be the distance from the top of the lighthouse to the sea.

$$\tan 20° = \frac{y}{150}$$

$y = 54.5955...\,\text{m}$

height of the lighthouse $= y - x$

$= 8.74\,\text{m (to 3 sf)}$

Use the unrounded value of x to find $y - x$.

Example 18

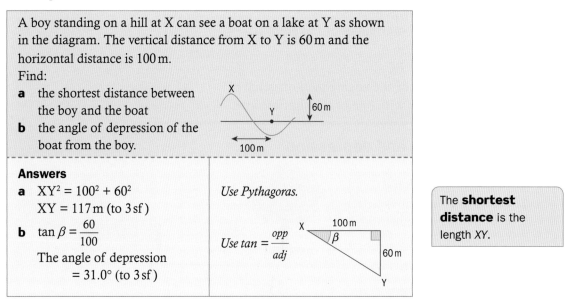

A boy standing on a hill at X can see a boat on a lake at Y as shown in the diagram. The vertical distance from X to Y is 60 m and the horizontal distance is 100 m.

Find:

a the shortest distance between the boy and the boat

b the angle of depression of the boat from the boy.

Answers

a $XY^2 = 100^2 + 60^2$

$XY = 117\,\text{m (to 3 sf)}$

Use Pythagoras.

b $\tan \beta = \frac{60}{100}$

The angle of depression

$= 31.0°\text{ (to 3 sf)}$

Use $\tan = \frac{opp}{adj}$

> The **shortest distance** is the length *XY*.

Exercise 3M

1 Find the angle of elevation of the top of a tree 13 m high from a point 25 m away on level ground.

Draw a diagram for each question.

2 A church spire 81 metres high casts a shadow 63 metres long. Find the angle of elevation of the sun.

3 The angle of depression from the top of a cliff to a ship at sea is 14°. The ship is 500 metres from shore. Find the height of the cliff.

4 Find the angle of depression from the top of a cliff 145 metres high to a ship at sea 1.2 kilometres from the shore.

5 A man whose eye is 1.5 metres above ground level stands 20 metres from the base of a tree. The angle of elevation to the top of the tree is 45°. Calculate the height of the tree.

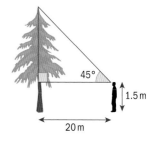

6 The height of a tree is 61.7 metres and the angle of elevation to the top of the tree from ground level is 62.4°. Calculate the distance from the tree to the point at which the angle was measured.

EXAM-STYLE QUESTION

7 The angle of depression of town B from town A is 12°.

a Find the angle of elevation of town A from town B.

The horizontal distance between the towns is 2 km.

b Find the vertical distance between the towns. Give your answer correct to the nearest metre.

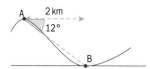

3.4 The sine and cosine rules

The sine and cosine rules are formulae that will help you to find unknown sides and angles in a triangle. They enable you to use trigonomentry in triangles that are **not** right-angled.

The formula and notation are simpler if you label triangles like this.

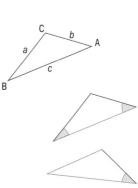

The sine rule

If you have this information about a triangle:

- two angles and one side, or
- two sides and a non-included angle,

then you can find the other sides and angles of the triangle.

- The side opposite $\hat{A}$ is a.
- The side opposite $\hat{B}$ is b.
- The side opposite $\hat{C}$ is c.

Also notice that

- $\hat{A}$ is between sides b and c.
- $\hat{B}$ is between sides a and c.
- $\hat{C}$ is between sides a and b.

> **→ Sine rule**
>
> In any triangle ABC with angles $\hat{A}$, $\hat{B}$ and $\hat{C}$, and opposite sides a, b and c respectively:
>
> $$\frac{a}{\sin A} = \frac{b}{\sin B} = \frac{c}{\sin C}$$
>
> or $\dfrac{\sin A}{a} = \dfrac{\sin B}{b} = \dfrac{\sin C}{c}$

The sine rule is in the Formula booklet.

Example 19

In triangle ABC, $b = 16\,\text{cm}$, $c = 10\,\text{cm}$ and $\hat{B} = 135°$.
- **a** Represent the given information in a labeled diagram.
- **b** Find the size of angle C.
- **c** Hence find the size of angle A.

Answers

a

b
$$\frac{16}{\sin 135°} = \frac{10}{\sin \hat{C}}$$

$16 \sin \hat{C} = 10 \sin 135°$

$\sin \hat{C} = \dfrac{10\sin 135°}{16}$

$\hat{C} = 26.2°$ (to 3 sf)

Substitute in the sine rule.

Cross multiply.

Make $\sin \hat{C}$ the subject of the formula.
Use your GDC.

c $\hat{A} + \hat{B} + \hat{C} = 180°$
$\hat{A} + 135° + 26.227... = 180°$
$\hat{A} = 18.8°$ (to 3 sf)

Use your GDC.

Cross multiply.
$$\frac{a}{b} = \frac{c}{d} \Leftrightarrow ad = bc$$

Example 20

In triangle PQR, find the length of RQ. Give your answer correct to two significant figures.

Answer

$\hat{P} = 78°$

$$\frac{RQ}{\sin 78°} = \frac{10}{\sin 82°}$$

$RQ = \dfrac{10\sin 78°}{\sin 82°}$

$= 9.9\,\text{km}$ (to 2 sf)

RQ is opposite angle P so first find the size of angle P.

Substitute in the sine rule.

Make RQ the subject of the formula.

Use your GDC.

Ptolemy (c. 90–168 CE), in his 13-volume work *Almagest*, wrote sine values for angles form 0° to 90°. He also included theorems similar to the sine rule.

Exercise 3N

1 Find the sides marked with letters.

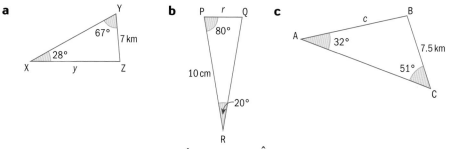

a

b

c

2 In triangle ABC, AC = 12 cm, Â = 30° and B̂ = 46°.
Find the length of BC.

3 In triangle ABC, Â = 15°, B̂ = 63° and AB = 10 cm. Find the
length of BC.

4 In triangle PQR, PR = 15 km, P̂ = 25° and Q̂ = 60°. Find the
length of QR.

5 In each triangle, find the angle indicated.

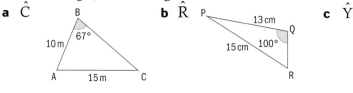

a Ĉ

b R̂

c Ŷ

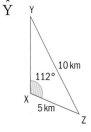

6 In triangle ABC, BC = 98 m, AB = 67 m and Â = 85°.
Find the size of Ĉ.

7 In triangle PQR, PQ = 5 cm, QR = 6.5 cm and P̂ = 70°.
Find the size of R̂.

EXAM-STYLE QUESTION

8 In the diagram, Â = 90°, CX = 10 m, AĈB = 30° and X̂ = 10°
a Write down the size of angle BCX.
b Find the length of BC.
c Find the length of AB.

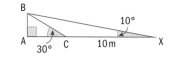

The cosine rule

If you have this information about a triangle:

- two sides and the included angle, or
- the three sides,

then you can find the other side and angles
of the triangle.

→ **Cosine rule**

In any triangle ABC with angles $\hat{A}$, $\hat{B}$ and $\hat{C}$, and opposite sides a, b and c respectively:

$$a^2 = b^2 + c^2 - 2bc\cos\hat{A}$$

This formula can be rearranged to

$$\cos\hat{A} = \frac{b^2 + c^2 - a^2}{2bc}$$

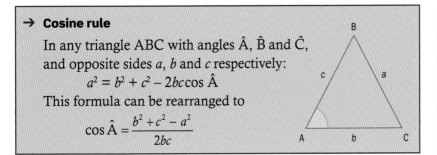

> These formulae are in the Formula booklet. The first version of the formula is useful when you need to find a side. The second version of the formula is useful when you need to find an angle.

Example 21

In triangle ABC, AC = 8.6 m, AB = 6.3 m and $\hat{A}$ = 50°.
Find the length of BC.

Answer

$BC^2 = 8.6^2 + 6.3^2 - 2 \times 8.6 \times 6.3 \times \cos 50°$

$BC^2 = 43.9975...$

$BC = 6.63$ m (to 3 sf)

Sketch the triangle.
Use $a^2 = b^2 + c^2$
$\quad\quad - 2bc\cos\hat{A}$

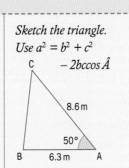

> The cosine rule applies to **any** triangle. For a right-angled triangle $A = 90°$. What does the formula look like? Do you recognize it? Is the cosine rule a generalization of Pythagoras' theorem?

Example 22

X, Y and Z are three towns. X is 20 km due north of Z.
Y is to the east of line XZ. The distance from Y to X is 16 km and the distance from Z to Y is 8 km.
a Represent this information in a clear and labeled diagram.
b Find the size of angle X.

Answers

a

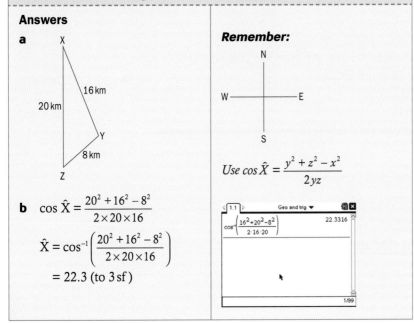

Remember:

N
W —————— E
S

Use $\cos\hat{X} = \dfrac{y^2 + z^2 - x^2}{2yz}$

b $\cos\hat{X} = \dfrac{20^2 + 16^2 - 8^2}{2 \times 20 \times 16}$

$\hat{X} = \cos^{-1}\left(\dfrac{20^2 + 16^2 - 8^2}{2 \times 20 \times 16}\right)$

$= 22.3$ (to 3 sf)

Exercise 30

1 Find the sides marked with letters.

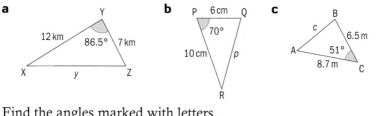

a

Y, 12 km, 86.5°, 7 km, X, y, Z

b

P, 6 cm, Q, 70°, 10 cm, p, R

c

B, c, 6.5 m, A, 51°, 8.7 m, C

2 Find the angles marked with letters.

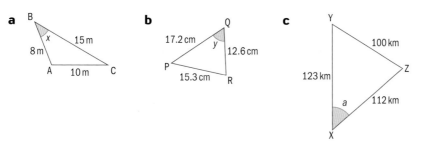

a

B, x, 15 m, 8 m, A, 10 m, C

b

Q, 17.2 cm, y, 12.6 cm, P, 15.3 cm, R

c

Y, 100 km, 123 km, Z, a, 112 km, X

3 In triangle ABC, CB = 120 m, AB = 115 m and $\hat{B}$ = 110°.
Find the length of side AC.

4 In triangle PQR, RQ = 6.9 cm, PR = 8.7 cm and $\hat{R}$ = 53°.
Find the length of side PQ.

5 In triangle XYZ, XZ = 12 m, XY = 8 m, YZ = 10 m.
Find the size of angle X.

EXAM-STYLE QUESTIONS

6 X, Y and Z are three towns. X is 30 km due south from Y.
Z is to the east of the line joining XY.
The distance from Y to Z is 25 km and the distance from X to Z is 18 km.
  a Represent this information in a clear and labeled diagram.
  b Find the size of angle Z.

7 Alison, Jane and Stephen are together at point A. Jane
walks 12 m due south from A and reaches point J. Stephen looks
at Jane, turns through 110°, walks 8 m from A and reaches
point S.
  a Represent this information in a clear and labeled diagram.
  b Find how far Stephen is from Jane.
  c Find how far **north** Stephen is from Alison.

8 The diagram shows a circle of radius 3 cm and center O.
A and B are two points on the circumference.
The length AB is 5 cm.
A triangle AOB is drawn inside the circle.
Calculate the size of angle AOB.

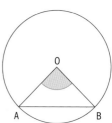

9 The diagram shows a crane PQR that carries a flat box W.
PQ is vertical, and the floor PM is horizontal.
Given that PQ = 8.2 m, QR = 12.3 m, PQ̂R =100° and
RW = 7.8 m, calculate

a PR

b the size of angle PRQ

c the height, h, of W above the floor, PM.

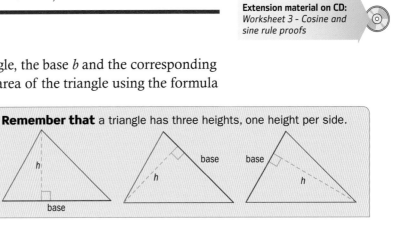

Extension material on CD:
*Worksheet 3 - Cosine and
sine rule proofs*

Area of a triangle

If you know one side of a triangle, the base b and the corresponding
height h, you can calculate the area of the triangle using the formula

$$A = \frac{1}{2}(b \times h)$$

If you do not know the
height, you can still calculate
the area of the triangle as
in the next example.

Remember that a triangle has three heights, one height per side.

Example 23

Calculate the area of triangle ABC.

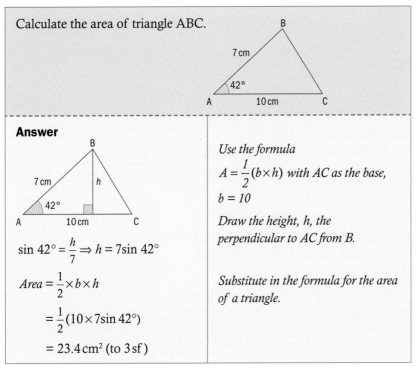

Answer

$\sin 42° = \dfrac{h}{7} \Rightarrow h = 7\sin 42°$

$Area = \dfrac{1}{2} \times b \times h$

$= \dfrac{1}{2}(10 \times 7\sin 42°)$

$= 23.4 \text{ cm}^2$ (to 3 sf)

Use the formula
$A = \dfrac{1}{2}(b \times h)$ *with AC as the base,*
$b = 10$

*Draw the height, h, the
perpendicular to AC from B.*

*Substitute in the formula for the area
of a triangle.*

You can use the same method for any triangle.

> → In any triangle ABC with angles Â, B̂ and Ĉ, and opposite sides a, b and c respectively, this rule applies:
> Area of triangle $= \frac{1}{2} ab \sin \hat{C}$

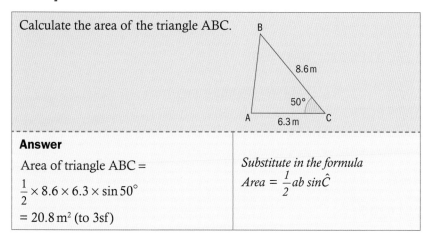

This formula is in the Formula booklet.

Example 24

Calculate the area of the triangle ABC.

B, 8.6 m, 50°, A, 6.3 m, C

Answer

Area of triangle ABC =

$\frac{1}{2} \times 8.6 \times 6.3 \times \sin 50°$

$= 20.8 \text{ m}^2$ (to 3sf)

Substitute in the formula
Area $= \frac{1}{2} ab \sin \hat{C}$

Exercise 3P

1 Calculate the area of each triangle.

a

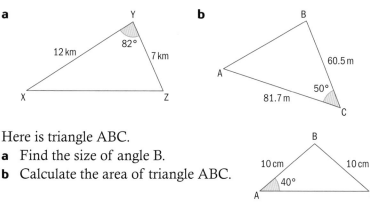

Y, 82°, 12 km, 7 km, X, Z

b

B, 60.5 m, A, 50°, 81.7 m, C

In the first century CE, Hero (or Heron) of Alexandria developed a different method for finding the area of a triangle using the lengths of the triangle's sides.

2 Here is triangle ABC.
  a Find the size of angle B.
  b Calculate the area of triangle ABC.

B, 10 cm, 10 cm, A, 40°, C

3 Here is triangle ABC.
  a Write down the size of angle C.
  b Find the area of triangle ABC.

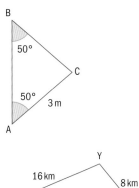

B, 50°, C, 50°, 3 m, A

4 Calculate the area of triangle XYZ.

Y, 16 km, 8 km, X, 20 km, Z

First find the size of one of the angles.

5 The diagram shows a triangular field XYZ.
XZ is 50 m, YZ is 100 m and angle X is 100°.
  a Find angle Z.
  b Find the area of the field. Give your answer correct to the nearest 10 m².

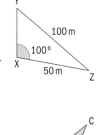

6 The area of an isosceles triangle ABC is 4 cm². Angle B
is 30° and AB = BC = x cm.
  a Write down, in terms of x, an expression for the area of the triangle.
  b Find the value of x.

7 In the diagram, AB = 5 cm, AD = 6 cm, BÂD = 90°,
BĈD = 30°, BD̂C = 70°.
  a Find the length of DB.   **b** Find the length of DC.
  c Find the area of triangle BCD.
  d Find the area of the quadrilateral ABCD.

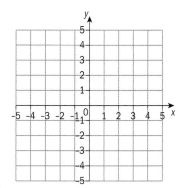

Review exercise

Paper 1 style questions

Give answers correct to 3 sf.

1 Line L_1 passes through the points A(1, 3) and B(5, 1).
  a Find the gradient of the line AB.

  Line L_2 is parallel to line L_1 and passes through the point (0, 4).
  b Find the equation of the line L_2.

2 Line L_1 passes through the points A(0, 6) and B(6, 0).
  a Find the gradient of the line L_1.
  b Write down the gradient of all lines perpendicular to L_1.
  c Find the equation of a line L_2 perpendicular to L_1 and
  passing through O(0, 0).

3 Consider the line L with equation $y = 2x + 3$.
  a Write down the coordinates of the point where
    i L meets the x-axis   **ii** L meets the y-axis.
  b Draw L on a grid like this one.
  c Find the size of the acute angle that L makes with the x-axis.

4 Consider the line L_1 with equation $y = -2x + 6$.
  a The point $(a, 4)$ lies on L_1. Find the value of a.
  b The point $(12.5, b)$ lies on L_1. Find the value of b.

  Line L_2 has equation $3x - y + 1 = 0$.
  c Find the point of intersection between L_1 and L_2.

5 The height of a vertical cliff is 450 m. The angle of elevation from a ship to
the top of the cliff is 31°. The ship is x metres from the bottom of the cliff.
  a Draw a diagram to show this information.
  b Calculate the value of x.

6 In the diagram, triangle ABC is isosceles.
AB = AC, CB = 20 cm and angle ACB is 32°.
Find   **a** the size of angle CAB
       b the length of AB
       c the area of triangle ABC.

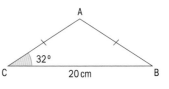

7 A gardener pegs out a rope, 20 metres long, to form a triangular
flower bed as shown in this diagram.
  a Write down the length of AC.
  b Find the size of the angle BAC.
  c Find the area of the flower bed.

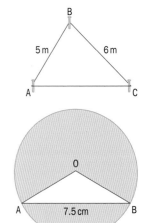

8 The diagram shows a circle with diameter 10 cm and center O.
Points A and B lie on the circumference and the length of
AB is 7.5 cm. A triangle AOB is drawn inside the circle.
  a Find the size of angle AOB.
  b Find the area of triangle AOB.
  c Find the shaded area.

Paper 2 style questions

1 **a** On a pair of axes plot the points A(–2, 5), B(2, 2) and C(8, 10).
    Use the same scale on both axes.
    The quadrilateral ABCD is a rectangle.
  b **i** Plot D on the pair of axes used in **a**.
    ii Write down the coordinates of D.
  c Find the gradient of line BC.
  d Hence write down the gradient of line DC.
  e Find the equation of line DC in the form $ax + by + d = 0$ where $a, b, d \in \mathbb{Z}$.
  f Find the length of   **i** DC   **ii** BC.
  g Find the size of the angle DBC.

2 The diagram shows a ladder AB. The ladder rests on the horizontal
ground AC. The ladder is touching the top of a vertical telephone pole CB.
The angle of elevation of the top of the pole from the foot of the
ladder is 60°. The distance from the foot of the ladder to the foot
of the pole is 2 m.
  a Calculate the length of the ladder.
  b Calculate the height of the pole.
  The ladder is moved in the same vertical plane so that its foot
  remains on the ground and its top touches the pole at a point P
  which is 1.5 m below the top of the pole.
  c Write down the length of CP.
  d Find the new distance from the foot of the ladder to the foot of the pole.
  e Find the size of the new angle of elevation of the top of
    the pole from the foot of the ladder.

3 The diagram shows a cross-country running course.
Runners start and finish at point A.

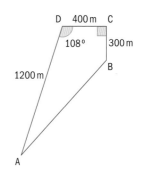

a Find the length of BD.

b Find the size of angle BDC, giving
your answer correct to two decimal places.

c Write down the size of angle ADB.

d Find the length of AB.

e **i** Find the perimeter of the course.

  ii Rafael runs at a constant speed of $3.8 \, \text{m s}^{-1}$.
  Find the time it takes Rafael to complete the course.
  Give your answer correct to the nearest minute.

f Find the area of the quadrilateral ABCD enclosed by the course.
Give your answer in km^2.

CHAPTER 3 SUMMARY

Gradient of a line

- If $A(x_1, y_1)$ and $B(x_2, y_2)$ are two points that lie on line L,
  the gradient of L is $m = \dfrac{y_2 - y_1}{x_2 - x_1}$.
- **Parallel lines** have the **same gradient**. This means that
  - if two lines are parallel then they have the same gradient
  - if two lines have the same gradient then they are parallel.
- Two lines are **perpendicular** if, and only if, they make
  an angle of 90°. This means that
  - if two lines are perpendicular then they make an angle of 90°
  - if two lines make an angle of 90° then they are perpendicular.
- Two lines are **perpendicular** if the product of their gradients is -1.

Equations of lines

- The equation of a straight line can be written in the form
  i $y = mx + c$, where m is the **gradient** and c is the y-**intercept**
    (the y-coordinate of the point where the line crosses the y-axis).
  ii $ax + by + d = 0$ where a, b and $d \in \mathbb{Z}$.
- The equation of any vertical line is of the form $x = k$
  where k is a constant.
- The equation of any horizontal line is of the form
  $y = k$ where k is a constant.
- If two lines are parallel then they have the same gradient
  and do not intersect.
- If two lines L_1 and L_2 are not parallel then they
  intersect at just one point. To find the point of intersection
  write $m_1 x_1 + c_1 = m_2 x_2 + c_2$ and solve for x.

point of intersection

Continued on next page

The sine, cosine, and tangent ratios

- Three trigonometric ratios in a right-angled triangle are defined as

$$\sin \alpha = \frac{\text{Opposite side}}{\text{Hypotenuse}}$$

$$\cos \alpha = \frac{\text{Adjacent side}}{\text{Hypotenuse}}$$

$$\tan \alpha = \frac{\text{Opposite side}}{\text{Adjacent side}}$$

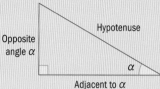

- The **angle of elevation** is the angle you lift your eyes through to look at something above.
- The **angle of depression** is the angle you lower your eyes through to look at something below.

The sine and cosine rules

- In any triangle ABC with angles A, B and C, and opposite sides a, b and c respectively:

$$\frac{a}{\sin A} = \frac{b}{\sin B} = \frac{c}{\sin C}$$

- In any triangle ABC with angles A, B and C, and opposite sides a, b and c respectively:

$$a^2 = b^2 + c^2 - 2bc \cos \hat{A}$$

This formula can be rearranged to

$$\cos \hat{A} = \frac{b^2 + c^2 - a^2}{2bc}$$

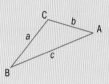

- In any triangle ABC with angles A, B and C, and opposite sides a, b and c respectively, this rule applies:

$$\text{Area of triangle} = \frac{1}{2} ab \sin \hat{C}$$

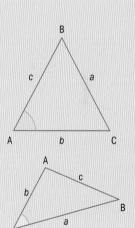

Making connections

Mathematics is often separated into different topics, or fields of knowledge.

- List the different fields of mathematics you can think of.
- Why do humans feel the need to categorize and compartmentalize knowledge?
- Does this help or hinder the search for more knowledge?

Algebra and geometry

Algebra and geometry are both mathematical disciplines with a very long history.

> algebra – generalizes arithmetical operations and relationships by using letters to represent unknown numbers. Possibly originated in solving equations, which goes back (at least) to Babylonian mathematics.

> geometry – studies the properties, measurement, and relationships of points, lines, planes, surfaces, angles, and solids. Origins in the very beginning of mathematics.

There was no common ground between algebra and geometry until René Descartes, the French philosopher and mathematician (1596–1650) showed that equations could be represented by lines on a graph, giving people an insight into what these equations mean, and where their solutions are found. Cartesian Geometry – representing equations for given values of x and y on a system of orthogonal (perpendicular) axes, is named after him.

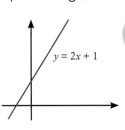

$y = 2x + 1$

It is said (although the story is probably a myth) that Descartes came up with the idea for his coordinate system while lying in bed and watching a fly crawl on the ceiling of his room.

Algebra and geometry are central to mathematics and school mathematics curricula around the world. Some schools run entirely separate courses on geometry and algebra, while others alternate mathematical topics throughout a course.

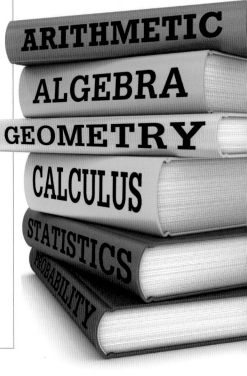

ARITHMETIC
ALGEBRA
GEOMETRY
CALCULUS
STATISTICS
PROBABILITY

Algebra and geometry are both useful in their own right but historically it is the interaction of these two areas that has led to many major mathematical developments and insights in the natural sciences, economics and of course other areas of mathematics.

As long as algebra and geometry have been separated, their progress have been slow and their uses limited, but when these two sciences have been united, they have lent each mutual forces, and have marched together towards perfection.

Joseph Louis Lagrange, 1736–1813, French mathematician

Fermat's Last Theorem

Fermat's Last Theorem states that no three positive integers a, b, and c can satisfy the equation $a^n + b^n = c^n$ for any integer value of n greater than 2. This theorem was first conjectured by Pierre de Fermat in 1637, in a note in a copy of *Arithmetica*, where he claimed he had a proof that was too large to fit in the margin. His proof, if it existed, was never found. It was not solved until 1995, when Andrew Wiles published a proof that he had been working on in secret for seven years.

▼ Andrew Wiles (1953–), British mathematician.

Wiles's complex proof uses the link between what were thought to be two separate areas of mathematics – modular forms and elliptic curves. Don't worry, these are not on the Mathematical Studies syllabus!

Many of the most famous proofs have needed input from different areas of mathematics.

Mathematical models

CHAPTER OBJECTIVES:

6.1 Concept of a function, domain, range and graph; function notation; concept of a function as a mathematical model

6.2 Drawing accurate graphs and sketch graphs; transferring a graph from GDC to paper; reading, interpreting and making predictions using graphs

6.3 Linear models: linear functions and their graphs

6.4 Quadratic models: quadratic functions and their graphs (parabolas); properties of a parabola; symmetry, vertex; intercepts; equation of the axis of symmetry

6.5 Exponential models: exponential functions and their graphs; concept and equation of a horizontal asymptote

6.6 Use of a GDC to solve equations involving combinations of the functions above

Before you start

You should know how to:

1 Substitute values into a formula, e.g. Given that $x = -1$, find the value of $y = 3x^2 + 2x$.
$y = 3(-1)^2 + 2(-1) \Rightarrow y = 1$

2 Use your GDC to solve quadratic equations and simultaneous equations in two unknowns, e.g. Solve
a $3x^2 + 9x - 30 = 0 \Rightarrow x = 2, x = -5$
b $\begin{cases} x + y = 4 \\ -2x + y = 1 \end{cases} \Rightarrow x = 1, y = 3$

3 Find the gradient, m, of a line joining two points, e.g. A(3, 5) and B(1, 4).
$y - y_1 = m(x - x_1)$
$m = \dfrac{5 - 4}{3 - 1}$
$m = \dfrac{1}{2}$

Skills check

1 a Find the value of $y = 2.5x^2 + x - 1$ when $x = -3$.
b Find the value of $h = 3 \times 2^t - 1$ when $t = 0$.
c Find the value of $d = 2t^3 - 5t^{-1} + 2$ when $t = \dfrac{1}{2}$.

2 Use your GDC to solve:
a $x^2 + x - 3 = 0$
b $2t^2 - t = 2$
c $\begin{cases} x - 2y = 3 \\ 3x - 5y = -2 \end{cases}$

> For help, see Chapter 12, Sections 1.1 and 1.2.

3 Find the gradient, m, of a line joining the two points:
a A(7, −2) and B(−1, 4)
b A(−3, −2) and B(1, 8)

The above photo shows the positions of a diver at various moments until he reaches the sea. Initially, the diver is at 40 m above sea level and it takes him 4.5 seconds to reach the sea. We can use mathematics to find a numerical relationship between the time in seconds, t, and the diver's height, h, in metres above sea level. The relationship linking the time, t, and the height, h, is a mathematical model. It can be described using a formula, a graph or a table of values.

To construct a mathematical model we usually begin by making some assumptions. Here, we assume that the diver is initially at 40 m above sea level and it takes him 4.5 seconds to reach the sea. The formula linking the variables t and h is

$$h = -1.97 \, (t^2 - 20.25) \quad \text{where } t \geq 0.$$

You can use this model to calculate the diver's height, h, above sea level at different times, t. Substitute the value of t into the formula to get the corresponding value for h. The table shows three pairs of values for t and h.

t (seconds)	h (metres)
0	40.0
1	38.0
4	8.37

The graph of $h = -1.97(t^2 - 20.25)$, $t \geq 0$, is shown.

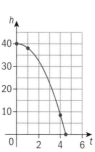

You can use the formula and/or the graph to answer questions such as:

At what height is the diver after 2 seconds?
How long does it take the diver to reach a height of 20 m above sea level?

The three pairs of values from the table are indicated with a ●.

In this chapter you will work with different types of mathematical models called **functions** to represent a range of practical situations. These functions help us to understand and predict the behavior of variables.

4.1 Functions

Mathematical models that link two variables are called functions.

> → A **function** is a relationship between two sets: a **first** set and a **second** set. Each element 'x' of the first set is related to **one and only one** element 'y' of the second set.

Example 1

Antonio and Lola are students at Green Village High School (GVHS). Miu is a student at Japan High School (JHS).
The set of students A = {Antonio, Lola, Miu}.
The set of schools B = {GVHS, JHS}.
Decide whether these relationships are functions. Justify your decisions.

a The relationship between the first set A and the second set B: 'x is a student at school y'.

b The relationship between the first set B and the second set A: 'x is the school where y is a student'.

In Chapter 1 we saw sets where the elements were numbers. However, the elements of a set may be any kind of object.

Answers

a This relationship is a function because each element of the first set A is related to one and only one element of the second set B; that is, each student studies at only one school.

Draw a mapping diagram to show how the elements of set A, Antonio, Lola and Miu, are related to the elements of set B, GVHS, JHS.

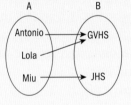

A mapping diagram is a drawing used to show how the elements in the first set are matched to the elements in the second set.

b This relationship is not a function because one element of the first set B, GVHS, is related to more than one element in the second set A, Antonio and Lola.

Draw a mapping diagram to show how the elements of set B are related to the elements of set A.

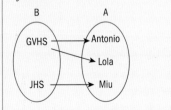

Example 2

Let $A = \{1, -1, 0, 2, 4\}$, $B = \{1, 0, 4\}$ and $C = \{1, 0, 4, 16\}$.
Decide whether these relationships are functions. Justify your decisions.

a The relationship between the first set A and the second set B: 'the square of x is y' or '$y = x^2$'.

b The relationship between the first set A and the second set C: 'the square of x is y' or '$y = x^2$'.

c The relationship between the first set C and the second set A: 'the square root of x is y' or '$y = \sqrt{x}$'.

In Example 2 as the elements of the sets are numbers, the relationships are **numerical**. In Mathematical Studies we work with numerical relationships that can be described using **equations**.

Answers

a It is not a function because one element of the first set A, 4, is not related to any element of the second set B.

Draw a table of values.
The elements of set A are the values of x. Use these to work out the corresponding values of y given $y = x^2$. Check that the values of y match the elements of set B.

A	B
x	$y = x^2$
1	1
−1	1
0	0
2	4
4	

$4^2 = 16$; 16 is not an element of set B.

b It is a function because each element of the first set A is related to one and only one element of the second set C.

A	C
x	$y = x^2$
1	1
−1	1
0	0
2	4
4	16

Think of everyday situations where you can define functions between two sets. For example, the relationship between a group of people and their names; the relationship between a tree and its branches; the relationship between the days and the mean temperature of each of these days, etc.

c It is a function because each element of the first set C is related to one and only one element of the second set A.

C	A
x	$y = \sqrt{x}$
1	1
0	0
4	2
16	4

1 Mrs. Urquiza and Mr. Genzer both teach mathematics.
Mick and Lucy are in Mrs. Urquiza's class. Lidia and Diana
are in Mr. Genzer's class.
Let the set of students A = {Mick, Lucy, Lidia, Diana}
and the set of teachers B = {Mrs. Urquiza, Mr. Genzer}.
Decide whether these relationships are functions. Justify your
decisions.
   a The relationship between the first set A and the second set B:
     'x is in y's mathematics class'.
   b The relationship between the first set B and the second set A:
     'x is y's mathematics teacher'.

2 Let A = {3, 7, 50}, B = {12, 16, 49, 100} and C = {49, 100}.
Decide whether these relationships are functions. Justify your
decisions.
   a The first set A, the second set B and the relationship 'x is a
     factor of y'.
   b The first set B, the second set A and the relationship 'x is a
     multiple of y'.
   c The first set C, the second set A and the relationship 'x is a
     multiple of y'.

3 Let A = {1, 2, 3, 4}, B = {2, 4, 6} and C = {1, 2, 4, 6}.
   a Decide whether these relationships are functions. Justify
     your decisions.
     i First set A, second set B and the relationship 'x is half of y'.
     ii First set A, second set C and the relationship 'x is half of y'.
     iii First set C, second set A and the relationship 'x is the
      double of y'.
     iv First set B, second set C and the relationship 'x is equal to y'.
     v First set C, second set A and the relationship 'x is equal to y'.
   b Draw a diagram to represent the relationships from part **a** that
     are functions.

4 Describe the following relationships between x and y using
equations.
   a y is double x.
   b Half of x is y.
   c The cubic root of x is y.
   d Half of the cube of x is y.

> The equation
> $y = x^2$ describes the
> relationship 'y is the
> square of x'.

5 Decide whether these relationships are functions. Explain your decision in the cases in which they are not functions.

   a The first set is $\mathbb{R}$, the second set is $\mathbb{R}$ and the relationship is defined by the equation $y = 3x + 1$.

   b The first set is $\mathbb{R}$, the second set is $\mathbb{R}$ and the relationship is defined by the equation $y = x^2$.

   c The first set is $\mathbb{R}$, the second set is $\mathbb{R}$ and the relationship is defined by the equation $y = \sqrt{x}$.

   d The first set is $A = \{x \geq 0, x \in \mathbb{R}\}$, the second set is $\mathbb{R}$ and the relationship is defined by the equation $y = \sqrt{x}$.

> $\mathbb{R}$ is the set of real numbers.

Domain and range of a function

A function is a relationship between two sets: a first set and a second set.

> → ● The first set is called the **domain** of the function. The elements of the domain, often thought of as the 'x-values', are the **independent variables**.
> ● For each value of 'x' (input) there is one and only one output. This value is called the **image** of 'x'. The set of all the images (all the outputs) is called the **range** of the function. The elements of the range, often thought of as the 'y-values', are the **dependent variables**.

Input   Output

equation

$x \longrightarrow y$

Domain   Range

> In Mathematical Studies the domain will always be the set of real numbers unless otherwise stated.

> We write domain and range values as sets inside curly brackets:
> Domain = {inputs}
> Range = {images or outputs}

Example 3

Consider the function $y = x^2$.

   a Find the image of   **i** $x = 1$ **ii** $x = -2$.

   b Write down the domain.

   c Write down the range.

- -

Answers

a **i** $y = 1$

    ii $y = 4$

b The domain is the set of real numbers, $\mathbb{R}$.

c The range is $y \geq 0$.

i *Substitute $x = 1$ into $y = x^2$*
$y = (-1)^2 \Rightarrow y = 1$
ii *Substitute $x = -2$ into $y = x^2$*
$y = (-2)^2 \Rightarrow y = 4$

Squaring any real number produces another real number. So the domain is the set of all real numbers.

The square of any positive or negative number will be positive and the square of zero is zero. So the range is the set of all real numbers greater than or equal to zero.

> The domain is assumed to be $\mathbb{R}$ unless there are any values which x cannot take.

Example 4

Consider the function $y = \dfrac{1}{x}$, $x \neq 0$.

a Find the image of: **i** $x = 2$ **ii** $x = -\dfrac{1}{2}$
b Write down the domain.
c **i** Decide whether $y = 0$ is an element of the range. Justify your decision.
   ii Decide whether $y = -5$ is an element of the range. Justify your decision.

Answers

a **i** $y = \dfrac{1}{2}$

   ii $y = \dfrac{1}{-\frac{1}{2}} = -2$

Substitute
i *$x = 2$ and*
ii *$x = -\dfrac{1}{2}$ into $y = \dfrac{1}{x}$.*

b The domain is the set of all real numbers except 0.

Since division by zero is not defined, the domain is the set of all real numbers except zero ($x \neq 0$).

c **i** $0 = \dfrac{1}{x}$

This equation has no solution. Therefore, $y = 0$ is not an element of the range.

Substitute **i** *$y = 0$ and*
   ii *$y = -5$ into $y = \dfrac{1}{x}$.*

Is there an input value (x) that gives an output (y) of 0?

   ii $-5 = \dfrac{1}{x}$

$x = -\dfrac{1}{5}$

So, $y = -5$ is an element of the range as it is the image of $x = -\dfrac{1}{5}$.

Is there an input value (x) that gives an output (y) of −5?

Gottfried Leibniz first used the mathematical term 'function' in 1673.

Exercise 4B

1 For each of the functions in **a**–**d**:

   i Copy and complete the table. Put a × in any cells that cannot be completed.

   ii Write down the domain.

   iii Decide if $y = 0$ is in the range of the function. Justify your decision.

a $y = 2x$

x	$-\dfrac{1}{2}$	0	1	3.5	
$y = 2x$					12

b $y = x^2 + 1$

x	-3	0	2	$\dfrac{1}{4}$		
$y = x^2 + 1$					5	5

c $y = \dfrac{1}{x+1}$, $x \neq -1$

x	-2	-1	0	$\dfrac{1}{2}$	3	
$y = \dfrac{1}{x+1}$						$\dfrac{1}{6}$

d $y = \sqrt{x}$, $x \geq 0$

x	-3	0	$\dfrac{1}{4}$		9	
$y = \sqrt{x}$				1		10

2 Decide whether each statement is true or false. Justify each of your decisions.

   a $y = 0$ is an element of the range of the function $y = \dfrac{2}{x}$.

   b The equation $y = x^2$ cannot take the value -1.

   c The equation $y = x^2 + 3$ cannot take the value 2.

   d For the function $y = x^2 - 1$ there are two values of x when $y = 3$.

   e For the function $y = \dfrac{x}{3} - 1$ the image of $x = -3$ is -2.

   f For the function $y = 2(-x + 1)$ the image of $x = -1$ is $y = 0$.

Graph of a function

A graph can represent a function.

> → The graph of a function f is the set of points (x, y) on the Cartesian plane where y is the image of x through the function f.

We use different letters to name functions: f, g, h, etc.

Drawing graphs

- Draw a table of values to find some points on the graph.
- On 2 mm graph paper, draw and label the axes with suitable scales.
- Plot the points.
- Join the points with a straight line or a smooth curve.

Cartesian coordinates and the Cartesian plane are named after the Frenchman René Descartes (1596–1650).

Example 5

a Draw the graph of the function $y = -x + 1$.

b Write down the coordinates of the point where the graph of this function intercepts the **i** x-axis **ii** y-axis.

c Decide whether the point A(200, −199) lies on the graph of this function.

d The point B(6, y) lies on the graph of this function. Find the value of y.

'Draw' means draw accurately on graph paper.

Answers

a

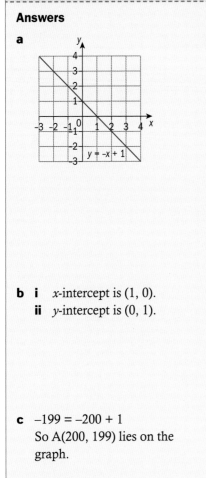

Draw a table of values. Use negative and positive values of x. Use the values of x to work out the corresponding values of y. When $x = -3$, $y = -(-3) + 1 = 4$.

x	−3	−1	0	1	3
y	4	2	1	0	−2

Use 2 mm graph paper. Let 1 cm = 1 unit. Label the x- and y-axes. Plot the coordinate points (−3, 4), (−1, 2), (0, 1), (1, 0) and (3, −2). Join the points with a straight line.

b **i** x-intercept is (1, 0).

ii y-intercept is (0, 1).

i To find the x-intercept, read off the point where the graph intersects the x-axis.

ii To find the y-intercept, read off the point where the graph intersects the y-axis.

A(200, −199)

c $-199 = -200 + 1$
So A(200, 199) lies on the graph.

Substitute the x- and y-values into the equation of the line to see if they 'fit'.

A point P lies on the graph of a function if and only if the point satisfies the equation of the function.

B(6, y)

d B(6, y) lies on the graph, so $y = -6 + 1 = -5 \Rightarrow y = -5$

Substitute x = 6 into the equation of the line to find the value of y at that point.

In the solution to the next example, the set notation $\{x \mid x \le 3\}$ is used. It is read as: the set of all x such that x is a real number less than or equal to 3.

Example 6

Here is the graph of a function f.
Use the graph to find
a the domain of f
b the range of f
c the points where the graph of f
    intersects the **i** x-axis and **ii** y-axis.

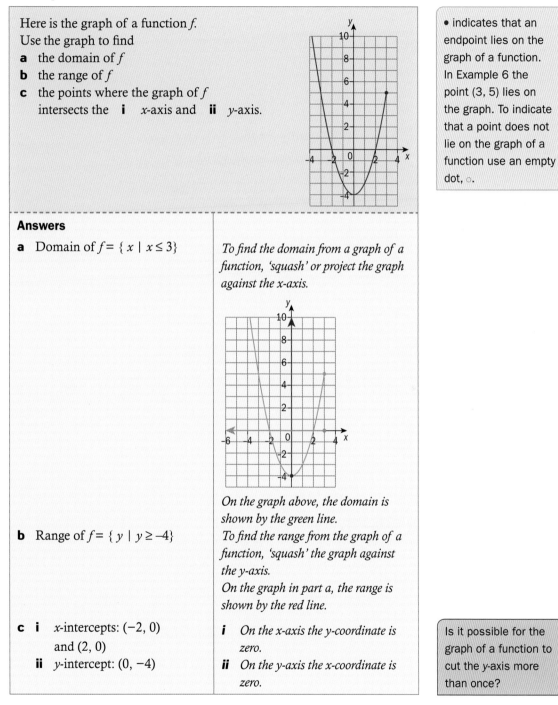

• indicates that an endpoint lies on the graph of a function. In Example 6 the point (3, 5) lies on the graph. To indicate that a point does not lie on the graph of a function use an empty dot, ○.

Answers

a Domain of $f = \{x \mid x \le 3\}$

To find the domain from a graph of a function, 'squash' or project the graph against the x-axis.

On the graph above, the domain is shown by the green line.

b Range of $f = \{y \mid y \ge -4\}$

To find the range from the graph of a function, 'squash' the graph against the y-axis.
On the graph in part a, the range is shown by the red line.

c i x-intercepts: $(-2, 0)$ and $(2, 0)$
  ii y-intercept: $(0, -4)$

i *On the x-axis the y-coordinate is zero.*
ii *On the y-axis the x-coordinate is zero.*

Is it possible for the graph of a function to cut the y-axis more than once?

Sketching a linear graph
- Draw and label the axes.
- Label the points where the graph crosses the x- or y-axis.

Example 7

Sketch a graph of the function $y = 3x - 1$.

> 'Sketch' means give a general shape of the graph.

Answer

The x-axis intercept is at $\left(\dfrac{1}{3}, 0\right)$

When $y = 0$, $x = \dfrac{1}{3}$

The y-axis intercept is at $(0, -1)$

When $x = 0$, $y = -1$

Draw the graph on your GDC.

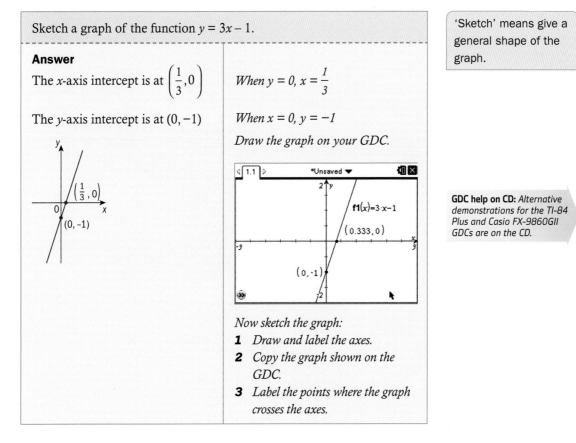

> **GDC help on CD:** *Alternative demonstrations for the TI-84 Plus and Casio FX-9860GII GDCs are on the CD.*

Now sketch the graph:
1. *Draw and label the axes.*
2. *Copy the graph shown on the GDC.*
3. *Label the points where the graph crosses the axes.*

Exercise 4C

EXAM-STYLE QUESTION

1. **a** Draw the graph of the function $y = 2x - 4$.
   b Write down the coordinates of the point where the graph of this function meets
      i the x-axis   **ii** the y-axis.
   c Decide whether the point A(250, 490) lies on the graph of this function. Justify your decision.
   d The point B(-3, y) lies on the graph of this function. Find the value of y.

2 For each of the graphs of functions in **a**–**d** write down
  i the domain **ii** the range
  iii the points where the graph meets the x-axis (where possible)
  iv the point where the graph meets the y-axis (where possible).

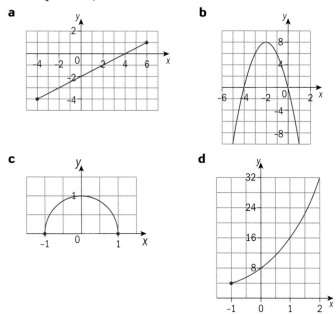

a   **b**   **c**   **d**

3 Decide whether these statements about the functions drawn in question 2 are true or false.

Function **a**
  i Point $(1, -1)$ lies on this graph.
  ii The image of $x = -2$ is 0. **iii** When $x = 6$, $y = 1$.

Function **b**
  i There are two values of x for which $y = 8$.
  ii There are two values of x for which $y = 4$.
  iii There is a value of x for which $y = 9$.

Function **c**
  i The line $x = 0.5$ intersects the graph of this function twice.
  ii The line $y = 0.5$ intersects the graph of this function twice.
  iii The image of $x = 0.2$ is the same as the image of $x = -0.8$.

Function **d**
  i The line $y = 1$ intersects the graph of this function once.
  ii When $x = 16$, $y = 1$.
  iii As the values of x increase, so do their corresponding values of y.

4 Sketch a graph for each of these functions.
  a $y = 2x + 3$ **b** $y = -x + 2$ **c** $y = 3x - 4$

Function notation

> → $y = f(x)$ means that the image of x through the function f is y.
> x is the independent variable and y is the dependent variable.

So, for example, if $f(x) = 2x - 5$
- $f(3)$ represents the image of $x = 3$.
  To find the value of $f(3)$ substitute $x = 3$: $f(3) = 2 \times 3 - 5 = 1$
- $f(-1)$ represents the image of $x = -1$.
  To find the value of $f(-1)$ substitute $x = -1$: $f(-1) = 2 \times -1 - 5 = -7$

We use different variables and different letters for functions; for example, $d = v(t)$, $m = C(n)$, etc.

$f(3) = 1$ can be read as: 'f at 3 equals 1' or 'f of 3 equals 1'.

Example 8

Consider the function $f(x) = -x^2 + 3x$.
a Find the image of $x = -2$.   **b** Find $f(1)$.
c Show that the point $(4, -4)$ lies on the graph of f.

- -

Answers

a $f(-2) = -(-2)^2 + 3 \times -2 = -10$	*Substitute $x = -2$ into $f(x) = -x^2 + 3x$.*
b $f(1) = -1^2 + 3 \times 1 = 2$	*Substitute $x = 1$ into $f(x)$.*
c $f(4) = -4^2 + 3 \times 4 = -4$ So $(4, -4)$ lies on the graph.	*If $(4, -4)$ lies on the graph of f then $f(4) = -4$. Substitute $x = 4$.*

One of the first mathematicians to study the concept of function was French Philosopher Nicole Oresme (1323–82). He worked with independent and dependent variable quantities.

Exercise 4D

1 Consider the function $f(x) = x(x - 1)(x + 3)$.
   a Calculate $f(2)$.   **b** Find the image of $x = \dfrac{1}{2}$.
   c Show that $f(-3) = 0$.
   d Decide whether the point $(-1, -4)$ lies on the graph of f. Justify your decision.

2 Consider the function $d(t) = 5t - t^2$.
   a Write down the independent variable of this function.
   b Calculate $d(2.5)$.
   c Calculate the image of $t = 1$.
   d Show that $d(1)$ and $d(4)$ take the same value.

3 Consider the function $C(n) = 100 - 10n$.
   a Calculate $C(2)$.
   b The point $(3, b)$ lies on the graph of the function C. Find the value of b.
   c The point $(a, 0)$ lies on the graph of the function C. Find the value of a.

EXAM-STYLE QUESTION
4 Here is the graph of the function $v(t) = -3t + 6$.
   a Write down the value of   **i** $v(1)$ **ii** $v(3)$.
   b The point $(m, 9)$ lies on the graph. Find the value of m.
   c Find the value of t for which $v(t) = 0$.
   d Find the set of values of t for which $v(t) < 0$.

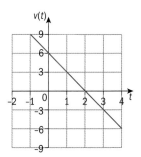

5 Consider the function $f(x) = 0.5(3 - x)$.

   a Draw the graph of f.

   b Find the point A where the graph of f meets the x-axis.

   c Find the point B where the graph of f meets the y-axis.

   d Solve the equation $f(x) = 2$.

6 Consider the function $h(x) = 3 \times 2^{-x}$.

   a Calculate **i** $h(0)$ **ii** $h(-1)$.   **b** Find x if $h(x) = 24$.

> $h(x) = 3 \times 2^{-x}$ is an exponential function. You will learn more about these in section 4.4.

Functions as mathematical models

We can use functions to describe real-life situations.

| Translate the situation into mathematical language and symbols. | → | Find the solution using mathematics. | → | Interpret the solution in the context of the problem. |

Example 9

A rectangular piece of card measures 20 cm by 10 cm. Squares of length x cm are cut from each corner. The remaining card is then folded to make an open box. Write a function to model the volume of the box.

Answer

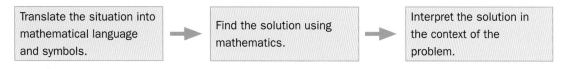

$V(x) = (20 - 2x)(10 - 2x)x$

First draw a diagram to represent the information given in the question. Carefully label the dimensions of the open box:
length $(20 - 2x)$ cm
width $(10 - 2x)$ cm
height x cm
The volume of the box, V, will depend on the value of x.
Volume of cuboid =
length × width × height.

> Look at Example 9.
> **1** What is the domain of the function $V(x)$? Can x take any value? Why? Try different values and draw a conclusion.
> **2** How could the function help you to find the maximum possible volume?

Exercise 4E

1 A rectangular piece of card measures 30 cm by 15 cm. Squares of length x cm are cut from each corner. The remaining card is then folded to make an open box of length l cm and width w cm.

   a Write expressions, in terms of x, for

     i the length, l **ii** the width, w.

   b Find an expression for the volume of the box, V, in terms of x.

     i Explain in words the meaning of $V(3)$.

     ii Find the value of $V(3)$.   **iii** Find the value of $V(3.4)$.

     iv Is $x = 8$ in the domain of the function $V(x)$? Justify your decision.

2 The perimeter of a rectangle is 24 cm and its length is x cm.

  a Find the width of the rectangle in terms of the length, x.

  b Find an expression for the area of the rectangle, A, in terms of x.

  c i Explain the meaning of $A(2)$.

   ii Calculate $A(2)$.

  d Is $x = 12$ in the domain of the function $A(x)$? Justify your decision.

> You can use mathematical functions to represent things from your own life. For example, suppose the number of pizzas your family eats depends on the number of football games you watch. If you eat 3 pizzas during every football game, the function would be 'number of pizzas' (p) = 3 times 'number of football games' (g) or $p = 3g$. Can you think of another real-life function? It could perhaps be about the amount of money you spend or the number of minutes you spend talking on the phone.

3 The Simpsons rent a holiday house costing 300 USD for the security deposit plus 150 USD per day. Let n be the number of days they stay in the house and C the cost of renting the house.

  a Write a formula for C in terms of n.

  b How much does it cost to rent the house for 30 days?

The Simpsons have 2300 USD to spend on the rent of the house.

  c i Write down an inequality using your answer to part **a** to express this condition.

   ii Hence, decide whether they have enough money to rent the house for two weeks.

   iii Write down the maximum number of days that they can rent the house.

4 An Australian company produces and sells books. The monthly **cost**, in AUD, for producing x books is modeled by $C(x) = 0.4x^2 + 1500$.
The monthly **income**, in AUD, for selling x books is modeled by $I(x) = -0.6x^2 + 160x$.

  a Show that the company's monthly profit can be calculated using the function
  $P(x) = -x^2 + 160x - 1500$.

  b What profit does the company make on producing and selling six books? Comment on your answer.

  c i What profit does the company make on producing and selling 40 books?

   ii Find the selling price of one book when 40 books are produced and sold.
   (Assume that all the books have the same price.)

  d Use your GDC to find the number of books for which $P(x) = 0$.

> Profit = Income – Cost

4.2 Linear models

Linear models of the form $f(x) = mx$

The straight line shown here has a positive gradient and the function $y = f(x)$ is increasing.

$f(0) = 0$ and the line passes through the origin $(0, 0)$.

The gradient of the line is given by $m = \dfrac{y_2 - y_1}{x_2 - x_1}$.

Using two points on the line, $(4, 6)$ and $(0, 0)$, the gradient is

$$m = \frac{6 - 0}{4 - 0} = \frac{3}{2} = 1.5.$$

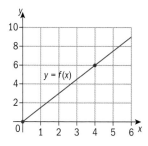

So $f(x) = 1.5x$.

This type of linear model is used in **conversion graphs**. The two variables which have a fixed relationship between them are in direct proportion, so their graphs are straight lines with a positive gradient passing through the origin.

> Conversion graphs can be used to convert one currency to another, or one set of units to another: for example, kilometres to miles, or kilograms to pounds.

Example 10

1 mile is equivalent to $1.6\,$km.
  a Draw a conversion graph of miles to km.
  b Find the gradient of the line.
  c Hence, write down a model for $k(x)$, where $k(x)$ is the distance in km and x is the distance in miles.

Answers

a

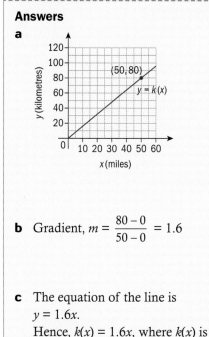

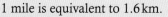

Use 2 mm graph paper.
Put miles on the x-axis.
Put kilometres on the y-axis.
Find two points to draw a straight line: 0 miles = 0 km so (0, 0) is on the line.
50 miles is equivalent to $1.6 \times 50 = 80$ km so (50, 80) is on the line.
Plot the two points and join with a straight line.

b Gradient, $m = \dfrac{80 - 0}{50 - 0} = 1.6$

*Use the two points from part **a** to find the gradient,*
$m = \dfrac{y_2 - y_1}{x_2 - x_1}.$

c The equation of the line is $y = 1.6x$.
Hence, $k(x) = 1.6x$, where $k(x)$ is the distance in km, and x is the distance in miles.

For a general linear function through the origin, f(x) = mx.
Here, the function is k (x) = mx.

> The equation $y = 1.6x$ can be rearranged to $x = \dfrac{y}{1.6}$ or $x = \dfrac{1}{1.6}\, y$ $= 0.625y.$ Use this to convert km to miles.

Exercise 4F

1 1 kg is equivalent to 2.2 pounds.

  a Convert 50 kg into pounds.

  b Draw a conversion graph of pounds to kilograms. Use x-values from 0 kg to 100 kg and y-values from 0 pounds to 250 pounds.

Plot the point you found in part **a**.

  c Find the gradient of the line. Hence, write down the model for $p(x)$, where $p(x)$ is the weight in pounds and x is the weight in kg.

  d Find $p(75)$ and $p(125)$.

  e Find the model for $k(x)$, where $k(x)$ is the weight in kg and x is the weight in pounds.

Write the formula as $y = \ldots$ and rearrange to make x the subject.

  f Calculate $k(75)$ and $k(100)$.

2 The exchange rate for pounds sterling (GBP) to Singapore dollars (SGD) is £1 = S$2.05.

  a Find the number of Singapore dollars equivalent to 50 GBP.

  b Draw a conversion graph of GBP to SGD. Use x-values from £0 to £100 and y-values from S$0 to S$250.

  c Find the gradient of the line. Hence, write down the model for $s(x)$, where $s(x)$ is the amount of money in SGD and x is the amount of money in GBP.

  d Find $s(80)$ and $s(140)$.

  e Find the model for $p(x)$, where $p(x)$ is the amount of money in GBP and x is the amount of money in SGD.

  f Calculate $p(180)$.

3 The exchange rate for pounds sterling (GBP) to US dollars (USD) is £1 = $1.55.

  a Find the number of US dollars equivalent to 60 GBP.

  b Draw a conversion graph of GBP to USD. Use the x-axis for GBP with $0 \le x \le 80$, and the y-axis for USD with $0 \le y \le 140$.

  c Find the gradient of the line. Hence, write down the model for $u(x)$, where $u(x)$ is the amount of money in USD and x is the amount of money in GBP.

  d Find $u(300)$ and $u(184)$.

  e Find the model for $p(x)$, where $p(x)$ is the amount of money in GBP and x is the amount of money in USD.

  f Calculate $p(250)$ and $p(7750)$.

Linear models of the form $f(x) = mx + c$

When two variables are not in direct proportion, their graphs are straight lines that do not pass through the origin, that is, **linear functions**.

> → A **linear function** has the general form
> $$f(x) = mx + c$$
> where m (the gradient) and c are constants.

You have seen the equation of a line in Chapter 3, Section 3.2.

Example 11

In a chemistry experiment, a liquid is heated and the temperature at different times recorded.
The table of results for one student is shown.

Time (x minutes)	2	4	6	9
Temperature (y°C)	30	40	50	65

a Draw a graph for this data.

b Find a model for $T(x)$, the temperature with respect to time, for these data.

c Use the model to predict:
  i the temperature of the liquid after 8 minutes
  ii the time taken for the liquid to reach 57 °C.

You can plot the graph on your GDC and find the model for $T(x)$. For help, see Chapter 12, Section 5.4.

Answers

a

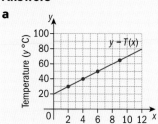

Time (x minutes)

Use 2 mm graph paper.
Put time on the x-axis.
Put temperature on the y-axis.
Plot the points from the table e.g. (2, 30) and join them with a straight line.

b Gradient, $m = \dfrac{65 - 40}{9 - 4}$

$= \dfrac{25}{5} = 5$

$T(x) = mx + c$
$T(x) = 5x + c$
$T(2) = 5 \times 2 + c = 30$
$10 + c = 30$
$c = 20$

Therefore, the model for the temperature is $T(x) = 5x + 20$.

The model will be in the form $T(x) = mx + c$. You need to find the constants m and c.
Use any two points from the table, e.g. (4, 40) and (9, 65), to find the gradient, $m = \dfrac{y_2 - y_1}{x_2 - x_1}$.

To find the value of c use any point from the table, e.g. (2, 30), which means T(2) = 30.

c i At 8 minutes:
$T(8) = 5 \times 8 + 20 = 60$
So, the temperature of the liquid after 8 minutes is 60 °C.

ii When $T(x) = 57$ °C:
$57 = 5x + 20$
$5x = 37$
$x = \dfrac{37}{5} = 7.4$
So, it takes 7.4 minutes for the liquid to reach 57 °C.

i *A time of 8 minutes means x = 8. Substitute x = 8 into the function in part **b**.*

ii *A temperature of 57°C means T(x) = 57. Substitute T(x) = 57 and solve for x.*

For Example 11, the equation of the model was $T(x) = 5x + 20$. Compare the equation of the model with

a the initial temperature

b the average rise in temperature every minute.

What conclusions can be drawn?

Exercise 4G

1 In a chemistry experiment, a liquid is heated and the temperature at different times is recorded. Here is a table of results.

Time (x minutes)	3	5	7	9
Temperature (y °C)	130	210	290	410

  a Draw a graph for these data.

  b What was the initial temperature of the liquid?

  c Find the linear model, $T(x)$, for the temperature of the liquid with respect to time.

> Use a scale up to 420 on the y-axis.

2 In a physics experiment, a spring is stretched by loading it with different weights, in grams. The results are given in the table.

Weight (x g)	40	50	75	90
Length of spring (y mm)	38	43	55.5	63

  a Draw a graph for these data.

  b Find the natural length of the spring.

  c By how many mm does the spring stretch when the weight increases from 50 g to 90 g?

  d Use the answer to part **c** to find the average extension of the spring in mm for each extra gram loaded.

  e Find the equation of the linear model, $L(x)$, for the length of the spring with respect to load.

> The natural length is the length of the spring with no loading.

3 The temperature of the water in a hot water tank is recorded at 15 minute intervals after the heater is switched on.

Time (x minutes)	15	30	45	60	75	90
Temperature (y °C)	20	30	40	50	60	70

  a Plot a graph of these data on your GDC.

  b Find the linear model, $T(x)$, for the temperature of the liquid with respect to time.

  c Find the temperature of the water after 85 minutes.

> Read off the value from your graph.

4 Different weights are suspended from a spring. The length of the spring with each weight attached is recorded in the table.

Weight (x g)	125	250	375	500
Length of spring (y cm)	30	40	50	60

  a Plot a graph of these data on your GDC.

  b Find the natural length of the spring.

  c By how many cm does the spring stretch when the weight is increased from 125 g to 375 g?

  d Find the weight that will stretch the spring to a length of 48 cm.

  e Find the equation of the linear model, $L(x)$, for the length of the spring with respect to load.

Linear models involving simultaneous equations

Sometimes you cannot find the model from the given data.
You may need to write equations to represent the situation
and solve them simultaneously.

For a reminder on
solving simultaneous
equations, see Chapter
13, Section 2.4.

Example 12

A carpenter makes wooden tables and chairs.
He takes 10 hours to make a table and 4 hours to make a chair.
The wood costs $120 for a table and $40 for a chair.
Find a model for
a the time required to make the tables and chairs
b the cost of making the tables and chairs.

The given values for
a model are called
constraints.

Answers

a Let t be the time required to
make the tables and chairs.
The model for the time
required is $t = 10x + 4y$.

*Let x be the number of tables and y
be the number of chairs.*
Total number of hours for the tables:
10 hours per table, x tables $\Rightarrow 10 \times x$
Total number of hours for the chairs:
4 hours per chair, y chairs $\Rightarrow 4 \times y$

b Let c be the cost of making
the tables and chairs.
The model for the cost is
$c = 120x + 40y$.

Total cost ($) for the tables:
$120 per table, x tables $\Rightarrow 120 \times x$
Total cost ($) for the chairs:
$40 per chair, y chairs $\Rightarrow 40 \times y$

The simultaneous equations arise when you are given values that the
model must satisfy.

Example 13

The carpenter in Example 12 works 70 hours one week and spends
$760 on wood.
How many tables and chairs can he make?

Answer

From Example 12:
$t = 10x + 4y$
$c = 120x + 40y$
$10x + 4y = 70$
$120x + 40y = 760$
Using a GDC: $x = 3$ and $y = 10$
The carpenter can make 3 tables
and 7 chairs.

The model must work for the values:
(time) $t = 70$ and (cost) $c = 760$.
*Write a set of simultaneous
equations.*
*Solve these either analytically or
using a GDC.*

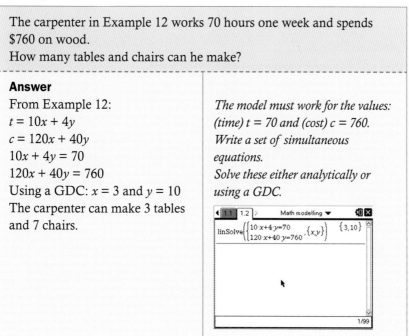

For help with using
a GDC to solve
simultaneous
equations see
Chapter 12,
Sections 1.1 and 3.4.

GDC help on CD: *Alternative
demonstrations for the TI-84
Plus and Casio FX-9860GII
GDCs are on the CD.*

1 To make a sponge cake you need 80 g of flour and 50 g of fat.
   To make a fruit cake you need 60 g of flour and 90 g of fat.
   Find a model for
   a the amount flour needed to make both cakes
   b the amount of fat needed to make both cakes.

   Peter has 820 g of flour and 880 g of fat.
   c How many of each type of cake can he make?

2 It takes 8 hours to make a table and 3 hours to make a chair.
   For a table the wood costs $100. For a chair the wood costs $30.
   A carpenter has 51 hours and $570.
   How many tables and chairs can she make?

3 A van carries up to 3 people and 7 cases.
   A car carries up to 5 people and 3 cases.
   How many vans and cars do you need for 59 people and
   70 cases?

4 A passenger plane carries 80 passengers and 10 tonnes of
   supplies.
   A transport plane carries 50 passengers and 25 tonnes of supplies.
   How many planes of each type do you need to carry 620 people
   and 190 tonnes of supplies?

5 A school mathematics department has 1440 euros to buy
   textbooks.
   Maths for All volume 1 costs 70 euros. *Maths for All* volume 2 costs
   40 euros.
   The department wants twice as many copies of volume 1 as
   volume 2.
   How many of each volume can they buy?

Extension material on CD:
Worksheet 4 - Equations

4.3 Quadratic models

Quadratic functions and their graphs

> → A **quadratic function** has the form
> $f(x) = ax^2 + bx + c$, where $a, b, c \in \mathbb{R}$ and $a \neq 0$.

Why $a \neq 0$? What kind
of function would you
get if $a = 0$?

The domain of a quadratic function can be the entire set of real
numbers ($\mathbb{R}$) or any subset of this.

Here are examples of some quadratic functions:

$f(x) = x^2 + 3x + 2$ $f(x) = x - 3x^2$ $f(x) = 3x^2 + 12$
$(a = 1, b = 3, c = 2)$ $(a = -3, b = 1, c = 0)$ $(a = 3, b = 0, c = 12)$

The simplest quadratic function is $f(x) = x^2$.

Here is a table of values for $f(x) = x^2$.

x	−3	−2	−1	0	1	2	3
f(x)	9	4	1	0	1	4	9

Plotting these values gives the graph shown here.

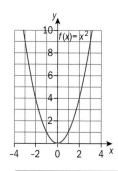

1 The graph is called a **parabola**.
2 The parabola has an **axis of symmetry** (the y-axis).
3 The parabola has a **minimum point** at (0, 0).
   The minimum point is called the **vertex** (or turning point) of the parabola.
4 The range of $f(x) = x^2$ is $y \geq 0$.

'Squash' the graph of $f(x) = x^2$ against the y-axis to confirm that the range is $y \geq 0$.

→ The graph of any quadratic function is a **parabola** – a ∪-shaped (or ∩-shaped) curve. It has an **axis of symmetry** and either a **minimum** or **maximum** point, called the **vertex** of the parabola.

The name 'parabola' was introduced by the Greek, Appollonius of Perga (c. 262–190 BCE) in his work on conic sections.

Investigation – the curve $y = ax^2$

1 Draw these curves on your GDC: $y = x^2$ and $y = -x^2$
   How are these two curves related?
2 Now draw: $y = 2x^2$ $y = 3x^2$ $y = 0.5x^2$
            $y = -2x^2$ $y = -3x^2$ $y = -0.5x^2$
   Compare each of these six graphs to $y = x^2$.
   Consider:
   a Is the curve still a parabola? Is the curve ∪-shaped or ∩-shaped?
   b Does it have a vertical line of symmetry?
   c What is its vertex? Is the vertex a minimum or maximum point?
3 What is the effect of changing the value of a?
   Draw a few more graphs to test your conjecture. (Remember to use positive and negative values of a and also use fractions.)

For help with drawing graphs on your GDC see Chapter 12, Section 4.1.

Without drawing the graph, how do you know that it will be ∩-shaped?

Investigation – the curve $y = x^2 + c$

Draw these curves on your GDC: $y = x^2$ $y = x^2 + 2$ $y = x^2 - 4$
                                $y = x^2 + 3$ $y = x^2 - 2$

Compare each graph to the parabola $y = x^2$. (Use the list of considerations in the preceding investigation as a guide.)
What is the effect of changing the value of c?

Sketching a quadratic graph (1)

- Draw and label the axes.
- Mark any points where the graph intersects the axes (the x- and y-intercepts). Label these with their coordinates.
- Mark and label the coordinates of any maximum or minimum points.
- Show one or two values on each axis to give an idea of the scale.

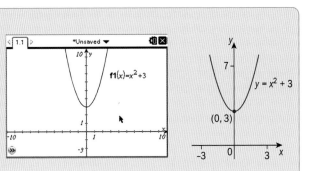

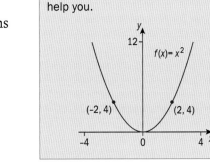

Exercise 4I

Use your results from the two previous investigations to help you sketch these graphs.

1 $y = 2x^2 + 1$

2 $y = -x^2 + 3$

3 $y = 3x^2 - 2$

4 $y = -2x^2 + 7$

Use this sketch graph of $f(x) = x^2$ to help you.

Investigation – the curves $y = (x + p)^2$ and $y = (x + p)^2 + q$

1 Use your GDC to draw these graphs:
$y = x^2$, $y = (x + 2)^2$, $y = (x + 3)^2$, $y = (x - 1)^2$, $y = (x - 0.5)^2$
Compare each graph to the graph of $y = x^2$.
What is the effect of changing the value of p?

2 Use your GDC to draw these graphs:
$y = (x + 2)^2 - 3$, $y = (x - 4)^2 + 2$, $y = (x - 1)^2 - 5$
What is the axis of symmetry of $y = (x + p)^2 + q$?
What are the coordinates of the vertex of $y = (x + p)^2 + q$?

Exercise 4J

For each graph, write down the coordinates of the vertex and the equation of the axis of symmetry.

1 $y = (x + 3)^2 - 2$

2 $y = (x + 5)^2 + 4$

3 $y = (x - 4)^2 - 1$

4 $y = (x - 5)^2 + 7$

5 $y = -(x + 3)^2 + 4$

The equation of the axis of symmetry must be given as 'x = . . .'.

Investigation – the curves $y = kx - x^2$ and $y = x^2 - kx$

Part A

1 Use your GDC to draw the graph of $y = 4x - x^2$.
What is the equation of its axis of symmetry?
What are the coordinates of the vertex?
What are the coordinates of the points at which the curve intersects the x-axis?

2 Draw these curves: $y = 2x - x^2$, $y = 6x - x^2$, $y = x - x^2$, $y = 5x - x^2$

3 What is the effect of varying the value of k?
What is the equation of the axis of symmetry of the curve $y = kx - x^2$?
What are the coordinates of the points at which the curve $y = kx - x^2$ intersects the x-axis?

Part B

Draw these curves: $y = x^2 - 2x$, $y = x^2 - 4x$, $y = x^2 - 6x$
Answer the same questions for these as you did for the curves in part A.

Investigation – curves of the form $y = (x - p)(x - q)$

1 Use your GDC to draw the graph of $y = (x - 1)(x - 3)$.
Where does it intersect the x-axis?
What is the equation of its axis of symmetry?
What are the coordinates of the vertex?

2 Answer the previous questions for the general curve $y = (x - p)(x - q)$.
(You may wish to draw more graphs of functions of this form.)

Exercise 4K

For each function, write down:
a the equation of the axis of symmetry
b the coordinates of the points at which the curve intersects the x-axis
c the coordinates of the vertex.

> Do not draw the graphs.

1 $y = x(x - 4)$ **2** $y = x(x + 6)$

3 $y = 8x - x^2$ **4** $y = 3x - x^2$

5 $y = x^2 - 2x$ **6** $y = x^2 - x$

> Factorize and then use the same method as in questions 1 and 2.

7 $y = x^2 + 4x$ **8** $y = x^2 + x$

9 $y = (x + 1)(x - 3)$ **10** $y = (x - 5)(x + 3)$

11 $y = (x - 2)(x - 6)$ **12** $y = (x + 2)(x - 4)$

Investigation – the general quadratic curve
$$y = ax^2 + bx + c$$

Part A: $a = 1$

1 Use your GDC to draw the graph of $y = x^2 - 4x + 3$.

Where does it intersect the x-axis?

What is the equation of its axis of symmetry?

What are the coordinates of the vertex?

2 Answer the previous questions for the general curve $y = ax^2 + bx + c$.

(You may wish to draw more graphs of functions of this form.)

Part B: varying a

Use your GDC to draw the graph of $y = 2x^2 - 4x + 3$ as a starting point.

Consider graphs of this form and answer the questions from Part A.

Exercise 4L

For each function, write down:

a the equation of the axis of symmetry

b the coordinates of the points at which the curve intersects the x-axis

c the coordinates of the vertex.

1 $y = x^2 - 2x + 3$ **2** $y = x^2 + 4x - 5$

3 $y = x^2 + 6x + 4$ **4** $y = 3x^2 - 6x + 2$

5 $y = 2x^2 - 8x - 1$ **6** $y = 2x^2 + 6x - 7$

7 $y = 0.5x^2 - x + 2$ **8** $y = 0.5x^2 + 3x - 4$

The general form of a quadratic function is $f(x) = ax^2 + bx + c$.

> A ∪-shaped graph is 'concave up'.
>
> A ∩-shaped graph is 'concave down'.

→ • If $a > 0$ then the graph is ∪-shaped; if $a < 0$ then the graph is ∩-shaped.
  • The curve intersects the y-axis at $(0, c)$.
  • The equation of the axis of symmetry is $x = -\dfrac{b}{2a}$, $a \neq 0$.
  • The x-coordinate of the vertex is $x = -\dfrac{b}{2a}$.

> This formula is in the Formula booklet. You should have found it in the investigation above.

→ The factorized form of a quadratic function is

$f(x) = a(x - k)(x - l)$.

  • If $a > 0$ then the graph is ∪-shaped; if $a < 0$ then the graph is ∩-shaped.
  • The curve intersects the x-axis at $(k, 0)$ and $(l, 0)$.
  • The equation of the axis of symmetry is $x = \dfrac{k + l}{2}$.
  • The x-coordinate of the vertex is also $x = \dfrac{k + l}{2}$.

> In a parabola, the axis of symmetry passes through the vertex.

→ **Finding the x-intercepts**

The function $f(x) = ax^2 + bx + c$ intersects the x-axis where $f(x) = 0$. The x-values of the points of intersection are the two solutions (or **roots**) of the equation $ax^2 + bx + c = 0$.
(The y-values at these points of intersection are zero.)

Example 14

Consider the function $f(x) = x^2 + 6x + 8$.

a Find
   i   the point where the graph intersects the y-axis
   ii   the equation of the axis of symmetry
   iii   the coordinates of the vertex
   iv   the coordinates of the point(s) of intersection with the x-axis.

b   Use the information from part **a** to sketch this parabola.

The Indian mathematician Sridhara is believed to have lived in the 9th and 10th centuries. He was one of the first mathematicians to propose a rule to solve a quadratic equation. Research why there is controversy about when he lived.

Answers

a **i**   The graph intersects the y-axis at $(0, 8)$.

General form: $f(x) = ax^2 + bx + c$.
Here: $f(x) = x^2 + 6x + 8$
So: $a = 1$, $b = 6$, $c = 8$
The curve intersects the y-axis at $(0, c)$.

   ii   The equation of the axis of symmetry is $x = -\dfrac{6}{2(1)} = -3$.

Use $x = -\dfrac{b}{2a}$, with $a = 1$ and $b = 6$.

   iii   The x-coordinate of the vertex is $x = -3$.
      The y-coordinate of the vertex is:
      $f(-3) = (-3)^2 + 6(-3) + 8$
           $= -1$
      So, the coordinates of the vertex are $(-3, -1)$.

*The x-coordinate of the vertex is $x = -\dfrac{b}{2a}$, which we found in part **b** so $x = -3$.*
To find the y-coordinate substitute $x = -3$ into the equation of the function.

   iv $x^2 + 6x + 8 = 0$
      $f(x) = 0$ when $x = -2$ or -4
      The graph intersects the x-axis at $(-2, 0)$ and $(-4, 0)$.

The curve intersects the x-axis where $f(x) = 0$, so put $x^2 + 6x + 8 = 0$ and solve using a GDC.

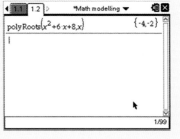

GDC help on CD: *Alternative demonstrations for the TI-84 Plus and Casio FX-9860GII GDCs are on the CD.*

▶ Continued on next page

b

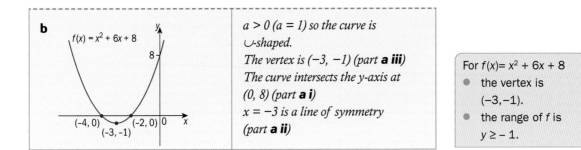

$a > 0 \ (a = 1)$ *so the curve is*
∪-shaped.
*The vertex is (−3, −1) (part **a iii**)*
The curve intersects the y-axis at
*(0, 8) (part **a i**)*
x = −3 is a line of symmetry
*(part **a ii**)*

> For $f(x) = x^2 + 6x + 8$
> ● the vertex is
> $(-3, -1)$.
> ● the range of f is
> $y \geq -1$.

Exercise 4M

For each function $f(x)$ in 1–8:

a Find
  i the coordinates of the point of intersection with the y-axis
  ii the equation of the axis of symmetry
  iii the coordinates of the vertex
  iv the coordinates of the point(s) of intersection with the x-axis
  v the range of f.
b Sketch the graph of the function.
c Use your GDC to draw the graph to check your results.

1 $f(x) = x^2 + 2x - 3$ **2** $f(x) = x^2 + 8x + 7$
3 $f(x) = x^2 - 6x - 7$ **4** $f(x) = x^2 - 3x - 4$
5 $f(x) = x^2 - 3x - 10$ **6** $f(x) = 2x^2 + x - 3$
7 $f(x) = 2x^2 + 5x - 3$ **8** $f(x) = 3x^2 - x - 4$

Sketching quadratic graphs

Example 15

a Sketch the graph of a parabola with vertex $(1, 2)$ and range $y \geq 2$.
b Sketch the graph of a parabola with x-intercepts at $x = -2$
  and $x = 3$ and y-intercept at $y = -1$.

Answers

a

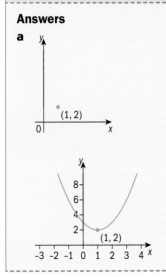

Draw and label the axes.
Use a vertical line to show the range
($y \geq 2$) of the function on the y-axis
(shown in red here).
Plot and label the vertex (1, 2).

> Is the parabola
> shown in Example 15
> part **a** the only one
> that satisfies the
> information given?
> If not, how many are
> there?

Draw a smooth curve through the
point (1, 2). The curve is symmetrical
about the vertical line through the
vertex, that is, x = 1.

▶ Continued on next page

b

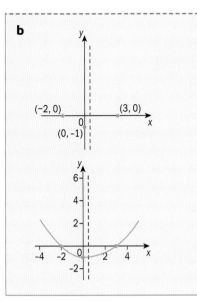

Draw and label the axes.
Mark the x-intercepts at $(-2, 0)$ and $(3, 0)$. The axis of symmetry is halfway between the two x-intercepts, at $x = \frac{1}{2}$. Draw it in (as shown by the dashed red line here).
Mark the y-intercept at $(0, -1)$.
Draw a smooth curve through the marked points.
The curve is symmetrical about $x = \frac{1}{2}$, and the axis of symmetry passes through the vertex.

The x-intercepts are the points where the graph crosses the x-axis. The y-values at these points are zero. The values at the x-intercepts are called the 'zeros' of the function.
The y-intercepts are the points where the graph crosses the y-axis. The x-values at these points are zero.

Sketching a quadratic graph (2)

● If you are given the function, use your GDC to draw the graph and then copy the information on to a sketch.
● If you are not given the function, use the information you are given and what you know about quadratic curves, that is:
  ■ They are ∪-shaped or ∩-shaped.
  ■ They have an axis of symmetry that passes through the vertex.

If a quadratic function only takes negative values between $x = m$ and $x = n$, what can you tell about $x = m$ and $x = n$? What happens at the points where x takes those negative values? Is the parabola ∪-shaped or ∩-shaped?
What can you tell about a quadratic function that only takes positive values between $x = m$ and $x = n$?

Exercise 4N

Sketch the graph of:

1 A parabola with vertex $(1, -3)$ and x-intercepts at -1 and 3.

2 A parabola with vertex $(-1, 2)$ and range $y \le 2$.

3 A parabola with an axis of symmetry at $x = 0$ and range $y \le -1$.

4 A parabola with x-intercepts at $x = 3$ and $x = 0$ and range $y \le 1$.

5 A parabola passing through the points $(0, -2)$ and $(4, -2)$ with a maximum value at $y = 2$.

6 A quadratic function f that takes negative values between $x = 2$ and $x = 5$, and $f(0) = 4$.

Intersection of two functions

→ Two functions $f(x)$ and $g(x)$ intersect at the point(s) where $f(x) = g(x)$.

To find the coordinates of the points of intersection
- use a GDC, or
- equate the two functions algebraically, rearrange to equal zero, and then solve on the GDC.

Example 16

Find the points of intersection of $f(x) = x^2 + x - 4$
and $g(x) = 3 - 4x - x^2$.

Answers

Method 1: Graphical
The points of intersection are
$(-3.5, 4.75)$ and $(1, -2)$.

For help with using a GDC to find the points of intersection of two curves see Chapter 12, Section 4.5.

Method 2: Algebraic

$f(x) = g(x)$ *Equate $f(x)$ and $g(x)$.*
$x^2 + x - 4 = 3 - 4x - x^2$
$2x^2 + 5x - 7 = 0$ *Rearrange to equal zero.*
$x = 1, x = -\dfrac{7}{2}$ *Solve using a GDC.*

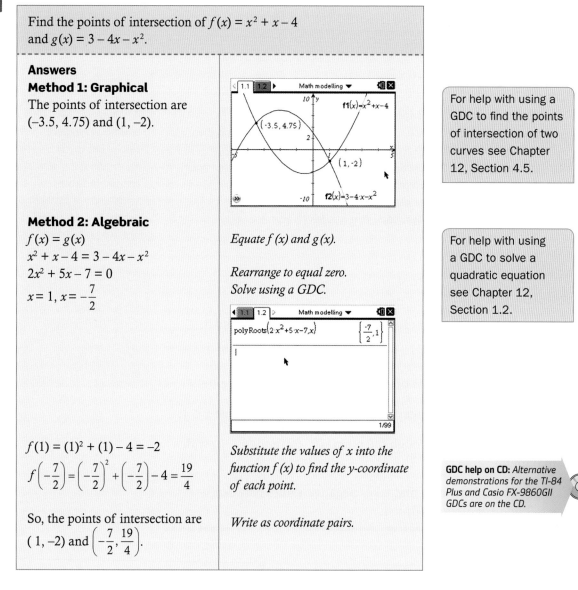

For help with using a GDC to solve a quadratic equation see Chapter 12, Section 1.2.

$f(1) = (1)^2 + (1) - 4 = -2$

$f\left(-\dfrac{7}{2}\right) = \left(-\dfrac{7}{2}\right)^2 + \left(-\dfrac{7}{2}\right) - 4 = \dfrac{19}{4}$

Substitute the values of x into the function $f(x)$ to find the y-coordinate of each point.

So, the points of intersection are
$(1, -2)$ and $\left(-\dfrac{7}{2}, \dfrac{19}{4}\right)$.

Write as coordinate pairs.

GDC help on CD: *Alternative demonstrations for the TI-84 Plus and Casio FX-9860GII GDCs are on the CD.*

Exercise 40

1 Here are two functions $f(x) = x^2 + 3x - 5$ and $g(x) = x - 2$ for the domain $-5 \leq x \leq 2$, $x \in \mathbb{R}$.

  a Use a GDC to draw the graphs of these functions and find the coordinates of their points of intersection.

  b Write down $f(x) = g(x)$ and solve it for x. Do you get the same answers as you did for part **a**?

  The function $h(x) = 2x - 3$ has the same domain.

  c Find the points of intersection of $f(x)$ and $h(x)$

    i algebraically   **ii** graphically.

> Find the points 'graphically' means draw the graphs on a GDC and read off the coordinates of the points of intersection.

2 Find the coordinates of the points of intersection of the graph of $f(x) = x^2 + 3x - 5$ for the domain $-5 \leq x \leq 2$, $x \in \mathbb{R}$, and the line $x + y + 5 = 0$.

> First rearrange the linear equation to make y the subject.

3 Find the points of intersection of the graphs of:

  a $f(x) = 5 + 3x - x^2$ and $g(x) = 1$

  b $f(x) = 5 + 3x - x^2$ and $h(x) = 2x + 3$

4 a Use a GDC to draw the graphs of the functions $f(x) = 2x^2 - x - 3$ and $g(x) = x + 1$ for the domain $-3 \leq x \leq 3$, $x \in \mathbb{R}$.

  b State the ranges of f and g on this domain.

  c Find the x-coordinates of the points of intersection of the two functions.

  d On the same axes, and for the same domain, draw the graph of the function $h(x) = 2x + 2$.

  e Solve the equation $f(x) = h(x)$ both graphically and algebraically.

  f Find the coordinates of the points of intersection of the graph of $y = f(x)$ and the line $x + y = 5$.

EXAM-STYLE QUESTION

5 The diagram shows the graphs of the functions $f(x) = x^2 - 3$ and $g(x) = 6 - x^2$ for values of x between -4 and 4.

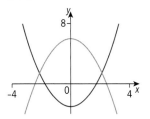

  a Find the coordinates of the points of intersection.

  b Write down the set of values of x for which $f(x) < g(x)$.

Finding the equation of a quadratic function from its graph

To find the equation of a graph of a quadratic function, use these facts:

For the graph of $f(x) = ax^2 + bx + c$

- the point of intersection of the function with the y-axis is $(0, c)$

- the equation of the axis of symmetry is $x = -\dfrac{b}{2a}$.

> You can use your GDC to find the equation of a quadratic function from its graph. For help see Chapter 12, Section 4.6.

Example 17

Find the equation of the quadratic function shown in the graph.

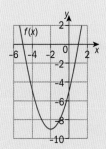

Answer

The general form of a quadratic function is given by $f(x) = ax^2 + bx + c$.

The function intersects the y-axis at the point $(0, -5)$. So $c = -5$.

$\Rightarrow f(x) = ax^2 + bx - 5$

The function intersects the y-axis at the point $(0, c)$. Read off the value of c from the graph.

The equation of the axis of symmetry is $x = -2$.

So: $-2 = -\dfrac{b}{2a}$

$\qquad -b = -4a$

$\qquad\ \ b = 4a$

The equation of the axis of symmetry is given by $x = -\dfrac{b}{2a}$. Substitute the value of x.

At the vertex, $x = -2$, $y = -9$.

So: $f(-2) = a(-2)^2 + b(-2) - 5 = -9$

$\qquad\qquad 4a - 2b - 5 = -9$

$\qquad\qquad 4a - 2b = -4$

Read off the coordinates of the vertex from the graph: $(-2, -9)$.
Substitute the x- and y-values into $f(x) = ax^2 + bx - 5$.

$\left.\begin{array}{l} b = 4a \\ 4a - 2b = -4 \end{array}\right\} \Rightarrow$

$\quad 4a - 2(4a) = -4$

$\qquad 4a - 8a = -4$

$\qquad\quad -4a = -4 \Rightarrow a = 1$

$b = 4a \Rightarrow b = 4$

Solve the simultaneous equations.

Therefore, the equation of the quadratic function is:

$f(x) = x^2 + 4x - 5$

Substitute the values of $a = 1$, $b = 4$ and $c = -5$ into $f(x) = ax^2 + bx + c$.

Exercise 4P

Find the equations of these quadratic functions.

1

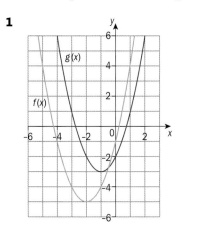

2

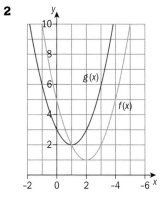

3

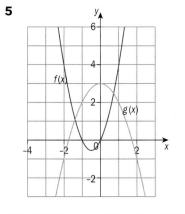

4

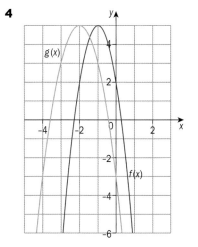

5

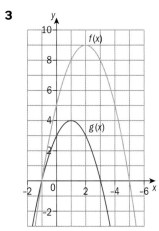

Quadratic models

Many real-life situations can be modeled using a quadratic function.

Example 18

A farmer wishes to fence off the maximum area possible to make a rectangular field.
She has 150 metres of fencing.
One side of the land borders a canal.
Find the maximum area of the field.

Canal

Width

Length

Ancient Babylonians and Egyptians studied quadratic equations like these thousands of years ago, to find solutions to problems involving areas of rectangles.

Answers

There are three variables:
● the length of the rectangle, l
● the width of the rectangle, w
● the area of the rectangle, A

Start by naming the variables in the problem.

The area of the rectangle $A = lw$.

Area = length × width.

As the total length of fencing is 150 m,

$$l + 2w = 150$$
$$l = 150 - 2w$$

Write down an equation for the perimeter of the field. Make l the subject.

So,

$$A = lw$$
$$A = (150 - 2w)w$$
$$A = 150w - 2w^2$$

Substitute the expression for l into the area formula.

Method 1: Using a GDC

The width, w, is 37.5 m.
$l = 150 - 2w = 150 - 75 = 75$ m
Maximum area,
$A = lw = 75 \times 37.5$
$\qquad = 2812.5$ m^2

Graph $A(x) = 150x - 2x^2$ on your GDC and read off the x-coordinate of the vertex: 37.5. This is the value of width, w, that gives the maximum value of A.

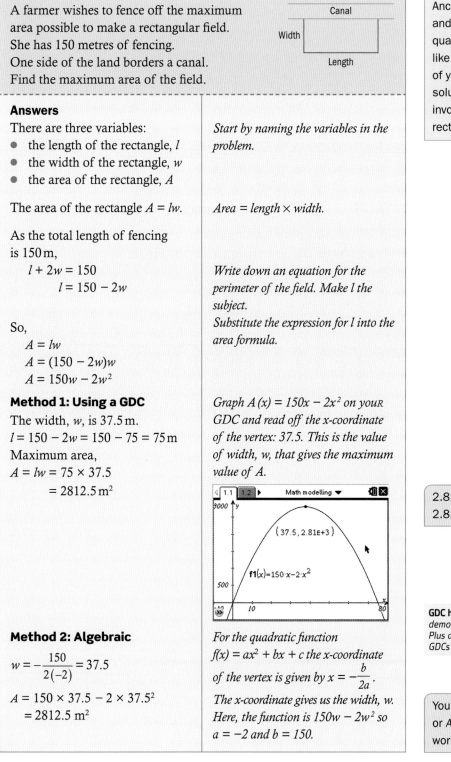

2.81E3 means
$2.81 \times 10^3 = 2810$.

GDC help on CD: *Alternative demonstrations for the TI-84 Plus and Casio FX-9860GII GDCs are on the CD.*

Method 2: Algebraic

$$w = -\frac{150}{2(-2)} = 37.5$$

$$A = 150 \times 37.5 - 2 \times 37.5^2$$
$$= 2812.5 \text{ m}^2$$

For the quadratic function $f(x) = ax^2 + bx + c$ the x-coordinate of the vertex is given by $x = -\dfrac{b}{2a}$.
The x-coordinate gives us the width, w. Here, the function is $150w - 2w^2$ so $a = -2$ and $b = 150$.

You can use $A = lw$ or $A = 150w - 2w^2$ to work out the area.

Exercise 4Q

1 a A farmer has 170 metres of fencing to fence off a rectangular area.

Find the length and width that give the maximum area.

> 1 Identify and name the variables.
> 2 Use the constraint to find a model for the 'length' (this model will be linear).
> 3 Find a model for the area (this model will be quadratic).

b A farmer has 110 metres of fencing to fence off a rectangular area.

Part of one side is a wall of length 15 m.

Find the dimensions of the field that give the maximum area.

2 A company's weekly profit, in riyals, is modeled by the function
$$P(u) = -0.032u^2 + 46u - 3000$$
where u is the number of units sold each week.
Find

> At break-even point there is no profit and no loss, so $P(u) = 0$.

a the maximum weekly profit
b the loss for a week's holiday period, where no units are sold
c the number of units sold each week at break-even point for the company.

EXAM-STYLE QUESTION

3 A rocket follows a parabolic trajectory.
After t seconds, the vertical height of the rocket above the ground, in metres, is given by
$$H(t) = 37t - t^2.$$

a Find the height of the rocket above the ground after 10 seconds.
b Find the maximum height of the rocket above the ground.
c Find the length of time the rocket is in the air.

> The trajectory is the path followed by an object.

4.4 Exponential models

Exponential functions and their graphs

→ In an **exponential function**, the independent variable is the **exponent** (or **power**).

Here are some examples of **exponential functions:**

$$f(x) = 2^x, \quad f(x) = 5(3)^x + 2, \quad g(x) = 5^{-x} - 3, \quad h(x) = \left(\frac{1}{3}\right)^x + 1$$

Investigation – exponential graphs

1 The number of water lilies in a pond doubles every week. In week one there were 4 water lilies in the pond. Draw a table and write down the number of water lilies in the pond each week up to week 12. Plot the points from the table on a graph of number of lilies against time. Draw a smooth curve through all the points.

> Time is the dependent variable, so it goes on the horizontal axis.

> The graph is an example of an increasing exponential graph.

2 A radioactive substance has a half-life of two hours. This means that every two hours its radioactivity halves.
A Geiger counter reading of the radioactive substance is taken at time $t = 0$. The reading is 6000 counts per second.
Two hours later ($t = 2$) the reading is 3000 counts per second.
What will the readings be at $t = 4$, $t = 6$, $t = 8$ and $t = 10$?
Plot the points on a graph of counts per second against time and join them with a smooth curve.

> Could the number of water lilies in a pond keep doubling forever? Will the radioactivity of the substance ever reach zero?

> This graph is an example of a decreasing exponential graph.

Does the shape of a ski slope form an exponential function? Investigate ski slopes on the internet to find out what the function is.

Graphs of exponential functions $f(x) = a^x$ where $a \in \mathbb{Q}^+$, $a \neq 1$

$\mathbb{Q}^+$ is the set of positive rational numbers.

Example 19

Draw a graph of the function $f(x) = 3^x$ for $-2 \leq x \leq 2$

Why can $a \neq 1$? What kind of function would you get if $a = 1$?

Answers

Method 1: By hand

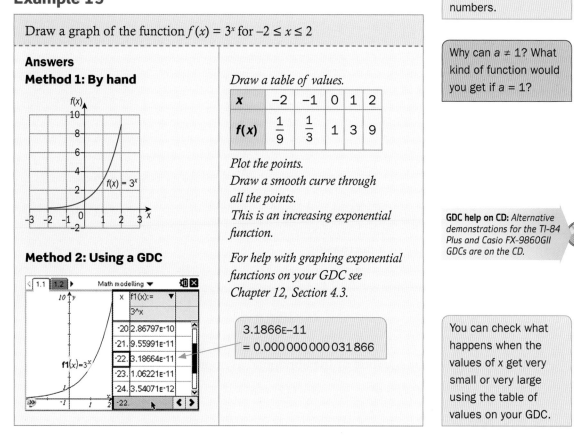

Draw a table of values.

x	−2	−1	0	1	2
f(x)	$\frac{1}{9}$	$\frac{1}{3}$	1	3	9

Plot the points.
Draw a smooth curve through all the points.
This is an increasing exponential function.

For help with graphing exponential functions on your GDC see Chapter 12, Section 4.3.

Method 2: Using a GDC

3.1866E−11
= 0.000 000 000 031 866

GDC help on CD: *Alternative demonstrations for the TI-84 Plus and Casio FX-9860GII GDCs are on the CD.*

You can check what happens when the values of x get very small or very large using the table of values on your GDC.

Look at the graph in Example 19. As the values of x get smaller, the curve gets closer and closer to the x-axis. The x-axis ($y = 0$) is a horizontal **asymptote** to the graph. At $x = 0$, $f(x) = 1$. As the values of x get very large, $f(x)$ gets larger even more quickly. We say that $f(x)$ tends to infinity. The function is an **increasing** exponential function.

An asymptote is a line that the curve approaches but never touches.

Here are some more graphs of **increasing** exponential functions.

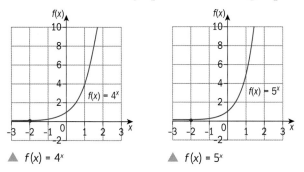

▲ $f(x) = 4^x$ ▲ $f(x) = 5^x$

All these graphs pass through the point $(0, 1)$ and have $y = 0$ (the x-axis) as a horizontal asymptote.

Graphs of exponential functions $f(x) = a^x$ where $0 < a < 1$

What happens if a is a positive proper fraction?

Here is the graph of $y = \left(\dfrac{1}{2}\right)^x$.

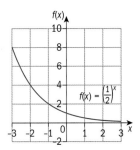

This graph also passes through the point $(0, 1)$ and has $y = 0$ (the x-axis) as a horizontal asymptote. However, this is an example of a **decreasing** exponential function.

> A **proper fraction** is a fraction where the numerator is smaller than the denominator.

> ● For an increasing exponential function, the y-values increase as the x-values increase from left to right.
> ● For a decreasing exponential function, the y-values decrease as the x-values increase from left to right.

Exercise 4R

Draw the graphs of these functions using your GDC.
For each, write down the coordinates of the point where the curve intersects the y-axis and the equation of the horizontal asymptote.

1 $f(x) = 2^x$ **2** $f(x) = 6^x$ **3** $f(x) = 8^x$

4 $f(x) = \left(\dfrac{1}{3}\right)^x$ **5** $f(x) = \left(\dfrac{1}{5}\right)^x$

Investigation – graphs of $f(x) = ka^x$ where $a \in \mathbb{Q}^+$ and $k \neq 0$ and $a \neq 1$

Use your GDC to draw the graphs of

1 $f(x) = 2(3)^x$ **2** $f(x) = 3\left(\dfrac{1}{2}\right)^x$ **3** $f(x) = -3(2)^x$

For each graph, write down
a the value of k in the equation $f(x) = ka^x$
b the point where the graph crosses the y-axis
c the equation of the horizontal asymptote.
What do you notice?

Investigation – graphs of $f(x) = ka^x + c$ where $a \in \mathbb{Q}^+$ and $k \neq 0$ and $a \neq 1$

Use your GDC to draw the graphs of

1 $f(x) = 2^x + 3$ **2** $f(x) = 3\left(\dfrac{1}{2}\right)^x - 4$ **3** $f(x) = -2(3)^x + 5$

for $-3 \leq x \leq 3$.

For each graph, write down

a the values of k and c in the equation $f(x) = ka^x + c$

b the point where the graph crosses the y-axis

c the equation of the horizontal asymptote.

   Work out $k + c$ for each graph. What do you notice?

→ In general, for the graph of $f(x) = ka^x + c$ where $a \in \mathbb{Q}^+$ and $k \neq 0$ and $a \neq 1$
   - the line $y = c$ is a **horizontal asymptote**
   - the curve passes through the point $(0, k + c)$.

Sketching an exponential graph
- Draw and label the axes.
- Label the point where the graph crosses the y-axis.
- Draw in the asymptotes.

Example 20

Sketch the graph of the function $f(x) = 3(2)^x - 1$

Answer

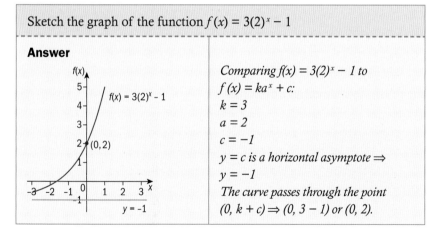

Comparing $f(x) = 3(2)^x - 1$ to
$f(x) = ka^x + c$:
$k = 3$
$a = 2$
$c = -1$
$y = c$ is a horizontal asymptote $\Rightarrow$
$y = -1$
The curve passes through the point
$(0, k + c) \Rightarrow (0, 3 - 1)$ or $(0, 2)$.

Exercise 4S

For each function, write down
   a the coordinates of the point where the curve cuts the y-axis
   b the equation of the horizontal asymptote.
Hence, sketch the graph of the function.

1 $f(x) = 2^x$ **2** $f(x) = 6^x$

3 $f(x) = \left(\dfrac{1}{3}\right)^x$ **4** $f(x) = \left(\dfrac{1}{5}\right)^x$

5 $f(x) = 3(2)^x + 4$

6 $f(x) = -2(4)^x - 1$

7 $f(x) = -1(2)^x + 3$

8 $f(x) = 4(3)^x - 2$

9 $f(x) = 0.5(2)^x + 3$

10 $f(x) = 2(0.5)^x + 1$

11 $f(x) = 0.4^x + 1$

12 $f(x) = 2(0.1)^x - 1$

Graphs of $f(x) = a^{-x} + c$ where $a \in \mathbb{Q}^+$ and $a \neq 1$

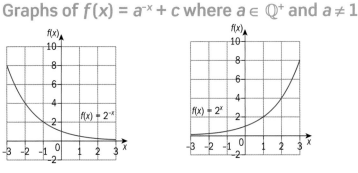

▲ Graph of $f(x) = 2^{-x}$.   ▲ Graph of $f(x) = 2^x$.

The graph of $f(x) = 2^{-x}$ is a reflection in the y-axis of the graph of $f(x) = 2^x$.

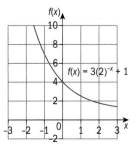   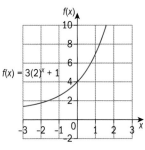

▲ Graph of $f(x) = 3(2)^{-x} + 1$.   ▲ Graph of $f(x) = 3(2)^x + 1$.

> $k = 3$ and $c = 1$.
> Notice that $3 + 1 = 4$.

The curves pass through the point $(0, 4)$ and the horizontal asymptote is $y = 1$.

→ In general, for the graph of $f(x) = ka^{-x} + c$ where $a \in \mathbb{Q}^+$ and $k \neq 0$ and $a \neq 1$
  • the line $y = c$ is a horizontal asymptote
  • the curve passes through the point $(0, k + c)$
  • the graph is a reflection in the y-axis of $g(x) = ka^x + c$.

Exercise 4T

For each function, write down
  a the coordinates of the point where the curve cuts the y-axis
  b the equation of the horizontal asymptote.
Hence, sketch the graph of the function.

1 $f(x) = 4(2)^{-x} + 2$ **2** $f(x) = -4^{-x} + 1$

3 $f(x) = -2(2)^{-x} + 3$ **4** $f(x) = 3(2)^{-x} - 2$

5 $f(x) = 0.5(3)^{-x} + 2$ **6** $f(x) = 0.5^{-x} + 1$

7 $f(x) = 2(0.1)^{-x} - 1$ **8** $f(x) = 0.4^{-x} + 2$

9 $f(x) = 3(0.2)^{-x} + 4$ **10** $f(x) = 5(3)^{-x} - 2$

Applications of exponential functions

Many real-life situations involving growth and decay can be modeled by exponential functions.

Example 21

The length, l cm, of a pumpkin plant increases according to the equation
$$l = 4(1.2)^t$$
where t is the time in days.

a Copy and complete the table. Give your answers correct to 3 sf.

t	0	2	4	6	8	10	12	14	16
l									

b Draw a graph of l against t for $0 \leq t \leq 20$ and $0 \leq l \leq 100$.

c How long is the pumpkin plant when $t = 0$?

d How long will the pumpkin plant be after 3 weeks?

Answers

a

t	0	2	4	6	8	10	12	14	16
l	4	5.8	8.3	11.9	17.2	24.8	35.7	51.4	74

Substitute each value of t into the equation to find the corresponding value of l.

b

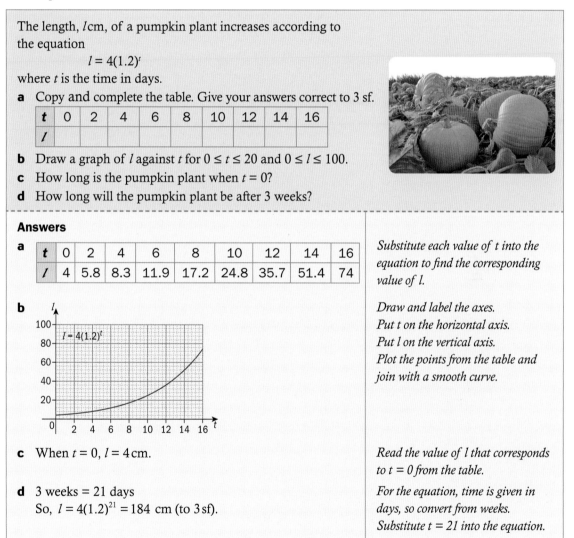

Draw and label the axes.
Put t on the horizontal axis.
Put l on the vertical axis.
Plot the points from the table and join with a smooth curve.

c When $t = 0$, $l = 4$ cm.

Read the value of l that corresponds to $t = 0$ from the table.

d 3 weeks = 21 days
So, $l = 4(1.2)^{21} = 184$ cm (to 3 sf).

For the equation, time is given in days, so convert from weeks.
Substitute $t = 21$ into the equation.

Example 22

Hubert invests 3000 euros in a bank at a rate of 5% per annum compounded yearly.

Let y be the amount he has in the bank after x years.

a Draw a graph to represent how much Hubert has in the bank after x years. Use a scale of 0–10 years on the x-axis and 2500–5000 euros on the y-axis.

b How much does he have after 4 years?

c How many years is it before Hubert has 4000 euros in the bank?

Answers

a The compound interest formula is:

$$y = 3000 \left(1 + \frac{5}{100}\right)^x$$

$y = 3000(1.05)^x$

where x = number of years.

This problem can be represented by a compound interest function.

Time (x years)	Amount (y euros)
0	3000
2	3307.50
4	3646.52
6	4020.29
8	4432.37
10	4886.68

Draw a table of values.

> The **compound interest** formula is an exponential (growth) function.

> You will learn more about compound interest in Chapter 7.

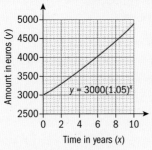

Draw and label the axes.

Plot the points and join them with a smooth curve.

b After 4 years Hubert has $3000(1.05)^4 = 3646.52$ euros.

Substitute $x = 4$ into the formula.

c Hubert has 4000 euros in the bank after 6 years.

You need to find the value of x for $y = 4000$ euros.

*From the table of values in part **a** you can see that after 6 years the amount is 4020.29.*

Check the amount after 5 years:

$y = 3000(1.05)^5 = 3828.84$

This is less than 4000 euros.

Exercise 4U

EXAM-STYLE QUESTIONS

1 Sketch the graphs of $f(x) = 2^x + 0.5$ and $g(x) = 2^{-x} + 0.5$
for $-3 \leq x \leq 3$.
  a Write down the coordinates of the point of intersection of the two curves.
  b Write down the equation of the horizontal asymptote to both graphs.

2 The value of a car decreases every year according to the function
$$V(t) = 26\,000x^t$$
where V is the value of the car in euros, t is the number of years after it was first bought and x is a constant.
  a Write down the value of the car when it was first bought.
  b After one year the value of the car is 22 100 euros. Find the value of x.
  c Calculate the number of years that it will take for the car's value to fall to less than 6000 euros.

3 The equation $M(t) = 150(0.9)^t$ gives the amount, in grams, of a radioactive material kept in a laboratory for t years.
  a Sketch the graph of the function $M(t)$ for $0 \leq t \leq 100$.
  b Write down the equation of the horizontal asymptote to the graph of $M(t)$.
  c Find the mass of the radioactive material after 20 years.
  d Calculate the number of years that it will take for the radioactive material to have a mass of 75 grams.

4 The area, $A\,\text{m}^2$, covered by a certain weed is measured at 06:00 each day.
On the 1st June the area was 50 m².
Each day the area of the weeds grew by the formula
$$A(t) = 50(1.06)^t$$
where t is the number of days after 1st June.
  a Sketch the graph of $A(t)$ for $-4 \leq t \leq 20$.
  b Explain what the negative values of t represent.
  c Calculate the area covered by the weeds at 06:00 on 15th June.
  d Find the value of t when the area is 80 m².

5 The graph shows the function
$f(x) = k(2)^x + c$.
Find the values of c and k.

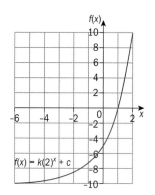

6 The temperature, T, of a cup of coffee is given by the function
$$T(t) = 18 + 60(2)^{-t}$$
where T is measured in °C and t is in minutes.
a Sketch the graph of $T(t)$ for $0 \le t \le 10$.
b Write down the temperature of the coffee when it is first served.
c Find the temperature of the coffee 5 minutes after serving.
d Calculate the number of minutes that it takes the coffee to reach a temperature of 40 °C.
e Write down the temperature of the room where the coffee is served. Give a reason for your answer.

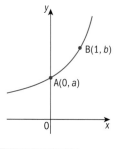

7 The value, in USD, of a piece of farm machinery depreciates according to the formula
$$D(t) = 18\,000(0.9)^t \quad \text{where } t \text{ is the time in years.}$$
a Write down the initial cost of the machine.
b Find the value of the machine after 5 years.
c Calculate the number of years that it takes for the value of the machine to fall below 9000 USD.

8 The graph of the function $f(x) = \dfrac{2^x}{a}$ passes through the points
$(0, b)$ and $(2, 0.8)$. Calculate the values of a and b.

9 The diagram shows the graph of $y = 2^x + 3$. The curve passes through the points A(0, a) and B(1, b).
a Find the value of a and the value of b.
b Write down the equation of the asymptote to the curve.

10 A function is represented by the equation $f(x) = 2(3)^x + 1$.
Here is a table of values of $f(x)$ for $-2 \le x \le 2$.

a Calculate the value a and the value of b.
b Draw the graph of $f(x)$ for $-2 \le x \le 2$.
c The domain of $f(x)$ is the real numbers. What is the range?

x	-2	-1	0	1	2
$f(x)$	1.222	a	3	7	b

4.5 Graphs of functions of the form $f(x) = ax^m + bx^n + ..., m, n \in \mathbb{Z}$

In Sections 4.2 and 4.3, you have seen examples of linear and quadratic functions. What happens when the power of x is an integer larger than 2 or smaller than 0?

Cubic functions

When the largest power of x is 3 the function is called a **cubic** function.

> → A cubic function has the form $f(x) = ax^3 + bx^2 + cx + d$, where $a \neq 0$. The domain is $\mathbb{R}$, unless otherwise stated.

Here are two examples of graphs of cubic functions.

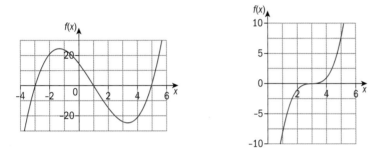

Example 23

The number of fish, F, in a pond from the period 1995 to 2010 is modeled using the formula
$$F(x) = -0.030x^3 + 0.86x^2 - 6.9x + 67$$
where x is the number of years after 1995.
  a Use your GDC to sketch the function for $0 \leq x \leq 18$.
  b Find the number of fish in the pond after 6 years.
  c Find the number of fish in the pond after 13 years.

Answers

a

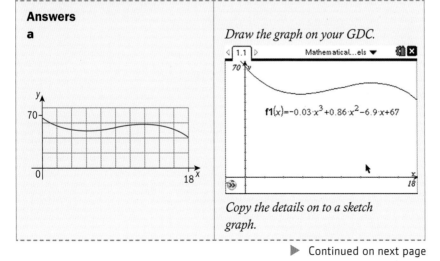

Draw the graph on your GDC.

$f1(x) = -0.03 \cdot x^3 + 0.86 \cdot x^2 - 6.9 \cdot x + 67$

Copy the details on to a sketch graph.

GDC help on CD: *Alternative demonstrations for the TI-84 Plus and Casio FX-9860GII GDCs are on the CD.*

▶ Continued on next page

b $F(6) = -0.030(6)^3 + 0.86(6)^2$ $\quad - 6.9(6) + 67$ $\quad = -6.48 + 30.96 - 41.4 + 67$ $\quad = 50.08$ So, after 6 years, there are 50 fish in the pond.	*Substitute x = 6 into the equation.* *Or, you can use your GDC table of values or the Trace function.*
c $F(13)$ $\quad = -0.030(13)^3 + 0.86(13)^2$ $\quad - 6.9(13) + 67$ $\quad = -65.91 + 145.34 - 89.7 + 67$ $\quad = 56.73$ So, after 13 years, there are 56 fish in the pond.	*Substitute x = 13 into the equation.* *Or, you can use your GDC table of values.*

Example 24

A pandemic is modeled using the equation
$$y = (x - 20)^3 + 5000$$
where x is the number of weeks after the outbreak started and y is the number of cases reported.
a Use your GDC to sketch the function for $0 \le x \le 30$.
b Find the number of cases after 10 weeks.
c Find the number of cases after 20 weeks.
d Is this a good model to represent the number of cases of a pandemic?

> A pandemic is an epidemic of an infectious disease that spreads over several continents.

Answers

a

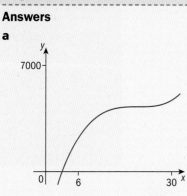

	Draw the graph on your GDC.
	Copy the details on to a sketch graph.
b $y = (10 - 20)^3 + 5000 = 4000$ So, after 10 weeks, there are 4000 cases.	*Substitute x = 10 into the equation.*
c $y = (20 - 20)^3 + 5000 = 5000$ So, after 20 weeks, there are 5000 cases.	*Substitute x = 20 into the equation.*
d No, because the number of cases starts to rise again after 20 weeks and will keep on rising.	*Consider:* *Does the graph keep increasing?* *Would you expect the pandemic to increase forever?*

> **GDC help on CD:** *Alternative demonstrations for the TI-84 Plus and Casio FX-9860GII GDCs are on the CD.*

> Can mathematical models accurately model the real world?

Investigation – quartic functions

When the largest power of x is 4 then the function is called a
quartic function.

A quartic function has the form
$f(x) = ax^4 + bx^3 + cx^2 + dx + e$, where $a \neq 0$. The domain is $\mathbb{R}$, unless
otherwise stated.

Substitute different values of a, b, c, d and e into the equation
$\quad f(x) = ax^4 + bx^3 + cx^2 + dx + e$.
Use your GDC to draw the functions.
What can you say about the shape of a quartic graph?

Exercise 4V

1 The times of high and low tides one day are modeled by the
function
$$f(x) = -0.0015x^4 + 0.056x^3 - 0.60x^2 + 1.65x + 4$$
where x is the number of hours after midnight.
 a Use your GDC to sketch the function for $0 \leq x \leq 20$.
 b Find the time of the low tides.
 c Find the times of the high tides.

2 Here is the graph of the function $f(x) = (x - 2)^4 + 6$

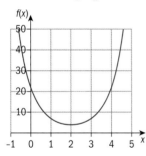

 a Find the value of $f(x)$ when $x = 2$.
 b Find the values of x when $y = 6$.
 c Write down the range of this function.

Graphs of functions when the power of x is a negative integer

Here is the graph of $y = x^{-1}$, $x \neq 0$, for $-10 \leq x \leq 10$

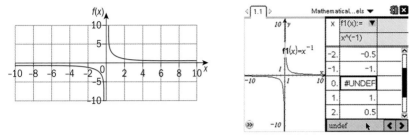

GDC help on CD: *Alternative demonstrations for the TI-84 Plus and Casio FX-9860GII GDCs are on the CD.*

The graph has two branches that do not overlap or cross the y-axis.

There is no value for y when $x = 0$. We call $x = 0$ a **vertical asymptote**.

When you look at the table of values on the GDC, you usually see UNDEF in the column for y whenever you have a vertical asymptote.

> A vertical asymptote occurs when the value of y tends to infinity as x tends to zero. This means that when x approaches 0 from either the negative side or the positive side then y approaches either a very big negative number or a very big positive number.

Here is the graph of $y = x^{-2}$, $x \neq 0$, for $-10 \leq x \leq 10$

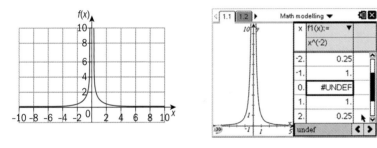

There is no value for y when $x = 0$, so $x = 0$ is a vertical asymptote. However, in this graph y tends to a very large positive number when x approaches 0 from either the negative side or the positive side.

Investigation – graphs of $y = ax^{-n}$

1 Use your GDC to draw the graphs of:
   - $y = x^{-3}$ for $-10 \leq x \leq 10$
   - $y = x^{-4}$ for $-10 \leq x \leq 10$

   Compare them to the graphs of $y = x^{-1}$ and $y = x^{-2}$.
   What do you notice?

2 Draw the graphs of:
   - $y = 2x^{-3}$ for $-10 \leq x \leq 10$
   - $y = 3x^{-4}$ for $-10 \leq x \leq 10$

   Compare these graphs to the others.
   What do you notice?

Example 25

A rectangle has an area of $1.5\,m^2$.

Let the length of the rectangle be y and the width be x.

a Show that $y = \dfrac{1.5}{x}$.

b Use your GDC to draw the graph of $y = \dfrac{1.5}{x}$ for $0 < x \le 10$.

c What happens when x gets closer to 0?

d What happens when x gets closer to 10?

e Write down the equations of the vertical and horizontal asymptotes.

> How many different rectangles could you draw with an area of $1.5\,m^2$?

Answers

a $x \times y = 1.5 \Rightarrow y = \dfrac{1.5}{x}$

Area = length × width.
Rearrange the formula to make y the subject.

b

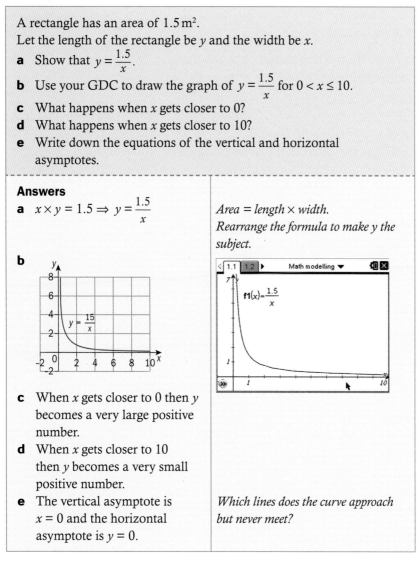

> **GDC help on CD:** *Alternative demonstrations for the TI-84 Plus and Casio FX-9860GII GDCs are on the CD.*

c When x gets closer to 0 then y becomes a very large positive number.

d When x gets closer to 10 then y becomes a very small positive number.

e The vertical asymptote is $x = 0$ and the horizontal asymptote is $y = 0$.

Which lines does the curve approach but never meet?

Exercise 4W

1 The temperature of water as it cools to room temperature is modeled by the function

$$f(x) = 21 + \frac{79}{x}, \; x \ne 0,$$

where x is the time in minutes and $f(x)$ represents the temperature in °C.

a Use your GDC to sketch the graph of the function for $0 < x \le 15$.

b Calculate the temperature of the water after 10 minutes.

c How many minutes does it take for the temperature to cool down to 50°C?

d Write down the equation of the vertical asymptote.

e Write down the equation of the horizontal asymptote.

f Write down the room temperature.

2 Oil is heated on a stove. The temperature is modeled by the function

$$f(x) = 100 - \frac{100}{x}, \; x \neq 0,$$

where x is the time in minutes from when the oil began to heat and $f(x)$ represents the temperature in °C.

a Use your GDC to sketch the graph of the function for $0 < x \leq 50$.

b Find the temperature of the oil after 10 minutes.

c Find the number of minutes that it takes the temperature to reach 30 °C.

d Write down the maximum temperature that the oil can reach.

3 a Use your GDC to sketch the graph of $f(x) = \dfrac{5}{x^2}, \; x \neq 0$.

b Write down the values of x when $y = 8$.

c Write down the equations of the vertical and horizontal asymptotes to the graph.

d Given that the domain of f is the real numbers, $x \neq 0$, write down the range of f.

4 a Use your GDC to sketch the graph of $f(x) = 3 + \dfrac{6}{x}, x \neq 0$, for $-10 \leq x \leq 10$.

b Find the value of $f(x)$ when $x = 8$.

c Find the value of x when $y = 5$.

d Write down the equations of the vertical and horizontal asymptotes to the graph.

e Given that the domain of f is the real numbers, $x \neq 0$, write down the range of f.

Graphs of more complex functions

Here is the graph of $f(x) = 3x^2 + \dfrac{2}{x}, \; x \neq 0$, for $-4 \leq x \leq 4$.

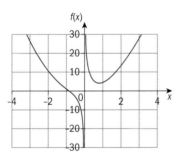

The graph has two separate branches.

$x = 0$ is a vertical asymptote.

The domain is $-\infty \leq x < 0, \; 0 < x \leq +\infty$.

> English mathematician John Wallis (1616–1703) introduced the symbol ∞ for infinity.

Example 26

A taxi company's fares depend on the distance, in kilometres, traveled.

The fares are calculated using the formula

$$f(x) = 2x + \frac{50}{x^2}$$

where x is the number of kilometres traveled ($x \neq 0$) and $f(x)$ is the fare in euros.

a Sketch the graph of the function for $0 < x \leq 20$.
b Find the cost for a journey of 10 kilometres.
c Find the number of kilometres traveled that gives the cheapest fare.

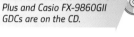

GDC help on CD: *Alternative demonstrations for the TI-84 Plus and Casio FX-9860GII GDCs are on the CD.*

Answers

a

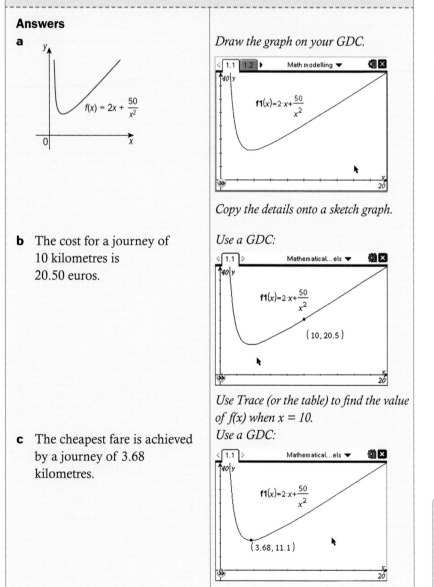

Draw the graph on your GDC.

Copy the details onto a sketch graph.

b The cost for a journey of 10 kilometres is 20.50 euros.

Use a GDC:

Use Trace (or the table) to find the value of f(x) when x = 10.

c The cheapest fare is achieved by a journey of 3.68 kilometres.

Use a GDC:

For help with finding the minimum value using a GDC see Chapter 12, Section 4.2, Example 20.

Example 27

A closed cuboid of height y cm has a square base of length x cm.
The volume of the cuboid is $500\,\text{cm}^3$.

a Write down an expression for the volume of the cuboid.

b Hence, find an expression for the surface area, A, of the cuboid in terms of x. Simplify your answer as much as possible.

c Use your GDC to draw a graph of the area function for $0 < x \le 30$.

d Use your GDC to find the dimensions that give a minimum surface area.

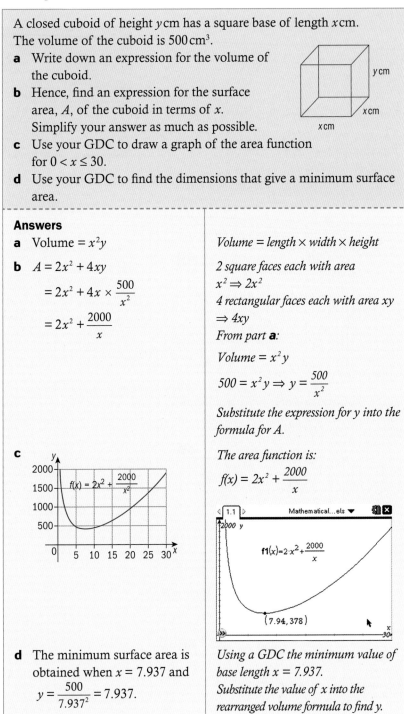

Answers

a Volume $= x^2 y$

b $A = 2x^2 + 4xy$

$\quad = 2x^2 + 4x \times \dfrac{500}{x^2}$

$\quad = 2x^2 + \dfrac{2000}{x}$

Volume $=$ length $\times$ width $\times$ height

2 square faces each with area
$x^2 \Rightarrow 2x^2$
4 rectangular faces each with area xy
$\Rightarrow 4xy$
*From part **a**:*

Volume $= x^2 y$

$500 = x^2 y \Rightarrow y = \dfrac{500}{x^2}$

Substitute the expression for y into the formula for A.

The area function is:

$f(x) = 2x^2 + \dfrac{2000}{x}$

c

d The minimum surface area is obtained when $x = 7.937$ and $y = \dfrac{500}{7.937^2} = 7.937$.

Using a GDC the minimum value of base length $x = 7.937$.
Substitute the value of x into the rearranged volume formula to find y.

GDC help on CD: *Alternative demonstrations for the TI-84 Plus and Casio FX-9860GII GDCs are on the CD.*

For help with finding a minimum value using a GDC see Chapter 12, Section 4.2, Example 20.

Exercise 4X

1 One section of a rollercoaster ride can be modeled by the equation

$$f(x) = \frac{20}{x} + 2x^2, \ x \neq 0,$$

where x is the time in seconds from the start of the ride and $f(x)$ is the speed in $m\,s^{-1}$.

a Use your GDC to sketch the graph of the function for $0 < x \leq 10$.

b Find the minimum value on the graph.

c Find the speed when $x = 6$.

d Find the times when the speed is $50\,m\,s^{-1}$.

2 An open box has the following dimensions:
length $= x\,cm$, breadth $= 2x\,cm$ and height $= y\,cm$.
The volume of the box is $300\,cm^3$.

a Write down an expression for the volume of the box.

b Find an expression for the surface area of the open box in terms of x only.

c Use your GDC to sketch the graph of the area function for $0 < x \leq 20$.

d Find the dimensions that make the surface area a minimum.

3 A pyramid has a square base of side x metres. The perpendicular height of the pyramid is h metres. The volume of the pyramid is $1500\,m^3$.

a Find an expression for the volume of the pyramid using the information given.

b Show that the height of each of the triangular faces is

$$\sqrt{\left(h^2 + \left(\frac{x}{2}\right)^2\right)}.$$

c Hence, find an equation for the total surface area of the pyramid.

d Write the equation in part **c** in terms of x only.

e Use your GDC to sketch the graph of this equation for $0 < x \leq 30$.

f Find the dimensions that make this surface area a minimum.

4 A fish tank in the shape of a cuboid has length 320 cm. Its length is twice that of its width.
To enhance viewing, the area of the four vertical faces should be maximized.
Find the optimum 'viewing area' of a fish tank that is fixed to the wall so that the area of three faces only should be considered.

Example 28

Consider the function $f(x) = \dfrac{3x - 12}{x}$, $x \neq 0$.

a Write down the domain of $f(x)$.

b Copy and complete the table of values for $f(x)$. Give your answers correct to two significant figures.

x	−24	−12	−4	−1	0	1	2	4	8	12	24
f(x)											

c Draw the graph of $f(x)$ for $-24 \leq x \leq 24$. Use a scale of 1 cm to represent 4 units on the horizontal axis and 1 cm to represent 2 units on the vertical axis.

d Write down the equation of the vertical asymptote to the graph of $f(x)$.

Answers

a The domain of f is the real numbers, $x \neq 0$.

The only value excluded is $x = 0$ (as division by zero is not defined).

Substitute each value of x into f(x) to find the corresponding value of f(x).

$x = 0$ has no image.

b

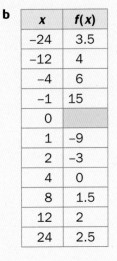

x	f(x)
−24	3.5
−12	4
−4	6
−1	15
0	
1	−9
2	−3
4	0
8	1.5
12	2
24	2.5

As x gets very large in absolute value the graph of f(x) gets closer and closer to a horizontal line. What is the equation of this line?

c

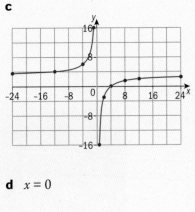

*Draw and label the axes. Plot the points from the table in part **b**. The graph has **two** branches. Points to the right of $x = 0$ are joined up with a smooth curve. Points to the left of $x = 0$ are joined up with another smooth curve.*

*Which vertical line does the curve approach but never meet? (Shown in blue on graph in part **c**.)*

Large in absolute value means very large positive numbers (1000, 10 000, etc.) or very large negative numbers (−1000, −10 000, etc.).

d $x = 0$

For more on absolute value, see Chapter 13, section 2.8.

Exercise 4Y

1 Consider the function $f(x) = 1 + \dfrac{2}{x}$, $x \neq 0$.

  a Write down the domain of $f(x)$.

  b Copy and complete the following table.

x	−10	−5	−4	−2	−1	−0.5	−0.2	0	0.2	0.5	1	2	4	5	10
f(x)															

  c Draw the graph for $-10 \leq x \leq 10$. Use a scale of 1 unit to represent 1 cm on each of the axes.

  d **i** Draw the vertical asymptote.

    ii Write down the equation of the vertical asymptote.

  e **i** Draw the horizontal asymptote.

    ii Write down the equation of the horizontal asymptote.

2 Consider the function $f(x) = 8x^{-1} + 3$, $x \neq 0$.

  a Write down the domain of $f(x)$.

  b Copy and complete the following table.

x	−10	−8	−5	−4	−2	−1	0	1	2	4	5	8	10
f(x)													

  c Draw the graph of $f(x)$ for $-10 \leq x \leq 10$. Use a scale of 1 cm to represent 2 units on both axes.

  d **i** Draw the vertical asymptote.

    ii Write down the equation of the vertical asymptote.

  e **i** Draw the horizontal asymptote.

    ii Write down the equation of the horizontal asymptote.

Sketching more complex graphs

Example 29

Sketch the graph of the function $f(x) = 2x - (x + 1)^2 + 13$ for $-5 \leq x \leq 5$.

Answer

Using a GDC:
Enter the function and adjust the window settings for x.

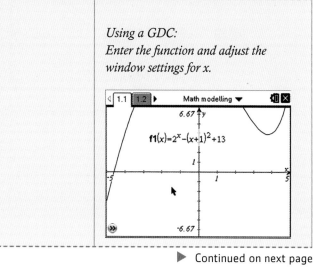

GDC help on CD: *Alternative demonstrations for the TI-84 Plus and Casio FX-9860GII GDCs are on the CD.*

▶ Continued on next page

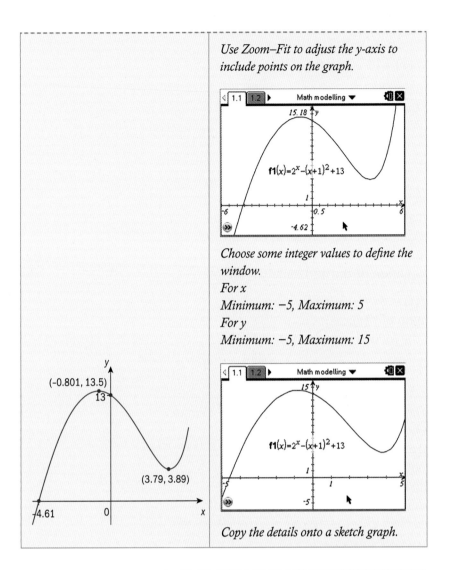

Use Zoom–Fit to adjust the y-axis to include points on the graph.

Choose some integer values to define the window.

For x

Minimum: −5, Maximum: 5

For y

Minimum: −5, Maximum: 15

Copy the details onto a sketch graph.

The range of the function in Example 29 is $\mathbb{R}$.

You can use a table on a GDC to give you an idea of the range of the function.

x	f1(x):= 2^x−(x+1)
-5.	-2.96875
-4.	4.0625
-3.	9.125
-2.	12.25
-1.	13.5

x	f1(x):= 2^x−(x+1)
-1.	13.5
0.	13.
1.	11.
2.	8.
3.	5.

Exercise 4Z

Use your GDC to help you sketch the graph of these functions.
Give the range for each function.

1 $f(x) = -0.5x + 1 + 3^x$

2 $f(x) = 2^x - x^2$

3 $f(x) = x(x-1)(x+3)$

4 $f(x) = x^4 - 3x^2 + 1$

5 $f(x) = 0.5^x - x^{-1}, x \neq 0$

4.6 Using a GDC to solve equations

Example 30

a Use your GDC to sketch the graphs of $f(x) = 2^x$ and
$g(x) = -x^2 + 3x + 2$.
b Hence, solve the equation $2^x + x^2 - 3x - 2 = 0$.

> 'Hence' means that
> you should try to use
> the previous part to
> answer this part of the
> question.

Answers

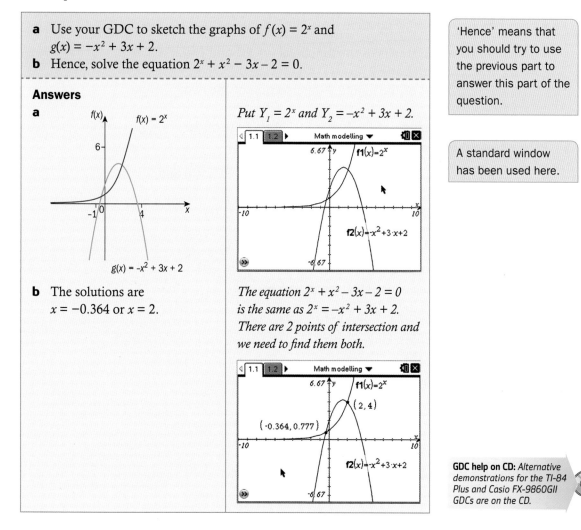

a

Put $Y_1 = 2^x$ and $Y_2 = -x^2 + 3x + 2$.

> A standard window
> has been used here.

b The solutions are
$x = -0.364$ or $x = 2$.

*The equation $2^x + x^2 - 3x - 2 = 0$
is the same as $2^x = -x^2 + 3x + 2$.
There are 2 points of intersection and
we need to find them both.*

> **GDC help on CD:** *Alternative
> demonstrations for the TI-84
> Plus and Casio FX-9860GII
> GDCs are on the CD.*

1 **a** On the same graph, sketch the curves $y = x^2$ and $y = 4 - \dfrac{1}{x}$ for values of x from -8 to 8 and values of y from -2 to 8. Show scales on your axes.

   b Find the coordinates of the points of intersection of these two curves.

EXAM-STYLE QUESTIONS

2 The functions f and g are defined by

$$f(x) = 1 + \frac{4}{x}, \; x \in \mathbb{R}, \; x \neq 0$$

   $g(x) = 3x, \; x \in \mathbb{R}$

   a Sketch the graph of f for $-8 \leq x \leq 8$.

   b Write down the equations of the horizontal and vertical asymptotes of the function f.

   c Sketch the graph of g on the same axes.

   d Hence, or otherwise, find the solutions of $1 + \dfrac{4}{x} - 3x = 0$.

   e Write down the range of function f.

3 The diagram shows the graphs of the functions $y = 5x^2$ and $y = 3^x$ for values of x between -2 and 2.

   a Find the coordinates of the points of intersection of the two curves.

   b Write down the equation of the horizontal asymptote of the exponential function.

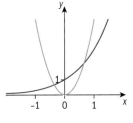

4 Two functions $f(x)$ and $g(x)$ are given by $f(x) = \dfrac{3}{x}, \; x \in \mathbb{R}, \; x \neq 0$ and $g(x) = x^3, \; x \in \mathbb{R}$.

   a On the same diagram sketch the graphs of $f(x)$ and $g(x)$ using values of x between -4 and 4, and values of y between -4 and 4. You must label each curve.

   b State how many solutions exist for the equation $\dfrac{3}{x} - x^3 = 0$.

   c Find a solution of the equation given in part **b**.

5 Sketch the graphs of $y = 3x - 4$ and $y = x^3 - 3x^2 + 2x$. Find all the points of intersection of the graphs.

6 Sketch the graphs of $y = 2^x$ and $y = x^3 + x^2 - 6x$. Find the coordinates of all the points of intersection.

7 Sketch the graphs of $y = x + 2$ and $y = \dfrac{5}{x}, \; x \neq 0$.

   a Find the solutions of the equation $\dfrac{5}{x} = x + 2$.

   b Write down the equation of the horizontal asymptote to $y = \dfrac{5}{x}$.

   c Write down the equation of the vertical asymptote to $y = \dfrac{5}{x}$.

4.7 Graphs of real-life situations

Linear and non-linear graphs can be used to represent a range of real-life situations.

Example 31

The graph below shows China's oil production and consumption from 1990 to 2010.

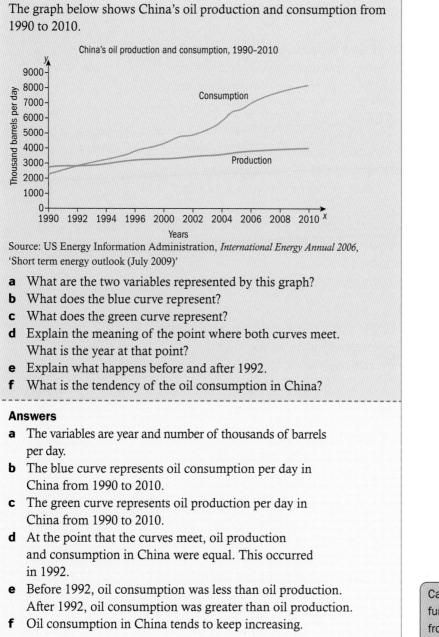

China's oil production and consumption, 1990–2010

Source: US Energy Information Administration, *International Energy Annual 2006*, 'Short term energy outlook (July 2009)'

a What are the two variables represented by this graph?
b What does the blue curve represent?
c What does the green curve represent?
d Explain the meaning of the point where both curves meet. What is the year at that point?
e Explain what happens before and after 1992.
f What is the tendency of the oil consumption in China?

- -

Answers
a The variables are year and number of thousands of barrels per day.
b The blue curve represents oil consumption per day in China from 1990 to 2010.
c The green curve represents oil production per day in China from 1990 to 2010.
d At the point that the curves meet, oil production and consumption in China were equal. This occurred in 1992.
e Before 1992, oil consumption was less than oil production. After 1992, oil consumption was greater than oil production.
f Oil consumption in China tends to keep increasing.

Can you deduce any further information from this graph?

Exercise 4AB

1 The water consumption in Thirsty High School is represented in the graph.

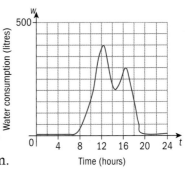

a Write down are the two variables represented by this graph.

b During what period of time is Thirsty High School open?

c During what intervals of time is consumption increasing?

d During what intervals of time is consumption decreasing?

e Find the time at which the consumption is at a maximum.

f Find the time at which the consumption is at a minimum.

2 The graph represents the temperature, in degrees Celsius, of some coffee after Manuela has heated it.

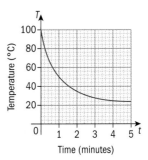

a Write down the two variables represented by this graph.

b Write down the initial temperature of the liquid after heating.

c Write down the temperature of the liquid 2 minutes after heating.

d Find the time it takes for the temperature to reach $68\,°C$.

e Decide whether the liquid reaches $22\,°C$ during the 5-minute period shown on the graph.

f Write down the room temperature.

EXAM-STYLE QUESTION

3 Under certain conditions the number of bacteria in a particular culture doubles every 5 seconds as shown by the graph.

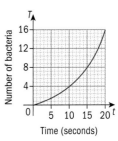

a Copy and complete the table below.

Time (*t* seconds)	0	5	10	15	20
Number of bacteria (*N*)	1				

b Write down the time it takes for the culture to have 6 bacteria.

c Calculate the number of bacteria in the culture after 1 minute if the conditions remain the same.

4 In a physics experiment a ball is projected vertically into the air from ground level.

The diagram represents the height of the ball at different times.

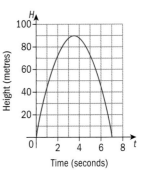

a Write down the height of the ball after one second.

b Find out how many seconds after being thrown the ball is at 60 metres.

c Write down the interval of time in which the ball is going up.

d Write down the interval of time in which the ball is coming down.

e Write down the maximum height reached by the ball and the time it takes the ball to reach that height.

f Explain what happens at $t = 7$.

5 The graph shows the tide heights, h metres, at time t hours after midnight for Blue Coast Harbor.

  a Use the graph to find
    i the height of the tide at 01:30
    ii the height of the tide at 05:30
    iii the times when the height of the tide is 3 metres.

The best time to catch fish in Blue Coast Harbor is when the tide is below 3 metres.

  b Find this best time, giving your answer as an inequality in t.

6 The temperature (°C) during a 24-hour period in a certain city is represented in the graph.

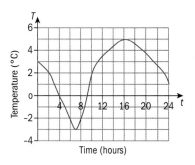

  a Determine how many times the temperature is exactly 0 °C during this 24-hour period.
  b Write down the interval of time in which the temperature falls below 0 °C.
  c Write down the time at which the temperature reaches its maximum value.
  d Write down the maximum temperature registered during this 24-hour period.
  e Write down the interval of time in which the temperature increases from 3 °C to 5 °C.
  f Write down the times at which the temperature is 4 °C.
  g Can you deduce from this graph whether the behavior of the temperature in the following day will be exactly the same as this day? Why?

7 The diagram represents a box with volume $16\,\text{cm}^3$. The base of the box is a square with sides x cm. The height of the box is y cm.

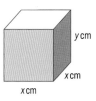

  a Write an expression for the height, y, in terms of x.
  b Copy and complete the table below for the function $y = f(x)$ from part **a**. Give your answers correct to two significant figures.

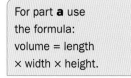

For part **a** use the formula:
volume = length × width × height.

x	0.5	1	2	4	8	10
$y = f(x)$						

  c Draw the graph of f for $0 < x \le 10$. Use a scale of 1 cm to represent 1 unit on the horizontal axis and 1 cm to represent 10 units on the vertical axis.
  d What happens to the height of the box as the values of x tend to infinity?

8 The diagram represents an open container with a capacity of 3 litres.

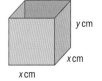

The base of the container is a square with sides x cm. The height of the container is y cm.

  a Write down the volume of the box in cm³.

  b Find an expression for the height, y, in terms of x.

  c Find an expression for the surface area of the container, A, in terms of x.

  d Copy and complete the table below. Give your answers correct to 2 significant figures.

x(cm)	5	10	15	20	25	30	35
$A(x)$(cm²)							

  e Draw the graph of A for $0 < x \le 35$. Use a scale of 2 cm to represent 5 units on the horizontal axis and 1 cm to represent 400 units on the vertical axis.

  f Use your graph to decide if there is a value of x that makes the surface area of the container a minimum. If there is, write down this value of x.

Review exercise

Paper 1 style questions

1 The graph represents the temperature in °C in a certain city last Tuesday.

  a Write down the interval of time in which the temperature was below 0 °C.

  b Write down the interval of time in which the temperature was above 11 °C.

  c Write down the maximum temperature last Tuesday. Give your answer correct to the nearest unit.

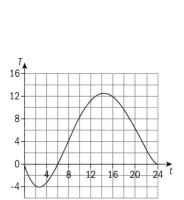

2 The cost c, in Singapore dollars (SGD), of renting an apartment for n months is a linear model

  $c = nr + s$

where s is the security deposit and r is the amount of rent per month.

Wan Ning rented the apartment for 6 months and paid a total of 35 000 SGD.

Tanushree rented the same apartment for 2 years and paid a total of 116 000 SGD.

Find the value of

  a r, the rent per month   **b** s, the security deposit.

3 Given that $f(x) = x^2 + 5x$

   a factorize $x^2 + 5x$

   b sketch the graph of $y = f(x)$. Show on your sketch

      i the coordinates of the points of intersection with the axes

      ii the equation of the axis of symmetry

      iii the coordinates of the vertex of the parabola.

4 A signal rocket is fired vertically from ground level by a gun. The height, in metres, of the rocket above the ground is a function of the time t, in seconds, and is defined by:

    $h(t) = 30t - 5t^2$, $0 \le t \le 6$.

   a Find the height of the rocket above the ground after 4 seconds.

   b Find the maximum height of the rocket above the ground.

   c Use your GDC to find the length of time, in seconds, for which the rocket is at a height of 25 m or more above the ground.

5 The graph of the function $f(x) = \dfrac{2^x}{m}$ passes through the points (3, 1.6) and (0, n).

   a Calculate the value of m.

   b Calculate the value of n.

   Find $f(2)$.

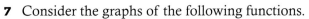

6 The diagram shows the graph of $y = x^2 - 2x - 15$. The graph crosses the x-axis at the point A, and has a vertex at B.

   a Factorize $x^2 - 2x - 15$.

   b Find the coordinates of the point

      i A   **ii** B.

7 Consider the graphs of the following functions.

      i $y = 8x + x^2$

      ii $y = (x - 3)(x + 4)$

      iii $y = x^2 - 2x + 5$

      iv $y = 5 - 4x - 3x^2$

   Which of these graphs

   a has a y-intercept below the x-axis

   b passes through the origin

   c does not cross the x-axis

   d could be represented by this diagram?

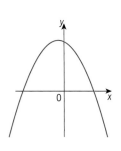

8 The figure shows the graphs of the functions

$$f(x) = (0.5)^x - 2 \quad \text{and} \quad g(x) = -x^2 + 4$$

for values of x between -3 and 3. The two graphs meet at the points A and B.

a Find the coordinates of

   i A   **ii** B.

b Write down the set of values of x for which $f(x) < g(x)$.

c Write down the equation of the horizontal asymptote to the graph of $f(x)$.

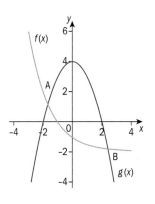

9 Gabriel is designing a rectangular window with a perimeter of 4.40 m. The length of the window is x m.

a Find an expression for the width of the window in terms of x.

b Find an expression for the area of the window, A, in terms of x.

Gabriel wants to make the amount of light passing through this window a maximum.

c Find the value of x that meets this condition.

10 a On the same graph sketch the curves $y = 3x^2$ and $y = \dfrac{1}{x}$ for values of x from -4 to 4 and values of y from -4 to 4.

b Write down the equations of the vertical and horizontal asymptotes of $y = \dfrac{1}{x}$.

c Solve the equation $3x^2 - \dfrac{1}{x} = 0$.

Paper 2 style questions

1 The number (n) of bacteria after t hours is given by the formula $n = 1500(1.32)^t$.

  a Copy and complete the table below for values of n and t.

Time (t hours)	0	1	2	3	4
Number of bacteria (n)	1500		2613	3450	

  b On graph paper, draw the graph of $n = 1500(1.32)^t$. Use a scale of 2 cm to represent 1 hour on the horizontal axis and 2 cm to represent 1000 bacteria on the vertical axis. Label the graph clearly.

  c Find

  i the number of bacteria after 2 hours 30 minutes. Give your answer to the nearest ten bacteria.

  ii the time it will take to form approximately 5000 bacteria. Give your answer to the nearest 10 minutes.

2 The functions f and g are defined by

$$f(x) = \frac{4}{x}, \ x \in \mathbb{R}, \ x \neq 0$$

$$g(x) = 2x, \ x \in \mathbb{R}$$

  a Sketch the graph of $f(x)$ for $-8 \leq x \leq 8$.

  b Write down the equations of the horizontal and vertical asymptotes of the function f.

  c Sketch the graph of g on the same axes.

  d Find the solutions of $\frac{4}{x} = 2x$.

  e Write down the range of function f.

3 A function is represented by the equation $f(x) = 2(1.5)^x + 3$. The table shows the values of $f(x)$ for $-3 \leq x \leq 2$.

x	−3	−2	−1	0	1	2
$f(x)$	3.59	3.89	a	5	6	b

  a Calculate the values for a and b.

  b On graph paper, draw the graph of $f(x)$ for $-3 \leq x \leq 2$, taking 1 cm to represent 1 unit on both axes.

The domain of the function $f(x)$ is the real numbers, $\mathbb{R}$.

  c Write down the range of $f(x)$.

  d Find the approximate value for x when $f(x) = 10$.

  e Write down the equation of the horizontal asymptote of $f(x) = 2(1.5)^x + 3$.

4 The graph shows the temperature, in degrees
Celsius, of Leonie's cup of hot chocolate t minutes
after pouring it. The equation of the graph is
$f(t) = 21 + 77(0.8)^t$ where $f(t)$ is the temperature
and t is the time in minutes after pouring the
hot chocolate out.

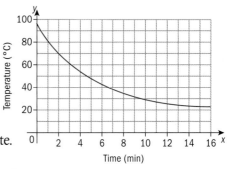

  a Find the initial temperature of the hot chocolate.
  b Write down the equation of the horizontal asymptote.
  c Write down the room temperature.
  d Find the temperature of the hot chocolate after 8 minutes.

5 Consider the functions
$$f(x) = x^2 - x - 6 \quad \text{and} \quad g(x) = -2x + 1$$
  a On the same diagram draw the graphs of $f(x)$ and $g(x)$
     for $-10 \le x \le 10$.
  b Find the coordinates of the local minimum of the graph
     of $f(x)$.
  c Write down the gradient of the line $g(x)$.
  d Write down the coordinates of the point where the graph of
     $g(x)$ cuts the y-axis.
  e Find the coordinates of the points of intersection of the
     graphs of $f(x)$ and $g(x)$.
  f Hence, or otherwise, solve the equation $x^2 + x - 7 = 0$.

6 a Sketch the graph of $f(x) = x^2 - \dfrac{3}{x}$, for $-4 \le x \le 4$.
  b Write down the equation of the vertical asymptote of $f(x)$.
  c On the same diagram draw the graph of $g(x) = -3(2)^x + 9$,
     for $-4 \le x \le 4$.
  d Write down the equation of the horizontal asymptote
     of $g(x)$.
  e Find the coordinates of the points of intersection of
     $f(x)$ and $g(x)$.

7 The profit (P) in euros made by selling homemade lemonade is modeled by the function

$$P = -\frac{x^2}{10} + 10x - 60$$

where x is the number of glasses of lemonade sold.

a Copy and complete the table.

x	0	10	20	30	40	50	60	70	80	90
P		30			180			150	100	

b On graph paper draw axes for x and $P(x)$, placing x on the horizontal axis and $P(x)$ on the vertical axis. Draw the graph of $P(x)$ against x by plotting the points.

c **Use your graph** to find

　i the maximum possible profit

　ii the number of glasses that need to be sold to make the maximum profit

　iii the number of glasses that need to be sold to make a profit of 160 euros

　iv the amount of money initially invested.

8 a Sketch the graph of the function $f(x) = x^2 - 7$, $x \in \mathbb{R}$, $-4 \le x \le 4$. Write down the coordinates of the points where the graph of $y = f(x)$ intersects the axes.

b On the same diagram sketch the graph of the function
$$g(x) = 7 - x^2, \, x \in \mathbb{R}, \, -4 \le x \le 4.$$

c Solve the equation $f(x) = g(x)$ in the given domain.

d The graph of the function $h(x) = x + c$, $x \in \mathbb{R}$, $-4 \le x \le 4$, where c is a positive integer, intersects twice with both $f(x)$ and $g(x)$ in the given domain. Find the possible values for c.

9 The functions f and g are defined by $f(x) = \frac{x^2}{2}$ and $g(x) = -\frac{x^2}{2} + 2x$, $x \in \mathbb{R}$.

a Calculate the coordinates of the points of intersection of the graphs $f(x)$ and $g(x)$.

b Find the equation of the axis of symmetry of the graph of $y = g(x)$.

c The straight line with equation $y = k$, $k \in \mathbb{R}$, is a tangent to the graph of g. Find the value of k.

d Sketch the graph of $f(x)$ and the graph of $g(x)$, using a rectangular Cartesian coordinate system with 1 cm as a unit. Show the coordinates of any points of intersection with the axes.

e Find the values of x for which $f(x) < g(x)$.

CHAPTER 4 SUMMARY
Functions

- A **function** is a relationship between two sets: a **first** set and a **second** set. Each element 'x' of the first set is related to **one and only one** element 'y' of the second set.
- The first set is called the **domain** of the function. The elements of the domain, the 'x-values', are the **independent variables**.
- For each value of 'x' (input) there is one and only one output. This value is called the **image** of 'x'. The set of all the images (all the outputs) is called the **range** of the function. The elements of the range, the 'y-values', are the **dependent variables**.
- The graph of a function f is the set of points (x, y) on the Cartesian plane where y is the image of x through the function f.
- $y = f(x)$ means that the image of x through the function f is y. x is the independent variable and y is the dependent variable.

Linear models

- A **linear function** has the general form $f(x) = mx + c$, where m (the gradient) and c are constants.
- When $f(x) = mx$ the graph passes through the origin, $(0, 0)$.

Quadratic models

- A **quadratic function** has the form $f(x) = ax^2 + bx + c$, where $a, b, c \in \mathbb{R}$ and $a \neq 0$.
- The graph of any quadratic function is a **parabola** – a $\cup$-shaped (or $\cap$-shaped) curve. It has an **axis of symmetry** and either a **minimum** or **maximum** point, called the **vertex** of the parabola.
- If $a > 0$ then the graph is $\cup$-shaped; if $a < 0$ then the graph is $\cap$-shaped.
- The curve intersects the y-axis at $(0, c)$.
- The equation of the axis of symmetry is $x = -\dfrac{b}{2a}$, $a \neq 0$.
- The x-coordinate of the vertex is $x = -\dfrac{b}{2a}$.
- The factorized form of a quadratic function is $f(x) = a(x - k)(x - l)$.
- If $a > 0$ then the graph is $\cup$-shaped; if $a < 0$ then the graph is $\cap$-shaped.
- A $\cup$-shaped graph is 'concave up'. A $\cap$-shaped graph is 'concave down'.
- The curve intersects the x-axis at $(k, 0)$ and $(l, 0)$.
- The equation of the axis of symmetry is $x = \dfrac{k+l}{2}$.
- The x-coordinate of the vertex is also $x = \dfrac{k+l}{2}$.
- The function $f(x) = ax^2 + bx + c$ intersects the x-axis where $f(x) = 0$. The x-values of the points of intersection are the two solutions (or **roots**) of the equation $ax^2 + bx + c$. (The y-values at these points of intersection are zero.)
- Two functions $f(x)$ and $g(x)$ intersect at the point(s) where $f(x) = g(x)$.

Continued on next page

Exponential models

- In an **exponential function,** the independent variable is the **exponent** (or **power**).
- In general, for the graph of $f(x) = ka^x + c$ where $a \in \mathbb{Q}^+$ and $k \neq 0$ and $a \neq 1$
  - the line $y = c$ is a **horizontal asymptote**
  - the curve passes through the point $(0, k + c)$.
- In general, for the graph of $f(x) = ka^{-x} + c$ where $a \in \mathbb{Q}^+$ and $k \neq 0$ and $a \neq 1$
  - the line $y = c$ is a horizontal asymptote
  - the curve passes through the point $(0, k + c)$
  - the graph is a reflection in the y-axis of $g(x) = ka^x + c$.

Cubic functions

- A cubic function has the form $f(x) = ax^3 + bx^2 + cx + d$, where $a \neq 0$. The domain is $\mathbb{R}$, unless otherwise stated.

The language of mathematics

Mathematics is described as a language. It has vocabulary (mathematical symbols with precise meaning) and grammar (an order in which we combine these symbols together to make them meaningful).

■ Mathematics is often considered a 'universal language'.
Can a language ever be truly universal?

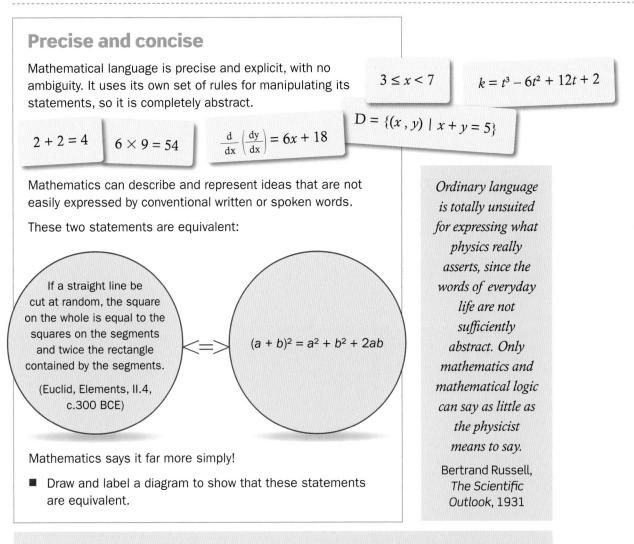

Precise and concise

Mathematical language is precise and explicit, with no ambiguity. It uses its own set of rules for manipulating its statements, so it is completely abstract.

$$3 \leq x < 7$$

$$k = t^3 - 6t^2 + 12t + 2$$

$$2 + 2 = 4$$

$$6 \times 9 = 54$$

$$\frac{d}{dx}\left(\frac{dy}{dx}\right) = 6x + 18$$

$$D = \{(x, y) \mid x + y = 5\}$$

Mathematics can describe and represent ideas that are not easily expressed by conventional written or spoken words.

These two statements are equivalent:

If a straight line be cut at random, the square on the whole is equal to the squares on the segments and twice the rectangle contained by the segments.

(Euclid, Elements, II.4, c.300 BCE)

$$\Longleftrightarrow$$

$$(a + b)^2 = a^2 + b^2 + 2ab$$

Ordinary language is totally unsuited for expressing what physics really asserts, since the words of everyday life are not sufficiently abstract. Only mathematics and mathematical logic can say as little as the physicist means to say.

Bertrand Russell, *The Scientific Outlook*, 1931

Mathematics says it far more simply!

■ Draw and label a diagram to show that these statements are equivalent.

'Mathematics is the abstract key which turns the lock of the physical universe.'
John Polkinghorne, *One World: The Interaction of Science and Theology*, 2007

Abstract language

■ What does '1' mean?

You can probably answer that with confidence. '1' is part of our language, we use it every day. Its meaning is clear to us. We can easily picture '1' banana.

But the language of mathematics has continued to expand to encompass more abstract concepts. Mathematicians call the square root of -1, 'i'.

■ What does this mean? Can you use i in everyday life?

What about pi (π)? Lots of people know this number.

It is the ratio $\dfrac{\text{circumference of a circle}}{\text{diameter of a circle}}$.

■ What does this 'mean'? Can you picture 'π' bananas?

■ Do π and i exist?

i is an abstract concept of mathematics that has also become part of our everyday, English language too. i or π are also abstract concepts of mathematics, but have not become part of everyday language. Mathematicians need and use these numbers. They are not any more abstract than the number 1. They appear in a mathematical context and allow us to think mathematically and communicate these ideas, to perform manipulations, to express results and model real-life occurrences in a simple way.

Simple and beautiful equations that model the world

Here are some famous equations

Einstein's equation: $E = mc^2$

Newton's second law: $F = ma$

Boyle's law: $V = \dfrac{k}{P}$

Schrödinger's equation: $\hat{H}\psi = E\psi$

Newton's law of universal gravitation: $F = G\dfrac{m_1 m_2}{r^2}$

These are simple equations (although they were not simple to derive!). Isn't it startling that so much of what happens in the universe can be described using equations like these?

These equations have helped to put a man on the moon and bring him back, develop wireless internet and understand the workings of the human body.

■ Do you think that mathematics and science will one day discover the ultimate 'theory of everything'? A theory that fully explains and links together all known physical phenomena? A theory that can predict the outcome of any experiment that could be carried out?

■ What will mathematicians and scientists do then?

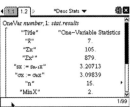

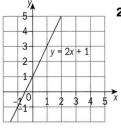

5 Statistical applications

CHAPTER OBJECTIVES:

4.1 The normal distribution; random variables; the parameters μ and σ; diagrammatic representation; normal probability calculations; expected value; inverse normal calculations

4.2 Bivariate data: correlation; scatter diagrams; line of best fit; Pearson's product-moment correlation coefficient, r

4.3 The regression line for y on x

4.4 The χ^2 test for independence: null and alternative hypotheses; significance levels; contingency tables; expected frequencies; degrees of freedom; p-values

Before you start

You should know how to:

1 Find the mean and standard deviation of a set of data and comment on the relationship between them, e.g. for the data set

4, 5, 6, 8, 12, 13, 2, 5, 6, 9, 10, 9, 8, 3, 5:

Mean =

$$\frac{(4+5+6+8+12+13+2+5+6+9+10+9+8+3+5)}{15}$$

$$=\frac{105}{15}=7$$

On a GDC, the mean is indicated by $\bar{x}$. Using a GDC, standard deviation $(\sigma_x) = 3.10$ (to 3 sf).

OneVar *number*, 1: *stat.results*	
"Title"	"One-Variable Statistics"
"$\bar{x}$"	7.
"Σx"	105.
"Σx^2"	879.
"$SX := Sn-1X$"	3.20713
"$\sigma X := \sigma n X$"	3.09839
"n"	15.
"MinX"	2.

The small standard deviation implies that the data are close to the mean.

2 Sketch the graph of the equation of a straight line, e.g. the straight line $y = 2x + 1$ passes through the point $(0, 1)$ and has gradient 2.

Skills check

1 Find the mean and standard deviation of these sets of data. Comment on your answers.

a 2, 4, 3, 6, 3, 2, 5, 3, 2, 5, 4, 4, 3, 5, 2, 3, 4, 5

b

x	Frequency
12	1
13	2
14	23
15	2
16	1

> For help, see Chapter 2, Sections 2.4 and 2.7.

2 Sketch the graphs of:

a $y = -3x + 4$

b $y = 2x - 6$

The people in this photograph are a sample of a population and a source of valuable data. Like a lot of data on natural phenomena, people's heights and weights fit a 'normal distribution', which you will study in this chapter. Medical statisticians use this information to plot height and weight charts, and establish guidelines on healthy weight.

The information can also be used to chart changes in a population over time. For example, the data can be analyzed to determine whether people, on the whole, are getting taller or heavier. These results may affect or even determine government health policy. Moreover, manufacturing and other industries may use the information to decide whether to, for example, make door frames taller or aircraft seats wider.

You may think that some data might be related, for example, people's height and shoe size, or perhaps a child's height and their later adult height. This chapter shows you how to investigate correlation and the strength of relationships between data sets.

Investigation – related data?

Do you think that height and shoe size are related?
Collect the height and shoe size of at least
60 students in your school.
Plot these data points on a graph. Use the *x*-axis for 'Height' and the *y*-axis for 'Shoe size'. Do not join up the points.
Does the data support your original hypothesis on height and shoe size?

The graph you will draw in this investigation is called a **scatter diagram**. You will find out more about scatter diagrams and the correlation between data sets in Section 5.2 of this chapter.

5.1 The normal distribution

For his Mathematical Studies Project, Pedro measures the heights of all the apple trees in his father's orchard. There are 150 trees.

If Pedro drew a diagram to represent the frequency of the heights of all 150 trees, what do you think it would look like?

Pedro then measures the heights of the apple trees in his uncle's orchard. If he drew a diagram of the frequencies of these heights, do you think that this diagram would look different to the previous one?

In both orchards there would probably be a few very small trees and a few very large trees – but those would be the exception. Most of the trees would fall within a certain range of heights. They would roughly fit a bell-shaped curve that is symmetrical about the mean. We call this a **normal distribution**.

Many events fit this type of distribution: for example, the heights of 21-year-old males, the results of a national mathematics examination, the weights of newborn babies, etc.

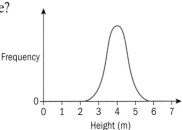

▲ Normal distribution diagram for the tree heights measured by Pedro

The properties of a normal distribution

> → The **normal distribution** is the most important continuous distribution in statistics. It has these properties:
> - It is a bell-shaped curve.
> - It is symmetrical about the mean, μ. (The mean, the mode and the median all have the same value.)
> - The x-axis is an asymptote to the curve.
> - The total area under the curve is 1 (or 100%).
> - 50% of the area is to the left of the mean and 50% to the right.
> - Approximately 68% of the area is within 1 standard deviation, σ, of the mean.
> - Approximately 95% of the area is within 2 standard deviations of the mean.
> - Approximately 99% of the area is within 3 standard deviations of the mean.

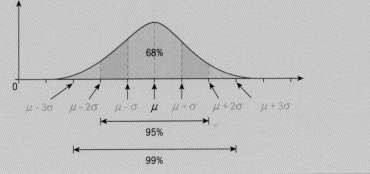

The normal curve is sometimes called the 'Gaussian curve' after the German mathematician Carl Friedrich Gauss (1777–1855). Gauss used the normal curve to analyze astronomical data in 1809. A portrait of Gauss and the normal curve appeard on the old German 10 Deutschmark note.

You can calculate the probabilities of events that follow a normal distribution.

Returning to Pedro and the apple trees, imagine that the mean height of the trees is 4 m and the standard deviation is 0.5 m.

Let the height of an apple tree be x.

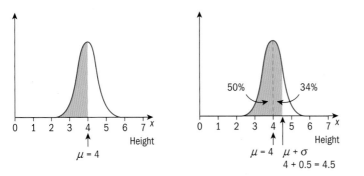

From the properties of the normal distribution:
Area to left of μ = 50%.
Area between μ and
$\mu + \sigma$ = 34% (68% ÷ 2).

The probability that an apple tree is less than 4 m is $P(x < 4) = 50\%$ or 0.5. And $P(x < 4.5) = 50\% + 34\% = 84\%$ or 0.84.

→ The **expected value** is found by multiplying the number in the sample by the probability.

For example, if we chose 100 apple trees at random, the expected number of trees that would be less than 4 m = $100 \times 0.5 = 50$.

Example 1

The waiting times for an elevator are normally distributed with a mean of 1.5 minutes and a standard deviation of 20 seconds.
a Sketch a normal distribution diagram to illustrate this information, indicating clearly the mean and the times within one, two and three standard deviations of the mean.
b Find the probability that a person waits longer than 2 minutes 10 seconds for the elevator.
c Find the probability that a person waits less than 1 minute 10 seconds for the elevator.

200 people are observed and the length of time they wait for an elevator is noted.
d Calculate the number of people expected to wait less than 50 seconds for the elevator.

Answers

a

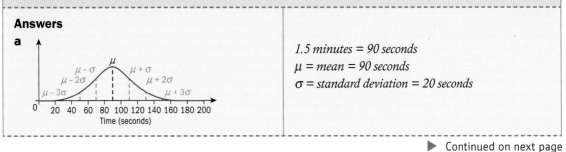

1.5 minutes = 90 seconds
μ = *mean = 90 seconds*
σ = *standard deviation = 20 seconds*

▶ Continued on next page

b

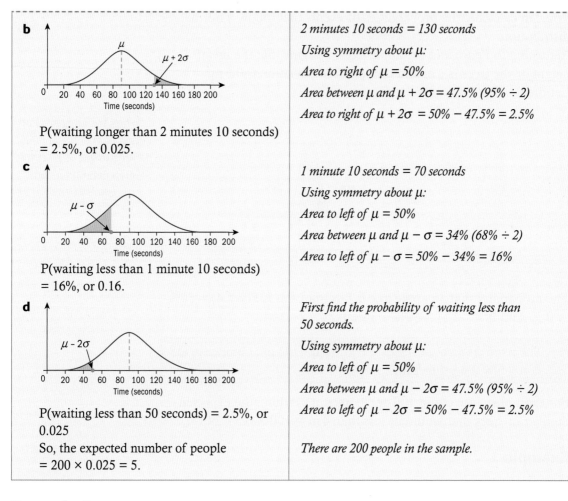

P(waiting longer than 2 minutes 10 seconds)
= 2.5%, or 0.025.

2 minutes 10 seconds = 130 seconds

Using symmetry about μ:

Area to right of μ = 50%

Area between μ and $\mu + 2\sigma$ = 47.5% (95% ÷ 2)

Area to right of $\mu + 2\sigma$ = 50% − 47.5% = 2.5%

c

P(waiting less than 1 minute 10 seconds)
= 16%, or 0.16.

1 minute 10 seconds = 70 seconds

Using symmetry about μ:

Area to left of μ = 50%

Area between μ and $\mu − \sigma$ = 34% (68% ÷ 2)

Area to left of $\mu − \sigma$ = 50% − 34% = 16%

d

P(waiting less than 50 seconds) = 2.5%, or
0.025
So, the expected number of people
= 200 × 0.025 = 5.

First find the probability of waiting less than 50 seconds.

Using symmetry about μ:

Area to left of μ = 50%

Area between μ and $\mu − 2\sigma$ = 47.5% (95% ÷ 2)

Area to left of $\mu − 2\sigma$ = 50% − 47.5% = 2.5%

There are 200 people in the sample.

Example 2

The heights of 250 twenty-year-old women are normally distributed with a mean of 1.68 m and standard deviation of 0.06 m.

a Sketch a normal distribution diagram to illustrate this information, indicating clearly the mean and the heights within one, two and three standard deviations of the mean.

b Find the probability that a woman has a height between 1.56 m and 1.74 m.

c Find the expected number of women with a height greater than 1.8 m.

Answers

a

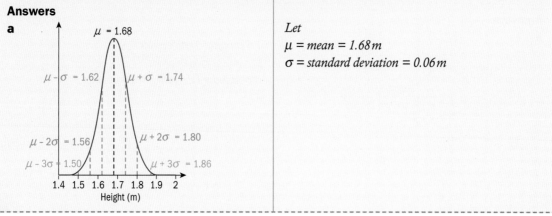

Let
μ = *mean* = *1.68 m*
σ = *standard deviation* = *0.06 m*

▶ Continued on next page

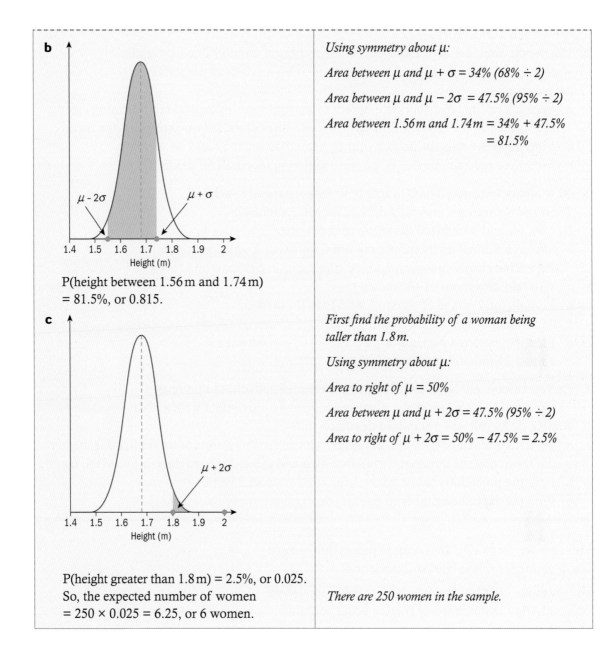

b

P(height between 1.56 m and 1.74 m)
= 81.5%, or 0.815.

Using symmetry about μ:

Area between μ and μ + σ = 34% (68% ÷ 2)

Area between μ and μ − 2σ = 47.5% (95% ÷ 2)

Area between 1.56 m and 1.74 m = 34% + 47.5%
= 81.5%

c

P(height greater than 1.8 m) = 2.5%, or 0.025.
So, the expected number of women
= 250 × 0.025 = 6.25, or 6 women.

First find the probability of a woman being taller than 1.8 m.

Using symmetry about μ:

Area to right of μ = 50%

Area between μ and μ + 2σ = 47.5% (95% ÷ 2)

Area to right of μ + 2σ = 50% − 47.5% = 2.5%

There are 250 women in the sample.

Exercise 5A

EXAM-STYLE QUESTION

1 The heights of 200 lilies are normally distributed with a mean of 40 cm and
a standard deviation of 3 cm.
   a Sketch a normal distribution diagram to illustrate this information.
     Indicate clearly the mean and the heights within one, two and three
     standard deviations of the mean.
   b Find the probability that a lily has a height less than 37 cm.
   c Find the probability that a lily has a height between 37 cm and 46 cm.
   d Find the expected number of lilies with a height greater than 43 cm.

2 100 people were asked to estimate the length of one minute. Their estimates were normally distributed with a mean time of 60 seconds and a standard deviation of 4 seconds.

a Sketch a normal distribution diagram to illustrate this information, indicating clearly the mean and the times within one, two and three standard deviations of the mean.

b Find the percentage of people who estimated between 52 and 64 seconds.

c Find the expected number of people estimating less than 60 seconds.

3 60 students were asked how long it took them to travel to school. Their travel times are normally distributed with a mean of 20 minutes and a standard deviation of 5 minutes.

a Sketch a normal distribution diagram to illustrate this information, indicating clearly the mean and the times within one, two and three standard deviations of the mean.

b Find the percentage of students who took longer than 25 minutes to travel to school.

c Find the expected number of students who took between 15 and 25 minutes to travel to school.

4 Packets of coconut milk are advertised to contain 250 ml. Akshat tests 75 packets. He finds that the contents are normally distributed with a mean volume of 255 ml and a standard deviation of 8 ml.

a Sketch a normal distribution diagram to illustrate this information, indicating clearly the mean and the volumes within one, two and three standard deviations of the mean.

b Find the probability that a packet contains less than 239 ml.

c Find the expected number of packets that contain more than 247 ml.

You can use your GDC to calculate values that are not whole multiples of the standard deviation.

For example, in question 4 of Exercise 5A, suppose we wanted to find the probability that a packet contains more than 250 ml.

First sketch a normal distribution diagram.

In a Calculator page 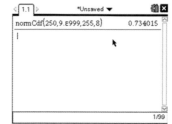 press MENU 5:Probability | 5:Distributions | 2:Normal Cdf and enter the lower bound (250), the upper bound (9×10^{999} – a *very* large number), the mean (255) and the standard deviation (8) in the wizard.

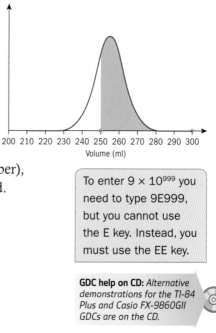

Volume (ml)

To enter 9×10^{999} you need to type 9E999, but you cannot use the E key. Instead, you must use the EE key.

GDC help on CD: *Alternative demonstrations for the TI-84 Plus and Casio FX-9860GII GDCs are on the CD.*

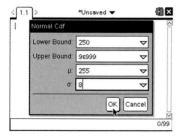

So, 73.4% of the packets contain more than 250 ml of coconut milk. Alternatively, enter normCdf, the lowest value, the highest value, the mean and the standard deviation directly into the calculator screen.

> For a very small number enter -9×10^{999}

Example 3

The lifetime of a light bulb is normally distributed with a mean of 2800 hours and a standard deviation of 450 hours.

a Find the percentage of light bulbs that have a lifetime of less than 1950 hours.
b Find the percentage of light bulbs that have a lifetime between 2300 and 3500 hours.
c Find the probability that a light bulb has a lifetime of more than 3800 hours.

120 light bulbs are tested.

d Find the expected number of light bulbs with a lifetime of less than 2000 hours.

Answers

a

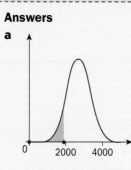

2.95% of the light bulbs have a lifetime of less than 1950 hours.

μ = mean = 2800 hours
σ = standard deviation = 450 hours

Lifetime less than 1950 hours:
lower bound = -9×10^{999}
upper bound = 1950

From GDC:
normCdf($-9E999$, 1950, 2800, 450) = 0.029 45
                                    = 2.95%

b

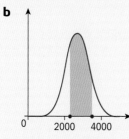

80.7% of the light bulbs have a lifetime between 2300 and 3500 hours.

Lifetime between 2300 and 3500 hours:
lower bound = 2300
upper bound = 3500

> Remember not to use $-9E999$ notations in an exam.

From GDC:
normCdf(2300, 3500, 2800, 450) = 0.8068 = 80.7%

c

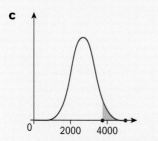

Only 1.31% of the light bulbs have a lifetime of more than 3800 hours.

Lifetime more than 3800 hours:
lower bound = 3800
upper bound = 9×10^{999}

GDC help on CD: *Alternative demonstrations for the TI-84 Plus and Casio FX-9860GII GDCs are on the CD.*

From GDC:
normCdf(3800, 9E999, 2800, 450) = 0.0131 = 1.31%

▶ Continued on next page

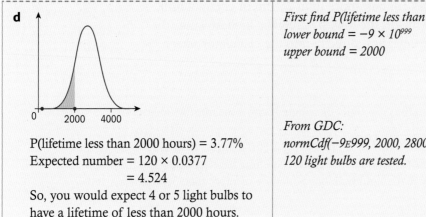

d

P(lifetime less than 2000 hours) = 3.77%
Expected number = 120 × 0.0377
= 4.524
So, you would expect 4 or 5 light bulbs to
have a lifetime of less than 2000 hours.

First find P(lifetime less than 2000 hours):
lower bound = −9 × 10⁹⁹⁹
upper bound = 2000

From GDC:
normCdf(−9ᴇ999, 2000, 2800, 450) = 0.0377 = 3.77%
120 light bulbs are tested.

Exercise 5B

EXAM-STYLE QUESTION

1 Jordi delivers daily papers to a number of homes in a village. The time taken
to deliver the papers follows a normal distribution with mean 80 minutes and
standard deviation 7 minutes.

a Sketch a normal distribution diagram to illustrate this information.

b Find the probability that Jordi takes longer than 90 minutes to deliver the papers.

Jordi delivers papers every day of the year (365 days).

c Calculate the expected number of days on which it would take Jordi longer
than 90 minutes to deliver the papers.

2 A set of 2000 IQ scores is normally distributed with a mean of 100 and a standard
deviation of 10.

a Calculate the probability that is represented by each of the following diagrams.

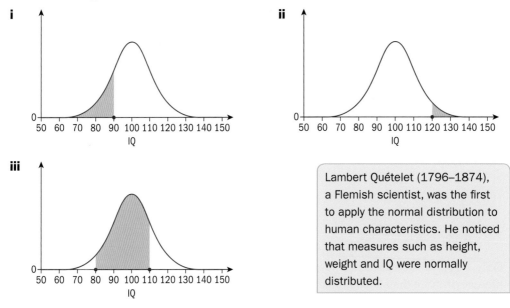

Lambert Quételet (1796–1874),
a Flemish scientist, was the first
to apply the normal distribution to
human characteristics. He noticed
that measures such as height,
weight and IQ were normally
distributed.

b Find the expected number of people with an IQ of more than 115.

3 A machine produces washers whose diameters are normally distributed with a mean of 40 mm and a standard deviation of 2 mm.

  a Find the probability that a washer has a diameter less than 37 mm.

  b Find the probability that a washer has a diameter greater than 45 mm.

Every week 300 washers are tested.

  c Calculate the expected number of washers that have a diameter between 35 mm and 43 mm.

EXAM-STYLE QUESTIONS

4 In a certain school, the monthly incomes of members of staff are normally distributed with a mean of 2500 euros and a standard deviation of 400 euros.

  a Sketch a normal distribution diagram to illustrate this information.

  b Find the probability that a member of staff earns less than 1800 euros per month.

The school has 80 members of staff.

  c Calculate the expected number of staff who earn more than 3400 euros.

5 The lengths of courgettes are normally distributed with a mean of 16 cm and a standard deviation of 0.8 cm.

  a Find the percentage of courgettes that have a length between 15 cm and 17 cm.

  b Find the probability that a courgette is longer than 18 cm.

The lengths of 100 courgettes are measured.

  c Calculate the expected number of courgettes that have a length less than 14.5 cm.

6 At a market, the weights of bags of kiwi fruit are normally distributed with a mean of 500 g and a standard deviation of 8 g.

A man picks up a bag of kiwi fruit at random.

Find the probability that the bag weighs more than 510 g.

EXAM-STYLE QUESTIONS

7 The scores in a Physics test follow a normal distribution with mean 70% and standard deviation 8%.

  a Find the percentage of students who scored between 55% and 80%.

30 students took the physics test.

  b Calculate the expected number of students who scored more than 85%.

8 A machine produces pipes such that the length of each pipe is normally distributed with a mean of 1.78 m and a standard deviation of 2 cm.

Any pipe whose length is greater than 1.83 m is rejected.

  a Find the probability that a pipe will be rejected.

500 pipes are tested.

  b Calculate the expected number of pipes that will be rejected.

Inverse normal calculations

Sometimes you are given the percentage area under the curve, i.e. the probability or proportion, and are asked to find the value corresponding to it. This is called an inverse normal calculation.

Always make a sketch to illustrate the information given.
You must always remember to use the area to the **left** when using your GDC. If you are given the area to the **right** of the value, you must subtract this from 1 (or 100%) before using your GDC.
For example, an area of 5% above a certain value means there is an area of 95% below it.

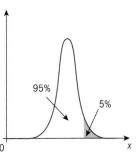

> In examinations, inverse normal questions will not involve finding the mean or standard deviation.

Example 4

The volume of cartons of milk is normally distributed with a mean of 995 ml and a standard deviation of 5 ml.
It is known that 10% of the cartons have a volume less than x ml.
Find the value of x.

- -

Answer

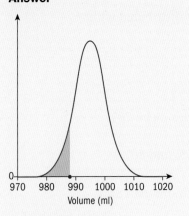

$x = 989$ ml (to 3 sf)

First sketch a diagram. The shaded area represents 10% of the cartons.
Using the GDC:

In a Calculator page ⊞ *press MENU 5:Probability | 5:Distributions | 3:Inverse Normal...*
Enter the percentage given (as a decimal, 0.1), the mean (995) and the standard deviation (5).

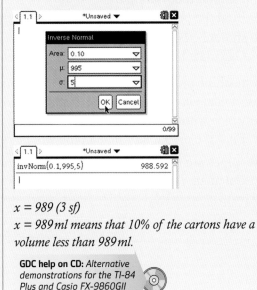

$x = 989$ (3 sf)
$x = 989$ ml means that 10% of the cartons have a volume less than 989 ml.

> **GDC help on CD:** *Alternative demonstrations for the TI-84 Plus and Casio FX-9860GII GDCs are on the CD.*

Example 5

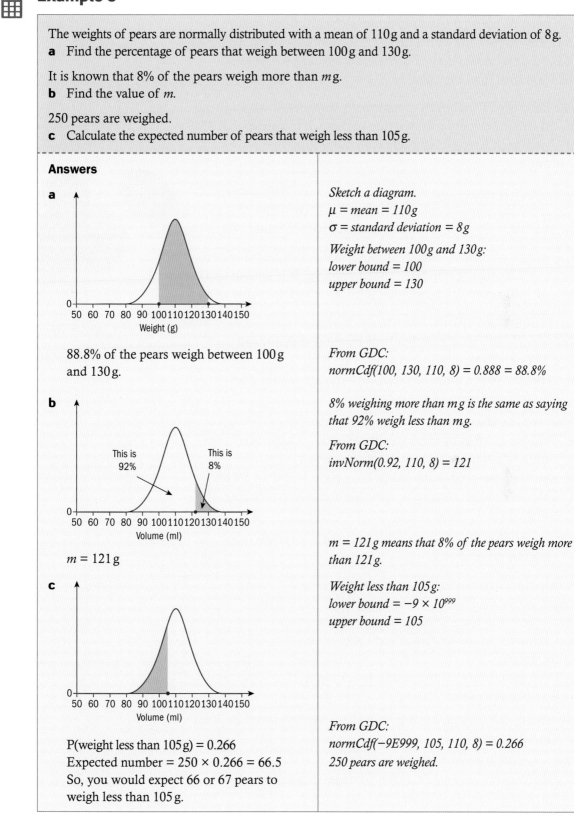

The weights of pears are normally distributed with a mean of 110g and a standard deviation of 8g.
a Find the percentage of pears that weigh between 100g and 130g.

It is known that 8% of the pears weigh more than m g.
b Find the value of m.

250 pears are weighed.
c Calculate the expected number of pears that weigh less than 105g.

Answers

a

88.8% of the pears weigh between 100g and 130g.

Sketch a diagram.
μ = mean = 110g
σ = standard deviation = 8g

Weight between 100g and 130g:
lower bound = 100
upper bound = 130

From GDC:
normCdf(100, 130, 110, 8) = 0.888 = 88.8%

b

This is 92%   This is 8%

$m = 121$ g

8% weighing more than m g is the same as saying that 92% weigh less than m g.

From GDC:
invNorm(0.92, 110, 8) = 121

m = 121g means that 8% of the pears weigh more than 121g.

c

P(weight less than 105g) = 0.266
Expected number = 250 × 0.266 = 66.5
So, you would expect 66 or 67 pears to weigh less than 105g.

Weight less than 105g:
lower bound = −9 × 10⁹⁹⁹
upper bound = 105

From GDC:
normCdf(−9E999, 105, 110, 8) = 0.266
250 pears are weighed.

1 The mass of coffee grounds in Super-strength coffee bags is normally distributed with a mean of 5 g and a standard deviation of 0.1 g. It is known that 25% of the coffee bags weigh less than p grams. Find the value of p.

2 The heights of Dutch men are normally distributed with a mean of 181 cm and a standard deviation of 5 cm. It is known that 35% of Dutch men have a height less than h cm. Find the value of h.

3 The weight of kumquats is normally distributed with a mean of 20 g and a standard deviation of 0.8 g. It is known that 15% of the kumquats weigh more than k grams. Find the value of k.

4 The weight of cans of sweetcorn is normally distributed with a mean of 220 g and a standard deviation of 4 g. It is known that 30% of the cans weigh more than w grams. Find the value of w.

EXAM-STYLE QUESTIONS

5 The weights of cats are normally distributed with a mean of 4.23 kg and a standard deviation of 0.76 kg.
  a Write down the weights of the cats that are within one standard deviation of the mean.

A vet weighs 180 cats.
  b Find the number of these cats that would be expected to be within one standard deviation of the mean.
  c Calculate the probability that a cat weighs less than 3.1 kg.
  d Calculate the percentage of cats that weigh between 3 kg and 5.35 kg.

It is known that 5% of the cats weigh more than w kg.
  e Find the value of w.

6 A manufacturer makes drumsticks with a mean length of 32 cm. The lengths are normally distributed with a standard deviation of 1 cm.
  a Calculate the values of a, b and c shown on the graph.
  b Find the probability that a drumstick has a length greater than 30.6 cm.

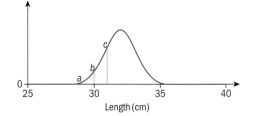

It is known that 80% of the drumsticks have a length less than d cm.
  c Find the value of d.

One week 5000 drumsticks are tested.
  d Calculate the expected number of drumsticks that have a length between 30.5 cm and 32.5 cm.

7 The average lifespan of a television set is normally distributed with a mean of 8000 hours and a standard deviation of 1800 hours.

   a Find the probability that a television set will break down before 2000 hours.

   b Find the probability that a television set lasts between 6000 and 12 000 hours.

   c It is known that 12% of the television sets break down before t hours.
   Find the value of t.

EXAM-STYLE QUESTIONS

8 The speed of cars on a motorway is normally distributed with a mean of $120 \, \text{km} \, \text{h}^{-1}$ and a standard deviation of $10 \, \text{km} \, \text{h}^{-1}$.

   a Draw a normal distribution diagram to illustrate this information.

   b Find the percentage of cars that are traveling at speeds of between $105 \, \text{km} \, \text{h}^{-1}$ and $125 \, \text{km} \, \text{h}^{-1}$.

It is known that 8% of the cars are traveling at a speed of less than $p \, \text{km} \, \text{h}^{-1}$.

   c Find the value of p.

One day 800 cars are checked for their speed.

   d Calculate the expected number of cars that will be traveling at speeds of between $96 \, \text{km} \, \text{h}^{-1}$ and $134 \, \text{km} \, \text{h}^{-1}$.

The speed limit is $130 \, \text{km} \, \text{h}^{-1}$.

   e Find the number of cars that are expected to be exceeding the speed limit.

9 The weights of bags of rice are normally distributed with a mean of 1003 g and a standard deviation of 2 g.

   a Draw a normal distribution diagram to illustrate this information.

   b Find the probability that a bag of rice weighs less than 999 g.

The manufacturer states that the bags of rice weigh 1 kg.

   c Find the probability that a bag of rice is underweight.

400 bags of rice are weighed.

   d Calculate the expected number of bags of rice that are underweight.

5% of the bags of rice weigh more than p g.

   e Find the value of p.

10 The weights of babies are normally distributed with a mean of 3.8 kg and a standard deviation of 0.5 kg.

  a Find the percentage of babies who weigh less than 2.5 kg.

  In a space of 15 minutes two babies are born. One weighs 2.34 kg and the other weighs 5.5 kg.

  b Calculate which event is more likely to happen.

  One month 300 babies are weighed.

  c Calculate the number of babies expected to weigh more than 4.5 kg.

  It was found that 10% of the babies weighed less than w kg.

  d Find the value of w.

5.2 Correlation

When two sets of data appear to be connected, that is, one set of data is dependent on the other, then there are various methods that can be used to check whether or not there is any **correlation**. One of these methods is the scatter diagram.

Data can be plotted on a scatter diagram with the **independent variable** on the horizontal axis and the **dependent variable** on the vertical axis. The pattern of dots will give a visual picture of how closely, if at all, the variables are related.

Types of correlation

→ In a **positive** correlation the dependent variable increases as the independent variable increases.

For example, fitness levels (dependent variable) increase as the number of hours spent exercising (independent variable) increase:

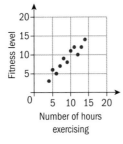

> → In a **negative** correlation the dependent variable decreases as the independent variable increases.

For example, the time taken to run a race (dependent variable) decreases as the training time (independent variable) increases:

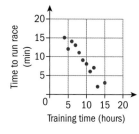

> → When the points are scattered randomly across the diagram there is **no** correlation.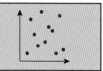

For example, the number of pairs of shoes that a person owns is not related to their age:

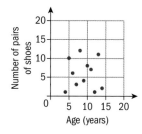

> → Correlations can also be described as strong, moderate or weak.

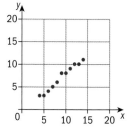

This is an example of a
strong positive correlation.

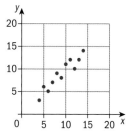

This is an example of a
moderate positive correlation.

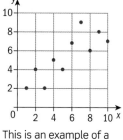

This is an example of a
weak positive correlation.

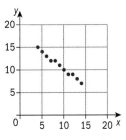

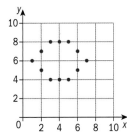

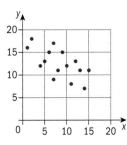

This is an example of a **strong negative** correlation.

This is an example of a **moderate negative** correlation.

This is an example of a **weak negative** correlation.

Correlations can also be classed as linear or non-linear.

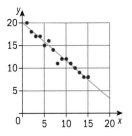

This is an example of a **linear** correlation.

This is an example of a **non-linear** correlation.

> For Mathematical Studies, you will only need to learn about **linear** correlations. However you may use other types of correlation in your project.

Example 6

The manager of a recreation park thought that the number of visitors to the park was dependent on the temperature.

He kept a record of the temperature and the numbers of visitors over a two-week period.

Plot these points on a scatter diagram and comment on the type of correlation.

Temperature (°C)	16	22	31	19	23	26	21	17	24	29	21	25	23	29
Number of visitors	205	248	298	223	252	280	233	211	258	295	229	252	248	284

Answer

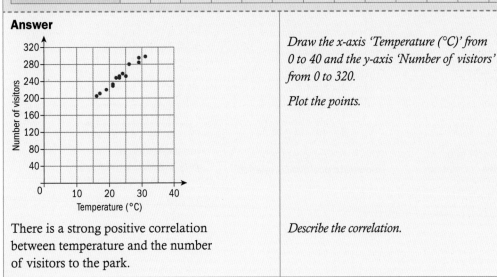

Draw the x-axis 'Temperature (°C)' from 0 to 40 and the y-axis 'Number of visitors' from 0 to 320.

Plot the points.

There is a strong positive correlation between temperature and the number of visitors to the park.

Describe the correlation.

Example 7

A Mathematical Studies student wanted to check if there was a correlation between the predicted heights of daisies and their actual heights.

Draw a scatter diagram to illustrate the data and comment on the correlation.

Predicted height (cm)	5.3	6.2	4.9	5.0	4.8	6.6	7.3	7.5	6.8	5.5	4.7	6.8	5.9	7.1
Actual height (cm)	4.7	7.0	5.3	4.5	5.6	5.9	7.2	6.5	7.2	5.8	5.3	5.9	6.8	7.6

Answer

(scatter diagram with Actual height (cm) on vertical axis 0–10 and Predicted height (cm) on horizontal axis 0–10)	*Draw x- and y-axes from 0 to 10.* *Plot 'Predicted height' on the horizontal axis and 'Actual height' on the vertical axis.*
There is a moderate positive correlation between predicted height and actual height.	*Describe the correlation.*

You can also use a GDC to plot a scatter diagram. For Example 7:

GDC help on CD: *Alternative demonstrations for the TI-84 Plus and Casio FX-9860GII GDCs are on the CD.*

First enter the data in two lists on a List & Spreadsheet page.

Then enter the variables onto the axes on a Data and Statistics page to draw the scatter diagram.

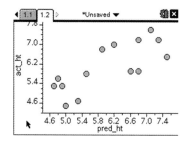

Exercise 5D

1 For each diagram, state the type of correlation (positive/negative and linear/non-linear) and the strength of the relationship (perfect/strong/moderate/weak/none).

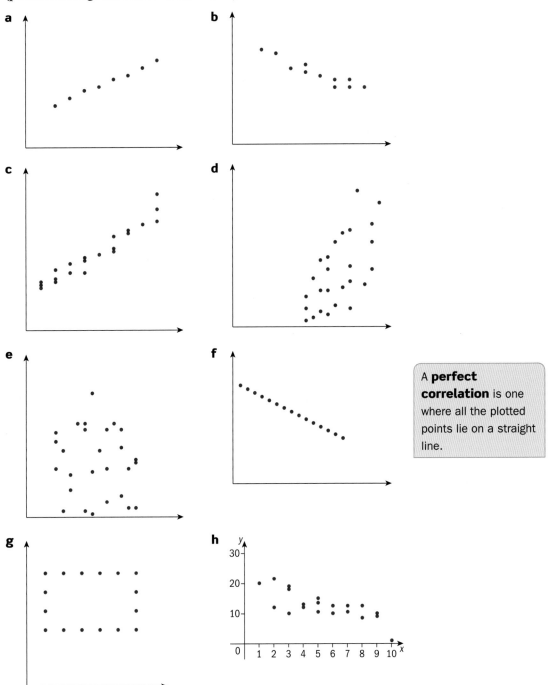

A **perfect correlation** is one where all the plotted points lie on a straight line.

2 For each set of data, plot the points on a scatter diagram and describe the type of correlation.

a

x	28	30	25	35	19	38	25	33	41	22	35	44
y	24	36	30	40	15	34	28	34	44	23	37	45

b

x	3	7	7	11	16	15	17	17	18	20
y	16	11	12	9	6	7	3	9	5	6

Line of best fit

A **line of best fit** is a line that is drawn on a graph of two sets of data, so that approximately as many points lie above the line as below it.

> → To draw the **line of best** fit by eye:
> - Find the mean of each set of data and plot this point on your scatter diagram.
> - Draw a line that passes through the mean point and is close to all the other points – with approximately an equal number of points above and below the line.

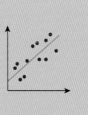

The line of best fit does not need to go through the origin and, in fact, in most cases it will not go through the origin.

Example 8

a For Example 6 draw the line of best fit on the diagram.
b For Example 7 draw the line of best fit on the diagram.

Geosciences use a line of best fit in
- flood frequency curves
- earthquake forecasting
- meteorite impact prediction
- climate change.

Answers

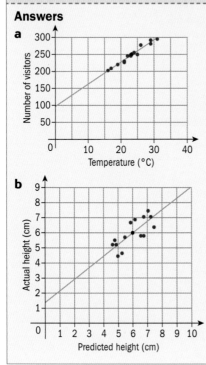

a *Calculate the means using your GDC. The mean temperature is 23.3, and the mean number of visitors is 251. Plot the mean point (23.3, 251) on the scatter diagram. Draw a line of best fit through the mean point so that there are roughly an equal number of points above and below the line.*

b *The mean predicted height is 6.03, and the mean actual height is 6.09. Plot the mean point (6.03, 6.09) on the scatter diagram and draw a straight line through it so that there are roughly an equal number of points above and below the line.*

You can also use a GDC to draw a line of best fit.
For Example 7:
Select MENU 4:Analyze | 6:Regression | 2:Show
Linear ($ax + b$).

Given a value of predicted height, use trace (MENU
4:Analyze | A:Graph Trace) to find the value of the
actual height from the graph.

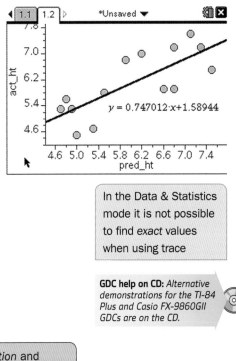

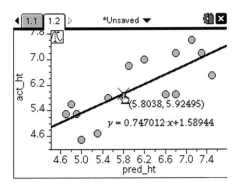

In the Data & Statistics mode it is not possible to find *exact* values when using trace

GDC help on CD: *Alternative demonstrations for the TI-84 Plus and Casio FX-9860GII GDCs are on the CD.*

There is often a lot of confusion between the concepts of *causation* and *correlation*. However, they should be easy enough to distinguish.
One action can *cause* another (such as smoking can cause lung cancer), or it can *correlate* with another (for example, blue eyes are correlated with blonde hair).

If one action *causes* another, then they are also *correlated*. But if two things are *correlated* it does *not* mean that one *causes* the other. For example, there could be a strong *correlation* between the predicted grades that teachers give and the actual grades that the students achieve. However, the achieved grades are not *caused* by the predicted grades.

Can you think of other examples?
Can you find articles in newspapers, magazines or online where *cause* is used incorrectly?

Exercise 5E

1 For each set of data:
  i Plot the points on a scatter diagram and describe the type of correlation.
  ii Find the mean of x and the mean of y.
  iii Plot the mean point on your diagram and draw a line of best fit by eye.

a

x	2	4	6	8	10	12	14	16	18	20	22	24
y	14	15	18	21	24	25	27	29	30	32	35	39

b

x	12	13	14	15	16	17	18	19	20	21
y	32	29	30	25	22	22	15	10	10	7

2 The following table gives the heights and weights of 12 giraffes.

Height (x m)	4.8	4.1	4.2	4.7	5.0	5.0	4.8	5.2	5.3	4.3	5.5	4.5
Weight (y kg)	900	600	650	750	1100	950	850	1150	1100	650	1250	800

a Plot the points on a scatter diagram and describe the correlation.

b Find the mean height and the mean weight.

c Plot the mean point on your diagram and draw a line of best fit by eye.

d Use your diagram to estimate the weight of a giraffe of height 4.6 m.

3 Fourteen students took a test in Chemistry and ITGS (Information Technology in a Global Society). The results are shown in the following table.

Chemistry (%)	45	67	72	34	88	91	56	39	77	59	66	82	96	42
ITGS (%)	42	76	59	44	76	88	55	45	69	62	58	94	85	58

a Plot the points on a scatter diagram and describe the correlation.

b Find the mean score for each test.

c Plot the mean point on your diagram and draw a line of best fit by eye.

d Use your diagram to estimate the result for an ITGS test when the chemistry score was 50%.

4 Twelve mothers were asked how many hours per day, on average, they held their babies and how many hours per day, on average, the baby cried. The results are given in the following table.

Baby held (hours)	1	2	3	3	4	4	5	6	6	7	8	9
Baby cried (hours)	6	6	5	5.5	4	3	3.5	2	2.5	2	1.5	1

a Plot the points on a scatter diagram and describe the correlation.

b Find the mean number of hours held and the mean number of hours spent crying.

c Plot the mean point on your diagram and draw a line of best fit by eye.

d Use your diagram to estimate the number of hours a baby cries if it is held for 3.5 hours.

5 The table shows the size of a television screen and the cost of the television.

Size (inches)	32	37	40	46	50	55	59
Cost ($)	450	550	700	1000	1200	1800	2000

a Plot the points on a scatter diagram and describe the correlation.

b Find the mean screen size and the mean cost.

c Plot the mean point on your diagram and draw a line of best fit by eye.

d Use your diagram to estimate the cost of a 52-inch TV.

Pearson's product-moment correlation coefficient

Karl Pearson (1857–1936) was an English lawyer, mathematician and statistician.

His contributions to the field of statistics include the product–moment correlation coefficient and the chi-squared test.

Pearson's career was spent largely on applying statistics to the field of biology.

He founded the world's first University statistics department at University College London in 1911.

▶ Karl Pearson

It is useful to know the **strength** of the relationship between any two sets of data that are thought to be related.

Pearson's product-moment correlation coefficient, r, is one way of finding a numerical value that can be used to determine the strength of a linear correlation between the two sets of data.

In examinations you will only be expected to use your GDC to find the value of r.

→ **Pearson's product-moment correlation coefficient**, r, can take all values between −1 and +1 inclusive.
- When $r = -1$, there is a **perfect negative** correlation between the data sets.
- When $r = 0$, there is **no** correlation.
- When $r = +1$, there is a **perfect positive** correlation between the data sets.
- A **perfect correlation** is one where **all** the plotted points lie on a straight line.

When r is between
- 0 and 0.25, the correlation is very weak
- 0.25 and 0.5, the correlation is weak
- 0.5 and 0.75, there is a moderate correlation
- 0.75 and 1, the correlation is strong.

The formula for Pearson's product–moment correlation coefficient for two sets of data, x and y, is: $r = \dfrac{s_{xy}}{s_x s_y}$

where s_{xy} is the covariance (beyond the scope of this course) and s_x and s_y are the standard deviations of x and y respectively.

You will be expected to use this formula to enhance your project.

Other formulae that you will need are:

$$s_{xy} = \sum \frac{(x - \bar{x})(y - \bar{y})}{n} \text{ or } \frac{\sum xy}{n} - \frac{\sum x}{n} \cdot \frac{\sum y}{n}$$

$$s_x = \sqrt{\frac{\sum (x - \bar{x})^2}{n}} \text{ or } \sqrt{\left(\frac{\sum x^2}{n} - \bar{x}^2 \right)} \qquad s_y = \sqrt{\frac{\sum (y - \bar{y})^2}{n}} \text{ or } \sqrt{\left(\frac{\sum y^2}{n} - \bar{y}^2 \right)}$$

Example 9

The data given below for a first-division football league show the position of the team and the number of goals scored.

Find the correlation coefficient, r, and comment on this value.

Position	1	2	3	4	5	6	7	8	9	10	11	12	13	14	15	16	17	18	19	20
Goals	75	68	60	49	59	50	55	46	57	49	48	39	44	56	54	37	42	37	40	27

GDC help on CD: *Alternative demonstrations for the TI-84 Plus and Casio FX-9860GII GDCs are on the CD.*

Answer

$r = -0.816$ (to 3 sf)

So, there is a **strong negative** correlation between the position of the team and the number of goals scored.

Using a GDC:

First enter 'Position' numbers and 'Goals' into two lists (X and Y respectively).

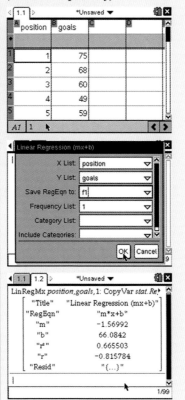

Your GDC also gives r^2, **the coefficient of determination**. This is an indication of how much of the variation in one set of data, y, can be explained by the variation in the other set of data, x. For example, if $r^2 = 0.821$, this means that 82.1% of the variation in set y is caused by the variation in set x. Here, either $r = 0.906$ which is a strong positive linear correlation, or $r = -0.906$ which is a strong negative linear correlation.

Example 10

The heights and shoe sizes of the students at Learnwell Academy are given in the table below. Find the correlation coefficient, r, and comment on your result.

Height (x cm)	145	151	154	162	167	173	178	181	183	189	193	198
Shoe size	35	36	38	37	38	39	41	43	42	45	44	46

Answer

$r = 0.964$ (to 3 sf)

This means that there is a **strong positive** correlation between height and shoe size.

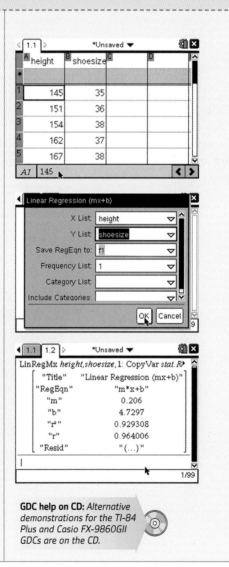

GDC help on CD: *Alternative demonstrations for the TI-84 Plus and Casio FX-9860GII GDCs are on the CD.*

Exercise 5F

1 The table gives the temperature (°C) at midday and the number of ice creams sold over a period of 21 days.

Temperature (°C)	22	23	22	19	20	25	23	20	17	18	23	24	22	26	19	19	20	22	23	22	20
Number of ice creams sold	59	61	55	40	51	72	55	45	39	35	59	72	63	77	37	41	44	50	59	48	38

Find the correlation coefficient, r, and comment on this value.

2 A chicken farmer selected a sample of 12 hens. During a two-week period, he recorded the number of eggs each hen produced and the amount of feed each hen ate. The results are given in the table.

a Find the correlation coefficient, r.

b Comment on the value of the correlation coefficient.

Number of eggs	Units of feed eaten
11	6.2
10	4.9
13	7.1
10	6.2
11	5.0
15	7.9
9	4.8
12	6.9
11	5.3
12	5.9
13	6.5
9	4.5

3 The table gives the average temperature for each week in December, January and February and the corresponding number of hours that an average family used their central heating.

Average temperature (°C)	4	1	3	−2	−9	−12	−8	−9	−2	1	3	5
Hours of heating	43	45	51	52	58	64	57	60	55	43	40	30

Find the correlation coefficient, r, and comment on this value.

4 Eight students complete examination papers in Economics and Biology. The results are shown in the table.

Student	A	B	C	D	E	F	G	H
Economics	64	55	43	84	67	49	92	31
Biology	53	42	44	79	75	52	84	29

Find the correlation coefficient, r, and comment on your result.

5 The table shows the age of a baby, measured in days, and the weight, in kilograms, at 08:00 on the corresponding day.

Age (days)	0	7	14	21	28	35	42
Weight (kg)	3.50	3.75	3.89	4.15	4.42	4.55	5.02

Find the correlation coefficient, r, and comment on your result.

6 The heights and weights of 10 students selected at random are shown in the table.

Height (x cm)	155	161	173	150	182	165	170	185	175	145
Weight (y kg)	50	75	80	46	81	79	64	92	74	108

Find the correlation coefficient, r, and comment on your answer.

7 The table shows the mock examination results and the actual results of 15 students at Top High College.

Mock	32	35	28	24	19	39	44	41	23	29	28	35	38	43	21
Actual	33	34	30	25	18	36	43	42	24	27	29	36	39	44	22

Find the correlation coefficient, r, and comment on your result.

8 The ages of 14 people and the times it took them to run 1 km are shown in the table.

Age (years)	9	12	13	15	16	19	21	29	32	43	48	55	61	66
Time (minutes)	7.5	6.8	7.2	5.3	5.1	4.9	5.2	4.6	4.9	6.8	6.2	7.5	8.9	9.2

Find the correlation coefficient, r, and comment on your result.

5.3 The regression line

> → The **regression line for y on x** is a more accurate version of a line of best fit, compared to best fit by eye.

The regression line for y on x, where y is the dependent variable, is also known as the least squares regression line. It is the line drawn through a set of points such that the sum of the squares of the distance of each point from the line is a minimum.

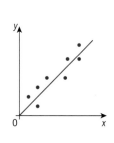

> → If there is a strong or moderate correlation, you can use the regression line for y on x to predict values of y for values of x within the range of the data.

You should only calculate the equation of the regression line if there is a moderate or strong correlation coefficient.

In your project you can work out the equation of the regression line for y on x using the formula:

$$(y - \bar{y}) = \frac{s_{xy}}{(s_x)^2}(x - \bar{x})$$

where $\bar{x}$ and $\bar{y}$ are the means of the x and y data values respectively, s_x is the standard deviation for the x data values, and s_{xy} is the covariance.

In examinations you will only be expected to use your GDC to find the equation of the regression line.

Example 11

Ten students train for a charity walk.
The table shows the average number of hours per week that each member trains and the time taken to complete the walk.

Training time (hours)	9	8	12	3	25	6	10	5	6	21
Time to complete walk (minutes)	15.9	14.8	15.3	18.4	13.8	16.2	14.1	16.1	16	14.2

a Find the correlation coefficient, r.
b Find the equation of the regression line.
c Using your equation, estimate how many minutes it will take a student who trains 18 hours per week to complete the walk.

British scientist and mathematician Francis Galton (1822–1911) coined the term 'regression'.

Answers

a $r = -0.767$ (to 3 sf)

First enter the data into two lists and then compute the results.

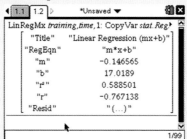

GDC help on CD: *Alternative demonstrations for the TI-84 Plus and Casio FX-9860GII GDCs are on the CD.*

b The equation of the regression line is:
$y = -0.147x + 17.0$

The general form of the equation is:
$y = mx + c$
From the GDC:
$m = -0.147$ (to 3 sf)
$c = 17.0$ (to 3 sf)

In this book we use $y = mx + c$ as the general form of a linear equation. The GDC uses $y = mx + b$ as the general form. Some people use $y = ax + b$.

c $y = -0.147(18) + 17.0 = 14.4$ (to 3 sf)
Therefore, the time taken is approximately 14.4 minutes.

Substitute 18 (hours) for x in the equation from part **b**.

Example 12

The table shows the number of mice for sale in a pet shop at the end of certain weeks.

Time (x weeks)	3	5	6	9	11	13
Number of mice (y)	41	57	61	73	80	91

a Find the correlation coefficient, r.

b Find the equation of the regression line for y on x.

c Use your regression line to predict the number of mice for sale after 10 weeks.

d Can you accurately predict the number of mice after 20 weeks?

> How do we know what we know? How sure can we be of our predictions? What predictions are made about population, or the climate?

Answers

a $r = 0.990$ (to 3 sf)

First enter the data into two lists.

```
◄ 1.1  1.2 ▷        *Unsaved ▼              ⊠
LinRegMx time,mice,1: CopyVar stat.RegEqn▸
      ⌈ "Title"    "Linear Regression (mx+b)" ⌉
      | "RegEqn"          "m*x+b"             |
      |   "m"             4.62929             |
      |   "b"             30.9039             |
      |   "r²"            0.979916            |
      |   "r"             0.989907            |
      ⌊ "Resid"          "{...}"             ⌋
                                        2/99
```

> **GDC help on CD:** *Alternative demonstrations for the TI-84 Plus and Casio FX-9860GII GDCs are on the CD.*

b The equation of the regression line is:

$y = 4.63x + 30.9$

The general form of the equation is:
$y = mx + c$
From the GDC:
$m = 4.63$ *(to 3 sf)*
$c = 30.9$ *(to 3 sf)*

c $y = 4.63(10) + 30.9 = 77.2$
   $= 77$
   After 10 weeks, the number of mice is 77.

*Substitute 10 (weeks) for x in the equation from part **b**.*

> Remember that you cannot use the regression line to predict values beyond the region of the given data.

d No, because it is too far away from the data in the table.

Exercise 5G

EXAM-STYLE QUESTION

1 The table shows the distance travelled by train between various places in India and the cost of the journey.

Distance (km)	204	1407	1461	793	1542	343	663	780
Cost (rupees)	390	2200	2270	1390	2280	490	1200	1272

a Find the correlation coefficient, r, and comment on your result.

b Find the equation of the regression line.

c Use your equation to estimate the cost of a 1000 km train journey.

2 Different weights were attached to a vertical spring and the length of the spring measured. The results are shown in the table.

Load (x kg)	0	2	3	5	6	7	9	11
Length (y cm)	15	16.5	17.5	18.5	18.8	19.2	20	20.4

a Find the correlation coefficient, r.
b Find the equation of the regression line.
c Use your equation to estimate the length of the spring when a weight of 8 kg is added.

3 Lijn is a keen swimmer. For his Mathematical Studies Project he wants to investigate whether or not there is a correlation between the length of the arm of a swimmer and the time it takes them to swim 200 m.
He selects 15 members of a swimming club to swim 200 m.
Their times (y seconds) and arm lengths (x cm) are shown in the table below.

Length of arm (x cm)	78	72	74	67	79	58	62	67	71	69	75	65	73	59	60
Time (y seconds)	130	135	132	143	133	148	140	139	135	145	129	140	130	145	142

a Calculate the mean and standard deviation of x and y.
b Calculate the correlation coefficient, r.
c Comment on your value for r.
d Calculate the equation of the regression line for y on x.
e Using your equation, estimate how many seconds it will take a swimmer with an arm length of 70 cm to swim 200 m.

4 Saif asked his classmates how many minutes it took them to travel to school and their stress level, out of 10, for this journey. The results are shown in the table.

Travel time (x minutes)	14	28	19	22	24	8	16	5	18	20	25	10
Stress level (y)	3	7	5	6	6	2	3	2	4	5	6	6

a Find the correlation coefficient, r.
b Find the equation of the regression line.
c Use your equation to estimate the stress level of a student who takes 15 minutes to travel to school.

5 The table shows the weight (g) and the cost (Australian dollars) of various candy bars.

Weight (xg)	62	84	79	65	96	58	99	48	73	66
Cost (y AUD)	1.45	1.83	1.78	1.65	1.87	1.42	1.82	1.15	1.64	1.55

   a Calculate the equation of the regression line for y on x.
   b Use your equation to estimate the cost of a candy bar weighing 70 g.

6 Ten students in Mr Craven's PE class did pushups and situps. Their results are shown in the following table.

Number of pushups (x)	23	19	31	53	34	46	45	22	39	27
Number of situps (y)	31	26	35	51	36	48	45	28	41	30

   a Find the equation of the regression line.

   A student can do 50 pushups.

   b Use your equation to estimate the number of situps the student can do.

7 Fifteen students were asked for their average grade at the end of their last year of high school and their average grade at the end of their first year at university. The results are shown in the table below.

High school grade (x)	44	49	53	47	52	58	67	73	75	79	82	86	88	91	97
University grade (y)	33	52	55	48	51	60	71	72	69	83	84	89	96	92	89

   a Find the equation of the regression line.

   A student scores 60 at the end of their last year of high school.

   b Use your equation to estimate the average university grade for the student.

8 A secretarial agency has a new computer software package. The agency records the number of hours it takes people of different ages to master the package. The results are shown in the table.

Age (x)	32	40	21	45	24	19	17	21	27	54	33	37	23	45
Time (y hours)	10	12	8	15	7	8	6	9	11	16	12	13	9	17

   a Find the equation of the regression line.
   b Using your equation, estimate the time it would take a 40-year-old person to master the package.

5.4 The chi-squared test

You may be interested in finding out whether or not certain sets of data are independent. Suppose you collect data on the favorite color of T-shirt for men and women. You may want to find out whether color and gender are independent or not. One way to do this is to perform a **chi-squared test (χ^2)** for independence.

To perform a chi-squared test (χ^2) there are four main steps.

Step 1: Write the **null (H_0)** and **alternative (H_1) hypotheses**.

H_0 states that the data sets are independent.
H_1 states that the data sets are not independent.

For example, the hypotheses for color of T-shirt and gender could be:
H_0: Color of T-shirt is independent of gender.
H_1: Color of T-shirt is not independent of gender.

Step 2: Calculate the chi-squared test statistic.

Firstly, you may need to put the data into a **contingency table**, which shows the frequencies of two variables. The elements in the table are the **observed** data. The elements should be frequencies (not percentages).

For the example above, the contingency table could be:

	Black	White	Red	Blue	Totals
Male	48	12	33	57	150
Female	35	46	42	27	150
Totals	83	58	75	84	300

If you are given the contingency table, you may need to extend it to include an extra row and column for the 'Totals'.

From the observed data, you can calculate the **expected frequencies**. Since you are testing for independence, you can use the formula for the probability of independent events to calculate the expected values. So:

The expected number of men who like black T-shirts is
$\frac{150}{300} \times \frac{83}{300} \times 300 = 41.5$.

The expected number of men who like white T-shirts is
$\frac{150}{300} \times \frac{58}{300} \times 300 = 29$ and so on.

The expected table of values would then look like this:

	Black	White	Red	Blue	Totals
Male	41.5	29	37.5	42	150
Female	41.5	29	37.5	42	150
Totals	83	58	75	84	300

> When two variables are independent, one does not affect the other. Here, you are finding out whether a person's gender influences their colour choice. You will learn more about mathematical independence in Chapter 8.

> The main entries in this table form a 2 × 4 **matrix** (array of numbers) - do not include the row and column for the totals.

> In examinations, the largest contingency table will be a 4 × 4.

> **Note:**
> - The expected values can **never** be less than 1.
> - The expected values must be 5 or higher.
> - If there are entries between 1 and 5, you can combine table rows or columns.

For calculations by hand, you need the expected frequencies to find the χ^2 value.

> → To calculate the χ^2 value use the formula $\chi^2_{\text{calc}} = \sum \dfrac{(f_o - f_e)^2}{f_e}$,
>
> where f_o are the observed frequencies and f_e are the expected frequencies.

For our example,

$$\chi^2_{\text{calc}} = \frac{(48-41.5)^2}{41.5} + \frac{(12-29)^2}{29} + \frac{(33-37.5)^2}{37.5} + \frac{(57-42)^2}{42} + \frac{(35-41.5)^2}{41.5}$$

$$+ \frac{(46-29)^2}{29} + \frac{(42-37.5)^2}{37.5} + \frac{(27-42)^2}{42}$$

$$= 33.8$$

In examinations, you will only be expected to use your GDC to find the χ^2 value.

Using your GDC to find the χ^2 value, enter the contingency table as a matrix (array) and then use the matrix with the χ^2 2-way test.

```
< 1.1 >            Stats Apps ▼         ⬛⬛
[48  12  33  57] →mat   [48  12  33  57]  ⬆
[35  46  42  27]        [35  46  42  27]

χ²2way mat: stat.results
            [ "Title"        "χ² 2-way Test" ]
            [ "χ²"           33.7615         ]
            [ "PVal"         2.22473E-7      ]
            [ "df"           3.              ]
            [ "ExpMatrix"    "[...]"         ]
            [ "CompMatrix"   "[...]"         ]
                                        ▼
                              ↖ 3/99
```

Your GDC calculates the expected values for you but you must know how to find them by hand in case you are asked to show one or two calculations in an exam question. To see the matrix for the expected values, type 'stat.' and then select 'expmatrix' from the menu that pops up.

From the screenshot, you can see that $\chi^2_{\text{calc}} = 33.8$ (to 3 sf). This confirms our earlier hand calculation.

Step 3: Calculate the critical value.

First note the **level of significance**. This is given in examination questions but you have to decide which level to use in your project. The most common levels are 1%, 5% and 10%.

Now you need to calculate the number of **degrees of freedom**.

> → To find the degrees of freedom for the chi-squared test for independence, use this formula based on the contingency table:
>
> Degrees of freedom = (number of rows − 1) (number of columns − 1)

GDC help on CD: *Alternative demonstrations for the TI-84 Plus and Casio FX-9860GII GDCs are on the CD.*

If the number of degrees of freedom is 1, you will be expected to use **Yates' continuity correction** to work out the chi-squared value. (In examinations the degrees of freedom will always be greater than 1.)

So, in our ongoing example, the number of degrees of freedom is
$(2 - 1) \times (4 - 1) = 3$

The level of significance and degrees of freedom can be used to find the critical value. However, in examinations, the **critical value** will always be given.

For our example, at the 1% level, the critical value is 11.345. At the 5% level, the critical value is 7.815. At the 10% level, the critical value is 6.251.

Step 4: Compare χ^2_{calc} against the critical value.

> → If χ^2_{calc} is **less than** the critical value then **do not reject** the null hypothesis.
> If χ^2_{calc} is **more than** the critical value then **reject** the null hypothesis.

In our example, at the 5% level, 33.8 > 7.815. Therefore, we reject the null hypothesis that T-shirt color is independent of gender.

Using a GDC, you can compare the p-value against the significance level.

> → If the p-value is **less** than the significance level then **reject** the null hypothesis.
> If the p-value is **more** than the significance level then **do not reject** the null hypothesis.

The *p*-value is the probability value. It is the probability of evidence against the null hypothesis.

Use the significance level as a decimal, so 1% = 0.01, 5% = 0.05 and 10% = 0.1.

So, for our example, p-value = 0.000 000 2 (see the GDC screenshot on page 234).

0.000 000 2 < 0.05, so we reject the null hypothesis.

> → **To perform a χ^2 test:**
> 1 Write the null (H_0) and alternative (H_1) hypotheses.
> 2 Calculate χ^2_{calc}:
>   **a** using your GDC (examinations)
>   **b** using the χ^2_{calc} formula (project work)
> 3 Determine:
>   **a** the p-value by using your GDC
>   **b** the critical value (given in examinations)
> 4 Compare:
>   **a** the p-value against the significance level
>   **b** χ^2_{calc} against the critical value

Investigation – shoe size and gender

Use the information that you collected at the beginning of this chapter to test if shoe size is independent of gender.

Example 13

One hundred people were interviewed outside a chocolate shop to find out which flavor of chocolate cream they preferred. The results are given in the table, classified by gender.

	Strawberry	Coffee	Orange	Vanilla	Totals
Male	23	18	8	8	57
Female	15	6	12	10	43
Totals	38	24	20	18	100

Perform a χ^2 test, at the 5% significance level, to determine whether the flavor of chocolate cream is independent of gender.

a State the null hypothesis and the alternative hypothesis.

b Show that the expected frequency for female and strawberry flavor is approximately 16.3.

c Write down the number of degrees of freedom.

d Write down the χ^2_{calc} value for this data.

The critical value is 7.815.

e Using the critical value or the p-value, comment on your result.

Answers

a H_0: Flavor of chocolate cream is independent of gender.
H_1: Flavor of chocolate cream is not independent of gender.

Write H_0 using 'independent of'.
Write H_1 using 'not independent of'.

b $\dfrac{43}{100} \times \dfrac{38}{100} \times 100 = 16.34$

So, the expected frequency for female and strawberry flavor is approximately 16.3.

From the contingency table:
Total for 'female' row = 43
Total for 'strawberry' column = 38
Total surveyed = 100

c Degrees of freedom = $(2 - 1)(4 - 1) = 3$

Degrees of freedom = (number of rows − 1) (number of columns − 1)
Here, there are 2 rows and 4 columns in the observed matrix of the contingency table.

d $\chi^2_{calc} = 6.88$

Using your GDC:
Enter the contingency table as a matrix. Use the matrix with χ^2 2-way test. Read off χ^2 value.
The p-value = 0.0758.

e 6.88 < 7.815; therefore, we do not reject the null hypothesis. There is enough evidence to conclude that flavor of chocolate cream is independent of gender.

Using the given critical value, check:
χ^2_{calc} < critical value → do not reject, or
χ^2_{calc} > critical value → reject.

Or, using the p-value, check:
p-value < significance level → reject, or
p-value > significance level → do not reject.
Significance level = 5% = 0.05. So, 0.0758 > 0.05 and we do not reject the null hypothesis.

Example 14

Members of a club are required to register for one of three games: billiards, snooker or darts.

The number of club members of each gender choosing each game in a particular year is shown in the table.

	Billiards	Snooker	Darts
Male	39	16	8
Female	21	14	17

Perform a χ^2 test, at the 10% significance level, to determine if the chosen game is independent of gender.

a State the null hypothesis and the alternative hypothesis.

b Show that the expected frequency for female and billiards is approximately 27.1.

c Write down the number of degrees of freedom.

d Write down the χ^2_{calc} value for this data.

The critical value is 4.605.

e Using the critical value or the *p*-value, comment on your result.

Answers

a H_0: The choice of game is independent of gender.
H_1: The choice of game is not independent of gender.

b $\left(\dfrac{52}{115}\right)\left(\dfrac{60}{115}\right)(115) = 27.130$

≈ 27.1

So, the expected frequency for female and billiards is approximately 27.1.

c Degrees of freedom =
$(2 - 1)(3 - 1) = 2$

d $\chi^2_{calc} = 7.79$

e 7.79 > 4.605; therefore, we reject the null hypothesis. There is enough evidence against H_0 to conclude that the choice of game is not independent of gender.

Expected value table from the GDC:

	Billiards	Snooker	Darts
Male	32.9	16.4	13.7
Female	27.1	13.6	11.3

The p-value = 0.0203
Or, using the p-value,
0.0203 < 0.10. Therefore, we reject the null hypothesis.

Exercise 5H

1 300 people were interviewed and asked which genre of books they mostly read. The results are given below in a table of observed frequencies, classified by age.

		Genre			
		Fiction	Non-fiction	Science fiction	Totals
Age	0–25 years	23	16	41	80
	26–50 years	54	38	38	130
	51+ years	29	43	18	90
	Totals	106	97	97	300

Perform a χ^2 test, at the 5% significance level, to determine whether genre of book is independent of age.

a State the null hypothesis and the alternative hypothesis.

b Show that the expected frequency for science fiction and the 26–50 age group is 42.

c Write down the number of degrees of freedom.

d Write down the χ^2_{calc} value for this data.

The critical value is 9.488.

e Using the critical value or the p-value, comment on your result.

2 Tyne was interested in finding out whether natural hair color was related to eye color. He surveyed all the students at his school. His observed data is given in the table below.

		Hair color			
		Black	Brown	Blonde	Totals
Eye color	Brown/Black	35	43	12	90
	Blue	8	27	48	83
	Green	9	20	25	54
	Totals	52	90	85	227

Perform a chi-squared test, at the 10% significance level, to determine if hair color and eye color are independent.

a State the null hypothesis and the alternative hypothesis.

b Find the expected frequency of a person having blonde hair and brown eyes.

c Write down the number of degrees of freedom.

d Write down the chi-squared value for this data.

The critical value is 7.779.

e Using the critical value or the p-value, comment on your result.

3 Three different flavors of dog food were tested on different breeds of dog to find out if there was any connection between favorite flavor and breed. The results are given in the table.

	Beef	Chicken	Fish	Totals
Poodle	13	11	8	32
Boxer	15	10	10	35
Terrier	16	12	9	37
Great Dane	17	11	8	36
Totals	61	44	35	140

A χ^2 test, at the 5% significance level, is performed to investigate the results.

a State the null hypothesis and the alternative hypothesis.

b Show that the expected frequency of a Boxer's favorite food being chicken is 11.

c Show that the number of degrees of freedom is 6.

d Write down the χ^2_{calc} value for this data.

The critical value is 12.59.

e Using the critical value or the p-value, comment on your result.

4 Eighty people were asked to identify their favorite film genre. The results are given in the table below, classified by gender.

	Adventure	Crime	Romantic	Sci-fi	Totals
Male	15	12	2	12	41
Female	7	9	18	5	39
Totals	22	21	20	17	80

A χ^2 test, at the 1% significance level, is performed to decide whether film genre is independent of gender.

a State the null hypothesis and the alternative hypothesis.

b Show that the expected frequency of a female's favorite film genre being crime is 10.2.

c Write down the number of degrees of freedom.

d Write down the chi-squared value for this data.

The critical value is 11.345.

e Using the critical value or the p-value, comment on your result.

5 Kyu Jin was interested in finding out whether or not the number of hours spent playing computer games per week had an influence on school grades. He collected the following information.

	Low grades	Average grades	High grades	Totals
0–9 hours	6	33	57	96
10–19 hours	11	35	22	68
> 20 hours	23	22	11	56
Totals	40	90	90	220

Perform a chi-squared test, at the 5% significance level, to decide whether the grade is independent of the number of hours spent playing computer games.

a State the null hypothesis and the alternative hypothesis.
b Show that the expected frequency of a high grade and 0–9 hours of playing computer games is 39.3.
c Show that the number of degrees of freedom is 4.
d Write down the χ^2_{calc} value for this data.

The critical value is 9.488.

e Using the critical value or the p-value, comment on your result.

6 The local authority conducted a survey in schools in Rotterdam to determine whether the employment grade in the school was independent of gender. The results of the survey are given in the table.

	Directors	Management	Teachers	Totals
Male	26	148	448	622
Female	6	51	1051	1108
Totals	32	199	1499	1730

Perform a χ^2 test, at the 10% significance level, to determine whether the employment grade is independent of gender.

a State the null hypothesis and the alternative hypothesis.
b Write down the table of expected frequencies.
c Write down the number of degrees of freedom.
d Write down the chi-squared value for this data.

The critical value is 4.605.

e Using the critical value or the p-value, comment on your result.

7 Ayako had a part-time job working at a sushi restaurant. She calculated the average amount of sushi sold per week to be 2000. She decided to find out if there was a relationship between the day of the week and the amount of sushi sold. Her observations are given in the table.

	< 1700	1700–2300	> 2300	Totals
Monday–Wednesday	38	55	52	145
Thursday–Friday	39	65	55	159
Saturday–Sunday	43	60	63	166
Totals	120	180	170	470

Perform a χ^2 test, at the 5% significance level, to determine whether the amount of sushi sold is independent of the day of the week.

a State the null hypothesis and the alternative hypothesis.
b Show that the expected frequency of selling over 2300 sushi on Monday–Wednesday is 52.4.
c Write down the number of degrees of freedom.
d Write down the χ^2_{calc} value for this data.

The critical value is 9.488.

e Using the critical value or the p-value, comment on your result.

8 Haruna wanted to investigate the connection between the weight of dogs and the weight of their puppies. Her observed results are given in the table.

		Puppy			
		Heavy	**Medium**	**Light**	**Totals**
	Heavy	23	16	11	50
Dog	**Medium**	10	20	16	46
	Light	8	15	22	45
	Totals	41	51	49	141

Perform a χ^2 test, at the 1% significance level, to determine whether a puppy's weight is independent of its parent's weight.

a State the null hypothesis and the alternative hypothesis.
b Show that the expected frequency of a medium dog having a heavy puppy is 13.4.
c Write down the number of degrees of freedom.
d Write down the χ^2_{calc} value for this data.

The critical value is 13.277.

e Using the critical value or the p-value, comment on your result.

Extension material on CD:
Worksheet 5 - Useful statistical techniques for the project

Review exercise

Paper 1 style questions

1 It is stated that the content of a can of drink is 350 ml. The content of thousands of cans is tested and found to be normally distributed with a mean of 354 ml and a standard deviation of 2.5 ml.

  a Sketch a normal distribution diagram to illustrate this information.

  b Find the probability that a can contains less than 350 ml.

  100 cans are chosen at random.

  c Find the expected number of cans that contain less than 350 ml.

2 6000 people were asked how far they lived from their work. The distances were normally distributed with a mean of 4.5 km and a standard deviation of 1.5 km.

  a Find the percentage of people who live between 2 km and 4 km from their work.

  b Find the expected number of people who live less than 1 km from their work.

3 The weights of bags of tomatoes are normally distributed with a mean of 1.03 kg and a standard deviation of 0.02 kg.

  a Find the percentage of bags that weigh more than 1 kg.

  It is known that 15% of the bags weigh less than p kg.

  b Find the value of p.

4 For each diagram, state the type of correlation.

  a   **b**

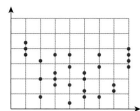

  c

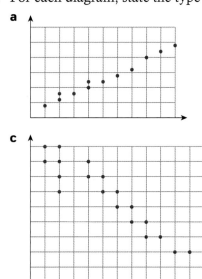

5 Plot these points on a diagram.

x	6	8	10	12	14	16
y	20	21	24	27	28	30

a State the nature of the correlation.
b Find the mean of the x-values and the mean of the y-values.
Plot this mean point on your diagram.
c Draw the line of best fit by eye.
d Find the expected value for y when x = 9.

6 The heights and arm lengths of 10 people are shown in the table.

Height (cm)	145	152	155	158	160	166	172	179	183	185
Arm length (cm)	38	42	45	53	50	59	61	64	70	69

a Find the correlation coefficient, r, and comment on your result.
b Write down the equation of the regression line.
c Use your equation to estimate the arm length of a person of height 170 cm.

7 The time taken to eat three doughnuts and the person's age is recorded in the table.

Age (years)	8	12	15	18	21	30	33	35	44	52	63	78
Time (seconds)	23	21	17	14	15	18	20	21	23	25	27	35

a Find the correlation coefficient, r, and comment on your result.
b Write down the equation of the regression line.
c Use your equation to estimate the time taken by a 40-year-old to eat three doughnuts.

8 100 people are asked to identify their favorite flavor of ice cream.
The results are given in the contingency table, classified by age (x).

	x < 25	25 ≤ x < 45	x ≥ 45	Totals
Vanilla	14	13	10	37
Strawberry	11	9	8	28
Chocolate	13	10	12	35
Totals	38	32	30	100

Perform a chi-squared test, at the 5% significance level, to determine whether flavor of ice cream is independent of age. State clearly the null and alternative hypotheses, the expected values and the number of degrees of freedom.

9 60 students go ten-pin bowling. They each have one throw with their right hand and one throw with their left. The number of pins knocked down each time is noted. The results are collated in the table.

	0–3	4–7	8–10	Totals
Right hand	8	28	24	60
Left hand	12	30	18	60
Totals	20	58	42	120

A χ^2 test is performed at the 10% significance level.

a State the null hypothesis.

b Write down the number of degrees of freedom.

c Show that the expected number of students who knock down 0–3 pins with their right hand is 10.

The p-value is 0.422.

d Write down the conclusion reached at the 10% significance level.

Give a clear reason for your answer.

10 Erland performs a chi-squared test to see if there is any association between the preparation time for a test (short time, medium time, long time) and the outcome (pass, does not pass). Erland performs this test at the 5% significance level.

a Write down the null hypothesis.

b Write down the number of degrees of freedom.

The p-value for this test is 0.069.

c What conclusion can Erland make?

Justify your answer.

Paper 2 style questions

1 The heights of Dutch men are normally distributed with a mean of 181 cm and a standard deviation of 9 cm.

a Sketch a normal distribution diagram to illustrate this information.

b Find the probability that a man chosen at random has a height less than 175 cm.

c Find the probability that a man chosen at random has a height between 172 cm and 192 cm.

Sixty men are measured.

d Find the expected number of men with a height greater than 195 cm.

It is known that 5% of the men have a height less than k cm.

e Find the value of k.

2 The weights of bags of sweets are normally distributed with a mean of 253 g and a standard deviation of 3 g.

   a Sketch a diagram to illustrate this information clearly.

   b Find the percentage of bags expected to weigh less than 250 g.

   Three hunderd bags are weighed.

   c Find the expected number of bags weighing more than 255 g.

3 The heights and weights of 10 students selected at random are shown in the table.

Height (x cm)	158	167	178	160	152	160	173	181	185	155
Weight (y kg)	50	75	80	46	61	69	64	86	74	68

   a Plot this information on a scatter graph. Use a scale of 1 cm to represent 25 cm on the x-axis and 1 cm to represent 10 kg on the y-axis.

   b Calculate the mean height.

   c Calculate the mean weight.

   d **i** Find the equation of the regression line.

      ii Draw the regression line on your graph.

   e Use your line to estimate the weight of a student of height 170 cm.

4 An employment agency has a new computer software package. The agency investigates the number of hours it takes people of different ages to reach a satisfactory level using this package. Fifteen people are tested and the results are given in the table.

Age (x)	33	41	22	46	25	18	16	23	26	55	37	34	25	48	17
Time (y hours)	8	10	7	16	8	9	7	10	12	15	11	14	10	16	7

   a Find the product-moment correlation coefficient, r, for these data.

   b What does the value of the correlation coefficient suggest about the relationship between the two variables?

   c Write down the equation of the regression line for y on x in the form $y = mx + c$.

   d Use your equation for the regression line to predict the time that it would take a 35-year-old person to reach a satisfactory level. Give your answer correct to the nearest hour.

5 Ten students were asked for their average grade at the end of their last year of high school and their average grade at the end of their first year at university. The results were put into a table as follows.

Student	High school grade, x	University grade, y
1	92	3.8
2	76	2.9
3	83	3.4
4	71	1.8
5	93	3.9
6	84	3.2
7	96	3.5
8	77	2.9
9	91	3.7
10	86	3.8

a Find the correlation coefficient, r, giving your answer to one decimal place.

b Describe the correlation between the high school grades and the university grades.

c Find the equation of the regression line for y on x in the form $y = mx + c$.

6 Several bars of chocolate were purchased and the following table shows the weight and the cost of each bar.

	Yum	Choc	Marl	Twil	Chuns	Lyte	BigM	Bit
Weight (x grams)	58	75	70	68	85	52	94	43
Cost (y euros)	1.18	1.45	1.32	1.05	1.70	0.90	1.53	0.95

a Find the correlation coefficient, r, giving your answer correct to two decimal places.

b Describe the correlation between the weight of a chocolate bar and its cost.

c Calculate the equation of the regression line for y on x.

d Use your equation to estimate the cost of a chocolate bar weighing 80 g.

7 The heights and dress sizes of 10 female students selected at random are shown in the table.

Height (x cm)	175	160	180	155	178	159	166	185	189	173
Dress size (y)	12	14	14	8	12	10	14	16	16	14

a Write down the equation of the regression line for dress size (y) on height (x), giving your answer in the form $y = ax + b$.

b Use your equation to estimate the dress size of a student of height 170 cm.

c Write down the correlation coefficient.

d Describe the correlation between height and dress size.

8 Members of a certain club are required to register for one of three games: badminton, table tennis or darts.

The number of club members of each gender choosing each game in a particular year is shown in the table.

	Badminton	Table tennis	Darts
Male	37	16	28
Female	32	10	19

Use a chi-squared test, at the 5% significance level, to test whether choice of game is independent of gender. State clearly the null and alternative hypotheses, the expected values and the number of degrees of freedom.

9 For his Mathematical Studies Project a student gave his classmates a questionnaire to find out which extra-curricular activity was the most popular. The results are given in the table below, classified by gender.

	Reading	Surfing	Skating	
Female	22	16	22	(60)
Male	14	18	8	(40)
	(36)	(34)	(30)	

The table below shows the expected values.

	Reading	Surfing	Skating
Female	p	20.4	18
Male	q	r	12

a Calculate the values of p, q and r.

The chi-squared test, at the 10% level of significance, is used to determine whether the extra-curricular activity is independent of gender.

b i State a suitable null hypothesis.

ii Show that the number of degrees of freedom is 2.

The critical value is 4.605.

c Write down the chi-squared statistic.

d Do you accept the null hypothesis? Explain your answer.

10 A company conducted a survey to determine whether position in upper management was independent of gender. The results of this survey are tabulated below.

	Managers	Junior executives	Senior executives	Totals
Male	135	90	75	300
Female	45	130	25	200
Totals	180	220	100	500

The table below shows the expected number of males and females at each level, if they were represented proportionally to the total number of males and females employed.

	Managers	Junior executives	Senior executives	Totals
Male	a	c	60	300
Female	b	d	40	200
Totals	180	220	100	500

a i Show that the expected number of male managers (a) is 108.
  ii Hence, write down the values of b, c and d.
b Write suitable null and alternative hypotheses for these data.
c i Find the chi-squared value.
  ii Write down the number of degrees of freedom.
  iii Given that the critical value is 5.991, what conclusions can be drawn regarding gender and position in upper management?

11 In the small town of Schiedam, population 8000, an election was held. The results were as follows.

	Urban voters	Rural voters
Candidate A	1950	1730
Candidate B	1830	1360
Candidate C	500	630

In **a–d** below, use a chi-squared test, at the 1% significance level, to decide whether the choice of candidate depends on where the voter lives.

H_0: The choice of candidate is independent of where the voter lives.

a Write down the alternative hypothesis.
b Show that the expected number of rural voters for candidate A is 1711.
c i Calculate the chi-squared value.
  ii Write down the number of degrees of freedom.

The critical value is 9.21.

d i State your conclusion.
  ii Explain why you reached your conclusion.

12 This table of observed results gives the number of candidates taking a Mathematics examination classified by gender and grade obtained.

	6 or 7	4 or 5	1, 2 or 3	Totals
Males	34	50	6	90
Females	40	60	10	110
Totals	74	110	16	200

The question posed is whether gender and grade obtained are independent.

a Show that the expected number of males achieving a grade of 4 or 5 is 49.5.

A chi-squared test is set up at the 5% significance level.

b i State the null hypothesis.

  ii State the number of degrees of freedom.

  iii Write down the chi-squared value.

The critical value is 5.991.

c What can you say about gender and grade obtained?

CHAPTER 5 SUMMARY
The normal distribution

- The **normal distribution** is the most important continuous distribution in statistics. It has these properties:
  - It is a bell-shaped curve.
  - It is symmetrical about the mean, μ. (The mean, the mode and the median all have the same value.)
  - The x-axis is an asymptote to the curve.
  - The total area under the curve is 1 (or 100%).
  - 50% of the area is to the left of the mean and 50% to the right.
  - Approximately 68% of the area is within 1 standard deviation, σ, of the mean.
  - Approximately 95% of the area is within 2 standard deviations of the mean.
  - Approximately 99% of the area is within 3 standard deviations of the mean.

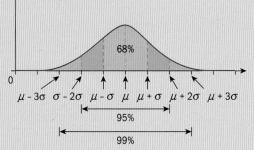

- The **expected value** is found by multiplying the number in the sample by the probability.

Continued on next page

Correlation

- In a **positive** correlation the dependent variable increases as the independent variable increases.

- In a **negative** correlation the dependent variable decreases as the independent variable increases.

- When the points are scattered randomly across the diagram there is **no** correlation.

- Correlations can also be described as strong, moderate or weak.
- To draw the **line of best** fit by eye:
  - Find the mean of each set of data and plot this point on your scatter diagram.
  - Draw a line that passes through the mean point and is close to all the other points – with approximately an equal number of points above and below the line.

- **Pearson's product-moment correlation coefficient**, r, can take all values between -1 and $+1$ inclusive.
  - When $r = -1$, there is a **perfect negative** correlation between the data sets.
  - When $r = 0$, there is **no** correlation.
  - When $r = +1$, there is a **perfect positive** correlation between the data sets.
  - A **perfect correlation** is one where **all** the plotted points lie on a straight line.

The regression line

- The **regression line for y on x** is a more accurate version of a line of best fit, compared to best fit by eye.
- If there is a strong or moderate correlation, you can use the regression line for y on x to predict values of y for values of x within the range of the data.

The chi-squared test

- To calculate the χ^2 value use the formula $\chi^2_{\text{calc}} = \sum \dfrac{(f_o - f_e)^2}{f_e}$, where f_o are the observed frequencies and f_e are the expected frequencies.

- To find the degrees of freedom for the chi-squared test for independence, use this formula based on the contingency table:
  Degrees of freedom = (number of rows − 1)(number of columns − 1)

Continued on next page

- If χ^2_{calc} is **less than** critical value, **do not reject** the null hypothesis.
  If χ^2_{calc} is **more than** critical value, **reject** the null hypothesis.
- If the p-value is **less than** significance level, **reject** the null hypothesis.
  If the p-value is **more than** significance level, **do not reject** the null hypothesis.
- To perform a χ^2 test:
  1 Write the null (H_0) and alternative (H_1) hypotheses.
  2 Calculate χ^2_{calc}: **a** using your GDC (examinations), or **b** using the χ^2_{calc} formula (project work).
  3 Determine: **a** the p-value using your GDC, or **b** the critical value (given in examinations).
  4 Compare: **a** the p-value against the significance level, or **b** χ^2_{calc} against the critical value.

Correlation or causation?

Correlation shows how closely two variables vary with each other.

Causation is when two variables directly affect each other.

Shaving less than once a day increases risk of stroke by 70%!

In 2003 British researchers found that there was a correlation between men's shaving habits and their risk of a stroke. This link emerged from a 20-year study of over 2,000 men aged 45–59 in Caerphilly, South Wales.

A strong **correlation** between two variables does not mean that one **causes** the other. There may be a cause and effect relation between the two variables, but you cannot claim this if they are only correlated. This is the **fallacy of correlation** – one of the most common logical fallacies.

Do you think a man could decrease his chance of having a stroke by shaving more? This seems silly, and suggests there might be a hidden intermediary variable at work.

In this case, the researchers think that shaving and stroke risk are linked by another variable – hormone levels. For example, testosterone has already been used to explain the link between baldness and a higher risk of heart disease.

If there is a correlation between two variables, be careful about assuming that there is a relationship between them. There may be no logical or scientific connection at all.

Analyse these examples of assumed correlation or causation.
Which illustrate the fallacy of correlation?

- Joining the military made me a disciplined and strong person
- I wore a hat today on my way to school and I was involved in a car accident; I will not be wearing that red hat again
- People who own washing machines are more likely to die in a car accident.

Anscombe's Quartet

Anscombe's Quartet is a group of four data sets that provide a useful caution against applying individual statistical methods to data without first graphing them. They have identical simple statistical properties (mean, variance, etc.) but look totally different when graphed.

▶ Francis Anscombe (1918–2001), British statistician.

- Find the mean of x, the mean of y, the variance of x and the variance of y and the r-value for each data set.

Set 1		Set 2		Set 3		Set 4	
x	y	x	y	x	y	x	y
4	4.26	4	3.1	4	5.39	8	6.58
5	5.68	5	4.74	5	5.73	8	5.76
6	7.24	6	6.13	6	6.08	8	7.71
7	4.82	7	7.26	7	6.42	8	8.84
8	6.95	8	8.14	8	6.77	8	8.47
9	8.81	9	8.77	9	7.11	8	7.04
10	8.04	10	9.14	10	7.46	8	5.25
11	8.33	11	9.26	11	7.81	8	5.56
12	10.84	12	9.13	12	8.15	8	7.91
13	7.58	13	8.74	13	12.74	8	6.89
14	9.96	14	8.1	14	8.84	19	12.5

1 Write down what you think the graphs and their regression lines will look like.

2 Using your GDC, sketch the graph of each set of points on a different graph.

3 Draw the regression line on each graph.

4 Explain what you notice.

6 Introducing differential calculus

CHAPTER OBJECTIVES:

7.1 Concept of the derivative as a rate of change; tangent to a curve

7.2 The principle that $f(x) = ax^n \Rightarrow f'(x) = anx^{n-1}$; the derivative of functions of the form $f(x) = ax^n + bx^{n-1} + \ldots$, where all exponents are integers

7.3 Gradients of curves for given values of x; values of x where $f'(x)$ is given; equation of the tangent at a given point; equation of the line perpendicular to the tangent at a given point (normal)

7.4 Values of x where the gradient of a curve is zero; solution of $f'(x) = 0$; stationary points; local maximum and minimum points

7.5 Optimization problems

Before you start

You should know how to:

1. Use function notation, e.g. If $f(x) = 3x + 7$ what is $f(2)$? $f(2) = 3 \times 2 + 7 = 13$

2. Rearrange formulae, e.g. Make x the subject of the formula:
$$y = 3x + 7$$
$$y - 7 = 3x \Rightarrow \frac{y-7}{3} = x$$

3. Use index notation, e.g. Write without powers
$$5^{-2} = \frac{1}{5^2} = \frac{1}{25}$$

4. Use the laws of indices, e.g. Simplify:
$$5^2 \times 5^4 = 5^{2+4} = 5^6$$
$$5^4 \div 5^6 = 5^{4-6} = 5^{-2}$$

5. Find the equation of a straight line given its gradient and a point, e.g. The line passing through the point $(2, 13)$ with gradient 3
$$(y - 13) = 3(x - 2)$$
$$y - 13 = 3x - 6$$
$$y = 3x + 7$$

Skills check

1. **a** $f(z) = 3 - 2z$, evaluate $f(5)$ and $f(-5)$
   b $f(t) = 3t + 5$, evaluate $f(2)$ and $f(-3)$
   c $g(y) = y^2$, evaluate $g(5)$ and $g\left(\frac{1}{2}\right)$
   d $g(z) = \frac{3}{z}$, evaluate $g(2)$ and $g(15)$
   e $f(z) = \frac{z^2}{z+1}$, evaluate $f(4)$ and $f(-3)$

2. Make r the subject of the formula:
   a $C = 2\pi r$ **b** $A = \pi r^2$ **c** $A = 4\pi r^2$
   d $V = \frac{\pi r^2 h}{3}$ **e** $V = \frac{2\pi r^3}{3}$ **f** $C = \frac{2A}{r}$

3. Write these without powers.
   a 4^2 **b** 2^{-3} **c** $\left(\frac{1}{2}\right)^4$

4. Write each expression in the form x^n:
   a $\frac{1}{x}$ **b** $\frac{1}{x^4}$ **c** $\frac{x^3}{x}$ **d** $\frac{x^2}{x^5}$ **e** $\frac{\left(x^2\right)^3}{x^5}$

5. Find the equation of the line that passes through
   a the point $(5, -3)$ with gradient 2
   b the point $(4, 2)$ with gradient -3.

The invention of the differential calculus, in the 17th century, was a milestone in the development of mathematics.

At its simplest it is a method of finding the gradient of a **tangent** to a curve. The gradient of the tangent is a measure of how quickly the function is changing as the x-coordinate changes.

All things move, for example, the hands on a clock, the sprinter in a 100 m race, the molecules in a chemical reaction, the share values on the stock market. Mathematics can be used to model all of these situations. Since each situation is dynamic, the models will involve differential calculus.

For more on the history of calculus, see pages 292–3.

In this chapter, you will investigate certain functions to discover for yourself the method of finding the gradient of a tangent to a curve, and check that this method can be applied to all similar curves. You will apply this technique in a variety of situations, to solve problems about graphs and to use mathematical models in 'real-world' problems.

In the photograph, all the cans have the same basic cylindrical shape. However, they are all different sizes. By the end of this chapter you will be able to determine the optimal design of a cylindrical can – one that uses the smallest amount of metal to hold a given capacity.

6.1 Introduction to differentiation

You have already met the concept of the gradient of a straight line. **Differentiation** is the branch of mathematics that deals with **gradient functions** of curves.

The gradient measures how fast y is increasing compared to the rate of increase of x.

The gradient of a straight line is constant, which means its direction never changes. The y-values increase at a constant rate.

> → If P is the point (a, b) and Q is (c, d) then the gradient, m, of the straight line PQ is $m = \dfrac{d - b}{c - a}$.

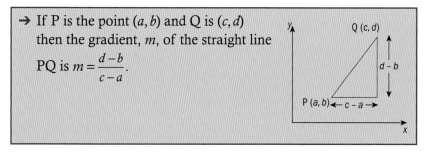

To calculate the gradient of a curve at a particular point you need to draw a tangent at that point. A tangent is a line that just touches the curve.

Here is the curve $y = x^2 - 4x + 7$.
It is a **quadratic** function. Its **vertex** is at the point $(2, 3)$.

The three tangents to the curve are shown in blue.

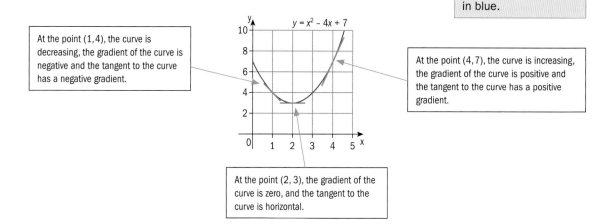

At the point $(1, 4)$, the curve is decreasing, the gradient of the curve is negative and the tangent to the curve has a negative gradient.

At the point $(4, 7)$, the curve is increasing, the gradient of the curve is positive and the tangent to the curve has a positive gradient.

At the point $(2, 3)$, the gradient of the curve is zero, and the tangent to the curve is horizontal.

The direction of a tangent to the curve changes as the x-coordinate changes. Therefore the gradient of the curve is not constant.

So, for any curve $y = f(x)$ which is not a straight line, its gradient changes for different values of x. The gradient can be expressed as a **gradient function**.

> → Differentiation is a method used to find the equation of the gradient function for a given function, $y = f(x)$.

Extension material on CD:
Worksheet 6 – More about functions

Investigation – tangents and the gradient function

The tangent to a graph at a given point is the straight line with its gradient equal to that of the curve **at that point**. If you find the gradient of the tangent, then you have also found the gradient of the curve at that point. Repeating this for different points, we can use the data obtained to determine the gradient function for the curve.

GDC instructions on CD:
These instructions are for the TI-Nspire GDC. Instructions for the TI-84 Plus and Casio FX-9860GII GDCs, and using a graph plotter, are on the CD.

1 **Plot the curve $y = x^2$ on your GDC**

Open a new document and add a Graphs page.

Save the document as 'Calculus'.

Enter x^2 into the function $f1(x)$.

Press enter del .

To get a better view of the curve, you should pan the axes in order to see more of it.

Click down on the touchpad in an area away from the axes, function or any labels.

The ➤ will change to 🖐.

Move the grasping hand with the touchpad. The window view will pan with it.

Click the touchpad when the window is in the required position.

2 **Add a tangent to the curve**

Press menu 7:Points & Lines | 7:Tangent

Press enter .

Move the ➤ with the touchpad towards the curve. It will change to a 🖐 and the curve will be highlighted.

Click the touchpad.

Choose a point on the curve by clicking the touchpad.

Now you have a tangent drawn at a point on the curve that you can move round to any point on the curve. To get some more information about the tangent, you need the coordinates of the point and the equation of the tangent.

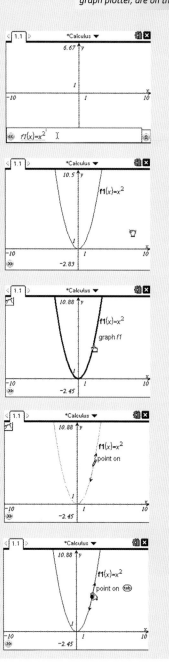

Continued on next page

Move the ↖ with the touchpad towards the point.
It will change to a ☜ and you will see 'point on tab '.
Press ctrl menu and select 7: Coordinates and Equations. Press enter .

3 Find the equation of the tangent.

Move the ↖ with the touchpad towards
the arrow at the end of the tangent.
It will change to a ☜ and you
will see 'line tab '.

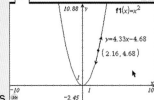

Press ctrl menu and select 7:Coordinates
and Equations. Press enter .
You should now have the coordinates
of the point and the equation of
the tangent labeled.

**4 Edit the *x*-coordinate so that the
point moves to (1, 1)**

Move the ↖ with the touchpad towards
the arrow at the *x*-coordinate of the
point. It will change to a ☜ and you
will see the numbers lighten and the
word 'text' appears.

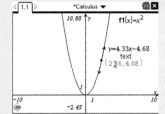

If you move the ↖ very slightly with the
touchpad it will change to a I. When
it does, click the touchpad.

> This is quite tricky
> and may take a bit
> of practice. If it does
> not work, press esc
> and start again.

The *x*-coordinate is now ready
for editing.

Use the del key to delete the current
value and type 1 . Press enter .

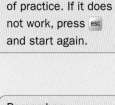

You have drawn the tangent to the
curve $y = x^2$ at the point (1, 1)

Its equation is $y = 2x - 1$, so
gradient of the tangent is 2.

> Remember:
> In the equation of
> a straight line
> $y = mx + c$, *m* is the
> gradient

5 Record this information in a table.

$y = x^2$

x-coordinate	−3	−2	−1	0	1	2	3	4	x
Gradient of tangent					2				

Worksheet on CD: *This table
is Worksheet 6.1 on the CD.*

6 Complete the table

Go back to the graph and edit the *x*-coordinate again. Change it to 2.
Write the gradient of the tangent at the point where the *x*-coordinate is
2 in your table. Repeat this until you have completed the table for all
values of *x* between −3 and 4.

Continued on next page

7   Look for a simple formula that gives the gradient of the tangent for any value of *x*

Write this formula in the bottom right cell in your copy of the table. Is this formula valid for all values of *x*? Try positive, negative and fractional values.

8   Repeat Steps 1–7 for the curve $y = 2x^2$

Draw the curve, then the tangents and complete this table.

$y = 2x^2$

x-coordinate	−3	−2	−1	0	1	2	3	4	x
Gradient of tangent									

Worksheet on CD: This table is Worksheet 6.1 on the CD.

Again, look for a simple formula that gives the gradient of the tangent for any value of x. Write it down.

You can repeat this process for other curves, but there is an approach that will save time. The formulae you found in the investigation are called the **gradient functions** of the curves. The gradient function can be written in several ways:

$$\frac{dy}{dx}, \quad \frac{d}{dx}(f(x)), \quad \text{or } f'(x).$$

You can use your GDC to draw a graph of the gradient function for any curve.

Investigation – GDC and the gradient function

1   Use the GDC to draw the gradient function of $y = 4x$

Add a new Graphs page to your document.

Enter 4x into the function *f*2(x).

Press enter .

GDC instructions on CD: These instructions are for the TI-Nspire GDC. Instructions for the TI-84 Plus and Casio FX-9860GII GDCs, and using a graph plotter, are on the CD.

2   Enter the gradient function in *f*3(*x*)

Click the ⟩⟩ symbol using the touchpad to open the entry line at the bottom of the work area.

Press ⬚|⬚ and use the ◀▶▲▼ keys to select the $\frac{d}{d\square}$ template.

Press enter .

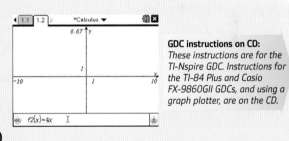

Enter x and *f*2 (x) in the template as shown.
Press enter .

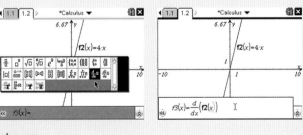

Continued on next page

You should have this diagram, with a horizontal line across the graph.

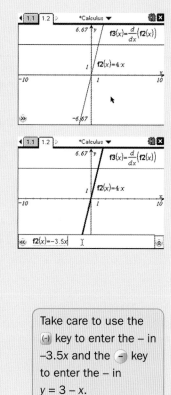

The graph plotter gives you a picture of the gradient function – you have to find the equation of this function.

The GDC drew the line $y = 4$.

The gradient of the line $y = 4x$ is '4'.

3 **Repeat for other functions**

Click the » symbol using the touchpad to open the entry line at the bottom of the work area.

Use the ▲ key to select $f2(x)$.

Enter a new function to replace $4x$.

In this way find the gradient functions for these straight lines.

a $y = -3.5x$

b $y = 2x + 4$

c $y = 5$

d $y = 3 - x$

e $y = -3.5$

f $y = 2 - \dfrac{1}{2}x$

> Take care to use the (-) key to enter the – in –3.5x and the ⊖ key to enter the – in $y = 3 - x$.

4 **Change the function to $y = x^2$**

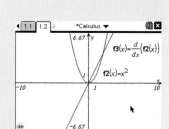

A straight line will appear on your screen as in the diagram on the right.

Write down the equation of this new straight line.

The GDC drew the line $y = 2x$.

The gradient function of the curve $y = x^2$ is '2x'.

This is the same result that you found by observation in the previous investigation.

Repeat for the curves $y = 2x^2$ and $y = 3x^2$ and write down the gradient functions for these curves.

5 **Tabulate your results**

You are now building up a set of results that you can use to generalize. To help with this, summarize your findings in a table. You should be able to see patterns in the results.

Continued on next page

Curve	$y = 4x$	$y = -3.5x$	$y = 2x + 4$	$y = 5$	$y = 3 - x$	$y = -3.5$	$y = 2 - \dfrac{1}{2}x$
Gradient function	4						

Curve	$y = x^2$	$y = 2x^2$	$y = 3x^2$	$y = 4x^2$	$y = -x^2$	$y = -2x^2$	$y = \dfrac{1}{2}x^2$
Gradient function	$2x$						

6 Extend your results

Complete this table for the curve $y = x^2 + 3x$ using the method from the first Investigation, on page 257.

Worksheet on CD: *This table is Worksheet 6.2 on the CD.*

$y = x^2 + 3x$

x-coordinate	−3	−2	−1	0	1	2	3	4
Gradient of tangent								

What is the algebraic rule that connects the answers for the gradient to the x-coordinates?

Check that your answer is correct by entering $x^2 + 3x$ in $f2(x)$ in the graphs page (Step 2 of this investigation) so that the GDC draws the gradient function.

What is the equation of this straight line?
Is its equation the same as the rule you found?

Use your GDC to find the gradient functions for the curves below. Look for a pattern developing.

> These should be the same! If they are not, check with your teacher.

a $y = x^2 + 3x$ **b** $y = x^2 - 5x$ **c** $y = 2x^2 - 3x$ **d** $y = 3x^2 - x$

e $y = 5x - 2x^2$ **f** $y = 2x - x^2$ **g** $y = x^2 + 4$ **h** $y = x^2 - 2$

i $y = 3 - x^2$ **j** $y = x^2 + x - 2$ **k** $y = 2x^2 - x + 3$ **l** $y = 3x - x^2 + 1$

Compare each curve to its gradient function and so determine the formula for the gradient function for the general quadratic curve

$$y = ax^2 + bx + c$$

Write down the gradient functions of the following curves **without using the GDC**.

1 $y = 5x^2 + 7x + 3$

2 $y = 5x + 7x^2 - 4$

3 $y = 3 + 0.5x^2 - 6x$

4 $y = 4 - 1.5x^2 + 8x$

> Do not proceed until you have answered these questions correctly.

Investigation – the gradient function of a cubic curve

GDC instructions on CD:
These instructions are for the TI-Nspire GDC. Instructions for the TI-84 Plus and Casio FX-9860GII GDCs, and using a graph plotter, are on the CD.

Now consider the simplest cubic curve $y = x^3$.

Change the function to $y = x^3$ using the GDC.

To enter x^3, press ⓧ ⌃ ③ ▶.

(You will need to press the ▶ key to get back to the base line from the exponent.)

This time a curve appears, instead of a straight line.

Find the equation of the curve.

This is the gradient function of $y = x^3$.

Once you have the equation of the curve, find the gradient function of $y = 2x^3$, $y = 3x^3$, …
Write down your answers in the worksheet copy of the table.

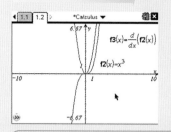

> Have a guess at the equation of the curve. Enter your guess to the gradient function. Adjust your equation until it fits. Then delete it.

Curve	$y = x^3$	$y = 2x^3$	$y = 3x^3$	$y = 4x^3$	$y = -x^3$	$y = -2x^3$	$y = \frac{1}{2}x^3$
Gradient function							

Extend your investigation so that you can find the gradient function of **any** cubic.
Be systematic, so try simple cubic curves first…

Worksheet on CD: *This table is Worksheet 6.3 on the CD.*

Curve	$y = x^3 + 4$	$y = 2x^3 - 3$	$y = x^3 + 5x$	$y = x^3 - 2x$	$y = x^3 + 2x^2$	$y = 2x^3 + \frac{1}{2}x^2$
Gradient function						

Then move on to more complicated cubic curves…

Curve	$y = x^3 + 3x^2 + 2$	$y = x^3 + 4x^2 + 3x$	$y = x^3 + 5x^2 - 4x + 1$	$y = x^3 - x^2 - 5x - 4$
Gradient function				

Generalize your results to determine the formula for the gradient function for the general cubic curve $y = ax^3 + bx^2 + cx + d$

You now have results for the gradient functions of linear functions, quadratic functions and cubic functions. Complete the worksheet copy of the table with these.

Function	Formula	Gradient function
Constant	$y = a$	
Linear	$y = ax + b$	
Quadratic	$y = ax^2 + bx + c$	
Cubic	$y = ax^3 + bx^2 + cx + d$	

Continued on next page

Investigation – the gradient function of any curve

In this investigation you find the gradient function of **any** curve.

Again, take a systematic approach.

1 Find the gradient function of $y = x^4$
2 Find the gradient function of $y = x^5$
3 Generalize these results to find the gradient function of $y = x^n$

GDC instructions on CD:
These instructions are for the TI-Nspire GDC. Instructions for the TI-84 Plus and Casio FX-9860GII GDCs, and using a graph plotter, are on the CD.

Up to this point, all the powers in your curve have been **positive**.

Consider the curves $y = \dfrac{1}{x}$, $y = \dfrac{1}{x^2}$, $y = \dfrac{1}{x^3}$, ... as well.

To enter $\dfrac{1}{x}$ on your GDC use the ⊞ key and select ⊕ from the template menu.

Remember
that $\dfrac{1}{x} = x^{-1}$

The final result

Function	Gradient function
$y = ax^n$	

Check this result with your teacher. Do not go on until you have done so.

The process of finding the gradient function of a curve is known as **differentiation.** In these investigations, you have learned for yourself how to differentiate.

Finding this result by investigation is not the same as *proving* it to be true. How, without proof, do we know that a result arrived at by pattern building is **always** true?

6.2 The gradient function

Differentiation is the algebraic process used to find the gradient function of a given function.

Two forms of notation are used for differentiation. The notation that you use will depend on the notation used in the question.

Calculus was discovered at almost the same time by both the British mathematician Isaac Newton (1642–1727), and the German mathematician Gottfried Leibniz (1646–1716). The controversy over the rival claims lasted for decades.

→ To differentiate a function, find the gradient function:

Function	Gradient function
$y = ax^n$	$\dfrac{dy}{dx} = nax^{n-1}$
$f(x) = ax^n$	$f'(x) = nax^{n-1}$

The process is valid for **all** values of n, both positive and negative.

The $\dfrac{dy}{dx}$ notation was developed by Leibniz. Newton's notation is now only used in physics. How important is mathematical notation in enhancing your understanding of a subject?

Example 1

Given $y = 4x^7$, find $\dfrac{dy}{dx}$.

Answer

$\dfrac{dy}{dx} = 7 \times 4x^{7-1}$

$\dfrac{dy}{dx} = 28x^6$

$y = ax^n$

$\dfrac{dy}{dx} = nax^{n-1}$

$a = 4,\ n = 7$

Example 2

Given $f(x) = 3x^5$, find $f'(x)$.

Answer

$f'(x) = 5 \times 3x^{5-1}$

$f'(x) = 15x^4$

$f(x) = ax^n$

$f'(x) = nax^{n-1}$

$a = 3,\ n = 5$

> The $f'(x)$ notation is from Euler (1707–83), who was perhaps the greatest mathematician of all.

Example 3

Given $f(x) = 3x - 4x^2 + x^3$, find $f'(x)$.

Answer

$f'(x) = 3x^{1-1} - 2 \times 4x^{2-1} + 3 \times x^{3-1}$

$f'(x) = 3 - 8x + 3x^2$

Differentiate each term separately.

> Remember that $x^1 = x$ and that $x^0 = 1$.

Exercise 6A

1 Find $\dfrac{dy}{dx}$.

   a $y = 4x^2$ **b** $y = 6x^3$ **c** $y = 7x^4$ **d** $y = 5x^3$

   e $y = x^4$ **f** $y = 5x$ **g** $y = x$ **h** $y = 12x$

   i $y = 9x^2$ **j** $y = \dfrac{1}{2}x^3$ **k** $y = \dfrac{1}{2}x^2$ **l** $y = \dfrac{3}{4}x^4$

2 Differentiate

   a $y = 7$ **b** $y = -3x^3$ **c** $y = -\dfrac{1}{4}x^4$ **d** $y = -\dfrac{2}{3}x^3$

   e $y = -x$ **f** $y = -3$ **g** $y = 5x^6$ **h** $y = -7x^9$

   i $y = \dfrac{1}{2}x^8$ **j** $y = \dfrac{3}{4}x^{12}$ **k** $y = -\dfrac{2}{3}x^9$ **l** $y = \dfrac{3}{4}$

3 Find $f'(x)$.

   a $f(x) = 3x^2 + 5x^3$ **b** $f(x) = 5x^4 - 4x$

   c $f(x) = 9x - 11x^3$ **d** $f(x) = x^4 + 3x + 2$

4 Find y'

   a $y = 8 - 5x + 4x^6$ **b** $y = 9x^2 - 5x + \dfrac{1}{2}$

   c $y = 7x + 4x^5 - 101$ **d** $y = x(2x + 3)$

> y' is another way of writing $\dfrac{dy}{dx}$.

You can use letters other than x and y for the variables. This changes the notation but not the process.

Example 4

Given $v = 3.5t^8$, find $\dfrac{dv}{dt}$.

Answer

$\dfrac{dv}{dt} = 8 \times 3.5t^{8-1}$

$\dfrac{dv}{dt} = 28t^7$

$v = at^n$

$\dfrac{dv}{dt} = nat^{n-1}$

$a = 3.5,\ n = 8$

Example 5

Given $f(z) = \dfrac{3z^4}{2}$, find $f'(z)$.

Answer

$f(z) = \dfrac{3z^4}{2} = \dfrac{3}{2} \times z^4$

$f'(z) = 4 \times \dfrac{3}{2} z^{4-1}$

$f'(z) = 6z^3$

$f(z) = az^n$

$f'(z) = naz^{n-1}$

$a = \dfrac{3}{2},\ n = 4$

Example 6

Given $f(t) = (3t - 1)(t + 4)$, find $f'(t)$.

Answer

$f(t) = 3t^2 + 12t - t - 4$
$f(t) = 3t^2 + 11t - 4$
$f'(t) = 6t + 11$

Multiply out the brackets.

Differentiate each term separately.

Exercise 6B

1 Find $\dfrac{dA}{dt}$.

   a $A = 4t(9 - t^2)$ 　　　　**b** $A = 6(2t + 5)$

   c $A = t^2(t - 5)$ 　　　　　**d** $A = (t + 2)(2t - 3)$

   e $A = (5 - t)(3 + 2t)$ 　　**f** $A = (6t + 7)(3t - 5)$

   g $A = (t^2 + 3)(t - 1)$ 　　**h** $A = 3(t + 3)(t - 4)$

2 Find $f'(r)$.

   a $f(r) = \dfrac{1}{2}(r + 3)(2r - 6)$ 　　**b** $f(r) = (r + 3)^2$

   c $f(r) = (2r - 3)^2$ 　　　　　　　　**d** $f(r) = (5 - 2r)^2$

   e $f(r) = 3(r + 5)^2$ 　　　　　　　　**f** $f(r) = 5(7 - r)^2$

You can also differentiate functions which have powers of x in the denominator of a fraction. First you must write these terms using negative indices.

Example 7

Given $y = \dfrac{4}{x^2}$, find $\dfrac{dy}{dx}$.

Answer

$y = 4 \times \dfrac{1}{x^2} = 4x^{-2}$	*Write the function in index form:* $\dfrac{1}{x^2} = x^{-2}$.
$\dfrac{dy}{dx} = -2 \times 4x^{-2-1}$	$a = 4$ *and* $n = -2$
$\dfrac{dy}{dx} = -8x^{-3}$	*Remember the rules for multiplying negative numbers.*
$\dfrac{dy}{dx} = \dfrac{-8}{x^3}$	*Rewrite in the original form.*

Example 8

Given $f(x) = \dfrac{12}{5x^3}$, find $f'(x)$.

Answer

$f(x) = \dfrac{12}{5} \times \dfrac{1}{x^3} = \dfrac{12}{5}x^{-3}$	*Write the function in index form.* $a = \dfrac{12}{5}$ *and* $n = -3$
$f'(x) = -3 \times \dfrac{12}{5} \times x^{-3-1}$	*Be **very** careful with minus signs.*
$f'(x) = \dfrac{-36}{5} \times x^{-4}$	*Simplify.*
$f(x) = \dfrac{-36}{5x^4}$	*Rewrite in the original form.*

Exercise 6C

Differentiate the following with respect to x.

Remember to use the same notation as the question.

1 $y = \dfrac{3}{x^2}$ **2** $f(x) = \dfrac{2}{x^4}$ **3** $y = \dfrac{7}{x}$

4 $f(x) = \dfrac{2}{x^8}$ **5** $y = \dfrac{5}{x^7}$ **6** $y = 9 + \dfrac{2}{x}$

7 $f(x) = 7x^2 + \dfrac{4}{x^5}$ **8** $y = 7 - 4x + \dfrac{5}{2x^2}$ **9** $g(x) = x^3 + \dfrac{3}{x^2}$

10 $y = 4x - \dfrac{3}{x}$ **11** $g(x) = 5x^3 - \dfrac{1}{x^4}$ **12** $y = \dfrac{x^4}{2} - \dfrac{3}{4x^8}$

13 $y = \dfrac{x^4}{8} + 3x^2 + \dfrac{5}{6x^4}$ **14** $g(x) = 2x^3 - x^2 + 2 - \dfrac{3}{2x^2}$ **15** $A(x) = x^2 - \dfrac{5}{2x} + \dfrac{3}{4x^2}$

6.3 Calculating the gradient of a curve at a given point

> → You can use the gradient function to determine the exact value of the gradient at any specific point on the curve.

Here is the curve $y = 2x^3 - x^2 - 4x + 5$ with **domain** $-2 \le x \le 2$. The curve intersects the y-axis at $(0, 5)$.

At $x = -2$ the function has a negative value.
It increases to a point A, then decreases to a point B and after $x = 1$ it increases again.

The gradient function of the curve will be negative between points A and B and positive elsewhere.

Differentiating, the gradient function is $\dfrac{dy}{dx} = 6x^2 - 2x - 4$.

At the y-intercept $(0, 5)$ the x-coordinate is 0. Substituting

this value into $\dfrac{dy}{dx}$: at $x = 0$, $\dfrac{dy}{dx} = 6(0)^2 - 2(0) - 4 = -4$

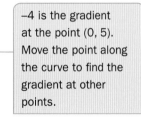

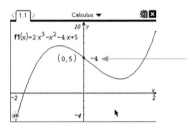

-4 is the gradient at the point $(0, 5)$.
Move the point along the curve to find the gradient at other points.

You can use this algebraic method to find the gradient of the curve at other points. For example,

at $x = -1$, $\dfrac{dy}{dx} = 6(-1)^2 - 2(-1) - 4$

$\dfrac{dy}{dx} = 4$

This result agrees with what can be seen from the graph.

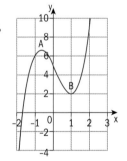

> **Will the gradient function be positive or negative at point A and at point B?**

> You can check this on your GDC. See Chapter 12, Section 6.1, Example 33.

> **GDC help on CD:** *Alternative demonstrations for the TI-84 Plus and Casio FX-9860GII GDCs are on the CD.*

> The gradient of the curve at $x = -1$ is 4 and at $x = 0$ it is -4

Exercise 6D

These questions can be answered using the algebraic method or using a GDC. Make sure you can do both.

1 If $y = x^2 - 3x$, find $\dfrac{dy}{dx}$ when $x = 4$.

2 If $y = 6x - x^3 + 4$, find $\dfrac{dy}{dx}$ when $x = 0$.

3 If $y = 11 - 2x^4 - 3x^3$, find $\dfrac{dy}{dx}$ when $x = -3$.

4 If $y = 2x(5x + 4)$, find the value of $\dfrac{dy}{dx}$ when $x = -1$.

5 Find the gradient of the curve $y = x^3 - 5x$ at the point where $x = 6$.

6 Find the gradient of the curve $y = 10 - \dfrac{1}{2}x^4$ at the point where $x = -2$.

7 Find the gradient of the curve $y = 3x\,(7 - 4x^2)$ at the point $(1, 9)$.

8 Find the gradient of the curve $y = 3x^2 - 5x + 6$ at the point $(-2, 28)$.

9 $s = 40t - 5t^2$

   Find $\dfrac{ds}{dt}$ when $t = 0$.

10 $s = t(35 + 6t)$

   Find $\dfrac{ds}{dt}$ when $t = 3$.

11 $v = 80t + 7$

   Find $\dfrac{dv}{dt}$ when $t = -4$.

12 $v = 0.7t - 11.9$

   Find $\dfrac{dv}{dt}$ when $t = 0.7$.

13 $A = 14h^3$

   Find $\dfrac{dA}{dh}$ when $h = \dfrac{2}{3}$.

14 $W = 7.25p^3$

   Find $\dfrac{dW}{dp}$ at $p = -2$.

15 $V = 4r^2 + \dfrac{18}{r}$.

   Find $\dfrac{dV}{dr}$ at $r = 3$.

16 $A = 5r + \dfrac{8}{r^2}$

   Find $\dfrac{dA}{dr}$ at $r = 4$.

17 $V = 7r^3 - \dfrac{8}{r}$

   Find $\dfrac{dV}{dr}$ at $r = 2$.

18 $A = \pi r^2 - \dfrac{2\pi}{r}$

   Find $\dfrac{dA}{dr}$ at $r = 1$.

19 $V = 6r + \dfrac{15}{2r}$

   Find $\dfrac{dV}{dr}$ at $r = 5$.

20 $C = 45r + \dfrac{12}{r^3}$

   Find $\dfrac{dC}{dr}$ at $r = 1$.

By working backwards you can find the coordinates of a specific point on a curve with a particular gradient.

Example 9

Point A lies on the curve $y = 5x - x^2$ and the gradient of the curve at A is 1. Find the coordinates of A.

Answer	
$\dfrac{dy}{dx} = 5 - 2x$	*First find $\dfrac{dy}{dx}$*
at A $\dfrac{dy}{dx} = 1$ so $5 - 2x = 1$ $\qquad x = 2$	*Solve the equation to find x.*
$y = 5(2) - (2)^2 = 6$ A is (2, 6)	*Substitute x = 2 into the equation of the curve to find y.*

Exercise 6E

1 Point P lies on the curve $y = x^2 + 3x - 4$. The gradient of the curve at P is equal to 7.
    a Find the gradient function of the curve.
    b Find the x-coordinate of P.
    c Find the y-coordinate of P.

2 Point Q lies on the curve $y = 2x^2 - x + 1$. The gradient of the curve at Q is equal to –9.
    a Find the gradient function of the curve.
    b Find the x-coordinate of Q.
    c Find the y-coordinate of Q.

3 Point R lies on the curve $y = 4 + 3x - x^2$ and the gradient of the curve at R is equal to –3.
    a Find the gradient function of the curve.
    b The coordinates of R are (a, b), find the value of a and of b.

EXAM-STYLE QUESTIONS

4 Point R lies on the curve $y = x^2 - 6x$ and the gradient of the curve at R is equal to 6.
    Find the gradient function of the curve.
    The coordinates of R are (a, b)
    Find the value of a and of b.

5 Find the coordinates of the point on the curve $y = 3x^2 + x - 5$ at which the gradient of the curve is 4.

6 Find the coordinates of the point on the curve $y = 5x - 2x^2 - 3$ at which the gradient of the curve is 9.

7 There are **two** points on the curve $y = x^3 + 3x + 4$ at which the gradient of the curve is 6.
    Find the coordinates of these two points.

8 There are **two** points on the curve $y = x^3 - 6x + 1$ at which the gradient of the curve is –3.
    Find the coordinates of these two points.
    Find the equation of the straight line that passes through these two points.

EXAM-STYLE QUESTION

9 There are **two** points on the curve $y = x^3 - 12x + 5$ at which the gradient of the curve is zero.
    Find the coordinates of these two points.
    Find the equation of the straight line that passes through these two points.

10 Point P $(1, b)$ lies on the curve $y = x^2 - 4x + 1$.
  a Find the value of b.
  b Find the gradient function of the curve.
  c Show that at P the gradient of the curve is also equal to b.
  d Q (c, d) is the point on the curve at which the gradient of the curve is equal to -2. Show that $d = -2$.

11 Point P $(5, b)$ lies on the curve $y = x^2 - 3x - 3$.
  a Find the value of b.
  b Find the gradient function of the curve.
  c Show that at P the gradient of the curve is also equal to b.
  d Q (c, d) is the point on the curve at which the gradient of the curve is equal to -3.
  Show that d is also equal to -3.

12 Consider the function $f(x) = 4x - x^2 - 1$.
  a Write down $f'(x)$.
  b Show that at $x = 5$, $f(x) = f'(x)$.
  c Find the coordinates of a second point on the curve $y = f(x)$ for which $f(x) = f'(x)$.

13 Consider the function $f(x) = 2x^2 - x + 1$.
  a Write down $f'(x)$.
  b Show that at $x = 2$, $f(x) = f'(x)$.
  c Find the coordinates of a second point on the curve $y = f(x)$ for which $f(x) = f'(x)$.

14 Consider the function $f(x) = 3x - x^2 - 1$.
  a Write down $f'(x)$.
  b Show that at $x = 1$, $f(x) = f'(x)$.
  c Find the coordinates of a second point on the curve $y = f(x)$ for which $f(x) = f'(x)$.

15 Consider the function $f(x) = 2x^2 - x - 1$.
  a Write down $f'(x)$.
  b Find the coordinates of the points on the curve $y = f(x)$ for which $f(x) = f'(x)$.

16 Consider the function $f(x) = x^2 + 5x - 5$.
  a Write down $f'(x)$.
  b Find the coordinates of the points on the curve $y = f(x)$ for which $f(x) = f'(x)$.

17 Consider the function $f(x) = x^2 + 4x + 5$.
  Find the coordinates of the point on the curve $y = f(x)$ for which $f(x) = f'(x)$.

6.4 The tangent and the normal to a curve

Here is a curve $y = f(x)$ with a point, P, on the curve.

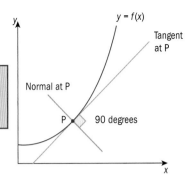

> → The tangent to the curve at any point P is the straight line which passes through P with gradient equal to the gradient of the curve at P.

The **normal** to the curve at P is the straight line which passes through P that is **perpendicular** to the tangent.

The tangent and the curve are closely related because, at P:

- the x-coordinate of the tangent is equal to the x-coordinate of the curve
- the y-coordinate of the tangent is equal to the y-coordinate of the curve
- the gradient of the tangent is equal to the gradient of the curve.

You can use differentiation to find the equation of the tangent to any curve at a point, P(a, b), provided that you know both the equation of the curve and the x-coordinate, a, of the point P.

> → To find the equation of the tangent to the curve at P(a, b):
>
> 1 Calculate b, the y-coordinate of P, using the equation of the curve.
>
> 2 Find the gradient function $\dfrac{dy}{dx}$.
>
> 3 Substitute a, the x-coordinate of P, into $\dfrac{dy}{dx}$ to calculate, m, the value of the gradient at P.
>
> 4 Use the equation of a straight line $(y - b) = m(x - a)$.

For more on the equation of a straight line, see Chapter 3.

Example 10

Point P has an x-coordinate 2. Find the equation of the tangent to the curve $y = x^3 - 3$ at P.
Give your answer in the form $y = mx + c$.

Answer

At $x = 2$, $y = (2)^3 - 3 = 5$	*Use $y = x^3 - 3$ to calculate the y-coordinate of P.*
$\dfrac{dy}{dx} = 3x^2$	*Find the gradient function $\dfrac{dy}{dx}$.*
At $x = 2$, $\dfrac{dy}{dx} = 3(2)^2 = 12$ $m = 12$	*Substitute 2, the x-coordinate at P, into $\dfrac{dy}{dx}$ to calculate m, the value of the gradient at P.*
At P $(2, 5)$ $(y - 5) = 12(x - 2)$ $y - 5 = 12x - 24$ $y = 12x - 19$	*Use the equation $(y - b) = m(x - a)$ with $a = 2$, $b = 5$, $m = 12$. Simplify.*

You can check the equation of the tangent using your GDC.

1 Find the equation of the tangent to the given curve at the stated point, P. Give your answers in the form $y = mx + c$.

   a $y = x^2$; P(3, 9)   **b** $y = 2x^3$; P(1, 2)

   c $y = 6x - x^2$; P(2, 8)   **d** $y = 3x^2 - 10$; P(1, -7)

   e $y = 2x^2 - 5x + 4$; P(3, 7)   **f** $y = 10x - x^3 + 5$; P(2, 17)

   g $y = 11 - 2x^2$; P(3, -7)   **h** $y = 5 - x^2 + 6x$; P(2, 13)

   i $y = 4x^2 - x^3$; P(4, 0)   **j** $y = 5x - 3x^2$; P(-1, -8)

   k $y = 6x^2 - 2x^3$; P(2, 8)   **l** $y = 60x - 5x^2 + 7$; P(2, 107)

   m $y = \dfrac{1}{2}x^4 - 7$; P(4, 121)   **n** $y = 17 - 3x + 5x^2$; P(0, 17)

   o $y = 2x(5 - x)$; P(0, 0)   **p** $y = \dfrac{1}{4}x^3 - 4x$; P(2, -6)

   q $y = \dfrac{3}{4}x^2 + 3$; P(-2, 6)   **r** $y = \dfrac{2}{3}x^3 + \dfrac{1}{3}$; $P\left(-1, -\dfrac{1}{3}\right)$

   s $y = \dfrac{1}{4}x^3 - 7x^2 + 5$; P(-2, -25)

2 Find the equation of the tangent to the given curve at the stated point. Give your answers in the form $ax + by + c = 0$.

   a $y = \dfrac{12}{x^2}$; (2, 3)   **b** $y = 5 + \dfrac{6}{x^3}$; (1, 11)

   c $y = 6x - \dfrac{8}{x^2}$; (-2, -14)   **d** $y = x^3 + \dfrac{6}{x^2}$; (-1, 5)

   e $y = 5x - \dfrac{8}{x}$; (4, 18)

To find the equation of the normal to a curve at a given point you need to do one extra step.

> → The normal is perpendicular to the tangent so its gradient, m', is found using the formula $m' = \dfrac{-1}{m}$, where m is the gradient of the tangent.

Example 11

Point P has x-coordinate -4.

Find the equation of the normal to the curve $y = \dfrac{12}{x}$ at P.

Give your answer in the form

$ax + by + c = 0$, where $a, b, c \in \mathbb{Z}$.

You learned about gradient of a perpendicular line in Chapter 3.

Answer

At $x = -4$, $y = \dfrac{12}{(-4)} = -3$

$\dfrac{dy}{dx} = -\dfrac{12}{x^2}$

Use $y = \dfrac{12}{x}$ to calculate the y-coordinate of P.

Find the gradient function $\dfrac{dy}{dx}$.

(Remember, $y = 12x^{-1}$.)

▶ Continued on next page

At $x = -4$, $\dfrac{dy}{dx} = -\dfrac{12}{(-4)^2} = -\dfrac{3}{4}$

The gradient of the **tangent**, $m = -\dfrac{3}{4}$

Substitute the value of x into $\dfrac{dy}{dx}$ to calculate, m, the value of the gradient at P.

Hence, the gradient of the normal, $m' = \dfrac{4}{3}$

The normal is perpendicular to the tangent.

The equation of the normal to $y = \dfrac{12}{x}$ at P($-4, -3$) is

$(y - (-3)) = \dfrac{4}{3}(x - (-4))$

Use the equation of a straight line $(y - b) = m(x - a)$ with $a = -4$, $b = -3$, $m = \dfrac{4}{3}$

$3(y + 3) = 4(x + 4)$

$3y + 9 = 4x + 16$

$4x - 3y + 7 = 0$

Simplify.
Rearrange to the form $ax + bx + c = 0$, where $a, b, c \in \mathbb{Z}$

> The gradient of a line perpendicular to a line whose gradient is m is $-\dfrac{1}{m}$.

> You cannot find the equation of a normal directly from the GDC.

Exercise 6G

Find the equation of the normal to the given curve at the stated point P. Give your answers in the form $ax + by + c = 0$

1 $y = 2x^2$; P(1, 2)

2 $y = 3 + 4x^3$; P(0.5, 3.5)

3 $y = \dfrac{x}{2} - x^2$; P(2, -3)

4 $y = \dfrac{3x^2}{2} + x$; P($-2, 4$)

5 $y = (x + 2)(5 - x)$; P(0, 10)

6 $y = (x + 2)^2$; P(0, 4)

7 $y = \dfrac{4}{x}$; P(2, 2)

8 $y = \dfrac{6}{x^2}$; P($-1, 6$)

9 $y = 6x + \dfrac{8}{x}$; P(1, 14)

10 $y = x^4 - \dfrac{3}{x^3}$; P($-1, 4$)

11 $y = 4 - 2x - \dfrac{1}{x}$; P(0.5, 1)

12 $y = 5x - \dfrac{9}{2x}$; P(3, 13.5)

Example 12

The gradient of the tangent to the curve $y = ax^2$ at the point P $(3, b)$ is 30. Find the values of a and b.

Answer

$\dfrac{dy}{dx} = 2ax$

$2a(3) = 30$

$\Rightarrow a = 5$

The equation of the curve is $y = 5x^2$.

$b = 5(3)^2 \Rightarrow b = 45$

As the gradient of the tangent is given, find $\dfrac{dy}{dx}$.

When $x = 3$, $\dfrac{dy}{dx} = 30$

Substitute $x = 3$ to find b.

Exercise 6H

1 Find the equation of the tangent to the curve $y = (x - 4)^2$ at the point where $x = 5$.

EXAM-STYLE QUESTIONS

2 Find the equation of the tangent to the curve $y = x(x^2 - 3)$ at the point where $x = -2$.

3 Find the equation of the normal to the curve $y = x + \dfrac{6}{x}$ at the point where $x = 4$.

4 Find the equation of the normal to the curve $y = x^2 - \dfrac{1}{x^2}$ at the point where $x = -1$.

5 Find the equations of the tangents to the curve $y = 3x^2 - 2x$ at the points where $y = 8$.

6 Find the equations of the tangents to the curve $y = 2x(3 - x)$ at the points where $y = -20$.

7 Find the equation of the normal to the curve $y = 7 - 5x - 2x^3$ at the point where it intersects the x-axis.

8 Find the equation of the normal to the curve $y = x^3 + 3x - 2$ at the point where $y = -6$.

9 **a** Find the value of x for which the gradient of the tangent to the curve $y = (4x - 3)^2$ is zero.
   b Find the equation of the tangent at this point.

EXAM-STYLE QUESTION

10 **a** Find the value of x for which the gradient of the tangent to the curve $y = x^2 + \dfrac{16}{x}$ is zero.
   b Find the equation of the tangent at this point.

11 **a** Find the value of x for which the gradient of the tangent to the curve $y = \dfrac{x^2}{2} + x - 3$ is 5.
   b Find the equation of the tangent at this point.

12 **a** Find the value of x for which the gradient of the tangent to the curve $y = x^4 + 3x - 3$ is 3.
   b Find the equation of the tangent at this point.
   c Find the equation of the normal at this point.

13 **a** Find the value of x for which the gradient of the tangent to the curve $y = 4x + \dfrac{3}{x^4}$ is 16.
   b Find the equation of the tangent at this point.
   c Find the equation of the normal at this point.

14 There are two points on the curve $y = 2x^3 + 9x^2 - 24x + 5$ at which the gradient of the curve is equal to 36. Find the equations of the tangents to the curve at these points.

EXAM-STYLE QUESTION

15 The gradient of the tangent to the curve $y = x^2 + kx$ at the point P (3, b) is 7.
Find the value of k and the value of b.

16 The gradient of the tangent to the curve $y = x^2 + kx$ at the point P (–2, b) is 1.
Find the value of k and that of b.

17 The gradient of the tangent to the curve $y = kx^2 - 2x + 3$ at the point P (4, b) is 2.
Find the value of k and that of b.

18 The gradient of the tangent to the curve $y = 4 + kx - x^3$ at the point P (–2, b) is –5.
Find the value of k and that of b.

19 The gradient of the tangent to the curve $y = px^2 + qx$ at the point P (2, 5) is 7.
Find the value of p and that of q.

20 The gradient of the tangent to the curve $y = px^2 + qx - 5$ at the point P (–3, 13) is 6.
Find the value of p and that of q.

6.5 Rates of change

The gradient function, $f'(x)$, of a function $f(x)$ is a measure of how $f(x)$ changes as x increases. We say that $f'(x)$ measures the **rate of change of _f_ with respect to _x_.**

> → For the graph $y = f(x)$, the gradient function $\dfrac{dy}{dx} = f'(x)$ gives the rate of change of y with respect to x.

In general, the **rate of change** of one variable with respect to another is the gradient function.

Other variables can also be used, for example:

if $A = f(t)$, then $\dfrac{dA}{dt} = f'(t)$ measures the **rate of change of _A_ with respect to _t_**.

If the variable t represents time, then the gradient function measures the rate of change with respect to the *time* that passes.

This is an important concept. If you measure how a variable changes as time is passing then you are applying mathematics to situations that are **dynamic** – to situations that are moving.

For example, if C represents the value of a car (measured on a day-to-day basis) we can say that C is a function of time: $C = f(t)$. Then, $\dfrac{dC}{dt} = f'(t)$ represents the rate at which the value of the car is changing – it measures the rate of change of C with respect to t, the rate of inflation or deflation of the price of the car.

Similarly, if s represents the distance measured from a fixed point to a moving object then s is a function of time: $s = g(t)$ and $\dfrac{ds}{dt} = g'(t)$ measures the rate of change of this distance, s, with respect to t.

$\dfrac{ds}{dt}$ measures the **velocity** of the object at time t.

> If v is the velocity of an object, what does $\dfrac{dv}{dt}$ represent?

Example 13

The volume of water in a container, $V\,\text{cm}^3$, is given by the formula $V = 300 + 2t - t^2$, where t is the time measured in seconds.

a   What does $\dfrac{dV}{dt}$ represent?

b   What units are used for $\dfrac{dV}{dt}$?

c   Find the value of $\dfrac{dV}{dt}$ when $t = 3$.

d   What does the answer to **c** tell you?

Answers

a $\dfrac{dV}{dt}$ represents the rate of change of the volume of water in the container. | *The rate at which the water is entering (or leaving) the container.*

b $\dfrac{dV}{dt}$ is measured in cm^3 per second (cm^3s^{-1}). | *The volume is measured in cm^3 and time is measured in seconds.*

c $\dfrac{dV}{dt} = 2 - 2t$

At $t = 3$,

$\dfrac{dV}{dt} = 2 - 2(3) = -4$ | $\dfrac{dV}{dt}$ *is negative, so*

d   Since this value is **negative**, the water is **leaving** the container at $4\,\text{cm}^3$ per second. | *the volume is decreasing.*

> How would you decide by considering $\dfrac{dv}{dt}$ whether the water was **entering** or **leaving** the container?

Example 14

A company mines copper, where the mass of copper, x, is measured in thousands of tonnes. The company's profit, P, measured in millions of dollars, depends on the amount of copper mined. The profit is given by the function $P(x) = 2.3x - 0.05x^2 - 12$

a Find $P(0)$ and $P(6)$ and interpret these results.

b Find $\dfrac{dP}{dx}$. What does $\dfrac{dP}{dx}$ represent?

c Find the value of P and $\dfrac{dP}{dx}$ when $x = 20$ and when $x = 25$.

d Interpret the answers to **c**.

e Find the value of x for which $\dfrac{dP}{dx} = 0$.

f Determine P for this value of x, and interpret this value.

> You can graph any function on the GDC. This could give you further insight into the problem.

Answers

a $P(0) = -12$; a loss of 12 million dollars.
$P(6) = 0$; there is no profit and no loss, this is the break-even point.

Substitute $x = 0$ in to $P(x)$.

b $\dfrac{dP}{dx} = -0.1x + 2.3$

$\dfrac{dP}{dx}$ represents the rate of change of the profit as the amount of copper mined increases.

$\dfrac{dP}{dx}$ measures the rate of change of P with respect to x.

c At $x = 20$, $P = 14$ and $\dfrac{dP}{dx} = 0.3$

At $x = 25$, $P = 14.25$ and $\dfrac{dP}{dx} = -0.2$

Substitute $x = 20$ and $x = 25$ into $P(x)$ and $\dfrac{dP}{dx}$.

d At both points the company is profitable.

At $x = 20$, $\dfrac{dP}{dx} > 0$ so a further increase in production will make the company **more profitable**.

At $x = 20$, $P(x)$ is increasing.

At $x = 25$, $\dfrac{dP}{dx} < 0$ so a further increase in production will make the company **less profitable**.

At $x = 25$, $P(x)$ is decreasing.

e $\dfrac{dP}{dx} = -0.1x + 2.3 = 0$

$0.1x = 2.3$

$x = \dfrac{2.3}{0.1} = 23$

Set $\dfrac{dP}{dx}$ equal to 0.

Solve for x.

23 000 tonnes of copper needs to be mined to maximize the company's profit.

x is measured in thousands of tonnes.

f $P(23) = 14.45$

14.45 million dollars is the **maximum** profit that the company can make.

Substitute $x = 23$ into $P(x)$.

Exercise 6I

EXAM-STYLE QUESTION

1 The volume of water in a container, $V\,\text{cm}^3$, is given by the formula
$V = 100 + 2t + t^3$, where t is the time measured in seconds.
  a How much water is there in the container initially?
  b How much water is there in the container when $t = 3$?
  c What does $\dfrac{\mathrm{d}V}{\mathrm{d}t}$ represent?
  d Find the value of $\dfrac{\mathrm{d}V}{\mathrm{d}t}$ when $t = 3$.
  e Use your answers to **b** and **d** to explain what is happening to
     the volume of water in the tank.

> Initially $t = 0$

2 The area, A, of a pool of water forming under a leaking pipe is
$A = 4t + t^2\ \text{cm}^2$ after t seconds.
  a What is the area of the pool initially?
  b What is the area of the pool when $t = 5$?
  c What does $\dfrac{\mathrm{d}A}{\mathrm{d}t}$ represent?
  d Find the value of $\dfrac{\mathrm{d}A}{\mathrm{d}t}$ when $t = 5$.
  e Use your answers to **b** and **d** to explain what is happening to
     the area of the pool.

3 The weight of oil in a storage tank, W, varies according to
the formula $W = 5t^2 + \dfrac{640}{t} + 40$ where W is measured in tonnes
and t is the time measured in hours, $1 \le t \le 10$.
  a Find the weight of oil in the tank at $t = 1$.
  b Find $\dfrac{\mathrm{d}W}{\mathrm{d}t}$.
  c Find the rate of change of the weight of the oil in the tank
     when
       i $t = 3$ **ii** $t = 5$.
  d What does your answer to **c** tell you?
  e Find the value of t for which $\dfrac{\mathrm{d}W}{\mathrm{d}t} = 0$.
  f Interpret your answer to **e**.

4 The volume of water, V, measured in m^3, in a swimming pool
after t minutes, where $t > 0$, is $V = 10 + 6t + t^2$.
  a Find the rate at which the volume is increasing when $t = 1$.
  b Find the rate at which the volume is increasing when there
     are $65\,\text{m}^3$ of water in the pool.

5 Water is flowing out of a tank. The depth of the water,
y cm, at time t seconds is given by $y = 500 - 4t - t^3$.
  a Find the rate at which the depth is decreasing at
     2 seconds and at 3 seconds.
  b Find the time at which the tank is empty.

6 The area, A cm², of a blot of ink is growing so that, after t seconds, $A = \dfrac{3t^2}{4} + \dfrac{t}{2}$.

    a Find the rate at which the area is increasing after 2 seconds.

    b Find the rate at which the area is increasing when the area of the blot is 30 cm².

7 The weight of oil in a storage tank, W, varies according to

    the formula $W = 10t + \dfrac{135}{t^2} + 4$ where W is measured in tonnes

    and t is the time measured in hours, $1 \le t \le 10$.

    a Find the rate at which the weight is changing after 2 hours.

    b Find the value of t for which $\dfrac{dW}{dt} = 0$.

8 The angle turned through by a rotating body, θ degrees, in time t seconds
is given by the relation $\theta = 4t^3 - t^2$.

    a Find the rate of increase of θ when $t = 2$.

    b Find the value of t at which the body changes direction.

9 A small company's profit, P, depends on the amount x of 'product' it makes.
This profit can be modeled by the function $P(x) = -10x^3 + 40x^2 + 10x - 15$.
P is measured in thousands of dollars and x is measured in tonnes.

    a Find $P(0)$ and $P(5)$ and interpret these results.

    b Find $\dfrac{dP}{dx}$.

    c Find the value of P and $\dfrac{dP}{dx}$ when   **i** $x = 2$ **ii** $x = 3$.

    d Interpret your answers to **c**.

    e Find the value of x and of P for which $\dfrac{dP}{dx} = 0$. What is the importance of this point?

6.6 Local maximum and minimum points (turning points)

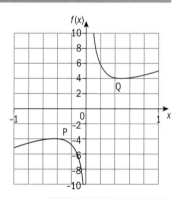

Here is the graph of the function

$$f(x) = 4x + \frac{1}{x}, \quad x \ne 0$$

The graph has two branches, because the function is **not defined** at the point $x = 0$.

First, look at the left-hand branch of the graph, for the **domain** $x < 0$.

As x increases, the curve increases to the point P. After point P, the curve decreases. P is said to be a local **maximum point**.

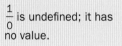

$\dfrac{1}{0}$ is undefined; it has no value.

You can determine that P is a local maximum point because just before P (for example, at A) the gradient of the curve is positive, and just after P (for example, at B) the gradient of the curve is negative.

At P itself, most importantly, the gradient of the curve is zero.

→ At a local maximum, the curve stops increasing and changes direction so that it 'turns' and starts decreasing. So, as x increases, the three gradients occur in the order: positive, zero, negative. Where the gradient is zero is the maximum point.

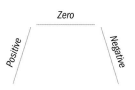

Now look at the right-hand branch of the graph, with the domain $x > 0$.

As x increases, the curve decreases to the point Q. After Q, the curve increases. Q is said to be a local **minimum point**.

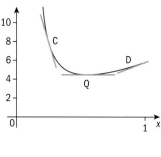

You can determine that Q is a local minimum point because just before Q (for example, at C) the gradient of the curve is negative and just after Q (for example, at D) the gradient of the curve is positive.

At Q itself, the gradient of the curve is zero.

→ At a local minimum, the curve stops decreasing and changes direction; it 'turns' and starts increasing. So, as x increases, the three gradients occur in the order: negative, zero, positive. Where the gradient is zero is the minimum point.

Local maximum and local minimum points are known as **stationary points** or **turning points**.

→ At any stationary or turning point – either local maximum or local minimum – $f'(x)$ is zero.

At a stationary point, if $y = f(x)$ then $\frac{dy}{dx} = 0$.

To find the coordinates of P (the local maximum) and of Q (the local minimum) for the function $f(x) = 4x + \dfrac{1}{x}$, use the fact that at each of these points $f'(x)$ is zero.

$f(x) = 4x + \dfrac{1}{x}$, so $f'(x) = 4 - \dfrac{1}{x^2}$

Remember that $\dfrac{1}{x} = x^{-1}$.

Set $f'(x) = 0$ which gives $4 - \dfrac{1}{x^2} = 0$

Adding $\dfrac{1}{x^2}$: $\qquad\qquad 4 = \dfrac{1}{x^2}$

Multiplying by x^2: $\qquad 4x^2 = 1$

Dividing by 4: $\qquad\qquad x^2 = \dfrac{1}{4}$

Taking square roots: $x = \dfrac{1}{2}$ or $x = -\dfrac{1}{2}$

Substitute each x-value into $f(x)$ to find the y-coordinate of each turning point.

At $x = \dfrac{1}{2}$, $f\left(\dfrac{1}{2}\right) = 4\left(\dfrac{1}{2}\right) + \dfrac{1}{\left(\frac{1}{2}\right)} = 4$

At $x = -\dfrac{1}{2}$, $f\left(-\dfrac{1}{2}\right) = 4\left(-\dfrac{1}{2}\right) + \dfrac{1}{\left(-\frac{1}{2}\right)} = -4$

You can find local maximum and local minimum points using a GDC, without using differentiation. See Chapter 12, Section 6.3.

So, the coordinates of the turning points are $\left(\dfrac{1}{2}, 4\right)$ and $\left(-\dfrac{1}{2}, -4\right)$

To determine which is the local maximum and which is the local minimum, look at the graph of the function: $\left(\dfrac{1}{2}, 4\right)$ is the local minimum and $\left(-\dfrac{1}{2}, -4\right)$ the local maximum.

> You cannot decide which is the maximum and which is the minimum simply by looking at the coordinates.

→ To find turning points, first set the gradient function equal to zero and solve this equation. This gives the *x*-coordinate of the turning point.

Exercise 6J

Find the values of x for which $\dfrac{\mathrm{d}y}{\mathrm{d}x} = 0$. Verify your answers by using your GDC.

1 $y = x^2 - 6x$

2 $y = 12x - 2x^2$

3 $y = x^2 + 10x$

4 $y = 3x^2 + 15x$

5 $y = x^3 - 27x$

6 $y = 24x - 2x^3$

7 $y = 4x^3 - 3x$

8 $y = 3x - 16x^3$

9 $y = 2x^3 - 9x^2 + 12x - 7$

10 $y = 5 + 9x + 6x^2 + x^3$

11 $y = x^3 - 3x^2 - 45x + 11$

12 $y = 12x^2 + x^3 + 36x - 8$

13 $y = 2x^3 - 6x^2 + 7$

14 $y = 17 + 30x^2 - 5x^3$

15 $f(x) = x + \dfrac{1}{x}$

16 $y = x + \dfrac{4}{x}$

17 $y = 4x + \dfrac{9}{x}$

18 $y = 8x + \dfrac{1}{2x}$

19 $y = 27x + \dfrac{4}{x^2}$

20 $y = x + \dfrac{1}{2x^2}$

Once you have found the *x*-coordinate of any turning point, you can then calculate the *y*-coordinate of the point and decide if it is a maximum or minimum.

Example 15

Find the coordinates of the turning points of the curve $y = 3x^4 - 8x^3 - 30x^2 + 72x + 5$. Determine the nature of these points.

'Determine the nature' means decide whether the point is a local maximum or a local minimum.

Answer

$y = 3x^4 - 8x^3 - 30x^2 + 72x + 5$

$\dfrac{\mathrm{d}y}{\mathrm{d}x} = 12x^3 - 24x^2 - 60x + 72$ *Differentiate.*

$\dfrac{\mathrm{d}y}{\mathrm{d}x} = 12x^3 - 24x^2 - 60x + 72 = 0$ *At each turning point $\dfrac{\mathrm{d}y}{\mathrm{d}x} = 0$.*

▶ Continued on next page

$x = -2, x = 1, x = 3$

At $x = -2$,

$y = 3(-2)^4 - 8(-2)^3 - 30(-2)^2 + 72(-2) + 5 = -95$

so $(-2, -95)$ is a turning point.

At $x = 1$, $y = 3(1)^4 - 8(1)^3 - 30(1)^2 + 72(1) + 5 = 42$

so $(1, 42)$ is a turning point.

At $x = 3$, $y = 3(3)^4 - 8(3)^3 - 30(3)^2 + 72(3) + 5 = -22$

so $(3, -22)$ is a turning point.

x-coordinate		-2		1		3	
Gradient		0		0		0	

$x = -10$ for $x < -2$ $f'(-10) = -12\,268$
$x = 0$ for $-2 < x < 1$ $f'(0) = 72$
$x = 2$ for $1 < x < 3$ $f'(2) = -48$
$x = 5$ for $x > 3$ $f'(5) = 672$

x-coordinate	-10	-2	0	1	2	3	5
Gradient	-12 268	0	72	0	-48	0	672

Gradient is zero
x = 1
Negative   Positive   Local maximum   Negative

x = -2
Gradient is zero

x = 3
Gradient is zero   Positive

Local minimum   Local minimum

$(-2, -95)$ is a local **minimum**.

$(1, 42)$ is a local **maximum**.

$(3, -22)$ is also a local **minimum**.

Solve this equation with your GDC.

Substitute the three values of x to find the y-coordinates.

To decide if points are maximum or minimum (without using the GDC) find the gradient at points on each side of the turning points. First, fill in the information on the turning points.

Now choose x-coordinates of points on each side of the turning points. Calculate the gradient at each point and enter them in the table.

Choose points close to the stationary point.

Sketch the pattern of the gradients from the table.

As the curve moves through $(-2, -95)$, the gradient changes negative → zero → positive.

As the curve moves through $(1, 42)$, the gradient changes positive → zero → negative.

As the curve moves through $(3, -22)$, the gradient changes negative → zero → positive.

Exercise 6K

Determine the coordinates of any turning points on the given curves.
For each, decide if it is a maximum or minimum.
Check your answers by using your GDC.

1 $y = x^3 - 9x^2 + 24x - 20$ **2** $y = x^3 + 6x^2 + 9x + 5$

3 $y = x(9 + 3x - x^2)$ **4** $y = x^3 - 3x^2 + 5$

5 $y = x(27 - x^2)$　　　　　**6** $y = x^2(9 - x)$

7 $f(x) = x + \dfrac{1}{x}$　　　　**8** $f(x) = x + \dfrac{9}{x}$

9 $f(x) = \dfrac{x}{2} + \dfrac{8}{x}$　　　　**10** $f(x) = \dfrac{9}{x} + \dfrac{x}{4}$

11 $f : x \to = x^2 - \dfrac{16}{x}$　　　**12** $f : x \to = 9x + \dfrac{1}{6x^2}$

> '$f : x \to$' is read as 'f such that x maps to' and means the same as '$f(x) =$'.

You can sometimes determine the nature of a turning point without checking points on either side.

Example 16

Find the coordinates of any turning points of the curve
$y = 9x - 3x^2 + 8$ and determine their nature.

Answer

At turning points:

$\dfrac{dy}{dx} = 9 - 6x = 0$ 　　　　*Solve for x.*

$\qquad x = 1.5$

$y = 9(1.5) - 3(1.5)^2 + 8$ 　　*Substitute x = 1.5 into*
$\quad = 14.75$ 　　　　　　　　*$y = 9x - 3x^2 + 8$.*

The turning point is (1.5, 14.75). 　*Quadratic graphs with a negative*
　　　　　　　　　　　　　　　coefficient of x^2 are this shape:
The turning point is a local
maximum.

> Quadratic graphs with a positive coefficient of x^2 are this shape:

Exercise 6L

Find the coordinates of the local maximum or local minimum point for each quadratic curve.
State the nature of this point.

1 $y = x^2 - 4x + 10$　　**2** $y = 18x - 3x^2 + 2$　　**3** $y = x^2 + x - 3$

4 $y = 8 - 5x + x^2$　　　**5** $y = 3x + 11 - x^2$　　**6** $y = 20 - 6x^2 - 15x$

7 $y = (x - 3)(x - 7)$　　**8** $y = x(x - 18)$　　　**9** $y = x(x + 4)$

6.7 Using differentiation in modeling: optimization

An introductory problem

In Chapter 4, you used quadratic functions to model various situations. One of the optimization problems was to maximize the area of a rectangular field that bordered a straight canal and was enclosed on three sides by 120 m of fencing.

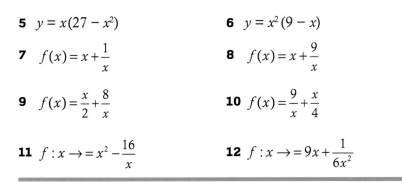

Canal

Width

Length

A model is a mathematical function that describes the situation. In this case, we need a model for the area of the field (the rectangle) for different widths.

First, identify the **variables** in the problem.
These are:
- the width of the field
- the length of the field
- the area of the field.

Second, identify any **constraints** in the problem. The constraint in this problem is that 120 m of fencing is used for three sides.

It often helps to try a few numerical examples in order to put the problem in context and to indicate the method. For example

1 If the width were 20 m, then the length would be $120 - 2(20) = 80$ m

the area would be $20 \times 80 = 1600$ m²

2 If the width were 50 m, then the length would be $120 - 2(50) = 20$ m

the area would be $50 \times 20 = 1000$ m²

> Note that, although the length of the fencing is constant, the size of the enclosed area varies.

Setting up the model

The model is for the area of the field and is a function of *both* its width and its length.

1 Define the variables.
Let A be the area of the field, x be the width of the field and y be the length of the field.
Then $A = xy$

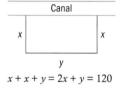

Canal

x x

y

$x + x + y = 2x + y = 120$

2 Write the constraint algebraically.
$120 = 2x + y$

> If you define the variables in a different way, you obtain a different function. Here you could have defined the **length** to be x and the width as y. The area A(x) would then have been a different – but correct – function.

3 Use the formula for the constraint to write the area function using just one variable.
Rearrange the constraint: $y = 120 - 2x$
Substitute in the area function: $A = xy = x(120 - 2x)$

So a model for the area of the field is $A(x) = x(120 - 2x)$, where x is the width of the field.

To determine the maximum area (the optimum solution) set the gradient function to zero.

The formula for the area is: $A(x) = x(120 - 2x)$

Expand the brackets: $A(x) = 120x - 2x^2$

Differentiate: $\dfrac{dA}{dx} = A'(x) = 120 - 4x$

Equate $\dfrac{dA}{dx}$ to zero: $120 - 4x = 0$

Solve: $4x = 120 \Rightarrow x = 30$

> The quadratic function A(x) has a negative coefficient of x² so the turning point is a maximum.

The width of the optimum rectangle is 30 m. To find the length substitute $x = 30$ into $y = 120 - 2x$.

$$120 - 2(30) = 60\,\text{m}$$

The dimensions of the rectangle are width 30 m and length 60 m.

To find the maximum area substitute $x = 30$ into $A(x) = x(120 - 2x)$.

The maximum area is $A(30) = (30)(120 - 2(30)) = 1800\,\text{m}^2$

> → In optimization problems, use differentiation to find an optimal value (either the maximum or the minimum) of a function as two variables interact.

You need to find an equation for this function in terms of these two variables and a constraint formula which links the variables. The constraint formula is used to remove one of the variables.

You can only use differentiation in functions with one variable.

Example 17

Optimize the function $A = 3xy$ subject to the constraint $x + y = 20$.

Answer

$y = 20 - x$	*Rearrange the constraint so y is the subject.*
$A = 3xy = 3x(20 - x)$	*Substitute y into the function.*
$A(x) = 60x - 3x^2$	*Simplify.*
$\dfrac{dA}{dx} = 60 - 6x$	*Differentiate.*
$60 - 6x = 0 \Rightarrow x = 10$	*Set $\dfrac{dA}{dx}$ to zero and solve for x.*
$A(10) = 60(10) - 3(10)^2 = 300$	*Substitute the value of x into A(x) to*
The optimal value of A is 300.	*find the optimal value of A.*

$A(x)$ is a quadratic function. Is the value 300 a maximum or a minimum?

Exercise 6M

1 $A = bh$, subject to the constraint $b - h = 7$.
  a Use the constraint to express b in terms of h.
  b Express A in terms of h.

2 $V = 3xt$ subject to the constraint $x + t = 10$.
  a Use the constraint to express x in terms of t.
  b Express V in terms of t.

3 $p = x^2y$ subject to the constraint $2x + y = 5$.
  a Use the constraint to express y in terms of x.
  b Express p in terms of x.

4 $R = \dfrac{1}{2}nr^2$ subject to the constraint $n - r = 25$.
  a Express R in terms of r.   **b** Express R in terms of n.

Choosing which variable to eliminate is an important skill. A bad choice will make the function more complicated.

5 $L = 2m(m + x)$ subject to the constraint $\frac{1}{2}(x + 5m) = 50$.

    a Express L in terms of m.   **b** Express L in terms of x.

6 $V = \pi r^2 h$ and $2r + h = 17$

    a Express V in terms of r.   **b** Express V in terms of h.

7 $y = 5x^2 + c$ and $12x - 2c = 3$

    a Express y in terms of x.

    b Use differentiation to find $\dfrac{dy}{dx}$.

    c Hence find the minimum value of y.

    d Find the value of c that corresponds to this minimum value.

> How do you know, without testing the gradient, that it is a minimum?

8 $N = 2n(5 - x)$ and $12n + 10x = 15$

    a Express N in terms of n.

    b Use differentiation to find $\dfrac{dN}{dn}$.

    c Hence find the minimum value of N.

    d Find the value of x that corresponds to this minimum value.

9 Given $A = \frac{1}{2}LB$ and $3L - 5B = 18$, express A in terms of L. Hence find the minimum value of A and the value of B that corresponds to this minimum value.

10 Given $C = \pi fr$ and $r = 30 - 3f$, express C in terms of either f or r. Hence find the maximum value of C and the values of f and of r that correspond to this maximum value.

11 Given $a - b = 10$ and $X = 2ab$, find the minimum value of X.

12 Given $x + 2t = 12$, find the maximum/minimum value of tx and determine the nature of this optimum value.

> Let $A = tx$

13 Given $3y + x = 30$, find the maximum/minimum value of $2xy$ and determine its nature.

14 Given $2M - L = 28$, find the values of L and M which give $3LM$ a maximum/minimum value. Find this optimum value, and determine its nature.

15 Given $c + g = 8$, express $c^2 + g^2$ in terms of g only. Hence find the minimum value of $c^2 + g^2$ subject to the constraint $c + g = 8$.

> Let $A = c^2 + g^2$

16 The sum of two numbers is 6. Find the values of these numbers such that the sum of their squares is a minimum.

17 Given that $r + h = 6$, express $r^2 h$ in terms of r only. Hence find the maximum value of $r^2 h$ subject to the constraint $r + h = 6$.

18 Given that $m + n = 9$, find the maximum/minimum values of $m^2 n$ and distinguish between them.

At the beginning of this chapter we defined the optimal design of a can as the one that uses the smallest amount of metal to hold a given capacity. Example 18 calculates the minimum surface area for a can holding $330\,cm^3$.

Example 18

Find the minimum surface area of a cylinder which has a volume of $330\,cm^3$.

Answer

Let

A be the total surface area of the cylinder.

Define the variables.

r be the radius of the base of the cylinder.

h be the height of the cylinder.

Then $A = 2\pi r^2 + 2\pi rh$

$\pi r^2 h = 330$

The constraint is that the volume of the cylinder is $330\ cm^3$.

$$h = \frac{330}{\pi r^2}$$

Rearrange to make h the subject.

$A = 2\pi r^2 + 2\pi rh$

$$= 2\pi r^2 + 2\pi r\left(\frac{330}{\pi r^2}\right)$$

Substitute the expression for h into the area function to reduce it to just one variable.

$$= 2\pi r^2 + \frac{660}{r}$$

Simplify.

$A = 2\pi r^2 + 660r^{-1}$

Write using indices.

$$\frac{dA}{dr} = 4\pi r + (-1)660r^{-2}$$

Differentiate.

$$\frac{dA}{dr} = 4\pi r - \frac{660}{r^2}$$

Simplify.

$$4\pi r - \frac{660}{r^2} = 0$$

Equate $\frac{dA}{dx}$ to zero to find the minimum.

$$4\pi r = \frac{660}{r^2}$$

Solve.

$4\pi r^3 = 660$

$$r^3 = \frac{660}{4\pi}$$

$$r^3 = \frac{165}{\pi} \Rightarrow r = \sqrt[3]{\frac{165}{\pi}}$$

You could solve this using a GDC.

$r = 3.74\,cm$ to 3 sf.

Is a drinks can perfectly cylindrical? What modeling assumptions do you need to make?

The surface area of a cylinder, $A = 2\pi r^2 + 2\pi rh$

The volume of a cylinder, $V = \pi r^2 h$

▶ Continued on next page

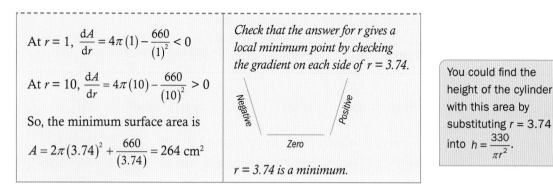

At $r = 1$, $\dfrac{dA}{dr} = 4\pi(1) - \dfrac{660}{(1)^2} < 0$

At $r = 10$, $\dfrac{dA}{dr} = 4\pi(10) - \dfrac{660}{(10)^2} > 0$

So, the minimum surface area is

$A = 2\pi(3.74)^2 + \dfrac{660}{(3.74)} = 264$ cm^2

Check that the answer for r gives a local minimum point by checking the gradient on each side of $r = 3.74$.

Negative   Positive

Zero

$r = 3.74$ is a minimum.

You could find the height of the cylinder with this area by substituting $r = 3.74$ into $h = \dfrac{330}{\pi r^2}$.

Exercise 6N

1 A gardener wishes to enclose a rectangular plot of land using a roll of wire-netting that is 40 m long. One side of the plot is to be the wall of the garden.
How should he bend the wire-netting to enclose the maximum area?

Draw a diagram first.

2 The sum of two numbers is 20. Let the first number be x. Write down an expression for the second number in terms of x.
Find the value of x given that twice the square of the first number added to three times the square of the second number is a minimum.

EXAM-STYLE QUESTIONS

3 An **open** rectangular box has its length double its width. The total surface area of the box is 150 cm^2.
The width of the box is x cm, and its height is h cm. Express the total surface area of the box in terms of x and h.
Use this expression (constraint) to find the volume of the box in terms of x only.
Hence, find the greatest possible volume of the box, and the width, length and height of the box required to give this volume.

4 A piece of wire 24 cm long is to be bent to form a rectangle with just one side duplicated for extra strength. Find the dimensions of the rectangle that give the maximum area.

5 A long strip of metal 120 cm wide is bent to form the base and two sides of a chute with a rectangular cross-section.
Find the width of the base that makes the area of the cross-section a maximum.

6 The sum of the height and the radius of the base of a cone is 12 cm. Find the maximum volume of the cone and the values of the height and the radius required to give this volume.

7 A closed box with a square base is to be made out of 600 cm^2 of metal. Find the dimensions of the box so that its volume is a maximum. Find the value of this maximum volume.

8 The total surface area of a closed cylindrical tin is to be 600 cm^2.
Find the dimensions of the tin if the volume is to be a maximum.

9 A square sheet of metal of side 24 cm is to be made into an open tray
of depth x cm by cutting out of each corner a square of side x cm and
folding up along the dotted lines as shown in the diagram.
Show that the volume of the tray is $4x(144 - 24x + x^2) \text{ cm}^3$.
Find the value of x for this volume to be a maximum.

10 A rectangular sheet of metal measures 16 cm by 10 cm. Equal squares
of side x cm are cut out of each corner and the remainder is folded up to
form a tray of depth x cm. Show that the volume of the tray is $4x(8 - x)$
$(5 - x) \text{ cm}^3$, and find the maximum volume.

11 A tin of soup is made in the shape of a cylinder so that the amount of metal
used in making the tin is a minimum. The volume of the tin is 350 cm^3.
  a If the radius of the base of the tin is 5 cm, find the height of the tin.
  b If the radius of the base of the tin is 2 cm, find the height of
    the tin.
  c **i** Use the volume of the tin to write down the constraint
      between the radius of the tin and its height.
    ii Show that the constraint can be written as $h = \dfrac{350}{\pi r^2}$

> The metal used in
> making the tin is the
> surface area of the
> cylinder.

  iii Find an expression for A, the total surface area of a cylinder,
      in terms of r only.
  iv Find the dimensions of the tin that minimize the total surface area
      of the tin.
  v Find the value of this minimum area.

12 The diagram shows a rectangular field with an area of
$50\,000 \text{ m}^2$. It has to be divided in half and also fenced in.
The most efficient way to enclose the area is to construct the fencing
so that the total length of the fence is minimized.
  a If the length (L) of the field is 200 m, what is the width?
  b Find the total length of the fencing in this case.
  c Use the fixed area to write down the problem constraint algebraically.
  d Find the dimensions of the field that make the length of fencing a
    minimum.
    Find the perimeter of the field in this case.

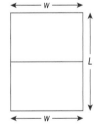

13 A second rectangular field is identical to that in question **12**.
The cost of the fencing around the perimeter is $3 per metre.
The cost of the dividing fence is $5 per metre. The most efficient
way to enclose the area minimizes the total **cost** of the fence.
  a Find the total cost of the fencing when the length (L) is 200 m.
  b Use the fixed area to write down the problem constraint algebraically.
  c Find the dimensions of the field that make the **cost** of the fencing
    a minimum. Find the cost in this case.

14 The page of a mathematics book is designed to have a printable area of $144\,cm^2$ plus margins of 2 cm along each side and 3 cm at the top and the bottom. The diagram for this is shown with the printable area shaded.

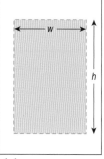

   a If the width of the printable area (w) is 9 cm, find its height (h). Using these values, find the area of the page.

   b If the width of the printable area is 14 cm, calculate the area of the page.

   c Write down an expression for the printable area in terms of w and h.

   d Write down an expression for P, the area **of the page** in terms of w and h.

   e Use the results of **c** and **d** to show that $P = 168 + 4h + \dfrac{864}{h}$.

   f Find the dimensions of the page that minimize the page area.

15 A fish tank is to be made in the shape of a cuboid with a rectangular base, with a length twice the width. The volume of the tank is fixed at 225 litres. The tank is to be made so that the total length of steel used to make the frame is minimized.

   a **i** If the length of the base is 100 cm, what is its width?

      ii Show that in this case, the height of the tank is 45 cm.

      iii Find the total length of the steel frame.

   b If the width of the tank is x, find an expression for the volume of the tank in terms of x and h, the height of the tank.

   c Show that L, the total length of the steel frame, can be written as $L = 6x + \dfrac{450\,000}{x^2}$.

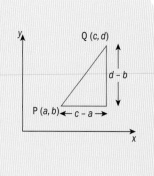

   d Find the dimensions of the tank that minimize the length of the steel frame. Find also the length of the frame in this case.

CHAPTER 6 SUMMARY

Introduction to differentiation

- If P is the point (a, b) and Q is (c, d) then the gradient, m, of the straight line PQ is $m = \dfrac{d - b}{c - a}$.

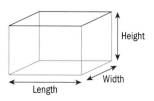

The gradient function

- To differentiate a function, find the gradient function:

Function	Gradient function
$y = ax^n$	$\dfrac{dy}{dx} = nax^{n-1}$
$f(x) = ax^n$	$f'(x) = nax^{n-1}$

The process is valid for **all** values of n, both positive and negative.

Continued on next page

Calculating the gradient of a curve at a given point

- You can use the gradient function to determine the exact value of the gradient at any specific point on the curve.

- At a local maximum or minimum, $f'(x) = 0$ $\left(\dfrac{\mathrm{d}y}{\mathrm{d}x} = 0 \right)$

The tangent and the normal to a curve

- The tangent to the curve at any point P is the straight line which passes through P with gradient equal to the gradient of the curve at P.
- To find the equation of the tangent to the curve at P(*a*, *b*):
  1. Calculate *b*, the *y*-coordinate of P, using the equation of the curve.
  2. Find the gradient function $\dfrac{\mathrm{d}y}{\mathrm{d}x}$.
  3. Substitute *a*, the *x*-coordinate of P, into $\dfrac{\mathrm{d}y}{\mathrm{d}x}$ to calculate, *m*, the value of the gradient at P.
  4. Use the equation of a straight line $(y - b) = m(x - a)$.

- The normal is perpendicular to the tangent so its gradient, m', is found using the formula $m' = \dfrac{-1}{m}$, where *m* is the gradient of the tangent.

Rates of change

- For the graph $y = f(x)$, the gradient function $\dfrac{\mathrm{d}y}{\mathrm{d}x} = f'(x)$ gives the rate of change of *y* with respect to *x*.

Local maximum and minimum points (turning points)

- At a local maximum, the curve stops increasing and changes direction so that it 'turns' and starts decreasing. So, as *x* increases, the three gradients occur in the order: positive, zero, negative. Where the gradient is zero is the maximum point.
- At a local minimum, the curve stops decreasing and changes direction; it 'turns' and starts increasing. So, as *x* increases, the three gradients occur in the order: negative, zero, positive. Where the gradient is zero is the minimum point.
- At any stationary or turning point – either local maximum or local minimum – $f'(x)$ is zero.

Using differentiation in modeling: optimization

- In optimization problems, use differentiation to find an optimal value (either the maximum or the minimum) of a function as two variables interact.

Mathematics - invention or discovery?

▶ The invention of the Hubble Telescope, which has been orbiting the Earth since 1990, has allowed astronomers to discover quasars, the existence of dark energy and the age of the universe.

■ Write down
  ● 3 'things' that have been **invented**
  ● 3 'things' that have been **discovered**.

Maybe under 'inventions' you have included such things as the wheel, the electric motor, the mp3 player. In 'discoveries' you could have included friction, electricity, magnetism, the fanged frog, the source of the Nile.

From these lists it appears that inventions are generally objects we can touch and feel, whereas discoveries are generally naturally occurring phenomena. People create inventions with their hands and with machinery. They seek new discoveries (often using new inventions to do so).

> *The laws of nature are but the mathematical thoughts of God.*
>
> Euclid

This brings us to one of the big questions in TOK about mathematics:

■ Is **mathematics** invented – just made-up, agreed-upon-conventions – or is it something humans somehow discover about the outside world?

We can use mathematics successfully to model real-world processes.

If mathematics is simply an invention of the human mind how can there be such wonderful applications in the outside world?

■ Is this because we create mathematics to mirror the world?

■ Or is the world intrinsically mathematical?

▼ In chapter 7 TOK you can see how the chambers of a nautilus shell relate to Fibonacci spirals.

From Euclidean to non-Euclidean geometry

Euclid formalized the rules of geometric shapes on flat planes. He began with a set of basic assumptions – his axioms and postulates – that seemed to come naturally from the observed world. For example, 'A straight line can be drawn between any two points'. Building upon these foundations, he proved properties of shapes, such as Pythagoras' Theorem and that interior angles in a triangle sum to 180°.

Other interesting geometrical properties are unknowable through Euclidean geometry.

For example, the angles of a triangle drawn with straight lines on the flat, 2-D surface of a sphere add up to more than 180°. Thus, non-Euclidean geometry was born, with different systems relying on new axioms.

■ Does this suggest that mathematics is an invention?

■ Can anyone start with any set of (non-contradictory) axioms that they want and create their very own mathematical system of rules, laws and theorems?

Axiomatic systems

You can create a system of axioms, but if they match the fundamental truths of the natural universe then the rules and laws arising from them are also bound by these fundamental principles. The conclusions (like Pythagoras' Theorem) already exist – whether you discover them or not. And if your system is consistent, no other conclusions are possible.

■ Does this suggest that mathematics is a discovery?

Axiomatic systems can be thought of as inventions, but they also reveal new truths about the nature of numbers – and that part is the discovery.

▶ An invention like the internal combustion engine is bound by the law of conservation of energy.

Newton vs. Leibniz

The development of calculus was truly a culmination of centuries of work by mathematicians all over the world.

The 17th century mathematicians Isaac Newton (English) and Gottfried Wilhelm Leibniz (German) are recognized for the actual development of calculus. One of the most famous conflicts in mathematical history is the argument over which one of them invented or discovered calculus first and whether any plagiarism was involved.

Today it is generally believed that Newton and Leibniz did develop calculus independently of one another.

Modern-day calculus emerged in the 19th century, due to the efforts of mathematicians such as Augustin-Louis Cauchy (French), Bernhard Riemann (German), Karl Weierstrass (German), and others.

■ What are some consequences when people seek personal acclaim for their work?

■ Suppose that Newton and Leibniz did develop calculus independently of one another. Would this offer support to the idea that calculus was discovered?

■ Did the work of these mathematicians arise from the need to solve certain real-world problems or purely from intellectual curiosity?

7 Number and algebra 2

Before you start

You should know how to:

1 Use and rearrange formulae, e.g. Given the formula $A = \pi r^2 + 2\pi rh$ find A when $r = 3$ and $h = 2$.

$A = \pi(3)^2 + 2\pi(3)(2) = 66.0$

Rearrange the formula to make h the subject.

$A - \pi r^2 = 2\pi rh \qquad h = \dfrac{A - \pi r^2}{2\pi r}$

2 Find percentages, e.g. Increase $4100 by 3%.

$\dfrac{3}{100} \times 4100 = \$123$

$\$4100 + \$123 = \$4223$ or

$\$4100 + 3\%$ of $\$4100 = \$4100(100\% + 3\%)$

$= \$4100(1.03) = \$4223$

3 Solve simultaneous equations either by hand or using GDC,

e.g.

$3x + 2y = 13$ ①

$x + 5y = 13$ ②

Eqn ② × 3: $3x + 15y = 39$ ③

Eqn ③ − ①: $13y = 26$

                  $y = 2$

Substitute $y = 2$ in ① to get $x = 3$

Skills check

1 Given the formula

$A = \pi r^2 + \pi rs$

  a find A when $r = 4$ and $s = 3$

  b rearrange the formula to make s the subject.

2 **a** Increase 630 GBP by 4%.

  b Increase 652 by 12%.

  c A shoe shop has a '20% off' sale. Find the sale price of a pair of shoes that originally cost 120 euros.

3 Solve simultaneously the equations:

$x - 2y = 11$

$3x + y = -2$

> For solving simultaneous equations using a GDC, see Chapter 12, Section 1.1.

Patterns are all around us – natural patterns and ones we build ourselves. This fence is made in sections – the first section has eight vertical pieces and two rails, the second section uses the right-hand vertical from the first one, so only has seven vertical pieces and two horizontal rails, and so on.

So the verticals give rise to the sequence 8, 15, 22, 29, 36, . . . and the horizontals are in the sequence 2, 4, 6, 8, 10, . . .

The fence isn't the only sequence pattern in this garden. Different varieties of flowers have different numbers of petals – and those numbers often occur in the Fibonacci sequence 1, 1, 2, 3, 5, 8, 13, . . .

At a microscopic level, bacteria in the soil are growing and reproducing so their total mass doubles every 8 hours. An initial mass of 0.2 grams gives the sequence 0.2, 0.4, 0.8, 1.6, 3.2, . . .

In this chapter you will learn about different types of sequences and how to use and define their rules, before exploring how these techniques are useful in a lot of different situations, including currency exchange and calculating bank interest.

> A number sequence is any pattern of numbers that follows a rule.

> What is the rule for generating the next term in the Fibonacci sequence?

> → A **sequence of numbers** is a list of numbers (finite or infinite) arranged in an order that obeys a certain rule.
> Each number in the sequence is called a **term**.

Pascal's triangle is named after French mathematician Blaise Pascal (1623–62). But the pattern was studied before Pascal was even born. Why was this number pattern named after Pascal? Who else knew about this pattern?

Investigation – number sequences

Here are three sequences of numbers.

1 2 3 4 5 6 7 …
1 8 27 64 125 216 343 …

The dots mean that the sequence carries on infinitely.

Use the internet to find out which sequences these are.
Write the next two terms in each sequence.
Look for some more sequences and see if your friends can continue the patterns.

Investigation – allowances

You have the following two options to choose from. Calculate how much money you will receive from your parents in total in each case (use 1 year = 52 weeks). Which option would you choose and why?

A Your parents give you an allowance of 5 euros a week when you are 5 years old. Every year your allowance is increased by 1 euro. You receive an allowance every week until your 21st birthday.

B Your parents give you an allowance of 5 euros a week when you are 5 years old. Every year your allowance is increased by 12%. You receive an allowance every week until your 21st birthday.

7.1 Arithmetic sequences

In an **arithmetic sequence** you can find each term by adding (or subtracting) the same number to the term before.

Here are some arithmetic sequences.

2	4	6	8	10	12	…	Add 2 each time.
3	3.5	4	4.5	5	5.5		
1	–2	–5	–8	–11	–14	…	Add –3 each time (adding –3
–5	–5.1	–5.2	–5.3	–5.4	–5.5		is the same as subtracting 3).

Here there are no dots – so the sequence has only six terms.

An arithmetic sequence goes up, or down, in equal-sized steps. The number added each time is called the **common difference**.

The first term is written as u_1; u_2 is the second term, u_3 is the third term, etc.

The common difference is written as d.

2	4	6	8	10	12	...	$u_1 = 2, d = 2$
3	3.5	4	4.5	5	5.5	...	$u_1 = 3, d = 0.5$
1	−2	−5	−8	−11	−14	...	$u_1 = 1, d = -3$
−5	−5.1	−5.2	−5.3	−5.4	−5.5	...	$u_1 = -5, d = -0.1$

d can be positive or negative.

Any arithmetic sequence can be written as:

u_1

$u_2 = u_1 + 1d$ ← Notice that 1 = (2 − 1)

$u_3 = u_2 + d = u_1 + 2d$ ← Notice that 2 = (3 − 1)

$u_4 = u_3 + d = u_1 + 3d$ ← Notice that 3 = (4 − 1)

...

Following the pattern,

> → The formula for the nth term of an arithmetic sequence is
> $$u_n = u_1 + (n − 1)d$$

The number of ds is always 1 less than the number of the term.

You can always find d by subtracting a term from the following one.
$$d = (u_2 − u_1) = (u_3 − u_2) = (u_4 − u_3) \text{ etc.}$$

Example 1

Here is a sequence of numbers.
2 5 8 11 14 17 ...
a Show that the sequence is an arithmetic sequence.
b Write down the common difference.
c Find the 10th term.
d Find the 25th term.

Answers

a 2 5 8 11 14 17 ...

The sequence goes up in equal-sized steps of 3 each time. So it is an arithmetic sequence.

b $d = 3$

c $u_{10} = 2 + (10 − 1) \times 3$
$= 2 + 27 = 29$

d $u_{25} = 2 + (25 − 1) \times 3$
$= 2 + 72 = 74$

Work out the differences between the terms.

Use the formula for the nth term with $n = 10$ and $d = 3$.

Here $n = 25$.

Example 2

The second term of an arithmetic sequence is 1 and the seventh term is 26.
a Find the first term and the common difference.
b Find the 100th term.

Answers

a $u_2 = u_1 + d = 1$
$u_7 = u_1 + 6d = 26$
$u_7 - u_2 = 6d - d = 26 - 1$
$5d = 25$
$d = 5$
$u_1 + d = 1$
$u_1 + 5 = 1$
$u_1 = -4$

Here you have two simultaneous equations. Solve them using algebra or a GDC.

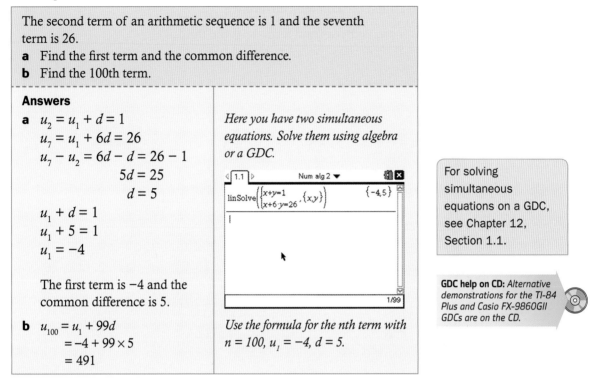

The first term is −4 and the common difference is 5.

For solving simultaneous equations on a GDC, see Chapter 12, Section 1.1.

GDC help on CD: *Alternative demonstrations for the TI-84 Plus and Casio FX-9860GII GDCs are on the CD.*

b $u_{100} = u_1 + 99d$
$= -4 + 99 \times 5$
$= 491$

Use the formula for the nth term with $n = 100$, $u_1 = -4$, $d = 5$.

Example 3

Here is a sequence of numbers   6   10   14   ...   50
a Write down the common difference.
b Find the number of terms in the sequence.

50 is the last term = u_n

Answers

a $d = 4$
b $u_n = 50 \Rightarrow u_1 + (n - 1)4 = 50$
$6 + (n - 1)4 = 50$
$(n - 1)4 = 44$
$(n - 1) = 11$
$n = 12$
So the sequence has 12 terms.

Use the formula for the nth term with $u_1 = 6$, $d = 4$. Solve for n.

Exercise 7A

EXAM-STYLE QUESTIONS

1 The first four terms of an arithmetic sequence are   3   7   11   15
  a Write down the eighth term in the sequence.
  b Find the 150th term.

2 The third term of an arithmetic sequence is 8 and the ninth term is 26.
  a Write down two equations in u_1 and d to show this information.
  b Find the values of u_1 and d.

3 The first term of an arithmetic sequence is –12 and the ninth term is 16.
Calculate the value of the common difference.

4 The first four terms of an arithmetic sequence are
3, 7, 11, 15, …
 a Write down the nth term of this sequence.
 b Calculate the 50th term of this sequence.

5 The nth term of an arithmetic sequence is $u_n = 42 - 3n$.
 a Calculate the values of the first two terms of this sequence.
 b Which term of the sequence is –9?
 c Two consecutive terms of this sequence, u_k and u_{k+1}, have a sum of 33. Find k.

> Consecutive means the two terms are next to one another.

6 The sixth term of an arithmetic sequence is 34. The common difference is 6.
 a Calculate the first term of the sequence.
The nth term is 316.
 b Calculate the value of n.

7 The first term of an arithmetic sequence is 8 and the common difference is 7. The nth term is 393. Find the value of n.

8 Here is a finite sequence.
 –5 –1 3 7 11 … 75
 a Write down the value of the common difference.
 b Find the 13th term.
 c Find the number of terms in the sequence.

9 Here is a finite sequence.
 8 10.5 13 15.5 … 188
 a Write down the value of the common difference.
 b Find the 12th term.
 c Find the number of terms that the sequence has.

10 The nth term of a sequence is given by the formula
 $u_n = 12 + 7n$.
 a Write down the first two terms.
 b Write down the common difference.
 c Find the 25th term.

The sum of the first n terms of an arithmetic sequence

The sum of the first n terms of an arithmetic sequence is called an **arithmetic series** and is written as S_n.

$$S_n = u_1 + u_2 + u_3 + u_4 + … + u_n$$

> **Carl Friedrich Gauss** (1777–1855) is often said to have been the greatest mathematician of the 19th century. Find out how Gauss worked out the sum of the first 100 integers.

Using the formula for the nth term, you can also write S_n as

$$S_n = u_1 + (u_1 + d) + (u_1 + 2d) + (u_1 + 3d) + \ldots + (u_1 + (n-1)d)$$

Writing the series backwards gives

$$S_n = (u_1 + (n-1)d) + (u_1 + (n-2)d) + (u_1 + (n-3)d) + (u_1 + (n-4)d) + \ldots + u_1$$

Adding these two series gives

$$2S_n = (2u_1 + (n-1)d) + (2u_1 + (n-1)d) + (2u_1 + (n-1)d) + (2u_1 + (n-1)d) + \ldots$$
$$+ (2u_1 + (n-1)d)$$

Since there are n terms

$$2S_n = n(2u_1 + (n-1)d)$$

> → The sum to n terms of an arithmetic sequence is given by
> the formula
> $$S_n = \frac{n}{2}(2u_1 + (n-1)d)$$

Use this form when you have the first term and the common difference.

You can rewrite this as

$$S_n = \frac{n}{2}(u_1 + u_1 + (n-1)d)$$

so, since $u_n = u_1 + (n-1)d$, this gives another formula.

> → Another formula for the sum to n terms of an arithmetic
> sequence is
> $$S_n = \frac{n}{2}(u_1 + u_n)$$

Use this form when you have the first term and the last term.

Example 4

The first four terms of an arithmetic sequence are
5   8   11   14   ...
Calculate the sum of the first 80 terms.

Answer

$$S_{80} = \frac{80}{2}(2 \times 5 + (80-1) \times 3)$$
$$= 9880$$

$n = 80$, $u_1 = 5$, $d = 3$

Use $S_n = \frac{n}{2}(2u_1 + (n-1)d)$.

Solution using GDC

Example 5

Find the sum of this series.
$$-3 \quad 1 \quad 5 \quad 9 \quad \dots \quad 81$$

Answer

$$u_n = 81 \Rightarrow u_1 + (n-1)d = 81$$
$$-3 + (n-1) \times 4 = 81$$
$$(n-1) \times 4 = 84$$
$$(n-1) = 21$$
$$n = 22$$

So, there are 22 terms.

$$S_{22} = \frac{22}{2}(-3 + 81) = 11 \times 78$$
$$= 858$$

First find out how many terms the series has. The last term is $81 = u_n$.

Use $S_n = \frac{n}{2}(u_1 + u_n)$.

1.1 ▷	Num alg 2 ▼	⬛❌
$\frac{22}{2} \cdot (-3+81)$		858

Exercise 7B

EXAM-STYLE QUESTIONS

1 Here are the first five terms of an arithmetic sequence.

  1, 6, 11, 16, 21, …

  a Write down the sixth number in the sequence.

  b Calculate the 50th term.

  c Calculate the sum of the first 50 terms of the sequence.

2 The first three terms of an arithmetic sequence are
$k + 4$, $5k + 2$ and $10k - 2$

  a Show that $k = 2$.

  b Find the values of the first three terms of the sequence.

  c Write down the value of the common difference.

  d Calculate the 25th term of the sequence.

  e Find the sum of the first 25 terms of the sequence.

3 The sixth term of an arithmetic sequence is 20 and the eleventh term is 50.

  a **i** Find the common difference.

    ii Find the first term of the sequence.

  b Calculate the sum of the first 100 terms.

4 The first four terms of an arithmetic sequence are   12, 8, 4, 0, …

  a Write down the nth term of this sequence.

  b Find the sum of the first 80 terms of this sequence.

5 The second term of an arithmetic sequence is 2 and the ninth term is -19.

  a **i** Find the common difference.

    ii Find the first term of the sequence.

  b Calculate the sum of the first 60 terms.

6 Find the sum of this arithmetic series:

$$-7 + -2 + 3 + 8 + \ldots + 238$$

7 Find the sum of this arithmetic series.

$$26 + 24.5 + 23 + 21.5 + \ldots - 17.5$$

8 The first three terms of an arithmetic sequence are

$4k - 2$, $3k + 4$ and $6k$

 a Show that $k = 2.5$.
 b Find the values of the first three terms of the sequence.
 c Write down the value of the common difference.
 d Calculate the 15th term of the sequence.
 e Find the sum of the first 15 terms of the sequence.

Applications of arithmetic sequences

You can use arithmetic sequences and series to solve problems in everyday life.

Example 6

Susan wants to buy a flat.
She has to pay for the flat in **20** yearly installments.
The first installment is 5500 euros. Each installment is 500 euros more than the one before.
 a Write down the values of the second and third installments.
 b Calculate the value of the final installment.
 c Show that the total amount that Susan would pay for the flat is 205 000 euros.

Answers

a Second installment = 6000 euros $d = 500$
Third installment = 6500 euros

b $u_{20} = 5500 + (20 - 1) \times 500$ *The final installment is u_{20}.*
$= 15\,000$ euros

c $S_{20} = \dfrac{20}{2}(2 \times 5500 + (20 - 1) \times 500)$ *The total she has to pay is S_{20}.*
Use your GDC to evaluate this.

$= 205\,000$ euros

Example 7

The sales of Smartphones are growing every year. At the end of 2006, the number sold was 25 000 000. At the end of 2010, the number sold was 35 800 000. Assuming that the sales figures follow an arithmetic sequence, calculate
 a the number of Smartphones sold at the end of 2008
 b the predicted number of Smartphones sold at the end of 2015.

Why might this **not** follow an arithmetic sequence in reality?

▶ Continued on next page

Answers

a $u_1 = 25\,000\,000$

$\quad u_5 = u_1 + 4d = 35\,800\,000$

$\quad 25\,000\,000 + 4d = 35\,800\,000$

$\qquad\qquad 4d = 10\,800\,000$

$\qquad\qquad\quad d = 2\,700\,000$

So, at the end of 2008

$\quad u_3 = 25\,000\,000 + 2 \times 2\,700\,000$

$\qquad = 30\,400\,000$

b At the end of 2015

$\quad u_{10} = 25\,000\,000 + 9 \times 2\,700\,000$

$\qquad = 49\,300\,000$

u_1 = sales for 2006
so u_5 = sales for 2010

Find d.

u_3 = sales for 2008

u_{10} = sales for 2010

u_1	2006
u_2	2007
u_3	2008
u_4	2009
u_5	2010
.	
.	
.	

Exercise 7C

EXAM-STYLE QUESTIONS

1 A woman deposits $50 into her daughter's savings account on her first birthday. On her second birthday she deposits $75, on her third birthday $100, and so on.

  a How much money will she deposit in her daughter's account on her 18th birthday?

  b How much in total will she have deposited after her daughter's 18th birthday?

2 Zain goes swimming. He swims the first length of the pool in 2.5 minutes. The time he takes to swim each length is 10 seconds more than he took to swim the previous length.

  a Find the time Zain takes to swim the third length.

  b Find the time taken for Zain to swim a total of 10 lengths of the pool.

3 Mr Zheng decides to increase the amount of money he gives to charities by p yen every year. In the first year he gives a yen. In the 6th year he gives twice as much as in the 3rd year. In the 10th year he gives 4000 yen. Find the value of p and a.

4 A lottery is offering prizes in a new competition. The winner may choose one of two options.

  Option one: $1200 each week for 10 weeks.

  Option two: $150 in the first week, $400 in the second week, $650 in the third week, increasing by $250 each week for a total of 10 weeks.

  a Calculate the amount you receive in the tenth week, if you select Option two.

  b What is the total amount you receive if you select Option two?

  c Which option has the greatest total value?

5 Jacky is playing a game. The first time that he passes the square 'Collect' he gets $100. The second time that he passes 'Collect' he gets $110. He gets $10 more each time he passes the square 'Collect'.

  a Find how much he gets when he passes the square 'Collect' for the 10th time.

  b Calculate how much he gets in total if he passes the square 'Collect' 15 times.

6 A small cinema has 25 rows of seats. The first row has 18 seats. Each row has two more seats than the previous row.

  a Find the number of seats in the 10th row.

  b Find the total number of seats in the cinema.

7.2 Geometric sequences

A sequence of numbers in which each term can be found by **multiplying** the preceding term by a **common ratio** is called a **geometric sequence**.

The first term is represented by u_1 and the common ratio by r.

Examples of geometric sequences are

2	4	8	16	32	...	$u_1 = 2$ and $r = 2$
6	3	1.5	0.75	0.375	...	$u_1 = 6$ and $r = 0.5$
3	–9	27	–81	243	...	$u_1 = 3$ and $r = -3$

> r can be positive or negative.

You can find r by dividing any term by the preceding term

i.e. $r = \dfrac{u_2}{u_1} = \dfrac{u_3}{u_2} = \dfrac{u_4}{u_3}$ etc.

Finding the nth term of a geometric sequence

The first term of a geometric sequence is u_1

The second term, $u_2 = u_1 \times r = u_1 r^1$ $1 = 2 - 1$

The third term, $u_3 = u_1 \times r \times r = u_1 r^2$ $2 = 3 - 1$

The fourth term, $u_4 = u_1 r^3$ $3 = 4 - 1$

...

The nth term is $u_n = u_1 r^{n-1}$

> The power of r is always 1 less than the number of the term.

> → The formula for the nth term of a geometric sequence is
> $$u_n = u_1 r^{n-1}$$

Example 8

Find the 8th term of this geometric sequence.
24   12   6   3   ...

Answer	
$u_1 = 24$ and $r = \dfrac{12}{24} = \dfrac{1}{2} = 0.5$	*Find r using $\dfrac{u_2}{u_1}$.*
So, $u_8 = 24(0.5)^7$	*Use $u_n = u_1 r^{n-1}$ with $n = 8$.*
$ = 0.1875$	*Use your GDC to evaluate this.*

Example 9

The second term of a geometric sequence is -15 and the fifth term is 405. Find the first term and the common ratio.

Answer	
$u_2 = u_1 \times r^1 = -15$	*Use $u_n = u_1 r^{n-1}$ to write equations for*
$u_5 = u_1 \times r^4 = 405$	*u_1 and u_5.*
$\dfrac{\cancel{u_1} \times r^4}{\cancel{u_1} \times r^1} = \dfrac{405}{-15}$	*Divide $\dfrac{u_5}{u_2}$ to cancel out u_1.*
$r^3 = -27$	
$r = -3$	
$u_2 = u_1 \times r$	*Use r to find u_1.*
$-15 = u_1 \times -3$	
$\dfrac{-15}{3} = u_1$	
$u_1 = 5$	

Exercise 7D

EXAM-STYLE QUESTIONS

1 A geometric sequence has the form 4, 8, 16, ...
   a State the common ratio for this sequence.
   b Calculate the 20th term of this sequence.

2 A geometric sequence has the form 6, 2, $\dfrac{2}{3}$, ...
   a State the common ratio for this sequence.
   b Calculate the 10th term of this sequence.

3 A geometric sequence has the form 1280, -640, 320, -160, ...
   a State the common ratio for this sequence.
   b Find the 8th term of this sequence.

4 A geometric sequence has all its terms positive.
   The first term is 5 and the third term is 20.
   a Find the common ratio.
   b Find the 7th term of this sequence.

5 The second term of a geometric sequence is 18 and the fourth term is $\frac{81}{2}$.

All the terms in the sequence are positive.

a Calculate the value of the common ratio.

b Find the 8th term in the sequence.

6 Consider the geometric sequence -16, a, -4, … for which the common ratio is $\frac{1}{2}$.

a Find the value of a.

b Find the value of the eighth term.

7 The second term of a geometric sequence is 18 and the fourth term is 8.

All the terms are positive.

Find the value of the common ratio.

8 A geometric sequence has all its terms positive. The first term is 12 and the third term is 48.

a Find the common ratio.

b Find the 12th term.

The sum of the first n terms of a geometric sequence, S_n

The sum of the terms of a geometric sequence is called a **geometric series**.

Investigation – grains of rice

An old Indian fable illustrates how the terms in a geometric sequence grow. A prince was so taken with the new game of chess that he asked its inventor to choose his reward. The man said he would like one grain of rice on the first square of the chess board, two grains on the second, four on the third, etc., doubling the number each time.

This seemed so little to ask that the prince agreed straight away. Servants started to bring in the rice – and to the prince's great surprise the grain soon overflowed the chess board to fill the palace.

How many grains of rice did the prince have to give to the man?

> This is a classic tale – you may find several different versions on the internet or in books.

Investigation – becoming a millionaire

Suppose your parents give you $0.01 the first month and double the amount every month after that.

How many months will it take you to become a millionaire?

The formula for the sum of n terms of a geometric sequence is
$$S_n = u_1 + u_1 \times r^1 + u_1 \times r^2 + u_1 \times r^3 + \ldots + u_1 \times r^{n-1}$$
$$rS_n = u_1 \times r^1 + u_1 \times r^2 + u_1 \times r^3 + \ldots + u_1 \times r^{n-1} + u_1 \times r^n$$
$$rS_n - S_n = u_1 \times r^n - u_1 \quad \text{as all the other terms cancel out.}$$
$$S_n(r - 1) = u_1(r^n - 1)$$

Multiply each term by r

Subtract the first line from the second line.

→ A formula for the sum to n terms of a geometric sequence is
$$S_n = \frac{u_1(r^n - 1)}{(r - 1)} \quad \text{where } r \neq 1$$
You can also write this as
$$S_n = \frac{u_1(1 - r^n)}{(1 - r)} \quad \text{where } r \neq 1$$

If $r = 1$ then the denominator would be zero and you cannot divide by zero.

Example 10

A geometric progression is another name for a geometric sequence.

A geometric progression has the form $6, 2, \frac{2}{3}, \ldots$

a State the common ratio for this sequence.
b Calculate the sum of the first 10 terms of this sequence.

Answers

a $r = \dfrac{2}{6} = \dfrac{1}{3}$

$r = \dfrac{u_2}{u_1}$

b $S_{10} = \dfrac{6\left(1 - \left(\frac{1}{3}\right)^{10}\right)}{\left(1 - \left(\frac{1}{3}\right)\right)} = 9.00\,(3\,\text{sf})$

Use $S_n = \dfrac{u_1(1 - r^n)}{(1 - r)}$

with $u_1 = 6$, $r = \dfrac{1}{3}$, $n = 10$.

Use your GDC to evaluate this.

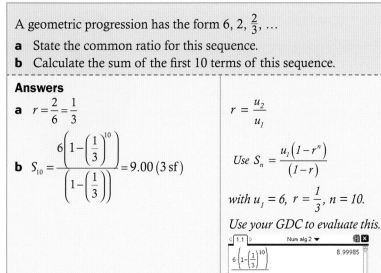

Exercise 7E

EXAM-STYLE QUESTIONS

1 The geometric sequence 16, 8, p, … has common ratio 0.5.
   a Find the value of p.   **b** Find the value of the seventh term.
   c Find the sum of the first 15 terms.

2 A geometric sequence has first term 2 and third term 32.
   a Find the common ratio.
   b Find the sum of the first 12 terms.

3 The first three terms of a geometric sequence are –2, 6, –18, …
   a State the common ratio for this sequence.
   b Calculate the sum of the first 10 terms of this sequence.

4 A geometric sequence has second term 21 and fourth term 5.25.
   a Find the common ratio.
   b Find the sum of the first 10 terms.

5 Find the sum of this geometric series.

$2 + 4 + 8 + 16 + \ldots + 8192$

6 Find the sum of this geometric series.

$-96 + 48 - 24 + 12 - \ldots - \dfrac{3}{8}$

Applications of geometric sequences

You can use geometric sequences and series to solve problems in everyday life.

> The example of the rice and the chess board in the investigation is a good example. Have you calculated the total number of grains of rice?

Example 11

Penelope is starting her first job. She will earn $24 000 in the first year and her salary will increase by 4% every year.
Calculate how much Penelope will earn in her 4th year of work.

Answer

$u_1 = \$24\,000$ and $r = 1.04$ Salary in fourth year will be $u_4 = 24\,000 \times 1.04^3$ $\quad = \$26\,996.74$	*Each year her salary increases by 4%* *so* *$r = 100\% + 4\%$* *$\quad = 104\%$* *$\quad = 1.04$*

Example 12

A ball is dropped vertically. It reaches a height of 1.6 m on the first bounce. The height of each subsequent bounce is 80% of the previous bounce.
a Find the height the ball reaches on the 6th bounce.
b Find the sum of the first seven terms of this sequence.

> Zeno van Elea (born about 490 BCE) is famous for 'Zeno's Paradox'. Research this.

Answers

a $u_1 = 1.6\,\text{m}$ and $r = 0.80$ $u_6 = 1.6 \times 0.80^5$ $\quad = 0.524\,\text{m}$	*80% = 0.80*
b $S_7 = \dfrac{1.6((0.80)^7 - 1)}{0.80 - 1}$ $\quad = 6.32 \ (3 \text{ sf})$	

Exercise 7F

1 A plant is 0.8 m tall. It grows 2% each week.
Find how tall the plant is after 10 weeks.

2 A sports car cost 75 000 GBP. Each year it loses 8% of its initial value.
Find how much the car is worth after 5 years.

> Remember that a loss of 8% means that the new value is
> 100% – 8% = 92% (or 0.92) of the original value.

3 A lottery prize is 10 Bulgarian Levs (BGN) in the first week, 20 BGN in the second week, 40 BGN in the third week, continuing to double for a total of 10 weeks.
Find the total prize money.

4 On her 21st birthday, Isabel received the following allowance from her parents: 80 Jordanian Dinar the first month and an increase of 5% every month for a full year.
a Find how much she received in month 8.
b Calculate how much Isabel received in total for the year.

5 The population of Growville is increasing at a rate of 4% each year. In 2010, the population was 210 000.
Calculate the population of Growville in 2013.

6 The population of Tomigu is growing each year. At the end of 2006, the population was 140 000. At the end of 2008, the population was 145 656. Assuming that these annual figures follow a geometric sequence, calculate
a the population of Tomigu at the end of 2007.
b the population of Tomigu at the end of 2012.

7 The tuition fees for the first three years of high school are given in the table.

Year	Tuition fees (in dollars)
1	6000
2	6300
3	6615

These tuition fees form a geometric sequence.
a Find the common ratio, r, for this sequence.
b If fees continue to rise at the same rate, calculate (to the nearest dollar) the total cost of tuition fees for the first six years of high school.

8 A mysterious illness is affecting the residents of Gezonddorp. The first day 8 people have the illness, the second day 24 new people have the illness and the third day 72 new people have the illness.
a Show that the number of new ill people forms a geometric sequence.
b Find how many new people have the illness on the 5th day.
c Find the total number of people who have the illness in the first week.

Extension material on CD:
Worksheet 7 - Infinite geometric sequences

7.3 Currency conversions

When you go on holiday from one country to another, you often have to change the currency you use at home to the currency of your holiday destination. Of course, different countries have different names for their currency.

You can change money at the airport, at a bank or at currency exchange bureaus. All of these places will advertise their **exchange rate**.

Sometimes you will be charged **commission** for changing from one currency to another.

Commission will either be a fixed amount or a percentage of the amount of money you are changing.

Sometimes there will be two exchange rates for changing currency – a **buying** rate and a **selling** rate. For example, if you live in Europe and are going on vacation in the USA then a bank will **sell** you US dollars (USD) at a certain rate. When you return, a bank will **buy** the dollars you have left and give you euros for them. Be aware, though, the bank always comes out best in any deal.

> Before coins and paper notes were invented, people used other things such as sharks' teeth, beans, sheep, tobacco, etc. to trade with.
> See if you can find other examples of unusual currencies.

> → To change one currency to another either multiply the amount by the appropriate exchange rate (if the bank is buying) or divide the amount by the appropriate exchange rate (if the bank is selling). If the bank charges a commission, calculate this first and subtract it from the amount before you change currency.

Example 13

Zoran wants to change 200 Singapore dollars (SGD) to euros.
The exchange rate is 1 SGD = 0.588 euros. The bank charges 2% commission.
Calculate how many euros Zoran receives.

Answer

2% of 200 = 0.02 × 200 = 4 SGD

So, Zoran changes 200 – 4 = 196 SGD to euros

196 SGD = 196 × 0.588 euros = 115.25 euros
Zoran receives 115.25 euros.

Calculate the commission.
Subtract the commission from the original amount.

Use 1 SGD = 0.588 euros.

Example 14

A French bank advertises the following conversion rates for buying and selling British pounds (GBP) and American dollars (USD).

BUY	SELL
1 USD = 0.761 euros	1 USD = 0.843 euros
1 GBP = 1.174 euros	1 GBP = 1.181 euros

a Write down the selling price for 1 USD.

b Peter has just returned from America and wants to exchange 250 USD back to euros. Calculate how many euros he receives.

c Jamai is going on holiday to Britain and wants to change 500 euros. Find how many GBP he receives.

> A poorer exchange rate may charge no commission. The bank can make money through the exchange rate and/or the commission.

▶ Continued on next page

Answers

a 1 USD = 0.843 euros

b 1 USD = 0.761 euros

So, 250 USD = 250 × 0.761

= 190.25 euros

Peter receives 190.25 euros.

*Peter wants the bank to **buy** his USD. Use the buying rate here.*

c 1 GBP = 1.181 euro

So, 1 euro = $\frac{1}{1.181}$ GBP

500 euros = 500 × $\frac{1}{1.181}$ GBP

= 423.37 GBP

*Jamai wants GBP. So, the bank will **sell** GBP to him.*

Why do we have different currencies? What effect do currency rate fluctuations have on trade?

Exercise 7G

EXAM-STYLE QUESTIONS

1 A family in Malaysia received a gift of 3500 USD from a cousin living in America.

The money was converted to Malaysian ringgit. One ringgit can be exchanged for 0.3236 USD.

Calculate how many ringgits the family received.

2 Joseph is spending a year traveling from America to France and Britain.

Here are the exchange rates.

> 1 US dollar (USD) = 0.783 euros
> 1 British pound (GBP) = 1.172 euros

Joseph changes 500 USD into euros.

a Calculate how many euros he receives.

He spends 328 euros in France and changes the remainder into GBP.

b Calculate how many GBP he receives.

3 A bank in Canada offers the following exchange rate between Canadian dollars (CAD) and euros (EUR). The bank sells 1 CAD for 0.821 EUR and buys 1 CAD for 0.758 EUR.

A customer wishes to exchange 800 Canadian dollars for euros.

a Find how many euros the customer will receive.

b The customer has to cancel his trip and changes his money back later when the rates are 'sells 1 CAD = 0.835 EUR, buys 1 CAD = 0.769 EUR'. Use the 'sell' information to find how many Canadian dollars he receives.

c How many Canadian dollars has he lost on the transaction?

*The bank will **buy** his CAD.*

4 Sjors is traveling to Europe. He withdraws 8000 Swedish kronor (SEK) from his savings and converts it to euros. The local bank buys SEK at 1 SEK = 0.111 euros and sells SEK at 1 SEK = 0.121 euros.

a Use the appropriate rate to calculate the amount of euros Sjors will receive.

b The trip is cancelled. How much will Sjors receive if the euros in part **a** are changed back to SEK?

c How much has Sjors lost after the two transactions?

5 1 Brazilian real (BRL) = 3.984 South African Rand (ZAR). Giving answers **correct to 2 decimal places**,

a convert 500 BRL to ZAR

b find how many BRL it costs to purchase 500 ZAR.

6 James, who lives in the UK, travels to Belgium. The exchange rate is 1.173 euros to one British pound (GBP) with a commission of 4 GBP, which is subtracted before the exchange takes place. James gives the bank 250 GBP.

a Calculate **correct to 2 decimal places** the number of euros he receives.

He buys 1 kilogram of Belgian chocolates at 2.25 euros per 100 g.

b Calculate the cost of his chocolates in GBP **correct to 2 decimal places**.

7 Jasmin plans to travel from Rotterdam to Los Angeles. She changes 2500 euros (EUR) to US dollars (USD) at an exchange rate of 1 EUR to 1.319 USD. Give all answers to this question **correct to 2 decimal places.**

a Calculate the number of USD Jasmin receives.

Jasmin spends 2050 USD and then decides to convert the remainder back to EUR at a rate of 1 EUR to 1.328 USD.

b Calculate the amount of EUR Jasmin receives.

If Jasmin had waited until she returned to Rotterdam she could have changed her USD at a rate of 1 EUR to 1.261 USD but the bank would have charged 0.6% commission.

c Calculate the amount of EUR Jasmin gained or lost by changing her money in Los Angeles.

8 The table shows some exchange rates for the Japanese yen (JPY).

Currency	1 JPY
Canadian dollar	0.01231
Chinese yuan	0.08086
Euro	0.009261
British pound	0.007897

Ayako has 2550 Japanese yen which she wishes to exchange for Chinese yuan.

a Calculate how many yuan she will receive. Give your answer to the nearest yuan.

Ross has 2150 Canadian dollars which he wishes to exchange for Japanese yen.

b Calculate how many yen he will receive. Give your answer to the nearest yen.

c Find how many British pounds there are to the euro. Give your answer correct to 2 decimal places.

9 Heindrich travels to England to buy some clothes. He changes 3000 euros to British pounds (GBP) at the rate of 1 euro = 0.8524 GBP.
The bank charges 1.5% commission.

a Find how many euros the bank charges for commission.

b Find how much Heindrich receives for his 3000 euros.

He spends 2100 GBP on clothes and changes the remaining money back into euros at an exchange rate of 1 GBP = 1.161 euro. This time the bank does not charge any commission.

c Find how many euros Heindrich receives.

10 A prize of 500 USD is awarded to three international students. Irma converts her prize money into Indian rupees (IDR) at an exchange rate of 1 USD = 44.95 IDR.

a Calculate how many IDR she receives.

Jose converts his prize money into Chilean pesos (CLP) at an exchange rate of 1 USD = 468.9 CLP. His bank charges 2% commission.

b Calculate how many CLP Jose receives.

11 Here is a currency conversion table:

	EUR	USD	JPY	GBP
Euros (EUR)	1	p	q	0.852
US dollars (USD)	0.759	1	81.92	0.647
Japanese yen (JPY)	0.00926	0.0122	1	0.0079
British pounds (GBP)	1.174	1.546	126.65	1

For example, from the table, 1 USD = 0.647 GBP. Use the table to answer these questions.

a Find the values of p and q.

b Muriel wants to change money at a bank in Bristol.

  i How many euros will she have to change to receive 150 British pounds (GBP)?

  ii The bank charges a 2.4% commission on all transactions. If she makes this transaction, how many British pounds will Muriel actually receive from the bank?

12 Minni visits Britain from America and exchanges 3000 US dollars (USD) into pounds (GBP) at an exchange rate of 1 USD = 0.652 GBP. The bank charges 2.5% commission.

a Calculate how many GBP she receives.

After Britain, Minni travels to Italy. She changes 550 GBP to euros at an exchange rate of 1 GBP = 1.18 euros.

The bank charges commission and then gives Minni 629 euros.

b Find the amount of the commission in GBP.

7.4 Compound interest

When you open a savings account at a bank, the bank pays you **interest**, depending on how much you have in your account.

The amount of money that you put in the bank at the start is called the **present value** (or **capital**).

The percentage that the bank offers is called the **rate**. You use the rate to calculate the interest on your capital over a certain time period. The interest can be compounded (added on) yearly, half-yearly, quarterly or monthly.

> Not everyone can borrow money. Research 'Credit worthiness'.

> 'Quarterly' means every quarter of a year, which is every 3 months.

Compound interest is when the interest calculated over the given time period is added to the capital and then this new capital is used to calculate the interest for the next time period.

Let PV = present value, FV = future value, r = rate and n = number of years.

If the interest is compounded yearly:

Start: PV

After 1 year: $PV + r\%$ of $PV = PV\left(1 + \dfrac{r}{100}\right)$

After 2 years: $PV\left(1 + \dfrac{r}{100}\right) + r\%$ of $PV\left(1 + \dfrac{r}{100}\right)$

$$= PV\left(1 + \frac{r}{100}\right) + \frac{r}{100}PV\left(1 + \frac{r}{100}\right)$$

$$= PV\left(1 + \frac{r}{100}\right)\left(1 + \frac{r}{100}\right)$$

$$= PV\left(1 + \frac{r}{100}\right)^2$$

and so on.

> $r\%$ of $PV = \dfrac{r}{100} \times PV$

After n years the **total** that you have in the bank is

$$FV = PV\left(1 + \frac{r}{100}\right)^n$$

If the interest is calculated **half-yearly** then the formula is

$$FV = PV\left(1 + \frac{r}{2(100)}\right)^{2n}$$

If the interest is calculated **quarterly** then the formula is

$$FV = PV\left(1 + \frac{r}{4(100)}\right)^{4n}$$

If the interest is calculated **monthly** then the formula is

$$FV = PV\left(1 + \frac{r}{12(100)}\right)^{12n}$$

> → The formula for calculating the future value of an investment with compound interest is
>
> $$FV = PV\left(1 + \frac{r}{k(100)}\right)^{kn}$$
>
> where FV is the future value, PV is the present value, r is the rate, n is the number of years and k is the number of compounding periods per year.

> You can use the **Finance Solver** on your GDC to work out number of years, rate, future value, etc. See Chapter 12, Section 7.1.

Example 15

Petra invests 6000 Costa Rican colóns (CRC) in a bank offering 4% interest compounded annually.

a Calculate the amount of money she has after 8 years.

Petra then withdraws all her money and places it in another bank that offers 4% interest per annum compounded monthly.

b Calculate the amount of money she has after 5 years.

Answers

a After 8 years

she has $6000 \times \left(1 + \dfrac{4}{100}\right)^8 = 8211.41\,\text{CRC}$

Or, using Finance Solver

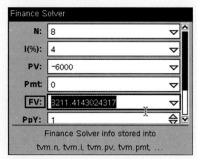

Use the formula for compounding annually.

$$FV = PV\left(1 + \frac{r}{100}\right)^n$$

with PV = 6000, r = 4, n = 8.
Round to 2 dp.

*Remember to put the capital as a **negative number** when using your Finance application.*

b $8211.41\left(1 + \dfrac{4}{12(100)}\right)^{12 \times 5}$

$= 10\,026.10\,\text{CRC}$

Or, using Finance Solver

Use the formula for compounding monthly.

$$FV = PV\left(1 + \frac{r}{k(100)}\right)^{kn}$$

where k = 12, r = 4, n = 5, PV = 8211.41

GDC help on CD: *Alternative demonstrations for the TI-84 Plus and Casio FX-9860GII GDCs are on the CD.*

Example 16

Ahmad invested 4000 Malaysian ringgits in a bank offering interest at a rate of 5% p.a. compounded quarterly.

a Calculate the amount of money that Ahmad has in the bank after 6 years.

b How long does it take for his money to double?

> p.a. stands for 'per annum' and means 'each year'.

Answers

a Ahmad will have $4000\left(1+\dfrac{5}{4(100)}\right)^{4\times6} = 5389.40$ ringgits after 6 years.

Or, using the GDC

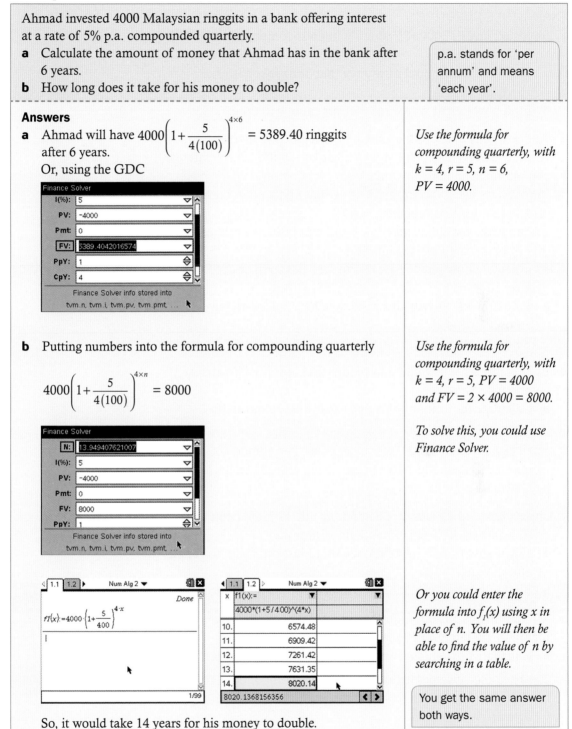

Use the formula for compounding quarterly, with k = 4, r = 5, n = 6, PV = 4000.

b Putting numbers into the formula for compounding quarterly

$$4000\left(1+\dfrac{5}{4(100)}\right)^{4\times n} = 8000$$

Use the formula for compounding quarterly, with k = 4, r = 5, PV = 4000 and FV = 2 × 4000 = 8000.

To solve this, you could use Finance Solver.

Or you could enter the formula into $f_i(x)$ using x in place of n. You will then be able to find the value of n by searching in a table.

> You get the same answer both ways.

So, it would take 14 years for his money to double.

> **GDC help on CD:** *Alternative demonstrations for the TI-84 Plus and Casio FX-9860GII GDCs are on the CD.*

Exercise 7H

1 Shunsuki invests 3000 JPY at 6.5% interest compounded annually for 15 years.

    a Calculate the amount of money that Shunsuki has after 15 years.

    b Find how long it would take for his money to double.

2 Andrew, Billy and Colin each have 2000 euros to invest.
Andrew invests his 2000 euros in a bank that offers 4.5% interest compounded annually.
Billy invests his 2000 euros in a bank that offers an interest rate of 4.4% p.a. compounded half-yearly.
Colin invests his 2000 euros in a bank that offers an interest rate of 4.3% p.a. compounded quarterly.

    a Calculate how much each of the men has in their accounts after 10 years.

    b Find how many years it would take for Andrew to have 3000 euros.

    c Find how many years it would take for Colin to double his money.

3 Brenda invests $5000 into an account that offers 3.4% interest compounded annually.

    a Calculate the amount of money that Brenda has in the account after 6 years.

Luke also invested $5000 into an account that offered r % interest compounded annually. After 6 years he had $6250 in his account.

    b Calculate r.

4 Hussein invests 20 000 Egyptian pounds (EGP) in a bank that offers a rate of 3.1% p.a. compounded monthly.

    a Calculate the amount of money that Hussein has in the bank after 5 years.

    b Find how many years that it will take for his money to double.

5 Minura invests 50 000 SGD (Singapore dollars) in an account that pays 7% interest per year, compounded yearly.

    a Calculate how much she will have in her account after 3 years.

The bank then changes the rate to 6.8% p.a. compounded monthly.

    b Calculate how much she will have in the account after another 3 years.

6 Mr Lin deposits 10 000 CNY (Chinese yuan) with Bank A that offers 8% interest p.a. compounded half-yearly.
Mr Lee deposits 10 000 CNY with Bank B that offers 8.2% interest p.a. compounded yearly.
Calculate who has earned the most interest after 2 years.

7 A bank is offering a rate of 6% per annum, compounded monthly.

Mrs Alcott invests 1000 GBP (British pounds) in this bank.

  a Calculate the amount of money she has in the account after 5 years.

Mr Bunt invests x GBP in this bank and the amount of money in his account after 5 years is 4000 GBP.

  b Calculate the value of x, correct to the nearest pound.

  c Calculate the number of years that it would take for Mrs Alcott's money to double.

8 Kelly has 8000 euros to invest. She invests a euros in Bank A which offers an interest rate of 6% compounded yearly. The remainder of her money she invests in Bank B which offers an interest rate of 5% compounded annually. She receives total interest of 430 euros at the end of the year.

  a Write an equation to represent this information.

  b Find the amount of money that Kelly invested in each bank.

Annual depreciation and inflation

An **increase** in monetary value is called **inflation**.

A **decrease** in monetary value is called **depreciation**.

To calculate inflation, you can use the formula for compound interest. To find depreciation you can use the same formula as for compound interest but the **rate will be negative** instead of positive.

Example 17

It is estimated that the value of a lump sum investment is 5% more than its value in the previous year.
Estimate the number of years that it will take for the investment to double.

Answer

Let a be the initial lump sum.

$$a\left(1+\frac{5}{100}\right)^n = 2a$$

$$\left(1+\frac{5}{100}\right)^n = 2$$

Using Finance Solver,

$n = 14.2$ years.

$$PV\left(1+\frac{r}{100}\right)^n = FV$$

Finance Solver

N:	14.206990082891
I(%):	5
PV:	-1
Pmt:	0.
FV:	2
PpY:	1

Finance Solver info stored into tvm.n, tvm.i, tvm.pv, tvm.pmt …

GDC help on CD: *Alternative demonstrations for the TI-84 Plus and Casio FX-9860GII GDCs are on the CD.*

Example 18

Lenny paid 32 000 USD for a new car. The car depreciates at a rate of 10% per annum.
Find the value of the car after 5 years.

Answer

$$32\,000\left(1-\frac{10}{100}\right)^5 = 18\,895.68\,\text{USD}$$

So, after 5 years his car will be worth 18 895.68 USD.

Exercise 7I

1 The rate of inflation is 2.3% per year. A bag of potatoes cost 3.45 euros in 2010.
   Find the cost of a bag of potatoes in 2013.

2 Pedro buys a house for 3 200 000 MXN (Mexican pesos).
   The house increases in value by 3.2% each year.
   Find how much the house is worth after 5 years.

3 Mauricio buys a car for 12 300 USD. The value of the car depreciates by 8% each year. Find the value of the car after 4 years.

4 Troy bought a gram of gold for 45 CAD (Canadian dollars).
   The price of gold increases by 2.03% each year.
   Find the value of the gold after 6 years.

5 Sangjae buys shares in a computer company for 18.95 KRW (Korean won) per share. The price of the shares depreciates by 15% per year for the following 2 years.
   Find the price of his shares after the 2 years.

6 Mrs Brash has a valuable antique vase that is worth 24 000 GBP (British pounds). The price of the vase increases each year by 1.8%.
   Find how much the vase is worth after 10 years.

7 Minna buys a new yacht for 85 000 USD. Each year the price of the yacht depreciates by 4.2%.
   Find the value of the yacht after 8 years.

8 Jenny has insured the contents of her house for 103 000 euros.
   The rate of inflation is 3.1% each year.
   How much should Jenny insure the contents of her house for in 5 years' time?

Review exercise

Paper 1 style questions

1 a At what interest rate, compounded annually, would you need to invest $500 in order to have $625 in 2 years?

b At this interest rate, how long would it take for your $500 to double in value?

2 a In a city, house prices have increased by 2.3% each year for the last three years. If a house cost USD 240 000 three years ago, calculate its value today, to the nearest dollar.

b In another city, a house worth USD 200 000 three years ago is now valued at USD 214 245. Calculate the yearly percentage increase in the value of this house.

3 Joseph decides to invest GBP 1200 of his money in a savings account which pays interest at 4.3%, compounded annually.

a How much interest will the GBP 1200 earn after 4 years?

b For how many years must Joseph invest his GBP 1200 in order to earn at least GBP 250 in interest?

c How long will it take for his money to double?

4 The exchange rate from US dollars (USD) to euros (EUR) is given by 1 USD = 0.753 EUR. Give the answers to the following correct to **two** decimal places.

a Convert 125 US dollars to euros.

b Roger receives 800 Australian dollars (AUD) for 610 EUR. Calculate the value of the US dollar in Australian dollars.

5 In 2010, Heidi joined a golf club. The fees were £1500 a year. Each year the fees increase by 3.5%.

a Calculate, **to the nearest £1**, the fees in 2012.

b Calculate the **total** fees for Heidi, who joined the golf club in 2010 and remained a member for five years.

6 Emma places €18 000 in a bank account that pays a nominal interest rate of 4.5% per annum, compounded quarterly.

a Calculate the amount of money that Emma would have in her account after 15 years. Give your answer correct to the nearest euro.

b After a period of time she decides to withdraw the money from this bank. There is €19 862.21 in her account. Find the number of months that Emma had left her money in the account.

7 The fourth term of an arithmetic sequence is 15 and the tenth term is 33.

   a Find the first term and the common difference.

   b Find the 50th term.

   c Find the sum of the first 50 terms.

8 Find the sum of this arithmetic series.

   $-15 - 13 - 11 - \ldots + 27$

9 A geometric sequence has second term 30 and fourth term 120.

   a Find the first term and the common ratio.

   b Find the 6th term.

   c Find the sum of the first 8 terms.

10 Here is a geometric sequence.

   54   18   6   2   …

   a Find the common ratio.

   b Find the 7th term.

   c Find the sum of the first 10 terms.

11 A geometric sequence has second term −4 and fourth term −1.

   a Find the first term and the common ratio.

   b Find the sixth term.

   c Find the sum of the first 6 terms.

12 Two students, Mary and John, play a game. Each time Mary passes START she receives $25. Each time John passes START he receives 15% of the amount he already has. Both students start with $200.

   a How much money will Mary have after she has passed START 10 times?

   b How much money will John have after he has passed START 10 times?

   c How many times will the students have to pass START for John to have more money than Mary?

13 The first term of an arithmetic sequence is 8 and the common difference is 8.

   a Find the value of the 36th term of the sequence.

   The first term of a geometric sequence is 3. The 6th term of the geometric sequence is equal to the 12th term of the arithmetic sequence given above.

   b Write down an equation using this information.

   c Calculate the common ratio of the geometric sequence.

Paper 2 style questions

1 A lottery is offering prizes in a new competition. The winner may choose one of three options.

Option one: $2000 each week for 10 weeks.

Option two: $1000 in the first week, $1250 in the second week, $1500 in the third week, increasing by $250 each week for a total of 10 weeks.

Option three: $15 in the first week, $30 in the second week, $60 in the third week continuing to double for a total of 10 weeks.

a Calculate the amount you receive in the eighth week, if you select

   i **option two**

   ii **option three**.

b What is the total amount you receive if you select **option two**?

c Which option has the greatest total value?

2 On Betty's 16th birthday she was given an allowance from her parents. She was given four choices.

Choice A: $150 every month of the year.

Choice B: A fixed amount of $1600 at the beginning of the year, to be invested at an interest rate of 10% per annum, compounded monthly.

Choice C: $105 the first month and an increase of $10 every month thereafter.

Choice D: $120 the first month and an increase of 5% every month.

a Assuming that Betty does not spend any of her allowance during the year, calculate, for each of the choices, how much money she would have at the end of the year.

b Which of the choices do you think Betty should choose? Give a reason for your answer.

c On her 17th birthday Betty invests $1500 in a bank that pays interest at r% per annum compounded annually. She would like to buy a car costing $1800 on her 20th birthday. What rate will the bank have to offer her to enable her to buy the car?

3 Cynthia wants to buy a house. She can choose between two different payment options. Both options require her to pay for the house in **20** yearly installments.

> Give all answers in this question correct to the **nearest** dollar.

Option 1: The first installment is $2000. Each installment is $250 more than the one before.

Option 2: The first installment is $2800. Each installment is 5% more than the one before.

a If Cynthia chooses **option 1,**
  i write down the values of the second and third installments
  ii calculate the value of the final installment
  iii show that the **total amount** that Cynthia would pay for the house is $87 500.

b If Cynthia chooses **option 2,**
  i find the value of the second installment
  ii show that the value of the fifth installment is $3403.42.

c Cynthia knows that the **total amount** she would pay for the house is not the same for both options. She wants to spend the least amount of money. Find how much she will save by choosing the cheaper option.

4 The first three terms of an arithmetic sequence are $3k + 1$, $5k$ and $6k + 4$
  a Show that $k = 5$.
  b Find the values of the first three terms of the sequence.
  c Write down the value of the common difference.
  d Calculate the 15th term of the sequence.
  e Find the sum of the first 20 terms of the sequence.

5 Arthur is starting his first job. He will earn a salary of 28 000 GBP in the first year and his salary will increase by 4% every year.
  a Calculate how much Arthur will earn in his 4th year of work.

  Arthur spends 24 000 GBP of his earnings in his first year of work. For the next few years, inflation will cause Arthur's living expenses to rise by 5% per year.
  b i Calculate the number of years it will be before Arthur is spending more than he earns.
  ii By how much will Arthur's spending be greater than his earnings in that year?

6 A geometric progression G_1 has 2 as its first term and 4 as its common ratio.
  a The sum of the first n terms of G_1 is 11 184 810. Find n.

  A second geometric progression G_2 has the form $2, \dfrac{2}{5}, \dfrac{2}{25}, \dfrac{2}{125}, \ldots$
  b State the common ratio for G_2.
  c Calculate the sum of the first 10 terms of G_2.

CHAPTER 7 SUMMARY

- A **sequence of numbers** is a list of numbers (finite or infinite) arranged in an order that obeys a certain rule.
- Each number in the sequence is called a **term**.

Arithmetic sequences

- The formula for the nth term of an arithmetic sequence is
$$u_n = u_1 + (n-1)d$$
- The sum to n terms of an arithmetic sequence is given by the formula
$$S_n = \frac{n}{2}(2u_1 + (n-1)d)$$
- Another formula for the sum to n terms of an arithmetic sequence is
$$S_n = \frac{n}{2}(u_1 + u_n)$$

Geometric sequences

- The formula for the nth term of a geometric sequence is
$$u_n = u_1 r^{n-1}$$
- A formula for the sum to n terms of a geometric sequence is
$$S_n = \frac{u_1(r^n - 1)}{(r-1)} \quad \text{where } r \neq 1$$
You can also write this as
$$S_n = \frac{u_1(1 - r^n)}{(1-r)} \quad \text{where } r \neq 1$$

Currency conversions

- To change one currency to another either multiply the amount by the appropriate exchange rate (if the bank is buying) or divide the amount by the appropriate exchange rate (if the bank is selling). If the bank charges a commission, calculate this first and subtract it from the amount before you change currency.

Compound interest

- The formula for calculating the future value of an investment with compound interest is
$$FV = PV\left(1 + \frac{r}{k(100)}\right)^{kn}$$
where FV is the future value, PV is the present value, r is the rate, n is the number of years and k is the number of compounding periods per year.

The nature of mathematics

Fibonacci: patterns in nature

The Italian mathematician, Fibonacci, Leonardo of Pisa, introduced the Fibonacci sequence in his book *Liber Abaci*, published in 1202.

In it he set this problem:

If you begin with a single pair of rabbits, and each month each pair produces a new pair which becomes productive from the second month on, how many pairs of rabbits will be produced in a year?

The diagram shows how the sequence grows:

Number of pairs

1st month: 1 pair of original two rabbits — 1

2nd month: still 1 pair as they are not yet productive — 1

3rd month: 2 pairs – original pair and the new pair they produce — 2

4th month: 3 pairs – original pair, pair they produced in 3rd month, pair they produced in 4th month — 3

The number of pairs gives the Fibonacci sequence

1, 1, 2, 3, 5, 8, 13, 21, 34, 55, 89, 144, 233, ...

— 5

where each term is the sum of the two preceding terms.

This is an idealized, hypothetical model of a situation.

- Criticize this model. What assumptions have been made? Are these assumptions reasonable?

- Do you think Fibonnacci was really trying to model a rabbit population? If not, how does this scenario help you to understand how the number pattern develops?

Flower power

The number of petals on plants are often numbers in the Fibonacci sequence

1   white calla lily
2   euphorbia
3   lilies, irises
5   buttercups, wild roses, larkspurs, columbines
8   delphiniums
13 corn marigolds, ragwort, cineraria, some daisies
21 asters, chicory
34 plantains, pyrethrum

Some daisies and flowers of the asteraceae family often have 34, 55, or 89 petals.

Fibonacci was not the only mathematician to work with this pattern.

Spirals

A **Fibonacci spiral** is formed by drawing a series of quarter-circle arcs within squares whose sides are in a Fibonacci sequence starting with a 1 × 1 square.

The resulting spiral figure is similar to the cross-section of a nautilus shell.

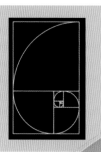

Fibonacci numbers and spirals have also been seen in:

- the arrangements of leaves around a stem
- the arrangement of seeds on flower heads
- the number of scales on the diagonals of a pineapple
- the pattern of sunflower seeds
- the number of spirals in pinecones
- the spiral of a chameleon's tail.

■ Research other examples of Fibonacci in nature.

Clearly the plants and animals do not know about this sequence – maybe they just grow in the most efficient ways.

Body facts

The human body has 2 hands each of which has 5 fingers, each of which has 3 parts separated by 2 knuckles.

All these numbers are in the Fibonacci sequence.

■ Is this a coincidence?

■ Could we be accused of looking for something where there is nothing?

A law of nature?

The Fibonacci number patterns occur so frequently in nature that the phenomenon is sometimes described as a 'law of nature'.

But there are deviations, sometimes even large ones, from Fibonacci patterns.

- There are many flowers with 4 petals (e.g. common primrose), 6 petals (hyacinth), 7 petals (starflower).

- Nautilus shells are not usually exact Fibonacci spirals.

■ How easy is it to find examples of patterns or mathematical relations in nature?

■ Does finding some examples of a pattern in nature reveal some mystical governing principle in nature?

■ Why do people ignore important cases that do not fit the pattern?

■ Are things that don't fit rationalized away as 'special cases'?

8 Sets and probability

CHAPTER OBJECTIVES:

3.5 Basic concepts of set theory: elements; subsets; intersection; union; complement; Venn diagram

3.6 Sample space; event A; complementary event A'; probability of an event; probability of a complementary event; expected value (a fair game)

3.7 Probability of combined events, mutually exclusive events, independent events; use of tree diagrams, Venn diagrams, sample space diagrams, and tables of outcomes; probability using 'with replacement' and 'without replacement'; conditional probability

Before you start

You should know how to

1 Use the terms integer, rational number, real number:

−2, 5, and 127 are integers

$\frac{1}{5}$ and $0.\dot{6} = \frac{2}{3}$ are rational numbers

$\sqrt{44}$ and $1.356\,724\,967\ldots$ are real numbers

2 Use and interpret inequalities such as $3 \le x \le 7$, $3 < x < 7$ or $3 \le x < 7$, e.g. if x is an integer and $3 \le x \le 7$, the possible values of x are 3, 4, 5, 6, 7.

3 Identify factors and prime factors, e.g.

List the factors of 18:

1, 2, 3, 6, 9, 18.

List the prime factors of 18:

2, 3.

Skills check

1 Determine whether each of the following is an integer, a rational number or a real number. If it is rational, write it as a fraction.

  a 5   **b** 1.875   **c** 0.333

  d 0.303 003 000 3...   **e** $\sqrt{0.5625}$

  f $\sqrt[3]{2.744}$ **g** π^2

2 x is an integer. For each inequality, write down the possible values of x.

  a $-2 \le x \le 3$ **b** $-3 < x \le 3$

  c $-2 \le x < 4$ **d** $-3 < x < 4$

3 **a** List the factors of

    i 12 **ii** 8 **iii** 17 **iv** 25 **v** 24

  b List the prime factors of

    i 12 **ii** 8 **iii** 17 **iv** 25 **v** 24

  c One of the numbers in part **b** is prime. Which?

  d How many factors does zero have? Is zero an integer? Rational? Real? Prime?

Here is one of the two 'integrated resorts' that have been built in Singapore. They are also known as casinos – big businesses that contribute over USD 1 000 000 000 to the country's tax revenues. Imagine the income for the companies when the **tax** they pay on it is one billion dollars!

Their business is gambling, and gambling is all about understanding the probability of winning and losing – and ensuring (as far as possible) that the casino always wins **overall**. The casino managers need to understand the laws of probability and be able to manipulate these in their favor, so that the casino makes a profit.

But if the casino **always** wins then the gamblers **always** lose, and this does not seem to be 'fair'.

In this chapter, you will investigate 'fair' games and how this idea of fairness relates to the probability of winning and losing. To do this you need to understand the fundamentals of probability theory. You will see that, although an intuitive approach is often helpful, sometimes intuition fails and you need the theory to fully understand the probability of an event.

The roots of probability lie in set theory, which can help you to visualize the problem, so this chapter begins with set theory and then goes on to apply it to probability theory.

Investigation – a contradiction?

A teacher asks her class how many of them study Chemistry. She finds
that there are 15. She then asks how many study Biology and finds
that there are 13.
Later, she remembers that there are 26 students in the class.
But 15 + 13 = 28.
Has she miscounted?

> How is it possible
> that the two totals
> are different?

What is the apparent contradiction in this problem?
How can you resolve it?

Once you have resolved the contradiction, try to answer these questions:

1 How many people study both Chemistry and Biology?
2 How many people study Chemistry but do not study Biology?
3 How many study neither subject?

Investigation – intuition

We all have a feeling about whether something is fair or not.

For example, in a football match the referee tosses a coin to decide
which team starts with possession of the ball. One team captain calls
heads and if the coin lands heads up, that team has possession.
This, we feel, is fair. But why?

> Write down *why*
> you think this
> is a fair way of
> deciding.

1 Are these scenarios 'fair'?

a To determine initial possession in a match between Team A and
Team B, the captain of Team A tosses a coin, then the captain of
Team B tosses the same coin. The team whose captain first tosses
'heads' gets possession.

b To determine initial possession in a match between Team A and
Team B, the captain of Team A chooses a number from 1 to 6. An
ordinary dice is rolled, and if that number turns up, Team A gets
possession. Otherwise, Team B gets possession.

c To determine initial possession in a match between Team A
and Team B, the captains of both teams roll an ordinary
dice once. The team whose captain rolled the higher score
gets possession.

> Is there a
> **guaranteed**
> higher score?
> What happens if
> there isn't?

2 The idea of what is fair changes when money is involved.
Are these situations involving two players, Paul and Jade, fair?
What makes a situation fair or not?

a Paul and Jade each place a bet of $1. Then an unbiased coin is
tossed. If the coin lands 'heads', Paul wins the $2; if the coin
lands 'tails', Jade wins the $2.

> An **unbiased** coin
> has an **equal**
> **probability** of
> landing on either
> 'heads' or 'tails'.

Continued on next page

b Paul and Jade each place a bet of $1. Then each takes turns to toss an unbiased coin, Paul going first. The first player to toss 'heads' wins the $2.

c Paul and Jade take turns to toss an unbiased coin, Paul going first. Each places a bet of $1 immediately before their turn. The first player to toss 'heads' wins the accumulated sum of money.

d Paul and Jade each place a bet of $1. Then an unbiased dice is rolled. If the dice shows a six, Paul wins the $2; if the dice does not show a six, Jade wins the $2.

e Paul and Jade each place a bet; Paul bets $1 and Jade bets $5. Then an unbiased cubical dice is rolled. If the dice shows a six, Paul wins the $6; if the dice does not show a six, Jade wins the $6.

> Casinos analyse all the gambling games played and ensure that they are **not** fair. Probability theory is the key to this. By the end of this chapter, you will be able to judge the merit of the statement: 'gambling is a tax on the mathematically ignorant'.

8.1 Basic set theory

> → A **set** is simply a collection of objects. The objects are called the **elements** of the set.

> We usually consider sets that have numbers as their elements; however, a set can be a collection of **any** type of object.

Some sets are so commonly used that they have their own symbols:

$\mathbb{Z}$ the set of integers $\{..., -3, -2, -1, 0, 1, 2, 3, ...\}$
$\mathbb{Z}^+$ the set of positive integers $\{1, 2, 3, ...\}$
$\mathbb{N}$ the set of natural numbers $\{0, 1, 2, 3, ...\}$
$\mathbb{Q}$ the set of rational numbers (fractions)
$\mathbb{R}$ the set of real numbers

> Is '5' a rational number? Is '–5'? Is zero a rational number?

We usually use capital letters for sets, and lower case letters for their elements.

There are several ways to describe which objects belong to a set:

$A = \{1, 2, 3, 4, 5\}$
$B = \{4, 5, 6, 7\}$
$C = \{$Names of the students absent from your school today$\}$

> You can explicitly list the elements of a set.

You can use **set builder notation**:

> You can describe the properties of the set's elements.

$D = \{x \mid 0 \leq x \leq 5\}$ The set of all numbers between 0 and 5, inclusive

$E = \{(x, y) \mid x + y = 5\}$ The set of pairs of numbers that add up to 5

$F = \{p \mid p$ is a prime number and a multiple of 10$\}$ The set of prime numbers which are a multiple of 10

> **Set builder notation** describes the properties of the elements of a set, using mathematical notation. The '|' symbol means 'such that', for example, the definition of set D is read as 'the set of all x such that x is greater than or equal to 0 and less than or equal to 5'.

$G = \{x \mid x$ is a square number less than 50$\}$ The set of square numbers less than 50.

$H = \{x \mid x \leq 200\}$ The set of all numbers less than or equal to 200.

As we shall see, the expressions used for sets D to H are not precise enough – they do not specify what sort of numbers the set elements should be. For example, if x in the definition for set D is an integer then D has six elements. If x is real, how many elements are there in D?

> → The number of elements in a set A is denoted as $n(A)$.

Set G has seven elements (assuming that x is an integer).

We write $n(G) = 7$, which is read as 'the number of elements in G is seven'.

Similarly, $n(A) = 5$ and $n(B) = 4$.
Set F has no elements, so $n(F) = 0$ and F is called the **empty set**.
The empty set is written as $\varnothing$ (or sometimes as { }).

Note that the set $\{0\}$ is **not** the empty set since it has **one** element – the number zero.

Sets A, B and G are examples of **finite sets**; they each contain a finite number of elements.

However, $n(\mathbb{Z}^+) = \infty$, so $\mathbb{Z}^+$ is an example of an **infinite set**.

Now consider set $D = \{x \mid 0 \leq x \leq 5\}$. This is read as '$x$ is any number that lies in between 0 and 5, inclusive'.

In this case it is impossible to list the elements of D, since, x has not been properly defined – it has not been stated whether x is an integer, a positive integer, a real number or a rational number.

1   If x is an **integer**, then $D = \{0, 1, 2, 3, 4, 5\}$ and $n(D) = 6$.
   In set builder notation, D is properly defined as
   $D = \{x \mid 0 \leq x \leq 5, x$ is an integer$\}$.

2   Suppose x is a **positive integer**. Then $D = \{1, 2, 3, 4, 5\}$ and $n(D) = 5$.
   In set builder notation, D is properly defined as
   $D = \{x \mid 0 \leq x \leq 5, x$ is a positive integer$\}$.

3   Suppose x is a **rational number**. Then D cannot be listed; it is an infinite set, $n(D) = \infty$.

Write down the elements of G to check how many there are.

Why is set F empty? Because, by definition, a prime number *cannot* be a multiple of 10.

Is $\{\varnothing\}$ the empty set?

Zero is an integer, but zero is not **positive**.

We can use mathematical notation to replace statements such as 'x is a positive integer' or, more precisely, 'x is an element of the set of positive integers'.

$\in$ means 'is an element of'

$\notin$ means 'is not an element of'

So $x \in \mathbb{Z}^+$ means 'x is a positive integer'.

$$1 \in A, \quad 49 \in G, \quad 8 \notin B, \quad (3,5) \notin E, \quad \pi \notin G,$$

using the sets on pages 331–332.

Example 1

Decide whether each set is well defined. Give reasons for your answer.
a $E = \{(x, y) \mid x + y = 5\}$
b $F = \{p \mid p$ is a prime number and a multiple of 10$\}$
c $H = \{x \mid x \le 200\}$

Answers

a E is **not** well defined, since we don't know which sets x and y belong to.

E becomes well defined if it is specified, for example, that $x \in \mathbb{Z}^+$, $y \in \mathbb{Z}^+$, so that $E = \{(x, y) \mid x + y = 5, x \in \mathbb{Z}^+, y \in \mathbb{Z}^+\}$.
Then $n(E) = 4$, since $E = \{(1, 4), (2, 3), (3, 2), (4, 1)\}$.

b $F = \{p \mid p$ is a prime number and a multiple of 10$\}$ **is** well defined since all multiples of 10 are integers and all prime numbers are positive integers.

$n(F) = 0$, however, since there is no prime multiple of 10.

c H is **not** well defined, since the set to which x belongs has not been specified.

$H = \{x \mid x \le 200, x \in \mathbb{N}\}$ is now well defined, and $n(H) = 201$.

Example 2

Write the set $\{5, 6, 7, 8, 9\}$ using set builder notation.

Answer

There are many different correct answers, including:

$\{x \mid 5 \le x \le 9, x \in \mathbb{Z}\}$ or $\{x \mid 5 \le x \le 9, x \in \mathbb{Z}^+\}$

$\{x \mid 5 \le x < 10, x \in \mathbb{Z}\}$ or $\{x \mid 5 \le x < 10, x \in \mathbb{Z}^+\}$

$\{x \mid 4 < x \le 9, x \in \mathbb{Z}\}$ or $\{x \mid 4 < x \le 9, x \in \mathbb{Z}^+\}$

$\{x \mid 4 < x < 10, x \in \mathbb{Z}\}$ or $\{x \mid 4 < x < 10, x \in \mathbb{Z}^+\}$

Exercise 8A

1 For each of the sets given below,
   a list its elements, if this is possible
   b state the number of elements in the set.
   $M = \{x \mid 2 \le x < 5, x \in \mathbb{Z}\}$
   $N = \{x \mid 0 < x \le 5, x \in \mathbb{Z}\}$
   $P = \{x \mid -2 \le x < 6, x \in \mathbb{Z}^+\}$
   $S = \{(x, y) \mid x + y = 5, x \in \mathbb{Z}^+, y \in \mathbb{Z}^+\}$
   $T = \{(x, y) \mid x + y = 5, x \in \mathbb{N}, y \in \mathbb{N}\}$
   $V = \{p \mid p \text{ is a prime number and a multiple of } 4\}$
   $W = \{x \mid x \text{ is a factor of } 20\}$
   $X = \{x \mid x < 200, x \in \mathbb{R}\}$

2 Here are three sets:
   $A = \{1, 2, 3, 4, 5, 6\}, B = \{2, 4, 6, 8, 10\}, C = \{3, 5, 7, 9, 11\}$.
   List the elements of the sets given by:
   a $\{x \mid x > 3, x \in A\}$ **b** $\{x \mid x \le 6, x \in B\}$
   c $\{x \mid 5 < x < 12, x \in C\}$ **d** $\{x \mid x = 2y + 1, y \in B\}$
   e $\{(x, y) \mid x = y, x, y \in B\}$ **f** $\{(x, y) \mid x = 2y, x \in B, y \in C\}$

3 Write these sets using set builder notation.
   a $\{2, 4, 6, 8, \ldots\}$ **b** $\{2, 3, 5, 7, 11, 13, \ldots\}$
   c $\{-2, -1, 0, 1, 2\}$ **d** $\{2, 3, 4, 5, 6, 7, 8\}$
   e $\{-2, 0, 2, 4, 6, 8\}$ **f** $\{3, 6, 9, 12, 15, 18\}$

8.2 Venn diagrams

Universal set

It is important to know what sort of elements are contained in a set.

In other words, in order to properly define a set, we need to define the **universal set**, those elements that are under consideration.

> → The **universal set** (denoted U), must be stated to make a set well defined.

The universal set is shown in diagrammatic form as a rectangle:

This type of set diagram is called a **Venn diagram**. Any set under consideration is shown as a circle inside the universal set.

The diagram is named after the English mathematician John Venn, who first used it.

Suppose that, as part of a problem, we were considering the months of the year that (in English) begin with the letter 'J'. Then the universal set, U, would be {January, June, July}.

Set A is defined as the set of all months which end in '...uary'.

Representing this on a Venn diagram, set A is a **subset** of U and is drawn inside the rectangle. This is written as $A \subset U$.

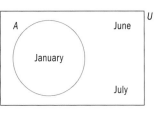

Since January $\in A$, it is written inside A. Since June, July $\notin A$, but June, July $\in U$, they are written inside the rectangle (U) but outside the circle (A).

For set $D = \{x \,|\, 0 \le x \le 5\}$ on page 331, its elements can only be defined properly when we define U. The three cases we considered were $U = \mathbb{Z}$, $U = \mathbb{Z}^+$, and $U = \mathbb{Q}$, respectively.

$\mathbb{Q}$, the set of rational numbers, is properly defined as
$$\left\{ \frac{p}{q} \,\middle|\, p, q \in \mathbb{Z}, q \ne 0 \right\}$$

Subsets

→ If every element in a given set, M, is also an element of another set, N, then M is a **subset** of N, denoted $M \subseteq N$

→ A **proper subset** of a given set is one that is **not identical** to the original set.

If M is a proper subset of N (denoted $M \subset N$) then
1 **every** element of M also lies in N and
2 there are some elements in N that do not lie in M.

If M is a proper subset of N then we write $M \subset N$.
If M could be equal to N then we write $M \subseteq N$.
Clearly, M and N are both subsets of the universal set U.

The Venn diagram on the right shows $M \subset N \subset U$.

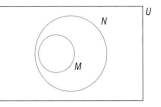

Example 3

Let $U = \{$months of the year that end (in English) with '...ber'$\}$
Let $A = \{$months of the year that begin with a consonant$\}$
Let $B = \{$months of the year that have exactly 30 days$\}$
Draw a Venn diagram to show
a sets U and A **b** sets U and B **c** sets U, A and B.

Answers

a

First, write down the sets
$U = \{$September, October, November, December$\}$
$A = \{$September, November, December$\}$
$B = \{$September, November$\}$
Note that since set A is not identical to U, we write $A \subset U$.

▶ Continued on next page

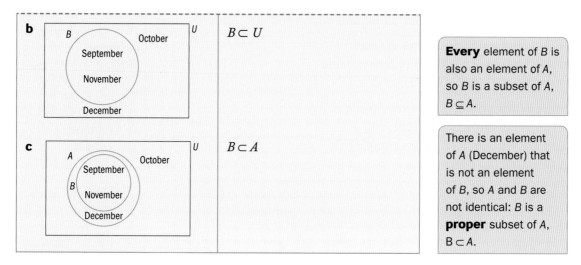

b

$B \subset U$

> **Every** element of B is also an element of A, so B is a subset of A, $B \subseteq A$.

c

$B \subset A$

> There is an element of A (December) that is not an element of B, so A and B are not identical: B is a **proper** subset of A, $B \subset A$.

Exercise 8B

Consider these sets:

$M = \{x \mid 2 \le x < 5, x \in \mathbb{Z}\}$
$N = \{x \mid 0 < x \le 5, x \in \mathbb{Z}\}$
$P = \{x \mid -2 \le x < 6, x \in \mathbb{Z}^+\}$
$S = \{(x, y) \mid x + y = 5, x \in \mathbb{Z}^+, y \in \mathbb{Z}^+\}$
$T = \{(x, y) \mid x + y = 5, x \in \mathbb{Z}, y \in \mathbb{Z}\}$
$V = \{p \mid p \text{ is a prime number and a multiple of 4}\}$
$W = \{x \mid x \text{ is a factor of 20}\}$
$X = \{x \mid x < 200, x \in \mathbb{R}\}$

State whether each statement is true or false:

1 $N \subseteq M$	**2** $S \subseteq T$	**3** $P \subseteq M$	**4** $W \subseteq X$
5 $N \subseteq P$	**6** $P \subseteq N$	**7** $\varnothing \subseteq W$	**8** $W \subseteq W$

In Exercise 8B you should have found that the last two examples were true:

For question 7, $\varnothing \subseteq W$ since every element of $\varnothing$ is in W. The fact that there are no elements in $\varnothing$ makes this certain!

Alternatively, there is no element in $\varnothing$ which is **not** in W, therefore W must contain $\varnothing$. Hence, $\varnothing$ is a subset of W.

This argument is valid for sets M, N, P, S too. In fact, it is valid for all sets.

> → The empty set $\varnothing$ is a subset of every set.

For question 8, since every element of W is in W, $W \subseteq W$. And the same argument is valid for all sets.

> → Every set is a subset of itself.

> When considering subsets you don't usually need to include either the empty set or the original set itself. The empty set and the set itself are not proper subsets of any set.

Exercise 8C

1 Consider these sets:

$M = \{x \mid 2 \leq x < 5, x \in \mathbb{Z}\}$

$N = \{x \mid 0 < x \leq 5, x \in \mathbb{Z}\}$

$P = \{x \mid -2 \leq x < 6, x \in \mathbb{Z}^+\}$

$S = \{(x, y) \mid x + y = 5, x \in \mathbb{Z}^+, y \in \mathbb{Z}^+\}$

$T = \{(x, y) \mid x + y = 5, x \in \mathbb{Z}, y \in \mathbb{Z}\}$

$V = \{p \mid p$ is a prime number and a multiple of 4$\}$

$W = \{x \mid x$ is a factor of 20$\}$

$X = \{x \mid x < 200, x \in \mathbb{R}\}$

State whether each statment is true or false:

a $N \subset M$ **b** $S \subset T$ **c** $P \subset M$ **d** $W \subset X$

e $M \subset P$ **f** $P \subset N$ **g** $\varnothing \subset T$ **h** $V \subset W$

2 a List all the subsets of
  i $\{a\}$ **ii** $\{a, b\}$ **iii** $\{a, b, c\}$ **iv** $\{a, b, c, d\}$
b How many subsets does a set with n members have?
c How many subsets does $\{a, b, c, d, e, f\}$ have?
d A set has 128 subsets. How many elements are there in this?

3 a List all the proper subsets of
  i $\{a\}$ **ii** $\{a, b\}$ **iii** $\{a, b, c\}$ **iv** $\{a, b, c, d\}$
b How many proper subsets does a set with n members have?
c How many proper subsets has $\{a, b, c, d, e, f\}$?
d A set has 254 subsets. How many elements are there in this?

Intersection

→ The **intersection** of set M and set N (denoted $M \cap N$) is the set of all elements that are in **both M and N**.

$M \cap N$ is the shaded region on the Venn diagram:

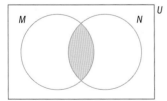

Example 4

Given the sets:
$A = \{1, 2, 3, 4, 5\}$
$B = \{x \mid 0 < x \leq 5, x \in \mathbb{Z}\}$
$C = \{p \mid p \text{ is a prime number and a multiple of } 10\}$
$D = \{4, 5, 6, 7\}$
$E = \{x \mid x \text{ is a square number less than } 50\}$
write down the sets
a $A \cap D$ **b** $A \cap B$ **c** $D \cap E$ **d** $C \cap D$

Answers	
	First, list the elements of each set:
	$A = \{1, 2, 3, 4, 5\}$
	$B = \{1, 2, 3, 4, 5\}$
	$C = \varnothing$
	$D = \{4, 5, 6, 7\}$
	$E = \{1, 4, 9, 16, 25, 36, 49\}$
	Compare the sets
a $A \cap D = \{4, 5\}$.	$A = \{1, 2, 3, 4, 5\}$ *and*
	$D = \{4, 5, 6, 7\}$.
b $A \cap B = \{1, 2, 3, 4, 5\}$.	*Sets A and B are identical.*
c The element 4 lies in both	$D = \{4, 5, 6, 7\}$ *and*
sets, hence $D \cap E = \{4\}$.	$E = \{1, 4, 9, 16, 25, 36, 49\}$.
d $C \cap D = \varnothing$.	*C does not contain any elements;*
	hence there is no element that lies in
	both sets.

Is it always true that for any set X:
$\varnothing \cap X = \varnothing$ and
$X \cap X = X$?

Union

→ The **union** of set M and set N (denoted $M \cup N$) is the set of all elements that are in **either M or N or both**.

$M \cup N$ is the shaded region on the Venn diagram:

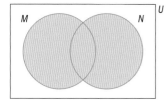

$M \cup N$ includes those elements that are in **both** M and N. This is important!

Example 5

Given the sets:
$A = \{1, 2, 3, 4, 5\}$
$B = \{1, 2, 3, 4, 5\}$
$C = \varnothing$
$D = \{4, 5, 6, 7\}$
$E = \{1, 4, 9, 16, 25, 36, 49\}$
Write down the sets
a $A \cup D$ **b** $A \cup B$ **c** $C \cup D$

- -

Answers

a $A \cup D = \{1, 2, 3, 4, 5, 6, 7\}$

b $A \cup B = \{1, 2, 3, 4, 5\}$

c $C \cup D = \{4, 5, 6, 7\}$

$A = \{1, 2, 3, 4, 5\}$ and
$D = \{4, 5, 6, 7\}$.
To write down $A \cup D$ list **every**
element of each set, but **only once**.
A and B are identical.
$C = \varnothing$ and $D = \{4, 5, 6, 7\}$.
$C \cup D = D$, since there are no extra
elements to list from C.

> Is it always true that
> for any set X:
> $\varnothing \cup X = X$ and
> $X \cup X = X$?

Complement

→ The **complement** of set M, denoted as M', is the set of all the
elements in the universal set that **do not** lie in M.

M' is the shaded part of this Venn diagram:

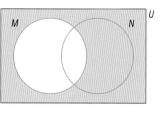

→ The complement of the universal set, U', is the empty set, $\varnothing$.

We can use Venn diagrams to
represent different combinations
of set complement, intersection
and union. For example, $M \cap N'$ is
shown here:

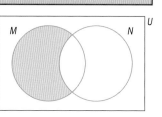

To see this in more detail, look at at the separate diagrams of M and N':

M

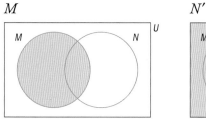

N'

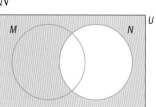

Combining these for the intersection $M \cap N'$ gives shading only in the area common to both diagrams.

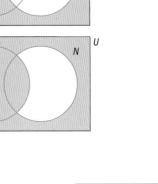

This diagram shows the set $M \cup N'$. Since it is the region that satisfies **either M or N'**, it includes the shading from both diagrams.

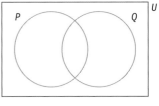

Exercise 8D

1 Copy the Venn diagram for sets P and Q. Shade the region that represents

  a $P \cup Q'$ **b** $P \cap Q'$ **c** $P' \cup Q'$

  d $P' \cap Q'$ **e** $(P \cup Q)'$ **f** $(P \cap Q)'$

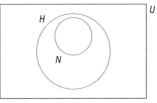

2 Copy the Venn diagram for sets H and N. Shade the region that represents

  a H' **b** $H \cap N'$ **c** N'

  d $H' \cup N'$ **e** $H' \cap N'$ **f** $H \cup N'$

3 Copy the Venn diagram for sets W and R. Shade the region that represents

  a W' **b** $W' \cap R'$ **c** $W' \cap R$

  d $W' \cup R'$ **e** $(W \cup R)'$ **f** $(W' \cap R)'$

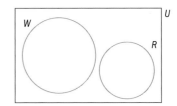

4 U is defined as the set of all integers. Consider the following sets:

$A = \{1, 2, 3, 4, 5\}$

$B = \{x \mid 0 \leq x < 5, x \in \mathbb{Z}\}$

$C = \{p \mid p \text{ is an even prime number}\}$

$D = \{4, 5, 6, 7\}$

$E = \{x \mid x \text{ is a square number less than } 50\}$

Write down the sets:

  a $A \cap B$ **b** $B \cap E$ **c** $C \cap D$ **d** $C \cap E$ **e** $B \cap D$

  f $A \cup B$ **g** $B \cup A$ **h** $C \cup D$ **i** $C \cup A$ **j** $B \cup D$

Decide whether each statment is true or false.

  k $A \subset B$ **l** $B \subset A$ **m** $C \subset A$ **n** $C \subset D$ **o** $(C \cap D) \subset E$

Venn diagrams can show individual set elements as well.

Example 6

$U = \{4, 5, 6, 7, 8, 9, 10\}$, $F = \{4, 5, 6, 7\}$ and $G = \{6, 7, 8, 9\}$.

a Draw a Venn diagram for F, G and U.

b Write down these sets:

   i F' **ii** $F \cap G'$ **iii** $(F \cap G)'$

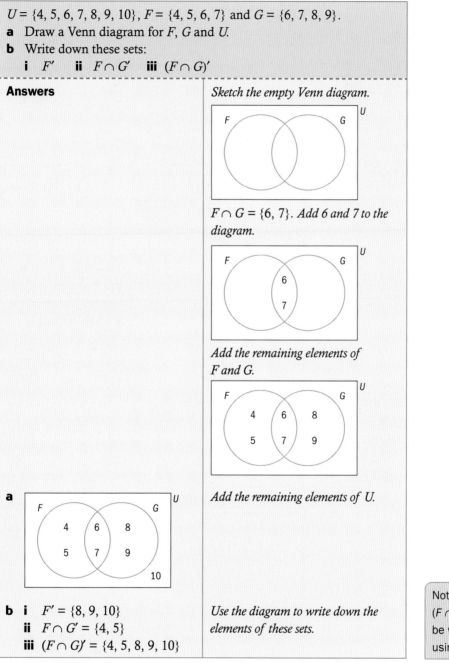

Answers

Sketch the empty Venn diagram.

$F \cap G = \{6, 7\}$. *Add 6 and 7 to the diagram.*

Add the remaining elements of F and G.

Add the remaining elements of U.

a

b **i** $F' = \{8, 9, 10\}$

   ii $F \cap G' = \{4, 5\}$

   iii $(F \cap G)' = \{4, 5, 8, 9, 10\}$

Use the diagram to write down the elements of these sets.

> Note that $F \cap G' \neq$ $(F \cap G)'$. You must be very precise when using brackets.

You can use Venn diagrams to work out the **number of elements** in each set without writing them all down.

Example 7

In this Venn diagram, each dot represents an element.
Write down:
a $n(G)$
b $n(F)$
c $n(G \cap F)$
d $n(H')$
e $n(F \cap H)$
f $n(G \cap H)$

Is each statement true or false?
g $n(F \cup H) = n(F) + n(H)$
h $n(G \cup H) = n(G) + n(H)$

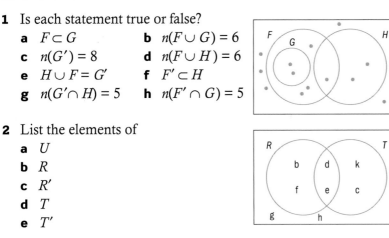

Answers

a $n(G) = 2$ *Count the dots in each set.*
b $n(F) = 6$
c $n(G \cap F) = 2$
d $n(H') = 10$
e $n(F \cap H) = 1$
f $n(G \cap H) = 0$
g The statement is false. $n(F \cup H) = 8, n(F) = 6, n(H) = 3$
h The statement is true. $n(G \cup H) = 5, n(G) = 2, n(H) = 3.$

> The statements in **e** and **f** help you decide whether statements **g** and **h** are true or false.

Exercise 8E

1 Is each statement true or false?
a $F \subset G$ **b** $n(F \cup G) = 6$
c $n(G') = 8$ **d** $n(F \cup H) = 6$
e $H \cup F = G'$ **f** $F' \subset H$
g $n(G' \cap H) = 5$ **h** $n(F' \cap G) = 5$

2 List the elements of
a U
b R
c R'
d T
e T'

3 List the elements of
a A
b A'
c $A \cup B'$
d $A \cap B'$
e $A' \cup B'$

8.3 Extending to three sets

This Venn diagram shows a general three-set problem.

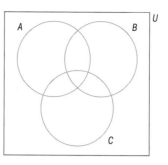

> Use the same notation for three sets. But take great care using brackets to describe the sets.

Example 8

Shade the region on a Venn diagram that shows the sets:

a $(A \cup B) \cap C$ **b** $A \cup (B \cap C)$

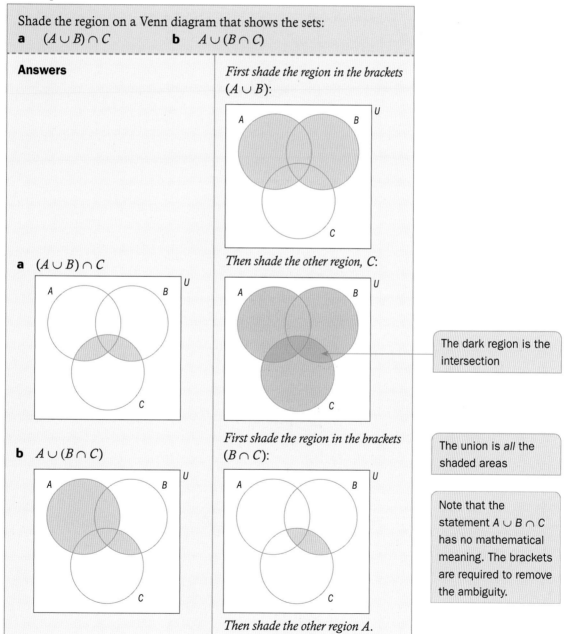

Answers

First shade the region in the brackets $(A \cup B)$:

a $(A \cup B) \cap C$

Then shade the other region, C:

> The dark region is the intersection

b $A \cup (B \cap C)$

First shade the region in the brackets $(B \cap C)$:

Then shade the other region A.

> The union is *all* the shaded areas

> Note that the statement $A \cup B \cap C$ has no mathematical meaning. The brackets are required to remove the ambiguity.

1 Shade the region on a three-set Venn diagram that shows each set:

a i $(A \cup B) \cup C$ **ii** $A \cup (B \cup C)$
b i $(A \cap B) \cap C$ **ii** $A \cap (B \cap C)$
c i $(A \cup C) \cap B$ **ii** $A \cup (C \cap B)$
d i $C \cap (A \cup B)$ **ii** $B \cup (C \cap A)$
e i $(A \cup B) \cup C'$ **ii** $A \cup (B \cup C')$
f i $(A \cap B') \cap C$ **ii** $A \cap (B' \cap C)$
g i $(A \cup C) \cap B'$ **ii** $A \cup (C \cap B')$

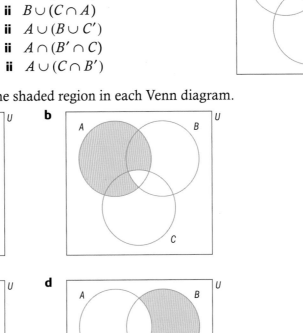

2 Use set notation to name the shaded region in each Venn diagram.

a

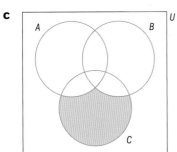

b

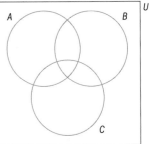

c

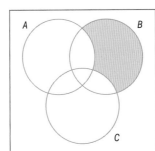

d

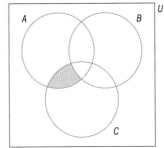

e

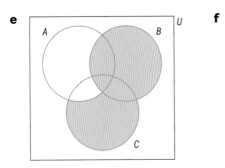

f

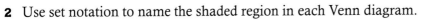

g

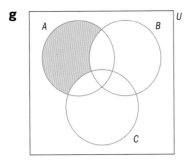

h

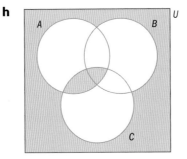

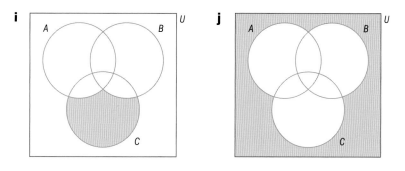

3 In this Venn diagram, $U = \{1, 2, 3, 4, 5, 6, 7, 8\}$.

List the elements of:

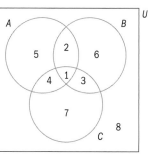

a $A \cap B \cap C$

b $A' \cap B \cap C$

c $A \cap B' \cap C$

d $A \cap B \cap C'$

e $A' \cap B' \cap C$

f $A' \cap B \cap C'$

g $A \cap B' \cap C'$

h $A' \cap B' \cap C'$

4 For the Venn diagram in question 3, list the elements of:

a $A \cap (B \cup C)$ **b** $A' \cap (B \cup C)$

c $(A \cup B') \cap C$ **d** $(A \cup B) \cap C'$

e $(A' \cup B') \cap C$ **f** $(A' \cup B) \cap C'$

g $B \cap (A' \cup C')$ **h** $B' \cap (A' \cup C)$

8.4 Problem-solving using Venn diagrams

Here is the problem from the first investigation in this chapter:

Investigation – a contradiction?

A teacher asks her class how many of them study Chemistry.
She finds that there are 15. She then asks how many study Biology and
finds that there are 13.
Later, she remembers that there are 26 students in the class.
But 15 + 13 = 28. Has she miscounted?

We can represent this problem on a Venn diagram.

Let B be the set of students studying Biology, and C be the set of students studying Chemistry. Then $n(B) = 13$, $n(C) = 15$ and $n(U) = 26$.

The teacher asks another question and finds out that 5 of the students study neither Biology nor Chemistry, so $n(B' \cap C') = 5$.

We can put what we know, and what we don't know, on a Venn diagram:

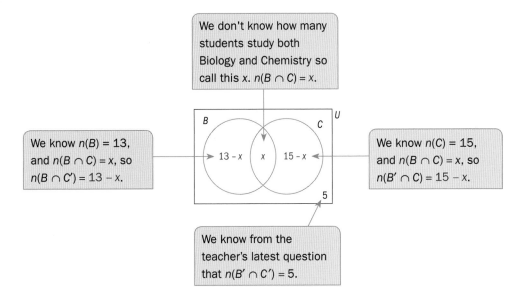

We don't know how many students study both Biology and Chemistry so call this x. $n(B \cap C) = x$.

We know $n(B) = 13$, and $n(B \cap C) = x$, so $n(B \cap C') = 13 - x$.

We know $n(C) = 15$, and $n(B \cap C) = x$, so $n(B' \cap C) = 15 - x$.

We know from the teacher's latest question that $n(B' \cap C') = 5$.

We also know that $n(U) = 26$. From the Venn diagram we can write

$$(13 - x) + x + (15 - x) + 5 = 26$$
$$33 - x = 26$$
$$x = 7$$

So now we can substitute for x on the Venn diagram, and answer questions like 'How many students study Chemistry but not Biology?'

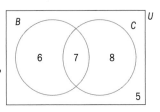

Exercise 8G

Use the Venn diagram to answer these questions:

1 How many students study Biology **only**? (That is, 'Biology, but not Chemistry'.)

2 How many students study **exactly** one science? (That is, 'Biology or Chemistry, but not **both**'.)

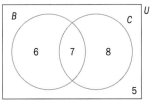

3 How many students study **at least** one science? (That is, 'Biology or Chemistry, or **both**'.)

4 How many students study one science? (That is, 'Biology or Chemistry, or **both**'.)

If you study two sciences, then you necessarily must study one!

5 How many students do not study Biology?

6 How many students do not study Chemistry?

7 How many Chemists study Biology?

8 How many Biologists do not study Chemistry?

9 How many science students do not study both Biology and Chemistry?

Example 9

In a class of 29 students, 19 study German, 14 study Hindi and 5 study both languages. Work out the number of students that study neither language.

Answer

Let G be the set of students who study German, and H the set of those studying Hindi.
From the information in the question

$$n(U) = 29 \qquad n(G) = 19 \qquad n(H) = 14 \qquad n(G \cap H) = 5$$

Draw a Venn diagram:

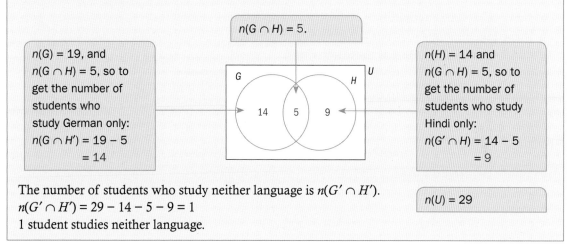

$n(G) = 19$, and $n(G \cap H) = 5$, so to get the number of students who study German only:
$n(G \cap H') = 19 - 5$
$\qquad = 14$

$n(G \cap H) = 5$.

$n(H) = 14$ and $n(G \cap H) = 5$, so to get the number of students who study Hindi only:
$n(G' \cap H) = 14 - 5$
$\qquad = 9$

The number of students who study neither language is $n(G' \cap H')$.
$n(G' \cap H') = 29 - 14 - 5 - 9 = 1$
1 student studies neither language.

$n(U) = 29$

Exercise 8H

1 There are 25 students in a class. 17 study French, 12 study Malay, and 10 study both languages.
   Show this information on a Venn diagram.
   Find the number of students who:
   a study French only **b** study Malay or French or both
   c study neither subject **d** do not study both subjects.

2 In a class 20 people take Geography, 17 take History, 10 take both subjects, and 1 person takes neither subject.
   Draw a Venn diagram to show this information.
   Find the number of students who:
   a are in the class **b** do not study History
   c study Geography but not History
   d study Geography or History but not both.

3 Of the 32 students in a class, 18 play the violin, 16 play the piano, and 7 play neither. Find the number of students who:
   a play the violin but not the piano **b** do not play the violin
   c play the piano but not the violin
   d play the piano or the violin, but not both.

4 There are 30 students in a mathematics class. 20 of the students have studied probability, 14 have studied set theory, and two people have studied neither.

Find the number of students who:

a have studied both topics

b have studied exactly one of these subjects

c have studied set theory, but not probability.

5 There are 25 girls in a PE group. 13 have taken aerobics before and 17 have taken gymnastics. One girl has done neither before.

Find the number of girls who:

a have taken both activities

b have taken gymnastics but not aerobics

c have taken at least one of these activities.

You can use the same ideas to draw Venn diagrams with more sets; see the following example.

Example 10

145 people answered a survey to find out which flavor of fruit juice, orange, apple or grape, they preferred.

The replies showed:

15 liked none of the three	35 liked orange and apple
55 liked grape	20 liked orange and grape
80 liked apple	30 liked apple and grape
75 liked orange	

Find the number of people who liked all three types of juice.

Answer

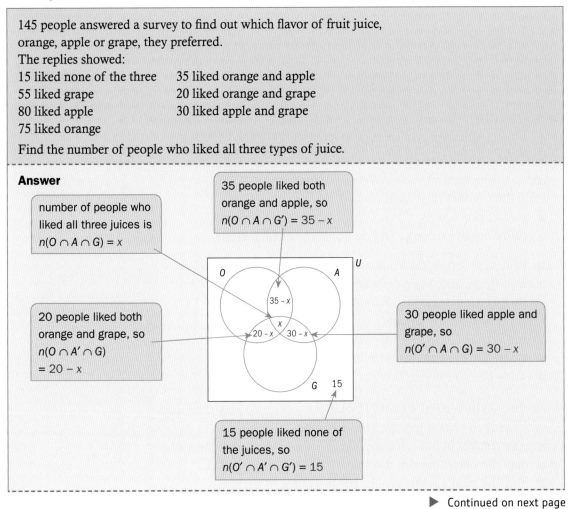

35 people liked both orange and apple, so
$n(O \cap A \cap G') = 35 - x$

number of people who liked all three juices is
$n(O \cap A \cap G) = x$

20 people liked both orange and grape, so
$n(O \cap A' \cap G) = 20 - x$

30 people liked apple and grape, so
$n(O' \cap A \cap G) = 30 - x$

15 people liked none of the juices, so
$n(O' \cap A' \cap G') = 15$

▶ Continued on next page

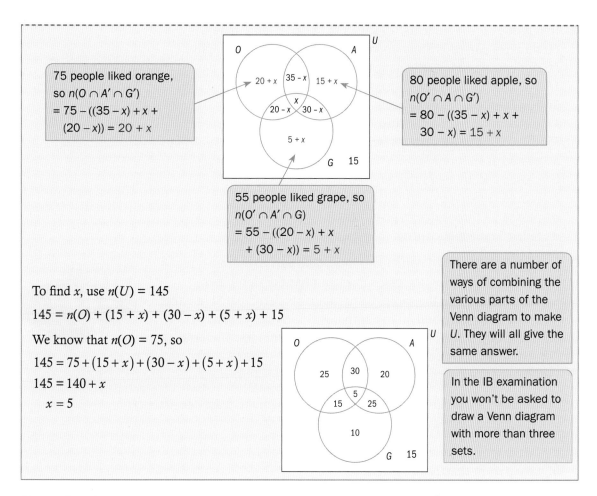

75 people liked orange, so $n(O \cap A' \cap G')$
$= 75 - ((35 - x) + x + (20 - x)) = 20 + x$

80 people liked apple, so $n(O' \cap A \cap G')$
$= 80 - ((35 - x) + x + 30 - x) = 15 + x$

55 people liked grape, so $n(O' \cap A' \cap G)$
$= 55 - ((20 - x) + x + (30 - x)) = 5 + x$

There are a number of ways of combining the various parts of the Venn diagram to make U. They will all give the same answer.

To find x, use $n(U) = 145$

$145 = n(O) + (15 + x) + (30 - x) + (5 + x) + 15$

We know that $n(O) = 75$, so

$145 = 75 + (15 + x) + (30 - x) + (5 + x) + 15$

$145 = 140 + x$

$x = 5$

In the IB examination you won't be asked to draw a Venn diagram with more than three sets.

Exercise 8I

Use the information from Example 10 to answer these questions.
1 Find the number in the survey above who
   a liked exactly two of the three flavors of juice
   b did not like orange juice
   c liked one flavor of juice only
   d did not like either orange or apple juice
   e did not like orange juice and did not like apple juice
   f liked at least two of the three flavors of juice
   g liked fewer than two of the three flavors of juice.

Find the number of orange juice drinkers who
   h liked apple juice
   i did not like grape juice
   j liked no other flavors of juice
   k liked exactly one other flavor of juice.

2 In a group of 105 students, 70 students passed Mathematics,
60 students passed History and 45 students passed Geography;
30 students passed Mathematics and History, 35 students passed
History and Geography, 25 passed Mathematics and Geography,
and 15 passed all three subjects.
Draw a Venn diagram to illustrate this information.
Find the number of students who

a passed at least one subject

b passed exactly two subjects

c passed Geography and failed Mathematics

d passed all three subjects given that they passed two

e failed Mathematics given that they passed History.

3 In a youth camp, each participant must take part in at least one
of the following activities: chess, backgammon or dominoes.
Of the total of 55 in the camp, 25 participants participated in chess,
24 in backgammon, and 30 in dominoes; 15 in both chess and
backgammon, 10 in both backgammon and dominoes, 5 in both
chess and dominoes, and 2 in all three events.

Draw a Venn diagram to show this information.

How many of the participants are not taking part in at least one activity?

Find the number of participants who

a take part in one activity only

b take part in exactly two activities

c do not take part in at least two activities

d take part in chess, given that they take part in dominoes

e take part in backgammon, given that they do not take
part in dominoes.

4 Fatty's Delight sells chicken, duck, and barbecued pork rice.
Of the 160 customers one day, 57 had chicken rice, 60 had
duck rice and 48 had barbecued pork rice. 30 customers ordered
chicken and duck rice, 25 ordered duck and barbecued pork rice,
35 ordered chicken and barbecued pork rice, and 20 ordered
all three types.
Draw a Venn diagram to show these data.

Find the number of customers who

a ordered more than one type of rice

b did not order a rice dish from Fatty's Delight

c did not order chicken rice

d ordered duck rice and one other rice dish.

5 In a community center in Buona Vista there are 170 youths. Of these, 65 take up climbing, 65 bouldering and 50 swimming; 15 take up climbing and bouldering, 10 bouldering and swimming, and 5 swimming and climbing. 17 youths take up other activities.

Let x be the number of youths who take up all three activities.
Show the above information in a Venn diagram.
Show clearly the number in each separate region in terms of x.
Form an equation satisfied by x, and hence find its value.

Find the number of youths who
  a take up one activity only
  b take up at least two activities
  c take part in fewer than two activities
  d take up bouldering given that they have already taken up climbing
  e take up one other activity given that they have already taken up swimming.

6 65 elderly men failed a medical test because of defects in at least one of these organs: the heart, lungs or kidneys. 29 had heart disease, 28 lung disease and 31 kidney disease. 8 of them had both lung and heart diseases, 11 had lung and kidney diseases, while 12 had kidney and heart diseases.

Draw a Venn diagram to show this information. You will need to introduce a variable.

Find the number of men who
  a suffer from all three diseases
  b suffer from at least two diseases
  c suffer from lung disease and exactly one other disease
  d suffer from heart disease and lung disease but not kidney disease
  e suffer from lung disease only.

7 Each of the 116 students in the Fourth Year of a school studies at least one of the subjects History, English and Art.
Of the 50 students who study Art,
      15 also study History and English,
      12 study neither History nor English, and
      17 study English but not History.
Of the 66 students who do not study Art,
      39 study both History and English,
      x study History only, and
      $2x$ study English only.

Draw a Venn diagram showing the number of students in each subset. Hence find

  a the value of x **b** the total number of students studying English.

8.5 Basic probability theory

Probability is the branch of mathematics that analyses random experiments. A **random experiment** is one in which we cannot predict the precise outcome. Examples of random experiments are 'tossing a coin' or 'rolling a dice' or 'predicting the gold, silver, and bronze medalists in a 100 m sprint'.

It is impossible to predict the outcome of a random experiment **precisely** but it is possible to
a   list the set of all possible outcomes of the experiment
b   decide how likely a particular outcome may be.

When tossing a coin, there are two possible outcomes: **heads** (H) and **tails** (T).

Also, the likelihood of getting a head is the same as getting a tail, so the probability of getting a head is one chance out of two. The probability of getting a tail is the same.

In other words, the set of equally likely outcomes is $\{H, T\}$ and $P(H) = P(T) = \dfrac{1}{2}$.

When rolling a dice, the set of equally likely possible outcomes has six elements and is $\{1, 2, 3, 4, 5, 6\}$.

As all six outcomes are equally likely, $P(1) = P(2) = \ldots = P(6) = \dfrac{1}{6}$.

Let event A be 'rolling an even number'.

To find $P(A)$, consider the set of equally likely outcomes $\{1, 2, 3, 4, 5, 6\}$.

There are six equally likely outcomes and three of these are even numbers, so $P(A) = \dfrac{3}{6}$.

Let B be the event 'rolling a prime number'.

To find $P(B)$, look again at the set of outcomes. There are three prime numbers: 2, 3, and 5 so, $P(B) = \dfrac{3}{6}$.

We can show the equally likely possible outcomes of rolling a dice on a Venn diagram using $U = \{1, 2, 3, 4, 5, 6\}$ and $A = \{\text{even numbers}\}$.

$$P(A) = \frac{n(A)}{n(U)} = \frac{3}{6}$$

Set B can be added to the Venn diagram to represent the event B.

$$P(B) = \frac{n(B)}{n(U)} = \frac{3}{6}$$

> There are some assumptions being made:
> **1** the coin is unbiased
> **2** the dice is unbiased
> **3** all sprinters are evenly matched

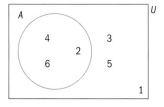

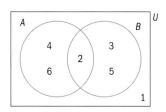

> → If all of the equally likely possible outcomes of a random experiment can be listed as U, the universal set, and an event A is defined and represented by a set A, then:
>
> $$P(A) = \frac{n(A)}{n(U)}$$
>
> There are three consequences of this law:
>
> **1** $\quad P(U) = \dfrac{n(U)}{n(U)} = 1 \quad$ (the probability of a **certain** event is 1)
>
> **2** $\quad P(\varnothing) = \dfrac{n(\varnothing)}{n(U)} = 0 \quad$ (the probability of an **impossible** event is 0)
>
> **3** $\quad 0 \leq P(A) \leq 1 \quad$ (the probability of any event **always** lies between 0 and 1)

Example 11

Find the probability that these events occur for the random experiment 'rolling a fair dice'.

a Rolling an odd number
b Rolling an even prime number
c Rolling an odd prime number
d Rolling a number that is either prime or even

> Unless stated otherwise, we will always be talking about a cubical dice with faces numbered 1 to 6.

Answers

a $\quad P(A') = \dfrac{n(A')}{n(U)} = \dfrac{3}{6}$

b $\quad P(A \cap B) = \dfrac{n(A \cap B)}{n(U)} = \dfrac{1}{6}$

c $\quad P(A' \cap B) = \dfrac{n(A' \cap B)}{n(U)} = \dfrac{2}{6}$

d $\quad P(A \cup B) = \dfrac{n(A \cup B)}{n(U)} = \dfrac{5}{6}$

Use the Venn diagram drawn earlier, where A is the event 'rolling an even number' and B is the event 'rolling a prime number'.
A is the event 'rolling an even number', so the probability of rolling an odd number is $P(A')$. From the Venn diagram, $A' = \{1, 3, 5\}$.

A is the event 'rolling an even number', and B is the event 'rolling a prime number', so the probability of rolling an even prime number is $P(A \cap B)$.

The probability of rolling an odd prime number is $P(A' \cap B)$.

The probability of rolling a number that is either prime or even is $P(A \cup B)$.

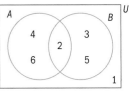

> This example illustrates the basics of probability theory: list the equally likely possible outcomes of a random experiment and count. Drawing a Venn diagram may clarify the situation.

Two further laws of probability:

→ • For complementary events, $P(A') = 1 - P(A)$
 • For combined events, $P(A \cup B) = P(A) + P(B) - P(A \cap B)$

> Use the Venn diagram to illustrate these laws.

Exercise 8J

1 A random experiment is: roll an unbiased six-faced dice.
   Let A be the event 'roll a square number' and let B be the event 'roll a factor of 6'.
   a List the elements of set A.
   b List the elements of set B.
   c Show sets A and B on a Venn diagram.
   d Write down $P(A)$.
   e Write down $P(B)$.
   f Find the probability that the number rolled is not a square number.
   g Find the probability that the number rolled is both a square number and a factor of 6.
   h Find the probability that the number rolled is either a square number or a factor of 6 or both.
   i Verify that both $P(A') = 1 - P(A)$ and
      $P(A \cup B) = P(A) + P(B) - P(A \cap B)$.

2 The numbers 3, 4, 5, 6, 7, 8, 9, 10 are written on identical pieces of card and placed in a bag. A random experiment is:
   a card is selected at random from the bag.
   Let A be the event 'a prime number is chosen' and
   let B be the event 'an even number is chosen'.
   a List the elements of set A.
   b List the elements of set B.
   c Show sets A and B on a Venn diagram.
   d Write down $P(A)$.
   e Write down $P(B)$.
   f Find the probability that the number rolled is composite (not a prime).
   g Find the probability that the number rolled is odd.
   h Find the probability that the number rolled is both even and prime.
   i Find the probability that the number rolled is either even or prime or both.
   j Verify that both $P(A') = 1 - P(A)$ and $P(B') = 1 - P(B)$.
   k Verify that $P(A \cup B) = P(A) + P(B) - P(A \cap B)$.
   l Find the probability that the number rolled is both odd and composite.
   m Find the probability that the number rolled is either odd or composite or both.
   n Verify that $P(A' \cup B') = P(A') + P(B') - P(A' \cap B')$

3 The numbers 2, 3, 4, 5, 6, 7, 8, 9 are written on identical pieces of card and placed in a bag. A random experiment is: a card is selected at random from the bag.

Let A be the event 'an odd number is chosen' and let B be the event 'a square number is chosen'.

a List the elements of set A.

b List the elements of set B.

c Show sets A and B on a Venn diagram.

d Write down P(A).

e Write down P(B).

f Find the probability that an odd square number is chosen.

g Find the probability that either an odd number or a square number is chosen.

h Verify that P($A \cup B$) = P(A) + P(B) − P($A \cap B$).

> The first book written on probability was *The Book of Chance and Games* by Italian philosopher and mathematician Jerome Cardan (1501–75). It explained techniquies on how to cheat and catch others at cheating.

4 A random experiment is: toss two unbiased coins.

a List the set of four equally likely possible outcomes.

b Find P(two heads show), P(one head shows), P(no heads show).

5 A random experiment is: toss three unbiased coins.

a List the set of eight equally likely possible outcomes.

b Find P(no heads), P(one head), P(two heads), P(three heads).

6 A random experiment is: toss four unbiased coins.

a Find P(no heads).

b Find P(four heads).

c Find P(one head).

d Find P(three heads).

e Use the answers **a** to **d** to deduce P(two heads).

f List the equally likely possible outcomes.

8.6 Conditional probability

In a class of 25 students, 16 students study French, 11 students study Malay and 4 students study neither language. This information can be shown in a Venn diagram.

Suppose a student is chosen at random from the class. We can use the techniques we have looked at already to find the probability that

a the student studies French and Malay

b the student studies exactly one language

c the student does not study two languages

d the student does not study French.

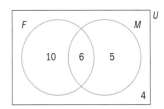

Using the Venn diagram on the right:

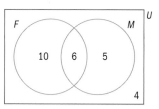

a $\dfrac{6}{25}$

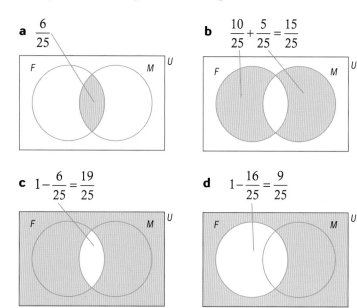

b $\dfrac{10}{25}+\dfrac{5}{25}=\dfrac{15}{25}$

c $1-\dfrac{6}{25}=\dfrac{19}{25}$

d $1-\dfrac{16}{25}=\dfrac{9}{25}$

What is the probability that a student chosen at random studies French, **given that** the student studies Malay?

The probability that a student studies French given that the student studies Malay is an example of a **conditional probability**. It is written $P(F|M)$.

Given that *M* has definitely occurred, then we are restricted to set M (the shaded area), rather than choosing from the universal set (the rectangle).

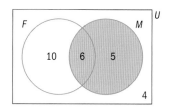

If we now want to determine the probability that F has also occurred, then we consider that part of F which also lies within M – the intersection of F and M (darkest shading).

The conditional probability, the probability that a student studies French given that the student studies Malay, is

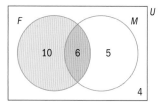

$$P(F|M) = \frac{n(F \cap M)}{n(M)} = \frac{6}{11}$$

> → The conditional probability that A occurs given that B has occurred is written as $P(A|B)$ and is defined as:
> $$P(A|B) = \frac{P(A \cap B)}{P(B)}$$

This requires a different approach because there is an extra condition: the student studies Malay.

Example 12

In a class of 29 students, 20 students study French, 15 students study Malay, and 8 students study both languages. A student is chosen at random from the class.

Find the probability that the student
a studies French
b studies neither language
c studies at least one language
d studies both languages
e studies Malay given that they study French
f studies French given that they study Malay
g studies both languages given that they study at least one of the languages.

Answers

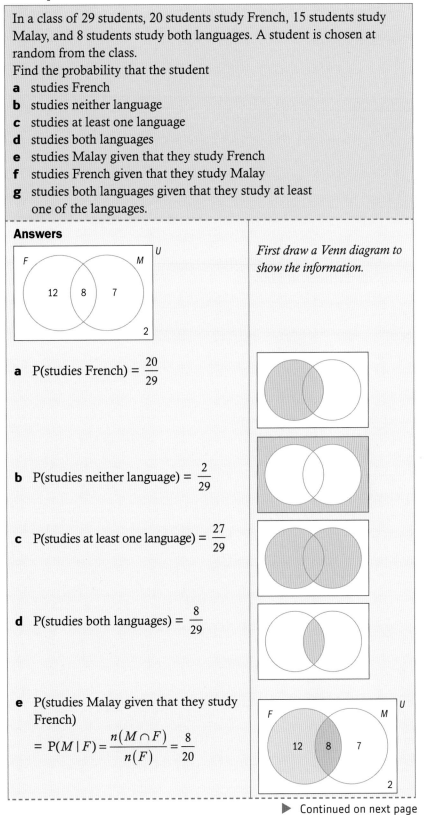

First draw a Venn diagram to show the information.

a P(studies French) = $\dfrac{20}{29}$

b P(studies neither language) = $\dfrac{2}{29}$

c P(studies at least one language) = $\dfrac{27}{29}$

d P(studies both languages) = $\dfrac{8}{29}$

e P(studies Malay given that they study French)

$= P(M \mid F) = \dfrac{n(M \cap F)}{n(F)} = \dfrac{8}{20}$

Probabilities **e** to **g** are conditional, and require more care

▶ Continued on next page

f P(studies French given that they study Malay)

$= P(F \mid M) = \dfrac{n(F \cap M)}{n(M)} = \dfrac{8}{15}$

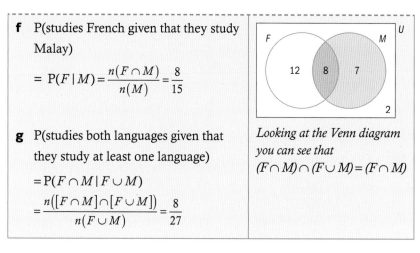

g P(studies both languages given that they study at least one language)

$= P(F \cap M \mid F \cup M)$

$= \dfrac{n([F \cap M] \cap [F \cup M])}{n(F \cup M)} = \dfrac{8}{27}$

Looking at the Venn diagram you can see that

$(F \cap M) \cap (F \cup M) = (F \cap M)$

Exercise 8K

The numbers in each set are shown on the Venn diagrams.

1 Find the probability that a person chosen at random:
 a is in A
 b is not in either A or B
 c is not in A and not in B
 d is in A, given that they are not in B
 e is in B, given that they are in A
 f is in both A and B, given that they are in A.

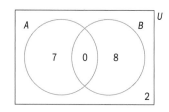

2 Find the probability that a person chosen at random:
 a is not in A
 b is neither in A nor in B
 c is not in both A and B given that they are in B
 d is not in A given that they are not in B
 e is in B given that they are in A
 f is in both A and B, given that they are not in A.

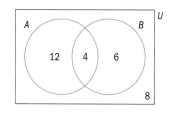

3 Find the probability that a person chosen at random:
 a is in B but not in A
 b is not in A or B
 c is in B and not in A
 d is in A given that they are not in B
 e is in B given that they are in A
 f is not in both A and B, given that they are in A.

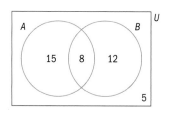

4 Find the probability that a person chosen at random:
 a is in A but not in both A and B
 b is not in A and not in both
 c is not in both A and B
 d is in A given that they are not in B
 e is in B given that they are in A
 f is not in A given that they are not in B.

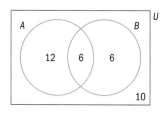

5 The Venn diagram shows the number of students who take Art and/or Biology in a class.

Use the Venn diagram to find the probability that a student chosen at random from the class:

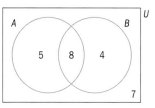

  a takes Art
  b takes Biology but not Art
  c takes both Art and Biology
  d takes at least one of the two subjects
  e takes neither subject
  f takes Biology
  g takes exactly one of the two subjects.

6 The Venn diagram shows the number of students who take Physics and/or Chemistry in a class.

Use the Venn diagram to find the probability that a student chosen at random from the class:

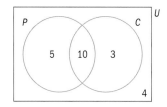

  a takes Physics but not Chemistry
  b takes at least one of the two subjects
  c takes Chemistry given that the student takes Physics
  d is a Chemist given that the student takes exactly one of the two subjects.

7 The Venn diagram shows the number of students who take Art and/or Drama in a class.

Use the Venn diagram to find the probability that a student chosen at random from the class:

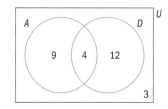

  a takes Drama but not Art
  b takes Drama given that they take Art
  c takes both subjects given that they take Drama
  d takes neither subject
  e takes Drama given that they take exactly one of the two subjects.

8 The Venn diagram shows the number of students who take Geography and/or History in a class.

Use the Venn diagram to find the probability that a student chosen at random from the class:

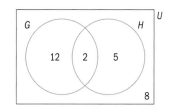

  a takes Geography but not History
  b takes Geography given that they do not take History
  c takes History given that they take at least one of the two subjects
  d takes Geography given they take History
  e takes Geography given that they take exactly one of the two subjects.

8.7 Two special cases: mutually exclusive and independent events

Two events, A and B, are **mutually exclusive** if whenever A occurs it is impossible for B to occur and, similarly, whenever B occurs it is impossible for A to occur.

Events A and A' are the most obvious example of mutually exclusive events – either one or the other must occur, but A and A' cannot occur at the same time.

> For example, in tossing a coin, the events 'a head is tossed' and 'a tail is tossed' are mutually exclusive.

Here is the Venn diagram for mutually exclusive events A and B.

As the two sets do not overlap, $A \cap B = \varnothing$.

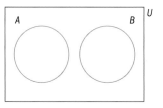

→ Events A and B are mutually exclusive if and only if $P(A \cap B) = 0$.

Example 13

The numbers 3, 4, 5, 6, 7, 8, 9, 10 are each written on an identical piece of card and placed in a bag. A random experiment is: a card is selected at random from the bag.

Let A be the event 'a prime number is chosen' and B the event 'an even number is chosen'.

a Draw a Venn diagram that describes the random experiment.

b Determine whether the events A and B are mutually exclusive.

Answers

a

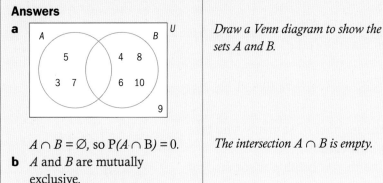

$A \cap B = \varnothing$, so $P(A \cap B) = 0$.

Draw a Venn diagram to show the sets A and B.

The intersection A ∩ B is empty.

b A and B are mutually exclusive.

In 1933, the Russian Mathematician Andrey Nikolaevich Kolmogorov (1903–1987) defined probability by these axioms:

- The probability of all occurrences is 1
- Probability has a value which is greater than or equal to zero
- When occurrences cannot coincide their probabilities can be added

The mathematical properties of probability can be deduced from these axioms. Kolmogorov used his probability work to study the motion of the planets and the turbulent flow of air from a jet engine.

> What is an axiom? Find out more about Euclid's axioms for geometry, written 2000 years ago.

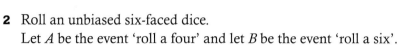

Exercise 8L

In each experiment, determine whether the events A and B are mutually exclusive.

1 Roll an unbiased six-faced dice.
Let A be the event 'roll a square number' and let B be the event 'roll a factor of six'.

2 Roll an unbiased six-faced dice.
Let A be the event 'roll a four' and let B be the event 'roll a six'.

3 Roll an unbiased six-faced dice.
Let A be the event 'roll a prime number' and let B be the event 'roll an even number'.

4 Roll an unbiased six-faced dice.
Let A be the event 'roll a square number' and let B be the event 'roll a prime number'.

5 Each of the numbers 3, 4, 5, 6, 7, 8, 9, 10 are written on identical pieces of card and placed in a bag. A card is selected at random from the bag.
Let A be the event 'a square number is chosen' and let B be the event 'an odd number is chosen'.

6 Each of the numbers 5, 6, 7, 8, 9, 10 are written on identical pieces of card and placed in a bag. A card is selected at random from the bag.
Let A be the event 'a square number is chosen' and let B be the event 'an even number is chosen'.

7 Each of the numbers 2, 3, 4, 5, 6, 7, 8, 9 are written on identical pieces of card and placed in a bag. A card is selected at random from the bag.
Let A be the event 'an even number is chosen' and let B be the event 'a multiple of three is chosen'.

8 Two unbiased coins are tossed.
Let A be the event 'two heads show' and let B be the event 'one head shows'.

If two events, A and B, are mutually exclusive, the effect of the first event, A, on the second, B, could not be greater – if A occurs, then it is impossible that B can occur (and vice versa). The occurrence of one event completely prevents the occurrence of the other.

The other extreme is when the occurrence of the one event does not affect in any way the occurrence of the other. Then the two events are **mathematically independent** of each other.

Another way to put this is that the probability that A occurs, $P(A)$, remains the same given that B has occurred. Writing this as an equation, A and B are independent whenever $P(A) = P(A \mid B)$.

The definition of $P(A \mid B)$ is:

$$P(A \mid B) = \frac{P(A \cap B)}{P(B)}$$

Thus whenever A and B are independent:

$$\frac{P(A \cap B)}{P(B)} = P(A)$$

Rearranging, $P(A \cap B) = P(A) \times P(B)$

> → A and B are independent if and only if $P(A \cap B) = P(A) \times P(B)$.

For example, if a one-euro coin is tossed and then a one-dollar coin is tossed, the fact that the euro coin landed 'heads' does not affect in any way whether the dollar coin lands 'heads' or 'tails'. The two events are independent of each other.

If you are asked to determine whether two events are independent, this is the test you must use.

Example 14

The numbers 2, 3, 4, 5, 6, 7, 8, 9 are each written on identical pieces of card and placed in a bag.
A card is selected at random from the bag.
Let A be the event 'an odd number is chosen' and let B be the event 'a square number is chosen'.
a Draw a Venn diagram to represent the experiment.
b Determine whether A and B are independent events.

Answers

a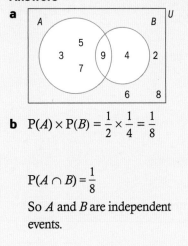

b $P(A) \times P(B) = \dfrac{1}{2} \times \dfrac{1}{4} = \dfrac{1}{8}$

$P(A \cap B) = \dfrac{1}{8}$

So A and B are independent events.

*The event $A \cap B$ is 'an odd number is chosen **and** a square number is chosen' or 'an odd square number is chosen'.*

From the Venn diagram,
$$P(A) = \frac{4}{8} = \frac{1}{2} \qquad P(B) = \frac{2}{8} = \frac{1}{4}$$

$A \cap B = \{9\}$, *hence* $P(A \cap B) = \dfrac{1}{8}$

Now, consider the definition for (mathematical) independence:
$P(A \cap B) = P(A) \times P(B)$.

This work links to the chi-squared test for independence that you studied in Chapter 5. Recall that to calculate the expected frequencies, the row total is multiplied by the column total and then divided by the overall total of frequencies. This is a direct consequence of the definition of mathematical independence.

Exercise 8M

For each experiment determine whether the events A and B are independent.

1   The numbers 1, 2, 3, 4, 5, 6, 7, 8, 9 are each written on identical cards and placed in a bag.
A card is selected at random from the bag.

Let A be the event 'an odd number is chosen' and let B be the event 'a square number is chosen'.

2   The numbers 1, 2, 3, 4, 5, 6 are each written on identical cards and placed in a bag.
A card is selected at random from the bag.

Let A be the event 'an even number is chosen' and let B be the event 'a square number is chosen'.

3   The numbers 2, 3, 4, 5, 6, 7, 8, 9, 10 are each written on identical cards and placed in a bag.
A card is selected at random from the bag.

Let A be the event 'a prime number is chosen' and let B be the event 'a multiple of three is chosen'.

4   The Venn diagram shows the number of students who take Art and/or Biology in a class.

Use the Venn diagram to determine whether taking Art and taking Biology are independent events.

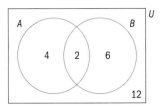

5   The Venn diagram shows the number of students who take Chemistry and/or Biology in a class.
Use the Venn diagram to determine whether taking Chemistry and taking Biology are independent events.

6   The Venn diagram shows the number of students who take Chemistry and/or Physics in a class.
Use the Venn diagram to determine whether taking Chemistry and taking Physics are independent events.

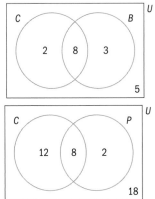

8.8 Sample space diagrams

A sample space diagram is a graphical way of showing the possible equally likely outcomes of an experiment rather than listing them.

One red dice and one blue dice are rolled together. Both dice are fair.

You can show all the possible outcomes on a grid.

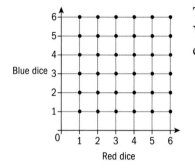

There are 36 possible outcomes, $n(U) = 36$.
You can use the sample space diagram to calculate probabilities.

Example 15

A red and a blue dice are rolled together. Calculate the probability that:
a The total score is 7.
b The same number comes up on both dice.
c The difference between the scores is 1.
d The score on the red dice is less than the score on the blue dice.
e The total score is a prime number.

Answers

a P(the total score is 7) $= \dfrac{6}{36}$

b P(the same number comes up on both dice) $= \dfrac{6}{36}$

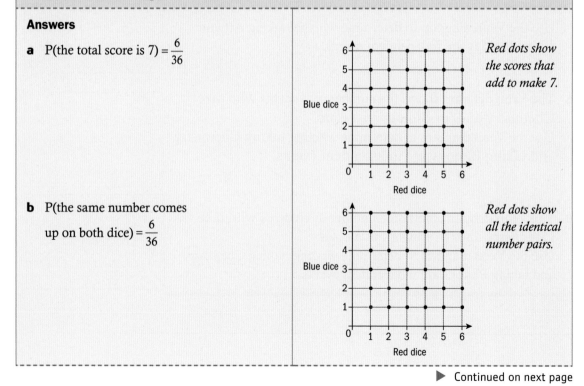

Red dots show the scores that add to make 7.

Red dots show all the identical number pairs.

▶ Continued on next page

c P(the difference between the scores is 1) $= \dfrac{10}{36}$

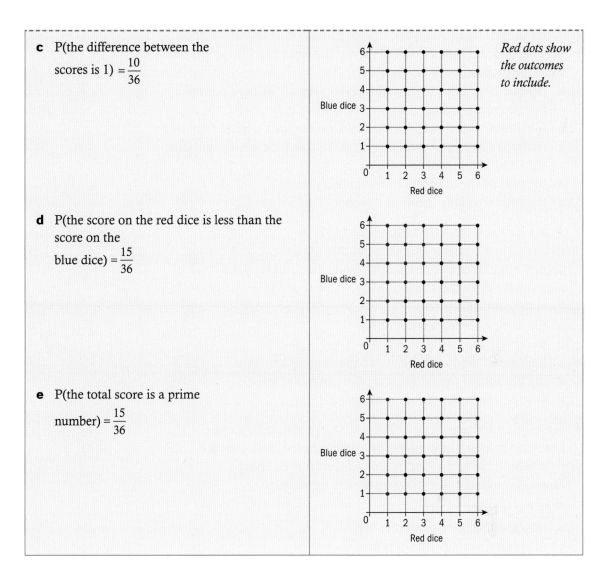

Red dots show the outcomes to include.

d P(the score on the red dice is less than the score on the blue dice) $= \dfrac{15}{36}$

e P(the total score is a prime number) $= \dfrac{15}{36}$

Exercise 8N

1 Draw the sample space diagram for this experiment: two tetrahedral dice, one blue and the other red, each numbered 1 to 4, are rolled. Find the probability that

a the number on the red dice is greater than the number on the blue dice

b the difference between the numbers on the dice is one

c the red dice shows an odd number and the blue dice shows an even number

d the sum of the numbers on the dice is prime.

2 A tetrahedral dice (numbered 1 to 4) and an ordinary six-sided dice are rolled. Draw the sample space diagram for this experiment. Find the probability that:
  a the number on the tetrahedral dice is greater than the number on the ordinary dice
  b the difference between the numbers on the dice is more than one
  c the ordinary dice shows an odd number and the tetrahedral dice shows an even number
  d the sum of the numbers on the dice is prime
  e the two dice show the same number.

3 A box contains three cards numbered 1, 2, 3.
A second box contains four cards numbered 2, 3, 4, 5. A card is chosen at random from each box. Draw the sample space diagram for the experiment.
Find the probability that:
  a the cards have the same number
  b the largest number drawn is 3
  c the sum of the two numbers on the cards is less than 7
  d the product of the numbers on the cards is at least 8
  e at least one even number is chosen.

4 Six cards, numbered 0, 1, 2, 3, 4, and 5, are placed in a bag. One card is drawn at random, its number noted, and then the card is replaced in the bag. A second card is then chosen. Draw the sample space diagram for the experiment.

Find the probability that:
  a the cards have the same number
  b the largest number drawn is prime
  c the sum of the two numbers on the cards is less than 7
  d the product of the numbers on the cards is at least 8
  e at least one even number is chosen.

5 Six cards, numbered 0, 1, 2, 3, 4, and 5, are placed in a bag. One card is drawn at random and is **not** replaced. A second card is then chosen. Draw the sample space diagram for the experiment.

Find the probability that:
  a the cards bear the same number
  b the larger number drawn is prime
  c the sum of the two numbers on the cards is less than 7
  d the product of the numbers on the cards is at least 8
  e at least one even number is chosen.

> Be careful: this is **not** the same sample space as for question 4.

8.9 Tree diagrams

Tree diagrams are another way of representing and calculating probabilities.

Example 16

Two fair dice are rolled, one red and one blue.
Using a tree diagram, find the probability that:
a a double six is rolled   **b** no sixes are rolled
c exactly one six is rolled   **d** at least one six is rolled.

Answers

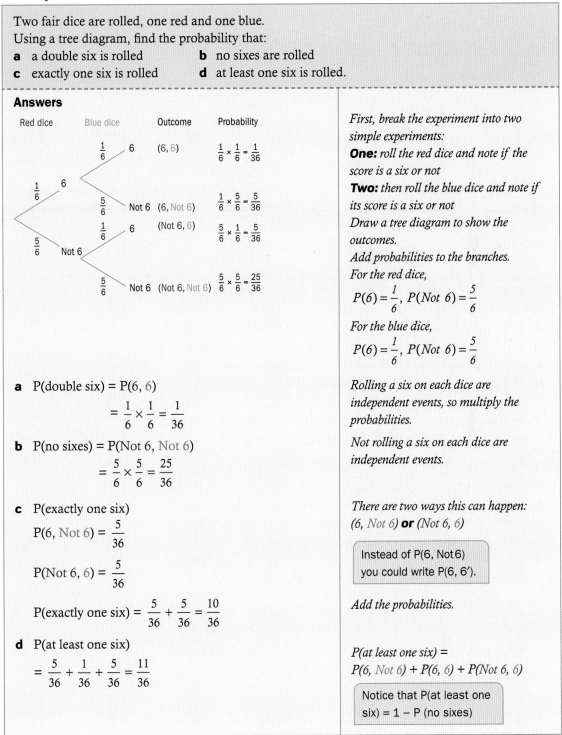

a P(double six) = P(6, 6)

$$= \frac{1}{6} \times \frac{1}{6} = \frac{1}{36}$$

b P(no sixes) = P(Not 6, Not 6)

$$= \frac{5}{6} \times \frac{5}{6} = \frac{25}{36}$$

c P(exactly one six)

$$P(6, \text{Not } 6) = \frac{5}{36}$$

$$P(\text{Not } 6, 6) = \frac{5}{36}$$

$$P(\text{exactly one six}) = \frac{5}{36} + \frac{5}{36} = \frac{10}{36}$$

d P(at least one six)

$$= \frac{5}{36} + \frac{1}{36} + \frac{5}{36} = \frac{11}{36}$$

First, break the experiment into two simple experiments:
One: *roll the red dice and note if the score is a six or not*
Two: *then roll the blue dice and note if its score is a six or not*
Draw a tree diagram to show the outcomes.
Add probabilities to the branches.
For the red dice,

$$P(6) = \frac{1}{6}, \ P(\text{Not } 6) = \frac{5}{6}$$

For the blue dice,

$$P(6) = \frac{1}{6}, \ P(\text{Not } 6) = \frac{5}{6}$$

Rolling a six on each dice are independent events, so multiply the probabilities.

Not rolling a six on each dice are independent events.

There are two ways this can happen:
(6, Not 6) **or** *(Not 6, 6)*

> Instead of P(6, Not 6) you could write P(6, 6′).

Add the probabilities.

P(at least one six) =
P(6, Not 6) + P(6, 6) + P(Not 6, 6)

> Notice that P(at least one six) = 1 − P (no sixes)

You can use tree diagrams for conditional probabilities too.

Example 17

For the experiment in Example 16, find the probability that, given that at least one six was rolled, then the red dice showed a six.

Answer	
P(six on red dice | at least one six rolled)	
$= \dfrac{\text{P(six on red dice and at least one six rolled)}}{\text{P(at least one six rolled)}}$	*Use the definition of conditional probability.*
$= \dfrac{\text{P}(6, \ 6) + \text{P}(6, \text{Not } 6)}{\text{P}(6, \ 6) + \text{P}(6, \text{Not } 6) + \text{P}(\text{Not } 6, 6)}$	
$= \dfrac{\left(\dfrac{1}{36} + \dfrac{5}{36}\right)}{\left(\dfrac{1}{36} + \dfrac{5}{36} + \dfrac{5}{36}\right)} = \dfrac{6}{11} = 0.545$	*Read the probabilities from the final column of the tree diagram in Example 16.* *Use your GDC to evaluate this – change to a fraction or give the answer correct to 3 significant figures.*

Exercise 80

1 A bag contains 6 red and 5 blue counters. One is chosen at random. Its color is noted and it is put back into the bag. Then a second counter is chosen.
   a Find the probability that exactly one red counter is chosen.
   b Find the probability that at least one blue counter is chosen.
   c Find the probability that one of each color is chosen.
   d If one of each color was chosen, what is the probability that the blue was chosen on the second pick?
   e If at least one blue was chosen, what is the probability that the blue was chosen on the first pick?

> These are conditional probabilities

2 A 5-sided dice is numbered 1, 2, 3, 4, 5. It is rolled twice.
   a Find the probability that exactly one prime number is rolled.
   b Find the probability that at least one prime number is rolled.
   c Given that at least one prime number was rolled, find the probability that two primes are rolled.
   d Given that at least one prime was rolled, find the probability that a prime was rolled on the first attempt.

3 To get to work I must go through two sets of traffic lights – first at Sixth Avenue and then at Dover Road. I get delayed at Sixth Avenue with probability $\dfrac{7}{10}$ and at Dover Road with probability $\dfrac{3}{5}$.
   Draw a tree diagram to show the possible delays on my journey to work.
   a Find the probability that I get delayed only once.
   b Find the probability that I do not get delayed at all.
   c Given that I get delayed exactly once, what is the probability that it was at Sixth Avenue?
   d Given that I get delayed, what is the probability that it was at Sixth Avenue?

4 On a journey to school a teacher has to pass through two sets of traffic lights (*A* and *B*). The probabilities that he will be stopped at these are $\frac{2}{7}$ and $\frac{1}{3}$ respectively. The corresponding delays are one minute and three minutes. Without these delays his journey takes 30 minutes. Draw a tree diagram to illustrate the possible delays.

   a Find the probability that the journey takes no more than 30 minutes.

   b Find the probability that the teacher has only one delay.

   c Given that the teacher gets delayed, what is the probability that it happened at *A*?

   d On a particular morning, the teacher has only 32 minutes to reach school on time. Find the probability that he will be late.

5 The probability that it will rain today is 0.2. If it rains today, the probability that it will rain tomorrow is 0.15. If it is fine today then the probability that it will be fine tomorrow is 0.9.

   a Find the probability that at least one day will be fine.

   b Given that at least one day is fine, what was the probability that it was today?

   c Given that at least one day is fine, what is the probability that both are fine?

> **Extension material on CD:**
> *Worksheet 8 - A game*

'Without replacement' problems

A classic probability problem involves picking a ball from a bag, noting its color and *not* replacing it, then picking another ball.

This means that the probability of choosing the next ball from the bag will be different from the probability of choosing the first.

You can use a tree diagram for this type of problem.

Example 18

There are 6 peppermints (*P*) and 2 liquorice (*L*) candies in a bag. A candy is picked **and not replaced in the bag**. Then a second candy is picked.

   a Find the probability that one of each type is chosen.

   b **Given that** one of each type was chosen, find the probability that the first one chosen was a peppermint.

> This 'without replacement' problem uses candies instead of balls.

Answers

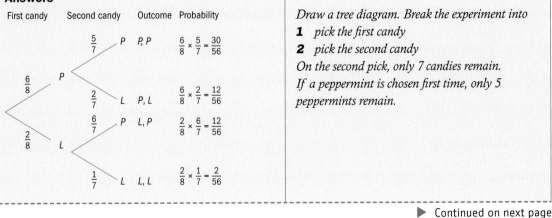

Draw a tree diagram. Break the experiment into

1 *pick the first candy*

2 *pick the second candy*

On the second pick, only 7 candies remain.

If a peppermint is chosen first time, only 5 peppermints remain.

▶ Continued on next page

a P(one of each type) = P(P, L) + P(L, P)

$$= \frac{12}{56} + \frac{12}{56} = \frac{24}{56} = \frac{3}{7}$$

b $P(A \mid B) = \dfrac{P(A \cap B)}{P(B)}$

$P(A \cap B) = P(P, L) = \dfrac{12}{56} = \dfrac{3}{14}$ and $P(B) = \dfrac{3}{7}$

So, $P(A \mid B) = \dfrac{\frac{3}{14}}{\frac{3}{7}} = \dfrac{1}{2}$

The outcomes corresponding to 'one of each type is chosen' are P, L and L, P.

Let A be the event that the first candy chosen is peppermint, and B be the event that one of each type is picked. Then we need P(A | B).

*P(B) is the probability from part **a**.*

Exercise 8P

1 A bag contains 6 red and 5 blue counters. One is chosen at random. Its color is noted, **and it is not put back in the bag**. Then a second counter is chosen.

   a Find the probability that exactly one red counter is chosen.

   b Find the probability that at least one blue counter is chosen.

   c Find the probability that one of each color is chosen.

   d If one of each color was chosen, what is the probability that the blue was chosen on the second pick?

   e If at least one blue was chosen, what is the probability that the blue was chosen on the first pick?

2 A bag contains 5 faulty and 7 working pens. A boy and then a girl each need to take a pen.

   a What is the probability that both take faulty pens?

   b Find the probability that at least one takes a faulty pen.

   c Given that exactly one faulty pen was taken, what is the probability that the girl took it?

3 To get to the school I can take one of two routes, via Kent Ridge or via Sunny Vale. I take the Kent Ridge route on average 3 times in a 5 day week. If I take this route the probability that I am delayed is 0.25. If I take the Sunny Vale route the probability that I am delayed is 0.5.

Draw a tree diagram that shows my journey to school.

   a Find the probability that I get delayed.

   b Find the probability that I go by Sunny Vale and I do not get delayed.

   c Given that I am delayed, what is the probability that I went via Kent Ridge?

   d Given that I am not delayed, what is the probability that I went via Sunny Vale?

4 The probability that it will snow today is 0.9. If it does snow today then the probability it will snow tomorrow is 0.7. However, if it does not snow today then the probability that it will snow tomorrow is 0.6.

Draw a tree diagram which shows the possible weather conditions for the two days.

  a Find the probability of two snowy days.
  b Find the probability of exactly one snowy day.
  c Given that there is exactly one snowy day, what is the probability that it is today?
  d Given that there is at least one snowy day, what is the probability that it is today?

5 There are eight identical discs in a bag, five of which are black and the other three are red. The random experiment is: choose a disc at random from the bag, do not return the disc to the bag, then choose a **second** disc from the bag. Find the probability that the second disc chosen is red.

Review exercise

Paper 1 style questions

1 The activities offered by a school are golf (G), tennis (T), and swimming (S). The Venn diagram shows the numbers of people involved in each activity.

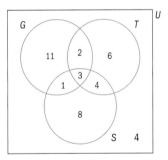

  a Write down the number of people who
   i play tennis only   **ii** play both tennis and golf
   iii play at least two sports   **iv** do not play tennis.
  b Copy the diagram and shade the part of the Venn diagram that represents the set $G' \cap S$.

2 A group of 40 children are surveyed to find out which of the three sports volleyball (A), basketball (B) or cricket (C) they play. The results are as follows:

7 children do not play any of these sports
2 children play all three sports
5 play volleyball and basketball
3 play cricket and basketball
10 play cricket and volleyball
15 play basketball
20 play volleyball.

  a Draw a Venn diagram to illustrate the relationship between the three sports played.
  b On your Venn diagram indicate the number of children that belong to each **disjoint** region.
  c Find the number of children that play cricket only.

3 The following Venn diagram shows the sets U, A, B and C.

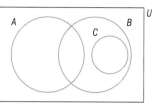

State whether the following statements are true or false for the information illustrated in the Venn diagram.

a $A \cup C = \varnothing$ **b** $C \subset (C \cup B)$

c $C \cap (A \cup B) = \varnothing$ **d** $C \subset A'$

e $C \cap B = C$ **f** $(A \cup B)' = A' \cap B'$

4 a Copy this diagram and shade $A \cup (B \cap C')$.

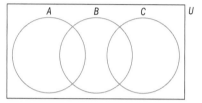

b In the Venn diagram on the right, the number of elements in each region is given.
Find $n((P \cup Q) \cap R)$.

c U is the set of positive integers, $\mathbb{Z}^+$.
E is the set of odd numbers.
M is the set of multiples of 5.

  i List the first four elements of the set M.

  ii List the first three elements of the set $E' \cap M$.

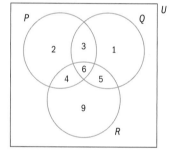

5 $\mathbb{Z}$ is the set of integers, $\mathbb{Q}$ is the set of rational numbers, $\mathbb{R}$ is the set of real numbers.

a Write down an element of $\mathbb{Z}$.   **b** Write down an element of $\mathbb{R} \cap \mathbb{Z}'$.

c Write down an element of $\mathbb{Q}$.   **d** Write down an element of $\mathbb{Q} \cup \mathbb{Z}'$.

e Write down an element of $\mathbb{Q}'$.   **f** Write down an element of $\mathbb{Q}' \cap \mathbb{Z}'$.

6 The table below shows the number of left- and right-handed tennis players in a sample of 60 males and females.

	Left-handed	Right-handed	Total
Male	8	32	40
Female	4	16	20
Total	12	48	60

If a tennis player was selected at random from the group, find the probability that the player is

a female and left-handed   **b** male or right-handed

c right-handed, given that the player selected is female.

7 A bag contains 3 red, 4 yellow and 8 green sweets.
You Jin chooses one sweet out of the bag at random and eats it.
She then takes out a second sweet.
  a Write down the probability that the first sweet chosen was red.
  b Given that the first sweet was not red, find the probability that
   the second sweet is red.
  c Find the probability that both the first and second sweets
   chosen were yellow.

8 Ernest rolls two cubical dice. One of the dice has three red faces and three
black faces. The other dice has the faces numbered from 1 to 6. By means
of a sample space diagram, or otherwise, find
  a the number of different possible combinations he can roll
  b the probability that he will roll a black and an even number
  c the probability that he will roll a number more than 4.

9 The table below shows the number of words in the extended
essays of an IB class.

Number of words	$3100 \leq w < 3400$	$3400 \leq w < 3700$	$3700 \leq w < 4000$	$4000 \leq w < 4300$
Frequency	7	20	18	5

  a Write down the modal group.
  b Write down the probability that a student chosen at random writes an
   extended essay with a number of words in the range: $4000 \leq w < 4300$.

The maximum word count for an extended essay is 4000 words.
Find the probability that a student chosen at random:
  c does not write an extended essay that is on or over the word count
  d writes an extended essay with the number of words in the range
   $3400 \leq w < 3700$ given that it is not on or over the word count.

Paper 2 style questions

1 Let $U = \{x \mid 8 \leq x < 13, x \in \mathbb{N}\}$.
P, Q and R are the subsets of U such that
$P = \{$multiples of four$\}$
$Q = \{$factors of 24$\}$
$R = \{$square numbers$\}$
  a List the elements of U.
  b i Draw a Venn diagram to show the relationship between sets P, Q and R.
   ii Write the elements of U in the appropriate places on the Venn diagram.
  c List the elements of:
   i $P \cap R$ **ii** $P' \cap Q \cap R$
  d Describe in words the set $P \cup Q$.
  e Shade the region on your Venn diagram that represents $(P \cup R) \cap Q'$.

2 In a club with 70 members, everyone attends either on Tuesday for Drama (D) or on Thursday for Sports (S), or on both days for Drama and Sports.

One week it is found that 48 members attend for Drama, 44 members attend for Sports, and x members attend for both Drama and Sports.

a i Draw and **label fully** a Venn diagram to illustrate this information.

　ii Find the number of members who attend for both Drama and Sports.

　iii Describe, in words, the set represented by $(D \cap S)'$.

　iv What is the probability that a member selected at random attends for Drama only or Sports only?

The club has 40 female members, 10 of whom attend for both Drama and Sports.

b What is the probability that a member of the club selected at random

　i is female and attends for Drama only or Sports only?

　ii is male and attends for both Drama and Sports?

3 On a particular day 50 children are asked to make a note of what they drank that day.

They are given three choices: water (P), fruit juice (Q) or coffee (R).

2 children drank only water.

4 children drank only coffee.

12 children drank only fruit juice.

3 children drank all three.

4 children drank water and coffee only.

5 children drank coffee and fruit juice only.

15 children drank water and fruit juice only.

a Represent the above information on a Venn diagram.

b How many children drank none of the above?

c A child is chosen at random. Find the probability that the child drank

　i fruit juice

　ii water or fruit juice but not coffee

　iii no fruit juice, given that the child did drink water.

d Two children are chosen at random. Find the probability that both children drank all three choices.

4 The sets P, Q, and R are subsets of U. They are defined as follows:

U = {positive integers less than 13}

P = {prime numbers}

Q = {factors of 18}

R = {multiples of 3}

a List the elements (if any) of

   i P **ii** Q **iii** R **iv** $P \cap Q \cap R$

b i Draw a Venn diagram showing the relationship between the sets U, P, Q and R.

   ii Write the elements of sets U, P, Q and R in the appropriate places on the Venn diagram.

c From the Venn diagram, list the elements of

   i $P \cup (Q \cap R)$ **ii** $(P \cup R)'$ **iii** $(P \cup Q)' \cap R'$

d Find the probability that a number chosen at random from the universal set U will be

   i a prime number

   ii a prime number, but **not** a factor of 18

   iii a factor of 18 or a multiple of 3, but **not** a prime number

   iv a prime number, given that it is a factor of 18.

5 There are two biscuit tins on a shelf. The **red** tin contains four chocolate biscuits and six plain biscuits. The **blue** tin contains one chocolate biscuit and nine plain biscuits.

A child reaches into the **red** tin and randomly selects a biscuit. The child returns that biscuit to the tin, shakes the tin, and then selects another biscuit.

a Draw a tree diagram that shows the possible outcomes. Place the appropriate probability on each branch of the tree diagram.

b Find the probability that

   i both biscuits chosen are chocolate

   ii one of the biscuits is plain and the other biscuit is chocolate.

c A second child chooses a biscuit from the **blue** tin. The child eats the biscuit and chooses another one from the **blue** tin. The tree diagram on the right represents the possible outcomes for this event.

   i Write down the value of a and of b.

   ii Find the probability that both biscuits are chocolate.

   iii What is the probability that *at least* one of the biscuits is plain?

d Suppose that, before the two children arrived, their brother randomly selected one of the biscuit tins and took out one biscuit. Calculate the probability that this biscuit was chocolate.

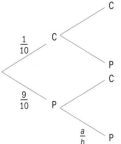

6 The data in the table below refer to a sample of 60 randomly chosen plants.

Growth rate	Classification by environment			
	desert	temperate	waterlogged	total
high	4	7	13	24
low	9	11	16	36
total	13	18	29	60

a i Find the probability of a plant being in a desert environment.
  ii Find the probability of a plant having a low growth rate and being in a waterlogged environment.
  iii Find the probability of a plant not being in a temperate environment.
b A plant is chosen at random from the above group.
  Find the probability that the chosen plant has
  i a high growth rate or is in a waterlogged environment, but not both
  ii a low growth rate, given that it is in a desert environment.
c The 60 plants in the above group were then classified according to leaf type. It was found that 15 of the plants had type A leaves, 36 had type B leaves and 9 had type C leaves.
  Two plants were randomly selected from this group. Find the probability that
  i both plants had type B leaves
  ii neither of the plants had type A leaves.

CHAPTER 8 SUMMARY

Basic set theory

- A **set** is simply a collection of objects. The objects are called the **elements** of the set.
- The number of elements in a set A is denoted as $n(A)$.

Venn diagrams

- The **universal set** (denoted U), must be stated to make a set well defined.
- If every element in a given set, M, is also an element of another set, N, then M is a **subset** of N, denoted $M \subseteq N$
- A **proper subset** of a given set is one that is **not identical** to the original set.
  If M is a proper subset of N (denoted $M \subset N$) then
  1 **every** element of M also lies in N and
  2 there are some elements in N that do not lie in M.
- The empty set $\varnothing$ is a subset of every set.
- Every set is a subset of itself.
- The **intersection** of set M and set N (denoted $M \cap N$) is the set of all elements that are in **both M and N**.

Continued on next page

- The **union** of set M and set N (denoted $M \cup N$) is the set of all elements that are in **either** M **or** N or **both**.
- The **complement** of set M, denoted as M', is the set of all the elements in the universal set that **do not** lie in M.
- The complement of the universal set, U', is the empty set, $\varnothing$.

Basic probability theory

- If all of the equally likely possible outcomes of a random experiment can be listed as U, the universal set, and an event A is defined and represented by a set A, then:

$$P(A) = \frac{n(A)}{n(U)}$$

There are three consequences of this law:

1 $P(U) = \dfrac{n(U)}{n(U)} = 1$ (the probability of a **certain** event is 1)

2 $P(\varnothing) = \dfrac{n(\varnothing)}{n(U)} = 0$ (the probability of an **impossible** event is 0)

3 $0 \leq P(A) \leq 1$ (the probability of any event **always** lies between 0 and 1)
- For complementary events, $P(A') = 1 - P(A)$
- For combined events, $P(A \cup B) = P(A) + P(B) - P(A \cap B)$

Conditional probability

- The conditional probability that A occurs given that B has occurred is written as $P(A \mid B)$ and is defined as:

$$P(A \mid B) = \frac{P(A \cap B)}{P(B)}$$

Two special cases: mutually exclusive and independent events

- Events A and B are mutually exclusive if and only if $P(A \cap B) = 0$.
- A and B are independent if and only if $P(A \cap B) = P(A) \times P(B)$.

It's not fair!

A certain degree of uncertainty

In mathematics we can be certain that we have the right answer – certain about what we know. Probability deals with situations that are uncertain.

- How can a theory involving uncertainty be quantified?
- Mathematics is an exact science.
  So how can probability considered to be mathematics?

Gambling dice

Probability theory began to develop in France in the 17th century, as mathematicians Blaise Pascal, Antoine Gombaud (also known as the Chevalier de Méré), and Pierre de Fermat discussed how to bet in a dice game.

The Chevalier de Méré asked: Which is more likely – rolling one 'six' in four throws of one dice, or rolling a 'double six' in 24 throws with two dice?

- Which option seems intuitively correct?
- Can you always trust your intuition?

At the time it was thought that the better option was to bet on the double six because so many more throws are allowed. The mathematicians analyzed the probabilities and proved rolling a six in four rolls of one dice is more likely.

- Can you prove this?

A moral question

While French mathematicians were developing probability theory, the English view was that 'as gambling is immoral, probability must not be studied'.

- How valid is this view?

- Discuss the statement 'Mathematics transcends morality; it cannot be immoral.'
- Why do people gamble? Why do people do things that they know will be self-destructive?

Fair games

In a game, X and Y toss a coin. If the coin lands heads then X wins. If the coin lands tails then Y wins.

- Is this a fair game?
- What do we mean by a fair game?

  - Are these games fair?
  1. Two people, X and Y, toss a coin. If the coin lands heads then X pays Y $5. If the coin lands tails then Y pays X $1.
  2. X and Y roll a dice. If the die shows 'one' then X pays Y $1. If the dice does not show 'one' then Y pays X $1.
  3. X and Y roll a dice. If the dice shows 'one' then X pays Y $5. If the dice does not show 'one' then Y pays X $1.

All the coins and dice in this section are 'unbiased'.

"Gambling is a tax on the mathematically ignorant."

- Do you agree or disagree?

Mathematical fairness

The mathematical definition of a fair game is a game in which the expected gain for each player is zero.

In a casino, **no game is fair**, from a mathematical standpoint. The casino has to make enough money to pay for the building, the electricity, its employees and taxes, as well as making a profit.

Who wins in roulette?

In roulette you can bet on single numbers, groups of numbers, rows, columns, whether the number that turns up is odd or even, red or black, or 'passé' or 'manqué'.

Passé means that the number is in the range 19 to 36 inclusive; manqué means that the number is in the range 1 to 18 inclusive.

Player X places a bet of $1 on manqué; if the ball lands in the range 1 to 18 inclusive, the casino pays X $1 (and X retains the original $1), if the ball lands outside this range, X loses the $1 to the casino. Is the game fair?

Intuitively you might think that the ball is equally likely to land on 'passé' or 'manqué', so the game is fair. But look at the picture of the roulette wheel: there are two outcomes colored green, labelled 0 and 00.

So there are 36 + 2 = 38 possible equally likely outcomes.

X would expect to lose $\$\frac{2}{38}$ per play (paying out $1 on 20 out of 38 plays and gaining $1 on 18 out of 38 plays)

The game is not fair. The $\$\frac{2}{38}$ is called the '**house edge**' and is the casino's profit margin. As a percentage this is $\frac{2}{38} \times 100 = 5.26\%$.

The house edge ensures that the casino can be run as any other business – to make money.

◄ A roulette table.

9 Logic

CHAPTER OBJECTIVES:

3.1 Basic concepts of symbolic logic: definition of a proposition; symbolic notation of propositions

3.2 Compound statements: implication, $\Rightarrow$; equivalence, $\Leftrightarrow$; negation, $\neg$; conjunction, $\wedge$; disjunction, $\vee$; exclusive disjunction, $\underline{\vee}$. Translation between verbal statements and symbolic form

3.3 Truth tables: concepts of logical contradiction and tautology

3.4 Converse, inverse, contrapositive; logical equivalence; testing the validity of simple arguments through the use of truth tables

Before you start

You should know how to:

1 Draw Venn diagrams to show the intersection of sets,
e.g. Draw a Venn diagram to show $A \cap B$.

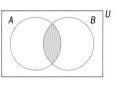

2 Draw Venn diagrams to show the union of sets,
e.g. Draw a Venn diagram to show $A \cup B$.

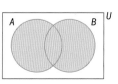

3 Draw Venn diagrams to show the complement of sets,
e.g. Draw a Venn diagram to show A'.

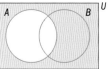

Skills check

1 Given the Venn diagram with sets A and B, draw Venn diagrams to show
  a $A \cap B$ **b** $A \cup B$
  c $A' \cup B$ **d** $(A \cup B)'$
  e $(A \cap B') \cup (A' \cap B)$

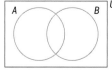

2 Given the Venn diagram with sets A, B and C, draw Venn diagrams to show
  a $A \cap B \cap C$ **b** $(A \cup B) \cap C$
  c $A \cup (B \cap C)$ **d** $(A \cup B) \cap C'$
  e $(A \cap B') \cup (A' \cap B)$

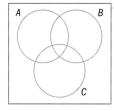

An ability to think logically is an important asset that is valued highly by all professionals. Studying mathematics is believed to promote this ability, and success in mathematics is often used to discriminate between candidates in non-mathematical situations, for example, law.

Mathematical logic can be used to analyze written statements in legal contracts, for example, to determine whether the arguments put forward are valid and precise, or not.

In this chapter you will study mathematical logic and its language. Its structure has many similarities with set theory:

'If you understand set theory, then you will understand logic.'

However, what if you **do not** understand set theory? Will you then understand logic ... or not?

Analyzing such statements (or arguments) and being able to understand what they really mean – as opposed to what they are intended to mean – is the purpose of mathematical logic.

> This statement has the following arguments related to it:
> 'If you do not understand set theory, then you will not understand logic.'
> 'If you understand logic, then you will understand set theory.'
> 'If you do not understand logic, then you will not understand set theory.'

Investigation – logical thinking

A sign on the door to the sports hall says:

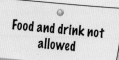

Food and drink not allowed

1 What does the writer of the sign
   (the Director of Sport) think the sign means?
   Are you allowed to bring food into the sports hall?
   Are you allowed to bring drink into the sports hall?

2 Are you allowed to bring food into the sports hall, without bringing
   in drinks?

3 Are you allowed to bring drinks into the sports hall, without bringing
   in food?

4 Does the sign say what the Director of Sport thinks it says?

5 What should the sign say if the Director of Sport wishes to stop
   any food from being brought into the sports hall and stop any drinks
   being brought in also?

9.1 Introduction to logic

Mathematical logic is taught as part of the syllabus for two IB
courses – MSSL and Further Mathematics. It is not taught for either
the HL or the SL course.

If you meet someone and she tells you that she studies
mathematical logic as part of her IB course, is it true that

> *she studies Further Mathematics?*

> Is this statement true?
> Is it **necessarily** true?

Determining whether a statement is **necessarily** true or false is part
of the analysis that we will undertake; statements form the basis of
logic.

Statements and connectives

A **statement** is a sentence or phrase and must have a precise
mathematical meaning.

'Awesome' and 'Cool' are not statements.

> → A (simple) **statement** has a truth value of **true** or **false**
> (but not both).

Here are some statements.

- Goh Chok Tong is the Prime Minister of Singapore.
- $2 + 2 = 5$
- Any square is also a rectangle.
- Any rectangle is also a square.
- If you do not do your homework, I will inform the school principal.

- I do not swim.
- I swim and play football.
- I do not swim or play football.
- I do not swim and I do not play football.
- To enter the race, you must be female or more than 45 years old.

All of the above phrases can be verified in some way and so are statements. Some of these statements are **simple** (and can easily be determined to be either true (T) or false (F)).

For example, Goh Chok Tong is **not** the Prime Minister of Singapore and so the **truth value** of the statement 'Goh Chok Tong is the Prime Minister of Singapore' is false (F).

Exercise 9A

Which of these are statements?

1	Rie studies Malay.	**2**	Is it snowing?
3	Make the tea.	**4**	The air-con is on.
5	5 > 2	**6**	7 < 3
7	$2 \leq 2$	**8**	Have a nice day!
9	The sun is not shining.	**10**	The glass is full.
11	The moon is made of green cheese.	**12**	A week consists of seven days.
13	A week consists of five days.	**14**	A month consists of 30 days.

9.2 Compound statements and symbols

> → A **compound statement** is made up of simple statements joined together by **connectives**.

The **five** connectives that we will use most commonly are:

NOT
AND
OR
OR
IF . . . THEN

> IF . . . THEN is one connective but there are actually five connectives in this list.

The 'OR' connective has two versions in everyday language and we must be very specific about which version we are using when analyzing arguments logically. Often the word 'either' is added in everyday language, though its use is not strictly necessary as part of the 'OR' connective.

These two examples show the two versions of 'OR'.

1 'You may study Mathematics at Higher Level or Standard Level in the IB diploma.'

2 'You may study Spanish or French as part of the IB diploma.'

In **1**: Mathematics can be studied either at Higher Level, or at Standard Level, but **not both**.

The version of 'OR' where the implied meaning is 'one or the other but not both' is known as '**exclusive or**'.

In **2**: Spanish can be studied as part of the IB diploma; so can French. A student can study **both** Spanish and French as part of the IB diploma.

This version of 'OR' where the implied meaning is 'one or the other or both' is known as '**inclusive or**'.

Exercise 9B

Decide which of the two versions of 'or' (exclusive/inclusive) is meant in the following compound statements.

1 Your mother is either Argentinean or Chilean.

2 Do you want either coffee or tea with your meal?

3 Do you take milk or sugar in your coffee?

4 Can you speak either Japanese or Korean?

5 He is allowed either food or drink after exercise.

6 I have injured either my knee or my ankle.

7 Under the terms of your lease, you may use the land for either residential or business purposes.

8 You can use one or two pillows when sleeping.

9 He is a captain in either the army or the navy.

10 x is odd or x is even.

11 Knowing that $(x - 2)(x + 1) = 0$, we know that x is equal to 2 or x is equal to -1.

12 $x \leq 5$

> $\leq$ means 'less than or equal to'

> → The **five** connectives have these names and symbols:
>
> | | | |
> |---|---|---|
> | NOT | **Negation** | ¬ |
> | AND | **Conjunction** | ∧ |
> | OR | **Inclusive disjunction** | ∨ |
> | OR | **Exclusive disjunction** | ∨̲ |
> | IF … THEN | **Implication** | ⇒ |

Here is question 7 from Exercise 9B again.

'Under the terms of your lease, you may use the land for either residential or business purposes.'

<aside>This is 'exclusive or'.</aside>

In a legal contract, both parties need to know exactly what that contract means. There is no place for ambiguity. The use of 'or' – whether inclusive or exclusive – in the above statement must be made precise.

Similarly, in mathematical logic, you need to make clear whether you are using inclusive or ($\vee$) or exclusive or ($\underline{\vee}$) in a statement.

The system used in mathematics and logic is:

1 If you use the word '**or**' in a statement it is **always** taken to be **inclusive**.
2 To use the exclusive version of 'or' you must add the extra phrase '**but not both**'.

So the statement

'you may use the land for either residential or business purposes'

means that you can use the land for either a residential development, a business development or a development that mixes **both** types.

To exclude the mixed development the statement has to be

'you may use the land for either residential or business purposes, **but not both**'.

9.3 Truth tables: negation

Mathematicians prefer symbols to words. You need to learn the symbols for logic.

Letters are used – usually p, q, r – to represent simple statements and these are combined with the connectives to form compound statements.

Each simple statement has a truth value associated with it (T or F – but not both) and these are tabulated in a **truth table.**

Each connective influences the overall truth value of the compound statement and has its own truth table associated with it.

Let p be the statement   'you may use the land for residential purposes'.

> → The **negation** of a statement p is written $\neg p$ (read as 'not-p').
> The relation between any statement p and its negation $\neg p$ is shown in a **truth table**:
>
> | p | $\neg p$ |
> |---|---|
> | T | F |
> | F | T |

Remember that a statement must take one of the **truth values** T (true) or F (false).

The table shows that $\neg p$ is false when p is true and that $\neg p$ is true when p is false.

It is impossible to have a statement which is both true and false at the same time; that would be a logical contradiction.

For statement p above, $\neg p$ is 'you may **not** use the land for residential purposes' or 'you may use the land for any purpose other than residential'.

The truth table above can be used as the definition of negation. **Negation** is an operation in logic; the corresponding operation in set theory is **complement**.

Take care when writing the negation of a statement. A common mistake is to state that the negation of a statement such as, 'Jan is the tallest in the class' is 'Jan is the smallest in the class'. This is not (necessarily) the case. For what circumstances would the second statement be the negation of the first?

Exercise 9C

1 Write negations of these statements.

  a The student is a council member. **b** She owns a mobile phone.

  c n is a prime number. **d** ABCD is a parallelogram.

  e Surabaya is the capital of Indonesia.

2 Write negations of these statements without using the word 'not'.

  a This word starts with a vowel.

  b There is an even number of pages in this book.

  c This price is inclusive of sales tax.

  d This shape is a quadrilateral.

  e He walked at a constant speed.

3 a In these statements, is q the negation of p? If not, say why it is not the negation.

   i p: Chihiro obtained the highest mark on the test.
   q: Chihiro obtained the lowest mark on the test.

   ii p: This test is difficult.
   q: This test is easy.

   iii p: Sahana scored more than 50% on the test.
   q: Sahana scored less than 50% on the test.

   iv p: Richard is inside the classroom.
   q: Richard is outside the classroom.

   v p: Nishad scored above average on the test.
   q: Nishad scored below average on the test.

b In all the above, q was obtained from p by using the opposite word. Does this method always give the correct negation?

c In **i** replace 'obtained' in p by 'did not obtain'. Is the resulting statement the negation of p? Does this method always give the correct negation?

4 Write negations of these statements.

   a x is greater than five.   **b** y is less than seven.
   c z is at least ten.   **d** b is at most 19.

5 The definitions of 'positive' and 'negative' are:

   x is positive whenever $x > 0$
   x is negative whenever $x < 0$

   a Is zero positive or negative?
   b Write down the negation of the statement 'x is negative' given that $x \in \{$real numbers$\}$.

6 Write down $\neg p$ for each statement p. Avoid the word 'not' if you can.

   a p: Courtney was absent from school on Friday.
   b p: This chair is broken.
   c p: The hockey team lost their match.
   d p: The soccer team won the tournament.
   e p: The hotel does not have running water.

7 Write negations of these statements. Avoid the word 'not' if you can.

   a p: His signature is illegible.
   b q: James is older than me.
   c r: The class contains fewer than eight boys.
   d s: Her family name begins with a P.
   e t: He has at least two sisters.

8 The wording of a negation may depend on the universal set that has been given.

If possible, write down the negations of these statements in the given domains, without using the word 'not'.

a X is a female doctor given that $U = \{\text{doctors}\}$.

b X is a female doctor given that $U = \{\text{females}\}$.

c X is a married man given that $U = \{\text{married people}\}$.

d X is a married man given that $U = \{\text{men}\}$.

e R is a positive rotation of more than 90° given that $U = \{\text{positive rotations}\}$.

f R is a positive rotation of more than 90° given that $U = \{\text{all rotations}\}$.

9.4 Truth tables: conjunction (and)

→ The **conjunction** of any two statements p and q is written $p \wedge q$. This **compound statement** is defined by this truth table.

p	q	$p \wedge q$
T	T	**T**
T	F	**F**
F	T	**F**
F	F	**F**

So, $p \wedge q$ is true only when **both** p and q are true. **Conjunction** corresponds to **intersection** in set theory.

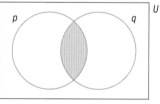

$p \wedge q$

Example 1

Let p represent 'It is at least 35° Celsius today' and let q represent 'It is Friday'. What does $p \wedge q$ represent?

- -

Answer

$p \wedge q$ represents 'It is 35° Celsius today and it is Friday'.

Example 2

Let p represent 'Dana got a 7 in HL Mathematics' and let q represent 'Yuri got a 5 in HL Mathematics'. What does $p \wedge q$ represent?

- -

Answer

$p \wedge q$ represents 'Dana got a 7 in HL Mathematics and Yuri got a 5 in HL Mathematics'.

The compound statement is true only when both the separate statements are true – only when it is Friday **and** only when the temperature is at least 35° Celsius. The statement is false on any other day of the week and it is also false on a cooler day.

Other compound statements related to $p \wedge q$ can be represented in terms of p and q, for example,

$\neg p \wedge q$ represents 'Dana did not get a 7 in HL Mathematics and Yuri got a 5 in HL Mathematics'.

$p \wedge \neg q$ represents 'Dana got a 7 in HL Mathematics and Yuri did not get a 5 in HL Mathematics'.

What does $\neg p \wedge \neg q$ represent?
What does $\neg(p \wedge q)$ represent?
Do they mean the same thing? Use a Venn diagram to investigate.

Exercise 9D

For questions 1 to 9, express each of these compound statements in words:

    a $p \wedge q$ **b** $\neg p \wedge q$ **c** $p \wedge \neg q$ **d** $\neg p \wedge \neg q$ **e** $\neg(p \wedge q)$

1 p: Susan speaks French. q: Susan speaks Spanish.

2 p: Jorge speaks Portuguese. q: Mei Ling speaks Malay.

3 p: All dogs bark. q: All flowers are yellow.

4 p: China is in Africa. q: Rwanda is in Asia.

5 p: Chicago is the largest q: Jakarta is the largest
       city in Canada.   city in Indonesia.

6 p: $x \leq 5$ q: $x \geq 5$
    Is it possible for both p and q to be true?

7 p: ABCD is a parallelogram. q: ABCD is a rectangle.
    Which of the statements **a** to **e** cannot possibly be true in this case?

8 p: Triangle ABC is right-angled at C. q: $AB^2 = AC^2 + BC^2 + 1$
    Which of the statements **a** to **e** cannot possibly be true in this case?
    Which of the statements **a** to **e** must be true in this case?

9 p: n is an odd integer. q: n is an even integer.
    Which of the statements **a** to **e** cannot possibly be true in this case?
    Which of the statements **a** to **e** must be true in this case?

10 Complete the truth table for $p \wedge \neg p$.

p	$\neg p$	$p \wedge \neg p$

    a First, enter the alternatives T/F in the column for p.
    b Next, use the definition of negation to complete
        the column for $\neg p$.
    c Finally, use the definition of conjunction to complete
        the column for $p \wedge \neg p$.
    If you have done this correctly, you will be able to deduce from
    the truth table that $p \wedge \neg p$ is a **logical contradiction**.

Often we use the word 'but' rather than 'and' in a statement that combines conjunction with negation. It is better to use 'and'.

Compound statements may combine statements that are related (as in question 1) or completely unrelated (as in question 3).

What is it about the final column that allows you to make this deduction?

11 Consider the statements

     p: Sei Jin came top in Mathematics.

     q: Sei Jin came top in English.

Write the statement

     r: Sei Jin came top in mathematics but not in English

in terms of p and q.

Construct a truth table showing how the truth value of r depends on the truth values of p and q.

12 Consider the statements

     p: n is divisible by 2. q: n is divisible by 5.

Write the statement

     r: n is divisible by 10

in terms of p and q.

Construct a truth table for r. For each row of the table, write down a value of n which gives that combination of truth values.

9.5 Truth tables: resolving an ambiguity - the 'or' connective

There are two versions of the 'or' connective: **inclusive or** and **exclusive or**.

Disjunction

> → The **disjunction** of any two statements p and q is written $p \vee q$. This is '**inclusive or**' and it is defined by this truth table.
>
> | p | q | $p \vee q$ |
> |:---:|:---:|:---:|
> | T | T | **T** |
> | T | F | **T** |
> | F | T | **T** |
> | F | F | **F** |
>
> $p \vee q$ is true if either p or q or possibly both are true.

Disjunction corresponds to **union** in set theory, where if x is an element of $p \cup q$, then x can be placed in either set p or set q or in the intersection of p and q.

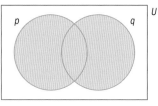

$p \vee q$

Example 3

Let p represent 'It is at least 35° Celsius today' and let q represent 'It is Friday'.
What does $p \lor q$ represent?

- -

Answer

$p \lor q$ represents 'It is at least 35° Celsius today or it is Friday or both'.

Note that to make the compound statement true it is only necessary that one of the individual statements is true – when it is Friday **anything** can be claimed about the temperature. For the compound statement to be true on any other day of the week the statement about the temperature must be true.

Example 4

Let p represent 'Dana got a 7 in HL Mathematics' and let q represent 'Yuri got a 5 in HL Mathematics'.
What does $p \lor q$ represent?

- -

Answer

$p \lor q$ represents 'Dana got a 7 in HL Mathematics or Yuri got a 5 in HL Mathematics or both'.

Exclusive disjunction

→ **Exclusive disjunction** is written $p \veebar q$ and is defined by this truth table.

p	q	$p \veebar q$
T	T	**F**
T	F	**T**
F	T	**T**
F	F	**F**

So, we exclude the possibility that both simple statements can be true simultaneously.

The equivalent in set theory of exclusive or is the **symmetric difference** and is shown on a Venn diagram as:

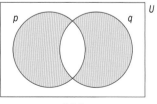

$p \veebar q$

Example 5

> The penalty for those found guilty of murder is death or life imprisonment.
> Explain why this is an example of exclusive or.
>
> **Answer**
> The penalties are alternatives. There is no possibility of both sentences being imposed and so it is clear that we are using exclusive or.

If there is some doubt about whether inclusive or exclusive or is intended, we always assume that **inclusive or** is being used. To make it clear that **exclusive or** is required the phrase '**but not both**' is added at the end.

Example 6

> What is the difference between these two statements?
> **a** The penalty for those found guilty of speeding is a $500 fine or 3 months in prison, but not both.
> **b** The penalty for those found guilty of speeding is a $500 fine or 3 months in prison.
>
> **Answers**
> In **a**, the person found guilty will either be fined or sent to prison, but not both.
> In **b**, it is possible the person found guilty of speeding will receive both a fine and a prison term.

Exercise 9E

1 Let p and q be the statements $\quad p$: $x < 36$ $\quad q$: $x = 36$
   a Express in words $\quad$ **i** $p \vee q$ $\qquad$ **ii** $p \veebar q$
   b The statement $x \le 36$ is **equivalent** to one of part **a**, **i** or **ii**. Which?

2 Three propositions p, q and r are defined as:

   p: The water is cold.
   q: The water is boiling.
   r: The water is warm.

   a Express in terms of p, q and r (as appropriate):
      i Either the water is cold or the water is warm.
      ii Either the water is cold or the water is warm, but not both.
      iii Either the water is boiling or the water is warm.
      iv Either the water is boiling or the water is warm, but (**and**) the water is not cold.
   b In the case of **iv** is it appropriate to use 'inclusive or' as a connective?

3 Let p, q and r be statements about the set of positive integers that are less than or equal to 36.

    p: x is a multiple of six.
    q: x is a factor of 36.
    r: x is a square number.

a Express in terms of p, q and r (as appropriate):
    i Either x is a multiple of six or x is a factor of 36.
    ii Either x is a multiple of six or x is a factor of 36, but not both.
    iii Either x is a multiple of six or x is a square number.
    iv Either x is a factor of 36 or x is a square number, but not both.
    v Either x is a multiple of six or x is a factor of 36 or x is a square number.
    vi Either x is a multiple of six or x is a factor of 36 but (**and**) x is not a square number.

b List the integers, x, that satisfy the statements in **a** (i.e. that make the statements about x true).

4 Let the propositions p, q and r be defined as:

    p: Matthew arrives home before six o'clock.
    q: Anna cooks dinner.
    r: Matthew washes the dishes.

Express in terms of p, q and r (as appropriate):

a Either Matthew arrives home before six o'clock or Anna cooks dinner.
b Either Matthew washes the dishes or Anna cooks dinner, but not both.
c Either Matthew arrives home before six o'clock or Matthew washes the dishes.
d Matthew washes the dishes and Anna cooks dinner.

5 The propositions p and q are defined as follows:

    p: You have understood this topic.
    q: You will be able to do this question.

Express in terms of p, q:

a You have understood this topic, or you will not be able to do this question.
b You have not understood this topic, and you will not be able to do this question.

6 Two propositions p and q are defined as follows where x is an element of the set of integers:

> p: x ends in zero.
> q: x is not divisible by 5.

Express in words:

a $p \lor q$

b $p \veebar q$

c $p \land q$

d $p \land \neg q$

e $\neg p \land q$

Write down a value of x that satisfies each of the above statements. Hence, determine which of the statements is necessarily false.

7 Consider these logic statements:

> p: I am studying French.
> q: I am studying Chinese.

a Express in terms of p and q:

　i I am studying French and I am studying Chinese.

　ii I am studying French or I am studying Chinese, but not both.

　iii I am studying French or I am studying Chinese.

　iv I am not studying French or I am not studying Chinese.

　v I am not studying either French or Chinese.

　vi I am not studying both French and Chinese.

　vii I am not studying French and I am not studying Chinese.

b In which of the statements in **a** can it be deduced that

　i It is **necessarily true** that I am studying both languages.

　ii It **may be true** that I am studying both languages.

　iii It is **necessarily true** that I am studying neither of the two languages.

　iv It **may be true** that I am studying neither of the two languages.

Logic was first studied in the ancient civilizations of India, China and Greece. Aristotle first defined the three essential subjects for study as logic, grammar and rhetoric – and these three subjects, known as the *trivium*, formed the basis of university education in Europe until the end of the Medieval period.

Nowadays logic is mainly studied in the contexts of philosophy, mathematics, semantics and computer science.

9.6 Logical equivalence, tautologies and contradictions

In logic, you must be careful to deduce only what is **necessarily** true from the statements. The best way to do this is to construct a truth table.

Example 7

Construct the truth table for the statement
¬*p* ∧ ¬*q*: I am not studying French and I am not studying Chinese
where *p* represents 'I am studying French' and *q* 'I am studying Chinese'.

Answer

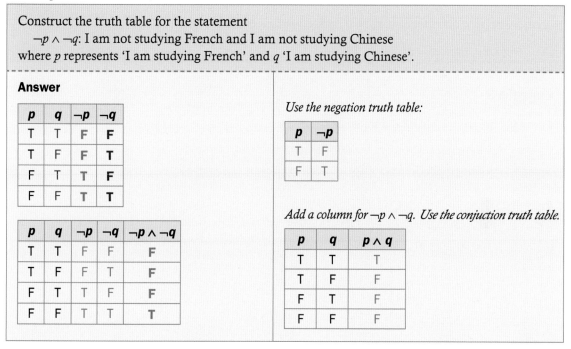

p	*q*	¬*p*	¬*q*
T	T	F	F
T	F	F	T
F	T	T	F
F	F	T	T

Use the negation truth table:

p	¬*p*
T	F
F	T

p	*q*	¬*p*	¬*q*	¬*p* ∧ ¬*q*
T	T	F	F	F
T	F	F	T	F
F	T	T	F	F
F	F	T	T	T

Add a column for ¬p ∧ ¬q. Use the conjuction truth table.

p	*q*	*p* ∧ *q*
T	T	T
T	F	F
F	T	F
F	F	F

Example 8

Construct the truth table for the statement
¬(*p* ∨ *q*): I am not studying either French or Chinese.
where *p* represents 'I am studying French' and *q* 'I am studying Chinese'.

Answer

p	*q*	(*p* ∨ *q*)
T	T	T
T	F	T
F	T	T
F	F	F

The brackets in the statement mean find p ∨ q first, then its negation.

Use the disjunction truth table.

p	*q*	(*p* ∨ *q*)
T	T	T
T	F	T
F	T	T
F	F	F

p	*q*	(*p* ∨ *q*)	¬(*p* ∨ *q*)
T	T	T	F
T	F	T	F
F	T	T	F
F	F	F	T

Use the negation truth table.

p	¬*q*
T	F
F	T

The entries in the last columns in the two truth tables in Examples 7 and 8 are exactly the same. These identical entries tell us that, **whatever the truth values of** p and q, the compound statements $\neg p \wedge \neg q$ and $\neg(p \vee q)$ have the same truth values. There is no *logical* difference between them.

> → The statements $\neg p \wedge \neg q$ and $\neg(p \vee q)$ are said to be **(logically) equivalent**. Equivalence is shown by the symbol $\Leftrightarrow$ so we write
>
> $$\neg p \wedge \neg q \Leftrightarrow \neg(p \vee q)$$

Equivalence depends only on the **structure** of the two compound statements. It does not depend on the meaning of the initial statements p and q.

Example 9

Construct the truth table for the statement

$\neg p \vee \neg q$: I am not studying French or I am not studying Chinese

where p represents 'I am studying French' and q 'I am studying Chinese'.

Answer

p	q	$\neg p$	$\neg q$	$\neg p \vee \neg q$
T	T	F	F	**F**
T	F	F	T	**T**
F	T	T	F	**T**
F	F	T	T	**T**

First use the truth table for negation, then the truth table for disjunction.

Note that the final column in this example is **not** the same as that for $\neg(p \vee q)$ in Example 8. Therefore:

> → $\neg p \vee \neg q$ is **not** equivalent to $\neg(p \vee q)$.

Tautologies and contradictions

> → • A **tautology** is a compound statement which is **true whatever** the truth values of the simple statements it is made up from.
> • A (logical) **contradiction** is a compound statement which is **false whatever** the truth values of its simple statements.

$[\neg(p \wedge q)] \vee q$ is a tautology because all of the entries in the column associated with that statement are true.

Here is the truth table for the statements

a $[\neg(p \wedge q)] \vee q$ and   **b** $[\neg(p \vee q)] \wedge p$

p	q	$p \wedge q$	$\neg(p \wedge q)$	$(p \vee q)$	$\neg(p \vee q)$	$[\neg(p \wedge q)] \vee q$	$[\neg(p \vee q)] \wedge p$
T	T	T	F	T	F	**T**	**F**
T	F	F	T	T	F	**T**	**F**
F	T	F	T	T	F	**T**	**F**
F	F	F	T	F	T	**T**	**F**

$[\neg(p \vee q)] \wedge p$ is a logical contradiction (or contradiction) since all of the entries in the column associated with that statement are false.

If there is a mix of true and false in the column, then the statement associated with that column is neither a tautology nor a contradiction.

So, for example, $p \wedge q$ is neither a tautology nor a contradiction; $p \vee q$ is neither a tautology nor a contradiction.

> If a compound statement is analyzed and found to be a tautology, then the logic behind the statement is valid. In a legal contract, all statements making up the contract should be tautologies. If the contract is drawn up from statements that are not all tautologies then there are 'loopholes', which could be challenged or exploited.

Exercise 9F

1 a Write down the truth tables for the statements of question 7 in Exercise 9E on page 394.
   b Determine which of the statements is equivalent to 'I am not studying French or I am not studying Chinese'.

2 Use truth tables to prove these logical equivalences.
   a $\neg(\neg p) \Leftrightarrow p$
   b $p \wedge p \Leftrightarrow p$
   c $p \vee (p \wedge q) \Leftrightarrow p$
   d $p \vee (\neg p \wedge q) \Leftrightarrow p \vee q$

3 Construct a truth table for the statement
   $$(p \wedge \neg q) \vee (\neg p \wedge q).$$
   This expression defines a logical operation on p and q which is similar to conjunction and disjunction. What operation is it?

4 Determine which of these are tautologies, which are contradictions and which are neither.
   a $p \vee \neg p$ b $p \wedge \neg p$
   c $p \wedge (p \wedge p)$ d $(p \vee q) \vee (\neg p \wedge \neg q)$
   e $(p \vee \neg q) \vee (\neg p \wedge q)$ f $(p \vee \neg q) \wedge (\neg p \wedge \neg q)$
   g $(\neg p \vee q) \wedge (p \wedge q)$ h $(p \wedge q) \wedge (\neg p \wedge \neg q)$

9.7 Compound statements made up from three simple statements

Compound statements made up from three simple statements need eight lines in the truth table.

The truth table for any compound statement that involves p, q and r begins like this:

Add the other columns according to the given rules. The only complication is that there are more entries.

p	q	r	
T	T	T	
T	T	F	
T	F	T	
T	F	F	
F	T	T	
F	T	F	
F	F	T	
F	F	F	

> How many lines are required in the truth table for a compound statement comprised of four simple statements? How many for five?

> You are not expected to deal with compound statements made up from more than three simple statements.

The same definitions apply:

- Equivalence is determined by looking at the final columns for each of the statements.
- A tautology is determined only if **all** the entries in the final column are T.
- A contradiction is determined only if **all** the entries in the final column are F.

Exercise 9G

Construct truth tables for these compound statements.
Determine whether each statement is

a a tautology **b** a contradiction **c** neither.

1 $p \vee (q \wedge r)$

p	q	r	$(q \wedge r)$	$p \vee (q \wedge r)$

2 $(p \vee \neg q) \vee r$

p	q	r	$\neg q$	$(p \vee \neg q)$	$(p \vee \neg q) \vee r$

3 $(p \wedge q) \vee (p \wedge \neg r)$

p	q	r	$\neg r$	$p \wedge q$	$p \wedge \neg r$	$(p \wedge q) \vee (p \wedge \neg r)$

4 $(p \vee q) \vee (r \wedge \neg q)$

5 $(p \wedge r) \wedge (q \wedge \neg r)$

6 $(\neg p \vee q) \vee (p \wedge r)$

7 $(\neg p \vee q) \wedge (p \vee r)$

8 $(p \vee q) \wedge (p \vee r)$ Which of questions 1–7 is this equivalent to?

Using brackets in statements

The statements $\neg p \wedge q$ and $\neg(p \wedge q)$ are not equivalent. Their meanings are not the same.

Take care when expressing written statements using logic notation to ensure you use brackets when necessary.

In general, you should use brackets every time you combine two simple statements using a connective.

Sometimes, however, brackets are not required.

> Write down the truth tables for $\neg p \wedge q$ and $\neg(p \wedge q)$.

> In general, it is better to use brackets rather than risk making a mistake by omitting them.

Example 10

Are brackets needed for the statements $(p \vee q) \vee r$ and $p \vee (q \vee r)$?

Answer

For $(p \vee q) \vee r$

p	*q*	*r*	$(p \vee q)$
T	T	T	T
T	T	F	T
T	F	T	T
T	F	F	T
F	T	T	T
F	T	F	T
F	F	T	F
F	F	F	F

Construct the truth table for $(p \vee q)$.

p	*q*	*r*	$(p \vee q)$	$(p \vee q) \vee r$
T	T	T	T	**T**
T	T	F	T	**T**
T	F	T	T	**T**
T	F	F	T	**T**
F	T	T	T	**T**
F	T	F	T	**T**
F	F	T	F	**T**
F	F	F	F	**F**

Add the $(p \vee q) \vee r$ column.

▶ Continued on next page

For $p \vee (q \vee r)$

p	q	r	(q ∨ r)
T	T	T	T
T	T	F	T
T	F	T	T
T	F	F	F
F	T	T	T
F	T	F	T
F	F	T	T
F	F	F	F

Construct the truth table for $(q \vee r)$.

p	q	r	(q ∨ r)	p ∨ (q ∨ r)
T	T	T	T	T
T	T	F	T	T
T	F	T	T	T
T	F	F	F	T
F	T	T	T	T
F	T	F	T	T
F	F	T	T	T
F	F	F	F	F

Add the $p \vee (q \vee r)$ column.

$(p \vee q) \vee r$ and $p \vee (q \vee r)$ are equivalent so brackets are not needed.

Compare the final columns of the two truth tables.

→ $(p \vee q) \vee r \Leftrightarrow p \vee (q \vee r)$

We can omit the brackets and write: $p \vee q \vee r$

Exercise 9H

1 Use truth tables to determine whether $(p \wedge q) \wedge r \Leftrightarrow p \wedge (q \wedge r)$ and thus whether brackets are needed.

2 Use Venn diagrams and the sets P, Q, R to show that $(P \cap Q) \cap R$ and $P \cap (Q \cap R)$ are equivalent.

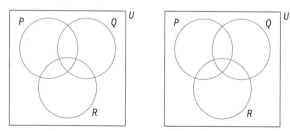

3 Use Venn diagrams and the sets P, Q, R to show that $(P \cup Q) \cup R$ and $P \cup (Q \cup R)$ are equivalent.

4 Use truth tables to determine whether $p \wedge (q \vee r) \Leftrightarrow (p \wedge q) \vee r$ and thus whether there is a need to use brackets.

5 Use Venn diagrams and the sets P, Q, R to determine whether $P \cap (Q \cup R)$ and $(P \cap Q) \cup R$ are equivalent.

6 Use truth tables to determine whether $p \vee (q \wedge r) \Leftrightarrow (p \vee q) \wedge r$ and thus whether there is a need to use brackets.

7 Use Venn diagrams and the sets P, Q, R to determine whether $P \cup (Q \cap R)$ and $(P \cup Q) \cap R$ are equivalent.

8 Are $(\neg p \wedge q) \vee (\neg q \wedge r) \vee (\neg r \wedge p)$ and $(\neg p \vee q) \wedge (\neg q \vee r) \wedge (\neg r \vee p)$ equivalent?

9.8 Arguments

> → A compound statement that includes **implication** is called an **argument**.

The analysis of an argument is perhaps the most important application of logic. Implication involves 'cause and effect' situations and can be summed up by the phrase 'if ... then ... '. It is perhaps the most important of all connectives.

Consider the statement 'If you do that again, I will inform your parents.' This compound statement uses:

the connective   if ... then ...
and the simple statements p: you do that again
and q: I will inform your parents.

This is written as $p \Rightarrow q$ where p is the **antecedent** and q the **consequent** (the cause and effect).

> → The truth table for implication is:
>
> | p | q | $p \Rightarrow q$ |
> |-----|-----|-------------------|
> | T | T | **T** |
> | T | F | **F** |
> | F | T | **T** |
> | F | F | **T** |

The implication truth table tends to confuse people. The best way to remember the implication is to remember that the only time we can say $p \Rightarrow q$ is FALSE is if p is TRUE and q is FALSE. All other times it is true.

The truth table for implication is a little strange since its two bottom lines have truth values that are not obvious.

The truth table has this form because the only way to **know** that the implication is **certainly** false (necessarily false) is if:

How is it possible that a false premise can imply a true statement?

- you do 'that' again (whatever 'that' may be) and
- as a consequence, I do **not** inform your parents.

Thus, in line 2 of the truth table, the implication is false. In line 1, the implication is obviously true and in the other cases, since the antecedent is false (you did not do that again) it cannot be proven that the consequent (and therefore the overall statement) is false.

Therefore, we take the overall statement as true.

Truth tables are used to analyze an argument to determine whether it makes sense (is valid) or makes no sense (is a contradiction) or may be applied in certain circumstances, but not others. Analyzing an argument in this way is known as **testing its validity**.

> → If the compound statement that represents an argument is a **tautology**, then the argument is **valid**.

If the compound statement that represents an argument is a not a tautology, then the argument is **invalid**. So, not only is a **contradiction** an invalid argument, but so is any statement that contains a truth value of **false** in any position.

When people make statements such as, 'If you do that again, I will inform your parents.', they consider the consequences only in the case where you do 'do that again'. However, for the truth table all possibilities must be considered.

Example 11

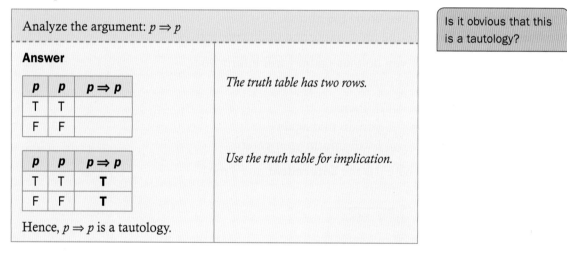

Analyze the argument: $p \Rightarrow p$

Answer

p	p	$p \Rightarrow p$
T	T	
F	F	

The truth table has two rows.

p	p	$p \Rightarrow p$
T	T	**T**
F	F	**T**

Use the truth table for implication.

Hence, $p \Rightarrow p$ is a tautology.

Is it obvious that this is a tautology?

Example 12

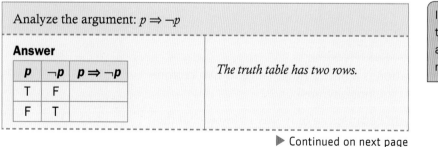

Analyze the argument: $p \Rightarrow \neg p$

Answer

p	$\neg p$	$p \Rightarrow \neg p$
T	F	
F	T	

The truth table has two rows.

Is it obvious whether this is a tautology, a contradiction or neither?

▶ Continued on next page

p	$\neg p$	$p \Rightarrow \neg p$
T	F	**F**
F	T	**T**

Use the truth table for implication.

Hence, $p \Rightarrow \neg p$ is neither a contradiction nor a tautology but is an invalid argument.

Is this a surprise?

Example 13

Analyze the argument: $(p \Rightarrow \neg p) \wedge (\neg p \Rightarrow p)$

Answer

p	$\neg p$	$p \Rightarrow \neg p$	$\neg p \Rightarrow p$	$(p \Rightarrow \neg p) \wedge (\neg p \Rightarrow p)$
T	F	**F**		
F	T	**T**		

$\neg p$	p	$\neg p \Rightarrow p$
F	T	**T**
T	F	**F**

Reverse the columns and use the truth table for implication.

Hence,

p	$\neg p$	$p \Rightarrow \neg p$	$\neg p \Rightarrow p$	$(p \Rightarrow \neg p) \wedge (\neg p \Rightarrow p)$
T	F	F	**T**	
F	T	T	**F**	

Take great care when completing the next column, since the initial columns are in reverse order.

p	$\neg p$	$p \Rightarrow \neg p$	$\neg p \Rightarrow p$	$(p \Rightarrow \neg p) \wedge (\neg p \Rightarrow p)$
T	F	F	T	**F**
F	T	T	F	**F**

Use the truth table for conjunction.

Hence, $(p \Rightarrow \neg p) \wedge (\neg p \Rightarrow p)$ is both a contradiction and an invalid argument.

Exercise 9I

1 Use truth tables to analyze the arguments $p \Rightarrow p \wedge q$ and $p \Rightarrow p \vee q$.

2 Use truth tables to analyze the arguments $p \wedge q \Rightarrow p$ and $p \vee q \Rightarrow p$.

3 Use a truth table to analyze the argument $(p \vee q \Rightarrow p) \wedge (p \Rightarrow p \wedge q)$.

4 Use a truth table to analyze the argument $(p \wedge q \Rightarrow p) \wedge (p \Rightarrow p \wedge q)$.

5 Use a truth table to analyze the argument $(p \wedge q \Rightarrow p) \vee (p \Rightarrow p \wedge q)$.

6 Use a truth table to analyze the argument $\neg(p \wedge q) \Rightarrow \neg p \vee \neg q$.

7 Use a truth table to analyze the argument $\neg(p \vee q) \Rightarrow \neg p \vee \neg q$.

8 Use a truth table to analyze the argument $\neg p \vee \neg q \Rightarrow \neg(p \wedge q)$.

9 Use a truth table to analyse the argument $\neg(p \vee q) \Rightarrow \neg p \wedge \neg q$.

Example 14

Analyze the argument:
'If the communists win the election, I will leave the country.
I am not leaving the country. Therefore, the communists will not win
the election.'

> Tenses (future, present, past) are not, in general, taken into account in analyzing statements.

Answer

p: The communists win the election.
q: I (will) leave the country.

Give variable names to the different simple statements.

If the communists win the election, I will leave the
country: $p \Rightarrow q$
I am not leaving the country: $\neg q$
The communists will not win the election: $\neg p$

Analyze the separate sentences.

So, the argument becomes $[(p \Rightarrow q) \land (\neg q)] \Rightarrow (\neg p)$

*Connect the simple statements with 'and' (conjunction) for the **antecedents** and 'if ... then' (implication) if they are **consequents**.*

Construct the truth table.

p	q	$(p \Rightarrow q)$	$\neg q$	$(p \Rightarrow q) \land (\neg q)$	$\neg p$	$[(p \Rightarrow q) \land (\neg q)] \Rightarrow (\neg p)$
T	T	T	F	F	F	**T**
T	F	F	T	F	F	**T**
F	T	T	F	F	T	**T**
F	F	T	T	T	T	**T**

Looking at the final column, $[(p \Rightarrow q) \land (\neg q)] \Rightarrow (\neg p)$ is a
tautology, so the argument is **valid**.

Note that when analyzing an argument, we are considering its form
only. We are not considering the meaning of each statement. When
the argument is a tautology, we conclude that the **structure** of the
argument is perfect.

Example 15

Analyze this argument.
'If the TOK lecture is too long, Kylie gets a headache. Kylie has a
headache. Therefore the TOK lecture is too long.'

▶ Continued on next page

Answer

p: The TOK lecture is too long.
q: Kylie gets a headache.

If the TOK lecture is too long, Kylie gets a headache: $p \Rightarrow q$
Kylie has a headache: q

Therefore the TOK lecture is too long. $\Rightarrow p$

$[(p \Rightarrow q) \wedge q] \Rightarrow p$

p	q	$(p \Rightarrow q)$	$(p \Rightarrow q) \wedge q$	$[(p \Rightarrow q) \wedge q)] \Rightarrow p$
T	T	T	T	T
T	F	F	F	T
F	T	T	T	F
F	F	T	F	T

This is not a tautology, so the argument is not valid (Kylie's headache was not caused by TOK **for certain**).
It is not a contradiction, so the TOK lecture **may or may not** have been the cause of Kylie's headache.

Give variable names to the different simple statements.

Then analyze the separate sentences.

Connect with 'and' (conjunction) for the **antecedents** *and 'if ... then' (implication) for the* **consequents**.

Construct the truth table.

The final column is TTFT.

→ There are four important types of argument:

- A **contradiction** is always false.   (last column of truth table all Fs)
- A **tautology** is always true.   (last column of truth table all Ts)
- A **valid** argument is always true.   (last column of truth table all Ts)
- An **invalid** argument is not always true.   (last column of truth table has at least one F)

These definitions mean that an invalid argument may (or may not) be a contradiction.
A contradiction, however, is always an invalid argument.

Example 16

Analyze this argument. Is it valid?
'If I kiss that frog, it will turn into a handsome prince. If that frog turns into a handsome prince I will marry him. Therefore, if I kiss that frog I will marry him.'

▶ Continued on next page

Chapter 9   405

Answer

p: I kiss that frog.

q: The frog turns into a handsome prince.

r: I marry it/him (the frog/handsome prince).

Give variable names to the statements.

Three statements will make an eight-line truth table.

The argument is: $[(p \Rightarrow q) \wedge (q \Rightarrow r)] \Rightarrow (p \Rightarrow r)$

The truth table is:

Write the argument in symbols.

Construct the truth table.

p	q	r	$p \Rightarrow q$	$q \Rightarrow r$	$(p \Rightarrow q) \wedge (q \Rightarrow r)$	$p \Rightarrow r$	$[(p \Rightarrow q) \wedge (q \Rightarrow r)] \Rightarrow (p \Rightarrow r)$
T	T	T	T	T	T	T	**T**
T	T	F	T	F	F	F	**T**
T	F	T	F	T	F	T	**T**
T	F	F	F	T	F	F	**T**
F	T	T	T	T	T	T	**T**
F	T	F	T	F	F	T	**T**
F	F	T	T	T	T	T	**T**
F	F	F	T	T	T	T	**T**

This is a tautology, so it is a valid argument.

The final column is all Ts.

Exercise 9J

Write each of these arguments in symbols and then test its validity.

1 If Eyala plugs the DVD player in, then it will blow a fuse. She does not plug the DVD player in. Therefore it will not blow a fuse.

2 If Muamar applies weed-killer to his garden then its yield will increase. The yield increases. Therefore Muamar did apply weed-killer.

3 Isaac will pass the Mathematics test or drop out of the IB diploma. He does not drop out of the IB diploma. Therefore he passed the Mathematics test.

4 If you like music then you will go to tonight's concert. If you go to tonight's concert then you will buy some CDs. You do not buy CDs. Therefore, you do not like music.

5 If a person has an annual medical, then many illnesses can be detected early. If illnesses are detected early then many lives can be saved. Therefore if people do not have annual medical, many lives will not be saved.

6 If you are involved in a car accident your insurance premiums will increase. If your insurance premiums increase then you will have to sell your car. Therefore if you are not involved in a car accident you will not have to sell your car.

7 If Doctor Underwood gives difficult tests then the students will fail. If the students fail then they will complain to Ms Smart. If they complain to Ms Smart then Doctor Underwood will be dismissed. Therefore, since he was not dismissed, he must give easy tests.

A related conditional

The statement $(p \Rightarrow q) \wedge (q \Rightarrow p)$ is an important statement. It has its own symbol ($\Leftrightarrow$) and its own name – **equivalence**, or the **biconditional**.

It describes cases where two statements are either true together or false together – they are equivalent to each other.

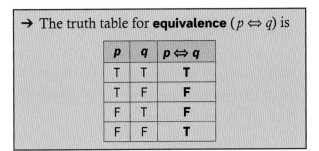

→ The truth table for **equivalence** ($p \Leftrightarrow q$) is

p	q	p ⇔ q
T	T	**T**
T	F	**F**
F	T	**F**
F	F	**T**

> Remember the difference between $\wedge$ and $\Leftrightarrow$
>
> $a \wedge b$ is TRUE only when both a AND b are TRUE.
>
> $a \Leftrightarrow b$ is TRUE when a is the same as b, so you can think of $\Leftrightarrow$ as meaning 'if and only if' or perhaps more usefully 'is the same as'.
>
> In particular: $a \wedge b$ is FALSE when a AND b are FALSE.
>
> $a \Leftrightarrow b$ is TRUE when a AND b are FALSE.
>
> Discuss with your teacher if you are still confused about this.

An example of the biconditional might be 'A parallelogram is a quadrilateral with two pairs of parallel sides'.

In this statement, p represents 'the shape is a parallelogram' and q represents 'the shape has two pairs of parallel sides'.

In English, we would say, 'the shape is a parallelogram **if and only if** it has two pairs of parallel sides'.

The two separate statements are equivalent to each other.

There are two possible ways to determine whether two statements are equivalent to each other.

1 a Write down the truth table for each of the two statements.
   b Compare the final two columns.
   c If these are the same, then the statements are equivalent.

2 a Write down a truth table for the two statements with an extra column for the biconditional.
   b Complete the truth table, including the biconditional column.
   c If the biconditional column shows a tautology, then the statements are equivalent.

Example 17

Determine whether the statements $(p \wedge \neg q) \vee (\neg p \wedge q)$ and $p \veebar q$ are equivalent.

Answer
Method 1

p	q	¬p	¬q	p ∧ ¬q	¬p ∧ q	(p ∧ ¬q) ∨ (¬p ∧ q)
T	T	F	F	F	F	**F**
T	F	F	T	T	F	**T**
F	T	T	F	F	T	**T**
F	F	T	T	F	F	**F**

Complete the truth table for $(p \wedge \neg q) \vee (\neg p \wedge q)$.

p	q	p ∨ q
T	T	**F**
T	F	**T**
F	T	**T**
F	F	**F**

Complete the truth table for $p \veebar q$.

The final columns are both **FTTF**, hence the two statements are equivalent.

Method 2

p	q	(p ∧ ¬q) ∨ (¬p ∧ q)	p ∨ q	[(p ∧ ¬q) ∨ (¬p ∧ q)] ⇔ (p ∨ q)
T	T	F	F	**T**
T	F	T	T	**T**
F	T	T	T	**T**
F	F	F	F	**T**

Complete a truth table for $(p \wedge \neg q) \vee (\neg p \wedge q)$ and $p \veebar q$ with an extra column for the biconditional ⇔.

The final column shows a tautology (all the truth values are T), hence the two statements are equivalent.

Exercise 9K

1 Use a truth table to determine whether the statement $\neg(p \wedge q) \Leftrightarrow (\neg p \vee \neg q)$ is a tautology.

2 Use a truth table to determine whether the statement $\neg(p \vee q) \Leftrightarrow (\neg p \wedge \neg q)$ is a tautology.

3 Use truth tables to analyze the statements $(p \wedge q) \Leftrightarrow p$ and $(p \vee q) \Leftrightarrow p$. Are the statements equivalent?

4 Determine whether the statements $\neg(p \wedge \neg q)$ and $\neg p \vee q$ are equivalent.

5 Determine whether the statements $\neg(p \vee \neg q)$ and $\neg p \wedge q$ are equivalent.

6 Determine the nature of the statement: $(p \vee \neg q) \Leftrightarrow (\neg p \wedge q)$.

7 Determine the nature of the statement: $\neg(p \vee q) \Leftrightarrow (p \wedge q)$.

8 Determine the nature of the statement: $(p \wedge \neg q) \Leftrightarrow (\neg p \vee q)$.

Three more conditional statements

When a person tries to formulate an argument, they often use imprecise language. Antecedents and consequents become mixed up and the implication connective can be reversed.

Although you might think you are arguing the direct statement $p \Rightarrow q$, if you use imprecise language, you may end up arguing $q \Rightarrow p$.

→ There are three commonly used arguments that are formed from the direct statement $p \Rightarrow q$:

$q \Rightarrow p$	the **converse** of the original statement
$\neg p \Rightarrow \neg q$	the **inverse** of the original statement
$\neg q \Rightarrow \neg p$	the **contrapositive** of the original statement.

Does it matter? Is $p \Rightarrow q$ the same as $q \Rightarrow p$?

Example 18

Find the converse, inverse and contrapositives for the direct argument
'*If the weather is sunny, then I will go for a swim.*'

Answer

p: The weather is sunny. q: I go swimming.

We have the statement $p \Rightarrow q$
The converse statement, $q \Rightarrow p$, is the argument

'*If I go for a swim, then the weather is sunny.*'

The inverse statement, $\neg p \Rightarrow \neg q$, is the argument

'*If the weather is not sunny, then I will not go for a swim.*'

The contrapositive, $\neg q \Rightarrow \neg p$, is the argument

'*If I do not go for a swim, then the weather is not sunny.*'

Write the statement in symbols.

Exercise 9L

1 Copy and complete the truth table for the converse, $q \Rightarrow p$.

Converse

p	q	p	$q \Rightarrow p$

Make sure that you construct the implication $q \Rightarrow p$.

2 Copy and complete the truth table for the inverse, $\neg p \Rightarrow \neg q$.

Inverse

p	q	$\neg p$	$\neg q$	$\neg p \Rightarrow \neg q$

3 Copy and complete the truth table for the contrapositive, $\neg q \Rightarrow \neg p$.

Contrapositive

p	q	$\neg q$	$\neg p$	$\neg q \Rightarrow \neg p$

> The columns for $\neg p$ and $\neg q$ have been reversed.

→ This table summarizes the truth values for the direct argument, $p \Rightarrow q$, and the related conditionals.

p	q	Statement $p \Rightarrow q$	Converse $q \Rightarrow p$	Inverse $\neg p \Rightarrow \neg q$	Contrapositive $\neg q \Rightarrow \neg p$
T	T	T	T	T	T
T	F	F	T	T	F
F	T	T	F	F	T
F	F	T	T	T	T

Remember that two statements are **logically equivalent** if they have the same truth values. This means that you could replace one statement by the other without changing the argument or its validity.

It is clear from the table above that a direct argument and its **contrapositive** are logically equivalent.

> They both have TFTT in the final column.

Hence, $(p \Rightarrow q) \Leftrightarrow (\neg q \Rightarrow \neg p)$ and the truth table for this statement has TTTT as its final column.

$(p \Rightarrow q) \Leftrightarrow (\neg q \Rightarrow \neg p)$ is a tautology. So, to make the argument $p \Rightarrow q$, you could make the argument $\neg q \Rightarrow \neg p$ with equal validity.

Example 19

Find the converse, inverse and contrapositives for the direct argument 'If it is raining, then the streets are wet'.

Answer

Direct:	*'If it is raining, then the streets are wet'*	$p \Rightarrow q$
Converse:	*'If the streets are wet, then it is raining'*	$q \Rightarrow p$
Inverse:	*'If it is not raining, then the streets are not wet'*	$\neg p \Rightarrow \neg q$
Contrapositive:	*'If the streets are not wet, then it is not raining'*	$\neg q \Rightarrow \neg p$

Example 20

Consider the argument:
  'if the final digit of an integer is zero, then the integer is divisible by five'.
Find the contrapositive, converse and inverse statements.

Answer

Contrapositive:

  'if an integer is not divisible by five, then its final digit is not zero'

This is equivalent to the original statement. Although the phrasing is different, it is the same argument.

> This argument is valid, however it is phrased.

Converse: 'if an integer is divisible by five, then the final digit of the integer is zero'

Inverse: 'if the final digit of an integer is not zero, then the integer is not divisible by five'

Clearly, neither the converse, nor the inverse statement, is **necessarily** true.

> Find an integer that makes the converse a false statement. This is known as a **counterexample**.

It is often the case that, given a valid argument, its converse is **not** valid. This is known as the **fallacy of the converse**.

> Express the phrase, 'an integer is divisible by five only if its final digit is zero', in terms of p and q.

Exercise 9M

For each of the arguments **a–s**:

1 Determine whether it is a valid argument. If it is invalid, give a counterexample.

2 Write down the converse, inverse and contrapositive statements.

3 Determine which of the converse, inverse and contrapositive statements are valid arguments. For each invalid argument, give a counterexample.

a If ABCD is a square, then ABCD is a quadrilateral.

b If ABCD is a rectangle, then ABCD is a parallelogram.

c If an integer is divisible by four then it is divisible by two.

d If an integer is divisible by three then it is an odd integer.

e If an integer is divisible by two then it is an even integer.

f If an integer is divisible by both four and three then it is divisible by twelve.

g If an integer is divisible by both four and two then it is divisible by eight.

h If the sum of two integers is even, then the two integers are both even.

i If the product of two integers is even, then the two integers are both even.

j If the sum of two integers is odd, then one of the integers is odd and the other is even.

k If the product of two integers is odd, then the two integers are both odd.

l If triangle ABC is right-angled, then $a^2 + b^2 = c^2$.

m The square of an odd integer is odd.

n If triangle ABC has three equal angles, then triangle ABC has three equal sides.

o If quadrilateral ABCD has four equal sides, then ABCD has four equal angles.

p If $x^2 = 25$, then $x = 5$.

q If $x^3 = 27$, then $x = 3$.

r If $x^2 > 25$, then $x > 5$.

s If $x^3 < 27$, then $x < 3$.

Extension material on CD:
Worksheet 9 - De Morgan's Laws

Review exercise

Paper 1 style questions

EXAM-STYLE QUESTION

1 a Copy and complete the truth table to show that
$\neg(p \lor q) \Rightarrow \neg p \land \neg q$ is a valid argument.

p	q	$p \lor q$	$\neg(p \lor q)$	$\neg p$	$\neg q$	$\neg p \land \neg q$	$\neg(p \lor q) \Rightarrow \neg p \land \neg q$
T	T			F	F		
T	F			F	T		
F	T			T	F		
F	F			T	T		

b Using the results of **a**, rewrite the following statement without using the phrase 'It is not true . . .'

'It is not true that she dances well, or sings beautifully.'

412   Logic

2 The following propositions are given.

      p: The train leaves from gate 2.

      q: The train leaves from gate 8.

      r: The train does not leave today.

  a Write a sentence, in words, for the following logic statement:

$$p \Rightarrow (\neg r \wedge \neg q).$$

  b Write the following sentence as a logic statement using *p*, *q*, *r* and logic notation:

    'The train leaves today if and only if it leaves from gate 2 or from gate 8.'

3 a Copy and complete the truth table.

p	*q*	$p \Rightarrow q$	$\neg p$	$\neg q$	$\neg q \vee p$	$\neg p \vee q$
T	T					
T	F					
F	T					
F	F					

  b What identity is shown by the truth table?

4 a Copy and complete the following truth table for

    p: $x > 3$

    q: $x^2 > 9$

p	*q*	$\neg p$	$\neg p \vee q$
T	T		
T	F		
F	T		
F	F		

  b Using the results of part **a**, and explaining your reasoning, is $\neg p \vee q$ true, or false, when

    i $x > 3$ and $x^2 \ngtr 9$?

    ii $x \ngtr 3$ and $x^2 > 9$?

    [Note: the symbol $\ngtr$ denotes '**not** greater than'.]

5 *p* and *q* are two statements:

    p: Ice creams are vanilla flavored.

    q: Ice creams are full of raisins.

  a Draw a Venn diagram to represent the statements above, carefully labeling all sets including the universal set.

    Shade the region that represents $p \vee q$.

  b On the Venn diagram, show

    i a point *x*, representing a vanilla flavored ice cream full of raisins

    ii a point *y*, representing a vanilla flavored ice cream not full of raisins.

  c Write each of the following using logic symbols.

    i If ice creams are not full of raisins, they are not vanilla flavored.

    ii Ice creams are not vanilla flavored or they are full of raisins.

    iii If ice creams are not full of raisins, they are vanilla flavored.

    iv Ice creams are vanilla flavored and they are not full of raisins.

  d State which one of the propositions in part **c** above is logically equivalent to:

    'If ice creams are vanilla flavored, they are full of raisins.'
    Give a reason.

6 The following propositions are given.

    p: Picasso painted picture A.
    q: Van Gogh painted picture A.

  a Write a sentence in words to define the logic statements

    i $p \vee \neg q$ **ii** $\neg p \wedge q$.

  b Copy and complete the following truth table.

p	q	$\neg p$	$\neg q$	$p \vee \neg q$	$\neg p \wedge q$
T	T				
T	F				
F	T				
F	F				

  c Draw two Venn diagrams and shade the area represented by $p \vee \neg q$ on the first diagram and $\neg p \wedge q$ on the second diagram.

  d Deduce the truth values of the logic statement
    $(p \vee \neg q) \Leftrightarrow (\neg p \wedge q)$

    i using the truth table

    ii using the Venn diagrams.
    Explain your answers clearly in words.

  e Write down the name given to a logic statement such as
    $(p \vee \neg q) \Leftrightarrow (\neg p \wedge q)$.

7 The following propositions are given.

    p: x is a multiple of 5.
    q: x is a multiple of 3.
    r: x is a factor of 90.

  a Write a sentence, in words, for the statement: $(q \vee r) \wedge \neg p$.

  b Write the following sentence as a logic statement using p, q, r and logic notation:

    If x is a factor of 90 then x is either a multiple of 5 or x is not a multiple of 3.

  c Use truth tables to determine the truth values of each of the following two statements:
    $(q \vee r) \wedge \neg p$ and $r \Rightarrow (p \vee \neg q)$.

d List the combinations of truth values of p, q and r that make
the statement $(q \vee r) \wedge \neg p$ true.
Write down a possible value of x for each of these
combinations of truth values.

e Construct a truth table to determine the conditions for
equivalence between the two statements $(q \vee r) \wedge \neg p$ and
$r \Rightarrow (p \vee \neg q)$
When the equivalence is true, describe in words the
conditions on the value of x.

CHAPTER 9 SUMMARY

Introduction to logic

- A (simple) **statement** has a truth value of **true** or **false** (but not both).

Compound statements and symbols

- A **compound statement** is made up of simple statements joined together by **connectives.**
- The five connectives have these names and symbols:

NOT	**Negation**	$\neg$
AND	**Conjunction**	$\wedge$
OR	**Inclusive disjunction**	$\vee$
OR	**Exclusive disjunction**	$\underline{\vee}$
IF ... THEN	**Implication**	$\Rightarrow$

Truth tables: negation

- The **negation** of a statement p is written $\neg p$ (read as 'not-p'). The relation between any statement p and its negation $\neg p$ is shown in a **truth table**.

p	$\neg p$
T	F
F	T

Truth tables: conjunction (and)

- The **conjunction** of any two statements p and q is written $p \wedge q$. This **compound statement** is defined by this truth table.

p	q	$p \wedge q$
T	T	**T**
T	F	**F**
F	T	**F**
F	F	**F**

Continued on next page

Truth tables: resolving an ambiguity – the 'or' connective

- The **disjunction** of any two statements p and q is written $p \vee q$. This is '**inclusive or**' and it is defined by this truth table.

p	q	p ∨ q
T	T	**T**
T	F	**T**
F	T	**T**
F	F	**F**

$p \vee q$ is true if either p or q or possibly both are true.

- **Exclusive disjunction** is written $p \underline{\vee} q$ and is defined by this truth table.

p	q	p ∨ q
T	T	**F**
T	F	**T**
F	T	**T**
F	F	**F**

Logical equivalence, tautologies and contradictions

- The statements $\neg p \wedge \neg q$ and $\neg(p \vee q)$ are said to be (**logically**) **equivalent**. Equivalence is shown by the symbol $\Leftrightarrow$ so we write

$$\neg p \wedge \neg q \Leftrightarrow \neg(p \vee q)$$

- $\neg p \vee \neg q$ is **not** equivalent to $\neg(p \vee q)$.
- A **tautology** is a compound statement which is **true whatever** the truth values of the simple statements it is made up from.
- A (logical) **contradiction** is a compound statement which is **false whatever** the truth values of its simple statements.

Compound statements made up from three simple statements

- $(p \vee q) \vee r \Leftrightarrow p \vee (q \vee r)$

Arguments

- A compound statement that includes **implication** is called an **argument**.
- The truth table for implication is:

p	q	p ⇒ q
T	T	**T**
T	F	**F**
F	T	**T**
F	F	**T**

Continued on next page

- If the compound statement that represents an argument is a **tautology**, then the argument is **valid.**
- There are four important types of argument:
  - A **contradiction** is always false.   (last column of truth table all Fs)
  - A **tautology** is always true.   (last column of truth table all Ts)
  - A **valid** argument is always true.   (last column of truth table all Ts)
  - An **invalid** argument is not always true. (last column of truth table has

       at least one F)

These definitions mean that an invalid argument may (or may not) be a contradiction. A contradiction, however, is always an invalid argument.

- The truth table for **equivalence** ($p \Leftrightarrow q$) is:

p	q	$p \Leftrightarrow q$
T	T	T
T	F	F
F	T	F
F	F	T

- There are three commonly used arguments that are formed from the direct statement $p \Rightarrow q$:

$q \Rightarrow p$ the **converse** of the direct statement

$\neg p \Rightarrow \neg q$ the **inverse** of the direct statement

$\neg q \Rightarrow \neg p$ the **contrapositive** of the direct statement.

- This table summarizes the truth values for the direct argument, $p \Rightarrow q$, and the related conditionals.

p	q	Statement $p \Rightarrow q$	Converse $q \Rightarrow p$	Inverse $\neg p \Rightarrow \neg q$	Contrapositive $\neg q \Rightarrow \neg p$
T	T	T	T	T	T
T	F	F	T	T	F
F	T	T	F	F	T
F	F	T	T	T	T

Logical conclusions

Discuss these three statements in the rectangle.

> **Statement 1:**
> $2 + 2 = 4$
>
> **Statement 2:**
> $2 \times 2 = 4$
>
> **Statement 3:**
> There are **exactly two** true statements inside this rectangle.

Electronic goods these days, such as video cameras, are said to have 'fuzzy logic'.

What is 'fuzzy logic'?

A close shave?

In a town there is (exactly) one male barber, and every man in the town is clean shaven. Some shave themselves every time. The others are shaved by the barber every time.

The barber obeys this rule: he shaves all and only those men in town who do not shave themselves.

■ Does the barber shave himself?

Crocodile tears

A (talking) crocodile that always lies has stolen a man's child. He says to the man

> I will return her to you if you guess correctly whether I will do so or not...

■ What should the father reply to ensure the return of his child?

'This statement is false.' Is it?

Testing logic

A teacher tells her Mathematical Studies class that she will give them a test on one school day this week (Monday to Friday) but that she will not tell them on which day it will be set: it will be a surprise.

The students consider the proposition and reason like this: if the surprise test were on Friday, then by the end of Thursday we would know that the test must be the next day; it would not be a surprise. Hence, the test cannot be on Friday.

The same argument holds for the remaining four days in the week (Monday through Thursday): if the surprise test does not take place by the end of Wednesday, it must be scheduled for Thursday, and so not a surprise. Hence, the test cannot be on Thursday.

Similarly, it cannot be on Wednesday or Tuesday so it must be on Monday, but this is no surprise, so no surprise test is possible.

The teacher then says, 'Close your books, here is your surprise test.'

The two guards

This is an ancient logic problem – dating back at least 2000 years.

On a walk, you come to a fork in the path. One path leads to Paradise, the other to Death. Both paths look the same, and each path has a guard. If you start down one path, you cannot turn back, so you need to choose the correct one first time.

One of the guards always tells the truth, and one always lies – but you don't know which guard is which.

You are allowed to ask one of the guards one question.

■ What should you ask, to make sure you can identify the path to Paradise?

"I know what you're thinking about," said Tweedledum; "but it isn't so, nohow." "Contrariwise," continued Tweedledee, "if it was so, it might be; and if it were so, it would be; but as it isn't, it ain't. That's logic."

From *Through the Looking Glass*, by Lewis Carroll

The two guards problem is a simpler version of Knights and Knaves problems, set on a fictional island where Knights always tell the truth and Knaves always lie.

■ Research some of these problems, and try to solve them.

10 Geometry and trigonometry 2

CHAPTER OBJECTIVES:

5.4 Geometry of three-dimensional solids; distance between two points; angle between two lines or between a line and a plane

5.5 Volumes and surface areas of three-dimensional solids

Before you start

You should know how to:

1 Use trigonometry in a right-angled triangle, e.g.

$$\sin 32° = \frac{3}{AC}$$

$$AC = \frac{3}{\sin 32°}$$

$$AC = 5.66 \text{ cm (3 sf)}$$

2 Find an angle, a side or the area of any triangle, e.g.

a using the cosine rule
$c^2 = a^2 + b^2 - 2ab \cos C$:

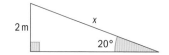

$$\cos \hat{C} = \frac{6.8^2 + 5^2 - 10^2}{2 \times 6.8 \times 5}$$

$$\cos \hat{C} = -0.4229...$$

$$\hat{C} = 115° \text{ (3 sf)}$$

b using the formula

$$A = \frac{1}{2} ab \sin C$$

$$A = \frac{1}{2} \times 6.8 \times 5 \times \sin 115°$$

$$= 15.4 \text{ cm}^2$$

Skills check

1 a Find x in this triangle.

b Find angle y.

2 In this triangle

a find the angle x

b find the area.

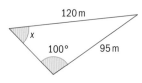

Goods are transported all around the world in containers like
this. These cuboid-shaped metal boxes come in uniform sizes,
so they can be moved from lorry to train to ship using standard
equipment.

A company using containers for transport needs to work out how
many of their products will fit into one container, and so how many
containers they will need. They might need to calculate the
maximum possible length of pipe that would fit into a container,
on the diagonal.

A company manufacturing containers needs to know how many
square metres of metal are needed to make each container.

In Chapter 5 you used geometry and trigonometry to solve
problems in two dimensions. In this chapter you will learn how
to calculate lengths and angles and solve problems in three
dimensions.

10.1 Geometry of three-dimensional solids

Geometry is the study of points, lines, planes, surfaces and solids.

No dimensions	One dimension	Two dimensions	Three dimensions
point	line	plane	solid

Can you draw a point with no dimensions?

A **plane** is a flat surface.

All the **faces** of a solid together make up the **surface** of the solid. A face of the solid may be **plane** or curved. A cuboid has 6 plane faces.

An **edge** is a line segment where two faces of a solid meet. This cuboid has 12 edges. The edges form the framework of the solid.

A **vertex** is a point where three or more edges meet. This cuboid has 8 vertices.

Euclid – the 'father of geometry'
Euclid (c325–c265 BCE) founded a school of mathematics in Alexandria, Egypt, and wrote thirteen volumes of *The Elements of Geometry*. These were the standard mathematics textbook for over 2000 years.

There are two groups of solids:

- Solids with all faces **plane**:
  - prisms
  - pyramids
- Solids with at least one **curved** face, e.g. cylinder, cone, sphere

Right prisms

→ In a **right prism** the end faces are the same shape and size and are parallel. All the other faces are rectangles that are **perpendicular** to the end faces.

Remember that two figures with the same shape and size are said to be **congruent**. In a prism the end faces are congruent.

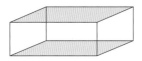

This is a right prism.

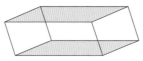

This is **not** a right prism. The end faces are not perpendicular to the other faces.

In Mathematical Studies you will only study right prisms.

> → If you cut parallel to the end faces of a right prism, the **cross-section** will always be the same shape and size.

Right prism   **Cross-section**

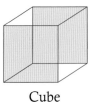

Cube

Square

> Can any cube be a cuboid?
> Can any cuboid be a cube?

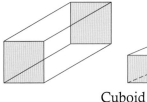

Cuboid

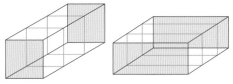

Square or rectangle

Triangular prism

Triangle

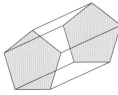

Pentagonal prism

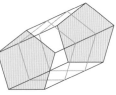

Pentagon

Pyramids

> → The base of a **pyramid** is a polygon. The other faces are triangles that meet at a point called the **apex**. In a **right pyramid** the apex is vertically above the center of the base.

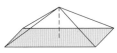

Rectangular-based pyramid
The base is a rectangle.

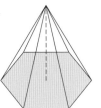

Hexagonal-based pyramid
The base is a hexagon.

> The pyramid at Giza in Egypt is the oldest of the seven wonders of the ancient world. It remained the tallest built structure for over 3800 years. What is the tallest structure now? How has mathematics been used in its design?

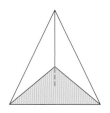

Triangular-based pyramid
The base is a triangle.

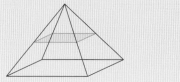

The **cross-sections** parallel to the base of a pyramid are the same shape as the base, but different sizes.

Solids with at least one curved face

In these solids the plane faces are shaded.

A **sphere** has one curved face.

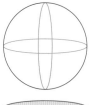

A **hemisphere** has two faces, one plane and one curved.

A **cylinder** has three faces, two plane and one curved.

Vertex or apex

A **cone** has two faces, one plane and one curved.

In Mathematical Studies you only study **right cones**. In a right cone the apex is vertically above the center of the base.

In a right cone:
- the **vertical height** h is the distance from the apex to the center of the base
- the **slant height** l is the distance from the apex to any point on the circumference of the base.

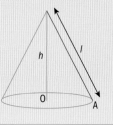

Example 1

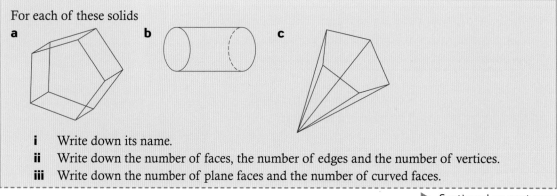

For each of these solids

a   b   c

i   Write down its name.
ii   Write down the number of faces, the number of edges and the number of vertices.
iii Write down the number of plane faces and the number of curved faces.

▶ Continued on next page

Answers

i		a Pentagonal prism	b Cylinder	c Pentagonal-based pyramid
ii	Faces	7	3	6
	Edges	15	2	10
	Vertices	10	0	6
iii	Plane faces	7	2	6
	Curved faces	0	1	0

Investigation – drawing a prism

Step 1 Draw one of the end faces.

Step 2 Draw the other end face. Remember that the end faces are congruent.

Step 3 Join up corresponding vertices with parallel lines.

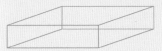

Now use this method to draw a triangular prism.

Exercise 10A

1 For each of these solids

a 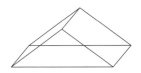   **b** 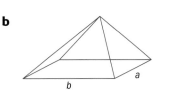   **c**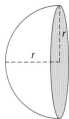

  i Write down its name.
  ii Write down the number of faces, the number of edges and
    the number of vertices.
  iii Write down the number of plane faces and the number of curved faces.

2 Draw prisms with these end faces.

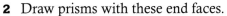
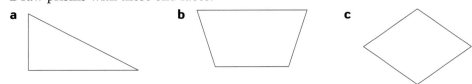
  a   **b**   **c**

10.2 Distance between points in a solid

You might need to calculate the distance between two vertices in a solid, or the distance between a vertex and the midpoint of an edge, or the distance between the midpoints of two lines. To do this you need first to identify right-angled triangles and then use Pythagoras' theorem.

Example 2

The diagram shows a cuboid ABCDEFGH, where AD = 7 cm, DC = 4 cm, and CG = 3 cm.

a Find the length of
   i AH **ii** AC **iii** DG **iv** AG.

b Find the distance between
   i the midpoint of CG and A
   ii the midpoint of AD and the midpoint of CG.

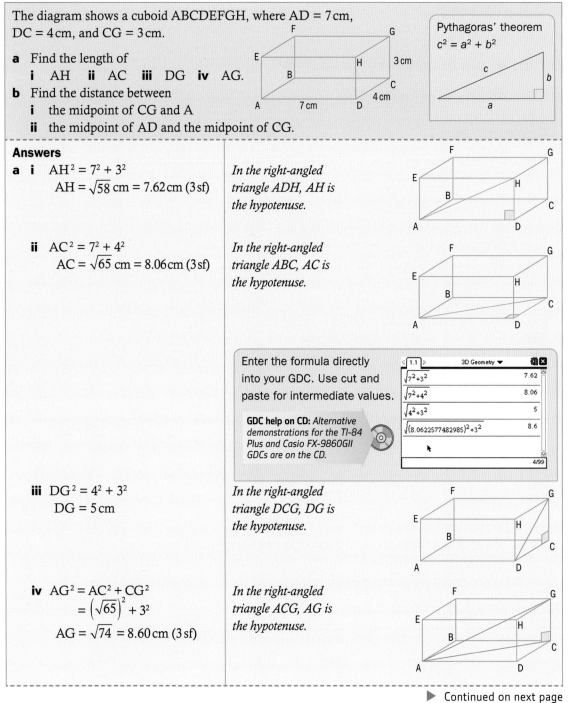

Pythagoras' theorem
$c^2 = a^2 + b^2$

Answers

a i $AH^2 = 7^2 + 3^2$
    $AH = \sqrt{58}$ cm = 7.62 cm (3 sf)

In the right-angled triangle ADH, AH is the hypotenuse.

ii $AC^2 = 7^2 + 4^2$
    $AC = \sqrt{65}$ cm = 8.06 cm (3 sf)

In the right-angled triangle ABC, AC is the hypotenuse.

Enter the formula directly into your GDC. Use cut and paste for intermediate values.

GDC help on CD: *Alternative demonstrations for the TI-84 Plus and Casio FX-9860GII GDCs are on the CD.*

3D Geometry	
$\sqrt{7^2+3^2}$	7.62
$\sqrt{7^2+4^2}$	8.06
$\sqrt{4^2+3^2}$	5
$\sqrt{(8.0622577482985)^2+3^2}$	8.6

iii $DG^2 = 4^2 + 3^2$
    $DG = 5$ cm

In the right-angled triangle DCG, DG is the hypotenuse.

iv $AG^2 = AC^2 + CG^2$
    $= \left(\sqrt{65}\right)^2 + 3^2$
    $AG = \sqrt{74} = 8.60$ cm (3 sf)

In the right-angled triangle ACG, AG is the hypotenuse.

▶ Continued on next page

b i $AM^2 = AC^2 + CM^2$

$\quad\quad = \left(\sqrt{65}\right)^2 + 1.5^2$

$\quad AM = 8.20\,\text{cm (3 sf)}$

Let M be the midpoint of CG. In the right-angled triangle ACM, AM is the hypotenuse.

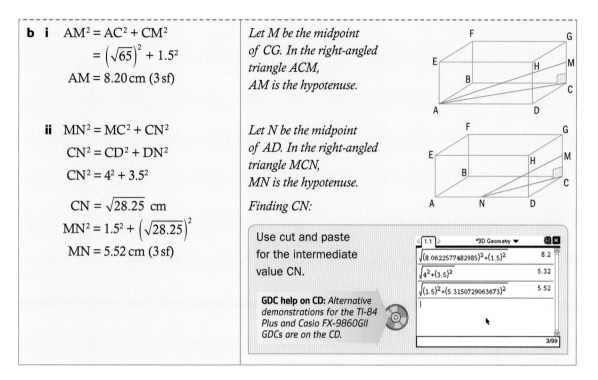

ii $MN^2 = MC^2 + CN^2$

$\quad CN^2 = CD^2 + DN^2$

$\quad CN^2 = 4^2 + 3.5^2$

$\quad CN = \sqrt{28.25}\,\text{cm}$

$MN^2 = 1.5^2 + \left(\sqrt{28.25}\right)^2$

$MN = 5.52\,\text{cm (3 sf)}$

Let N be the midpoint of AD. In the right-angled triangle MCN, MN is the hypotenuse.

Finding CN:

Use cut and paste for the intermediate value CN.

GDC help on CD: *Alternative demonstrations for the TI-84 Plus and Casio FX-9860GII GDCs are on the CD.*

1.1	*3D Geometry ▼
$\sqrt{(8.0622577482985)^2+(1.5)^2}$	8.2
$\sqrt{4^2+(3.5)^2}$	5.32
$\sqrt{(1.5)^2+(5.3150729063673)^2}$	5.52
I	
	3/99

Example 3

In the diagram, ABCD is the rectangular base of a right pyramid with apex E. The sides of the base are 8 cm and 5 cm, and the height OE of the pyramid is 7 cm.
Find the length of

a AC

b EC

c EM, where M is the midpoint of CD.

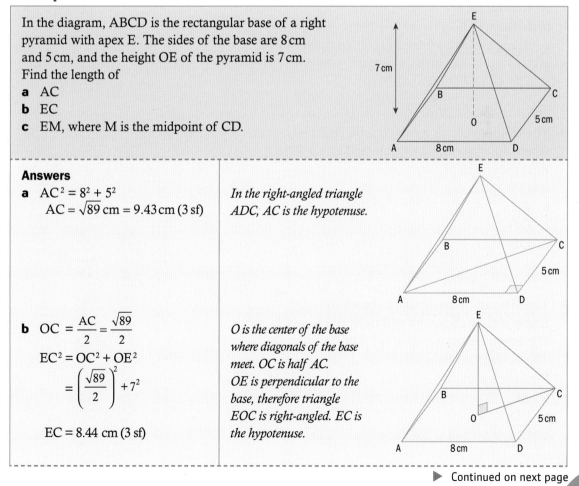

Answers

a $AC^2 = 8^2 + 5^2$

$\quad AC = \sqrt{89}\,\text{cm} = 9.43\,\text{cm (3 sf)}$

In the right-angled triangle ADC, AC is the hypotenuse.

b $OC = \dfrac{AC}{2} = \dfrac{\sqrt{89}}{2}$

$\quad EC^2 = OC^2 + OE^2$

$\quad\quad = \left(\dfrac{\sqrt{89}}{2}\right)^2 + 7^2$

$\quad EC = 8.44\,\text{cm (3 sf)}$

O is the center of the base where diagonals of the base meet. OC is half AC. OE is perpendicular to the base, therefore triangle EOC is right-angled. EC is the hypotenuse.

▶ Continued on next page

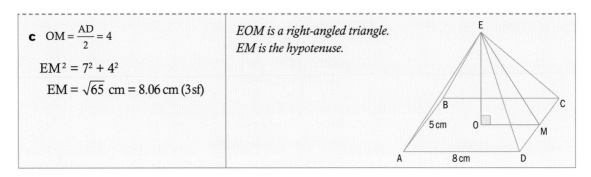

c $OM = \dfrac{AD}{2} = 4$

$EM^2 = 7^2 + 4^2$

$EM = \sqrt{65}$ cm $= 8.06$ cm (3sf)

EOM is a right-angled triangle.
EM is the hypotenuse.

Exercise 10B

1 Copy the cuboid shown in the diagram.
In different sketches mark clearly these right-angled triangles:
  a triangle ACD
  b triangle AGH
  c triangle HBA
  d triangle MCD, where M is the midpoint of EH.

2 Copy the right pyramid shown in the diagram.
In different sketches mark clearly:
  a triangle BCD
  b triangle EOC
  c triangle EOM, where M is the midpoint of CD.

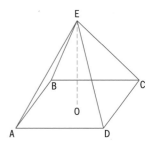

3 The diagram shows a cuboid ABCDEFGH,
where AD = 4 cm, CD = 6 cm and CG = 9 cm.
Find the length of
  a DB
  b ED
  c DG
  d DF.

4 The diagram shows a square-based pyramid.
E is vertically above the middle of the base, O.
The height of the pyramid is 1.5 m.
The sides of the base are 0.6 m.
Find the length of
  a AC
  b ED
  c EM, where M is the midpoint of CD.

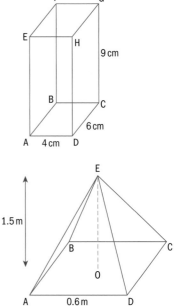

5 The diagram shows a cone with base center O and radius 4 cm. The slant height of the cone is 9 cm. Find OV, the height of the cone.

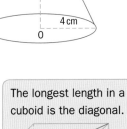

: EXAM-STYLE QUESTIONS

6 The diagram represents a cupboard in a gym. It has the dimensions shown.
  a Calculate the length of AC.
  b Find the length of the longest fitness bar that can fit in the cupboard.

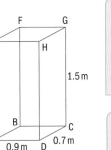

> The longest length in a cuboid is the diagonal.

7 The Great pyramid of Giza has a square base. At the present time the length of one side of the base is 230.4 m and the height is 138.8 m.
  a Calculate the length of the base diagonal.
  b Calculate the distance from the apex to the midpoint of a side of the base.
  c Calculate the length of one sloping edge of the pyramid.

> Sketch the pyramid and label the lengths you know.

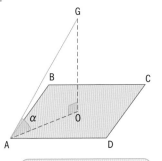

10.3 Angles between two lines, or between a line and a plane

To calculate angles start by identifying right-angled triangles. Then use trigonometry.

In the diagram, ABCD is a plane and AG is part of a line. To find the angle α that AG makes with the plane ABCD:

1 Drop a perpendicular from G to the plane.
2 Label the point where the perpendicular meets the plane.
3 Draw the right-angled triangle AOG. Angle α is opposite OG.
4 Use trigonometry to find α.

> The angle between the plane ABCD and the line AG is also the angle between the lines OA and AG.

Example 4

Copy the cuboid shown in the diagram.
Mark the angles described. Use a different
diagram for each angle.

a The angle that the plane ADHE makes
with the line AG

b The angle that the plane ADHE makes
with the line AC

c The angle that the plane ABCD makes with the line CE

d The angle between the lines BH and HA

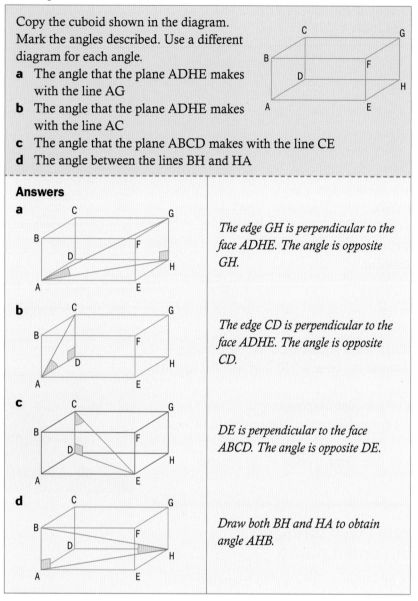

Answers

a

The edge GH is perpendicular to the face ADHE. The angle is opposite GH.

b

The edge CD is perpendicular to the face ADHE. The angle is opposite CD.

c

DE is perpendicular to the face ABCD. The angle is opposite DE.

d

Draw both BH and HA to obtain angle AHB.

Example 5

Copy the diagram of a rectangular-based pyramid.
E is vertically above the middle of the base, O. Mark the
angles described. Use a different diagram for each angle.

a The angle that the base ABCD makes with the edge DE

b The angle that the base ABCD makes with ME, where M is
the midpoint of CD

c The angle between the lines BE and ED

d The angle between the lines DE and EC

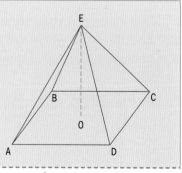

▶ Continued on next page

Answers

a

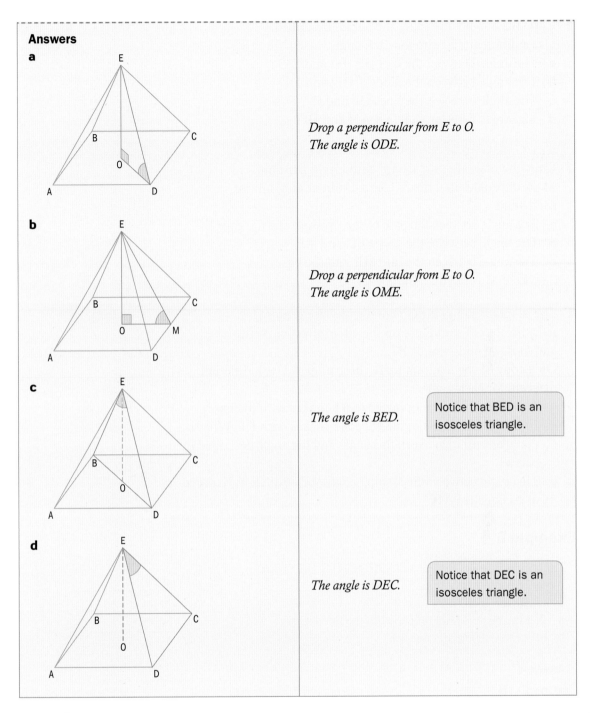

Drop a perpendicular from E to O.
The angle is ODE.

b

Drop a perpendicular from E to O.
The angle is OME.

c

The angle is BED.

> Notice that BED is an isosceles triangle.

d

The angle is DEC.

> Notice that DEC is an isosceles triangle.

Exercise 10C

1 Copy the cuboid shown and mark the angles described.
Use a different diagram for each angle.

   a The angle that the face ADHE makes with the line EG
   b The angle that the face ADHE makes with the line EC
   c The angle that the face EFGH makes with the line CE
   d The angle between the lines CE and CF
   e The angle between the lines CE and EA

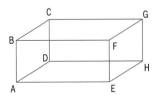

2 Copy the cuboid and mark the angles described.
Use a different diagram for each angle.
  a The angle between the face AEHD and DG
  b The angle between the face AEHD and DF
  c The angle between the lines CF and CA
  d The angle between the lines AH and HG

3 Copy the diagram of a square-based right pyramid.
Mark the angles described. Use a different diagram for
each angle.
  a The angle between the base of the pyramid and the edge EC
  b The angle between the edges EC and AE
  c The angle between the line ME and the base,
      where M is the midpoint of CD

4 The diagram shows a right cone with base center O.
A is the apex. T and P are on the circumference of the base
and O is the midpoint of PT.
On a copy of the diagram mark these angles. Use a different
diagram for each angle.
  a The angle that the sloping edge AT makes with the base
  b The angle that the sloping edge AT makes with PT. What is
      the relationship between this angle and the angle described in
      part **a**?
  c The angle between the sloping edges AT and AP. What type
      of triangle is PAT?

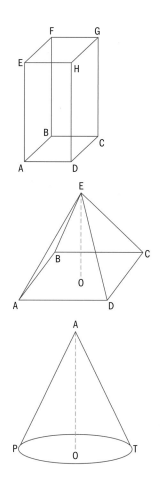

⊞ Example 6

The diagram shows the cuboid ABCDEFHG. AE is 9 cm,
AB is 2 cm and EH is 5 cm.
Calculate the angle
  a that the plane ADHE makes with the line AG
  b between the lines BH and HE.

Answers

a $\tan G\hat{A}H = \dfrac{GH}{HA}$

  $HA^2 = 9^2 + 5^2$
  $HA = \sqrt{106}$ cm
  $\tan G\hat{A}H = \dfrac{2}{\sqrt{106}}$
  $G\hat{A}H = 11.0°\,(3\,sf)$

AGH is a right-angled triangle with
$G\hat{H}A = 90°$ and GH = 2 cm.
Calculate GÂH.

The tangent links the sides GH and HA.
Find HA using Pythagoras. Keep
the exact value $\sqrt{106}$ for the next
calculation to get the final answer as
accurate as possible.
Substitute for HA in the tangent.
Round to 3 sf in the last step.

$\tan = \dfrac{\text{opposite}}{\text{adjacent}}$

▶ Continued on next page

b $\tan B\hat{H}E = \dfrac{BE}{EH}$

As BCHE is a rectangle, triangle BEH is right-angled.
So BEH is a right-angled triangle with $B\hat{E}H = 90°$. Calculate angle BHE.
The tangent links the sides BE and EH.

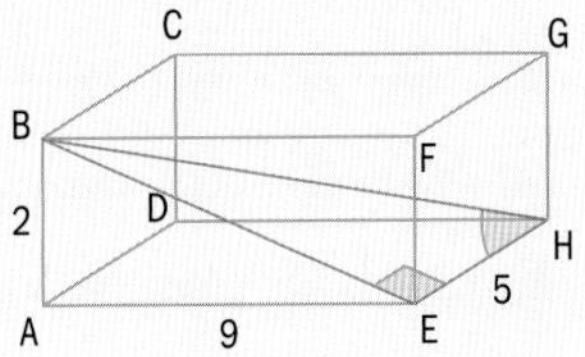

$BE^2 = 2^2 + 9^2$

$BE = \sqrt{85}$ cm

Find BE using Pythagoras.

$\tan B\hat{H}E = \dfrac{\sqrt{85}}{5}$

Substitute for BE in the tangent.

$B\hat{H}E = 61.5°$ (3 sf)

Example 7

The diagram shows the right pyramid ABCDE.
The base is a rectangle with AD = 6 cm and CD = 4 cm.
The height of the pyramid is 7 cm.
a i Calculate angle AEO.
  ii Calculate angle AEC.
b Calculate angle EMO, where M is the midpoint of CD.
c i Calculate the length of ED.
  ii Hence calculate angle DEC.

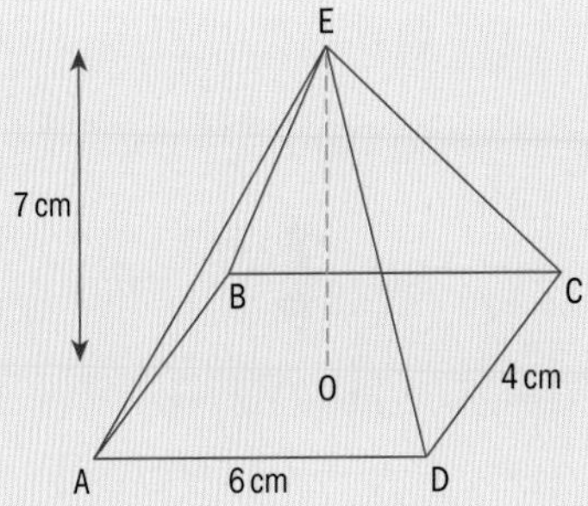

Answers

a i $\tan A\hat{E}O = \dfrac{AO}{EO}$

AOE is a right-angled triangle with $\hat{O} = 90°$. We are looking for angle AEO.
The tangent links AO (half of AC) and EO, the height.
Find AC using Pythogoras and then halve it.

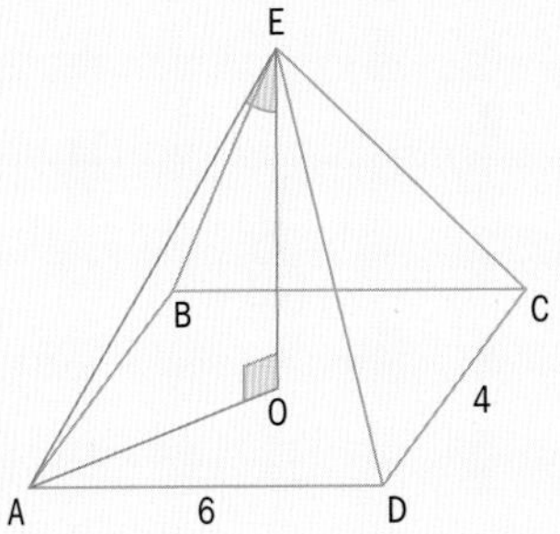

$AC^2 = 6^2 + 4^2$

$AC = \sqrt{52}$ cm

$AO = \dfrac{\sqrt{52}}{2}$ cm

$\tan A\hat{E}O = \dfrac{\dfrac{\sqrt{52}}{2}}{7}$

Substitute for AO and EO in the tangent.

$A\hat{E}O = 27.3°$ (3 sf)

ii $A\hat{E}C = 2 \times A\hat{E}O$

      $= 2 \times 27.252...$

$A\hat{E}C = 54.5°$ (3 sf)

Triangle AEC is isosceles, so EO is a line of symmetry.
EO bisects angle AEC.
So $A\hat{E}C$ is twice $A\hat{E}O$.

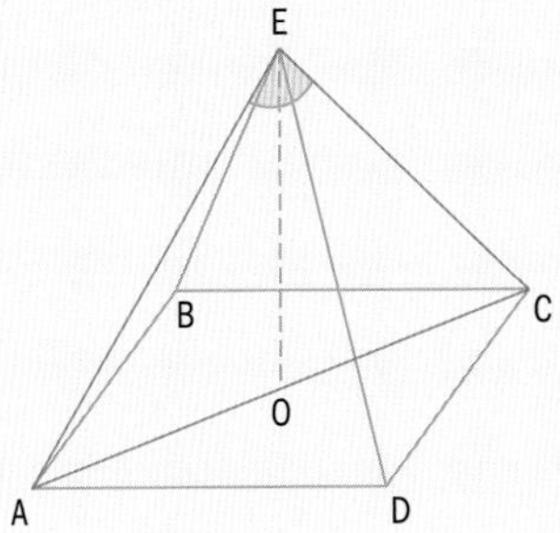

▶ Continued on next page

b $\tan \hat{EMO} = \dfrac{EO}{OM}$

$\tan \hat{EMO} = \dfrac{7}{3}$

$\hat{EMO} = 66.8°$ (3 sf)

EMO is a right-angled triangle with $\hat{O} = 90°$.
The tangent links EO and OM.
OM is half of $AD = \dfrac{6}{2} = 3$.

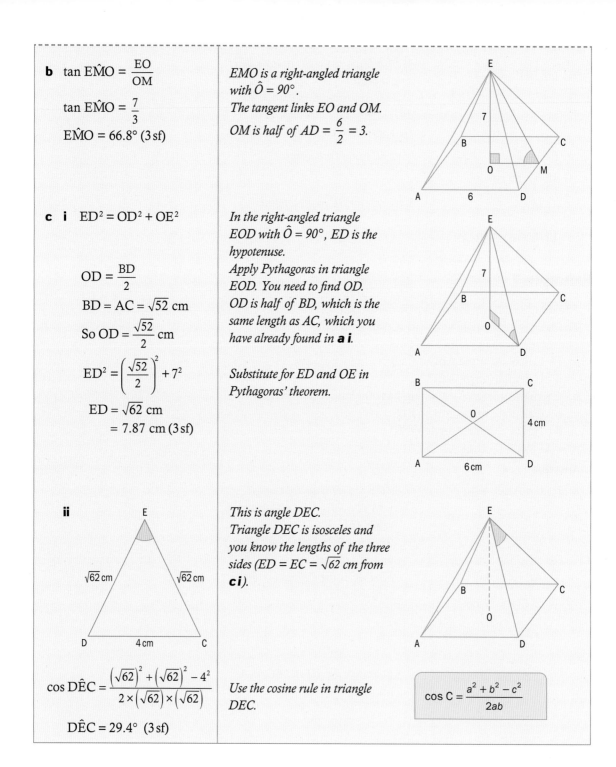

c i $ED^2 = OD^2 + OE^2$

$OD = \dfrac{BD}{2}$

$BD = AC = \sqrt{52}$ cm

So $OD = \dfrac{\sqrt{52}}{2}$ cm

$ED^2 = \left(\dfrac{\sqrt{52}}{2}\right)^2 + 7^2$

$ED = \sqrt{62}$ cm

$\quad = 7.87$ cm (3 sf)

In the right-angled triangle EOD with $\hat{O} = 90°$, ED is the hypotenuse.
Apply Pythagoras in triangle EOD. You need to find OD.
*OD is half of BD, which is the same length as AC, which you have already found in **a i**.*

Substitute for ED and OE in Pythagoras' theorem.

ii

This is angle DEC.
*Triangle DEC is isosceles and you know the lengths of the three sides (ED = EC = $\sqrt{62}$ cm from **c i**).*

$\cos \hat{DEC} = \dfrac{\left(\sqrt{62}\right)^2 + \left(\sqrt{62}\right)^2 - 4^2}{2 \times \left(\sqrt{62}\right) \times \left(\sqrt{62}\right)}$

$\hat{DEC} = 29.4°$ (3 sf)

Use the cosine rule in triangle DEC.

$$\cos C = \dfrac{a^2 + b^2 - c^2}{2ab}$$

Exercise 10D

EXAM-STYLE QUESTIONS

1 In the cuboid ABCDEFGH, AD = 10 cm, CD = 4 cm and AE = 3 cm.
  a i Calculate the length of AC.
    ii Calculate the angle that AG makes with the face ABCD.
  b i Calculate the length of AF.
    ii Find the angle that the face AEFB makes with the line AG.

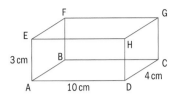

2 The diagram shows cube ABCDEFGH with side length 2 m.
  a Calculate the length of BD.
  b Find the angle that DF makes with the face ABCD.

  Let M be the midpoint of BF.
  c Find the angle that MD makes with the face ABCD.

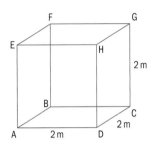

3 The diagram shows a cuboid ABCDEFGH, where AD = 4 cm, CD = 6 cm and CG = 9 cm.
  a i Calculate the length of BD.
    ii Find the angle that AF makes with the face BFGC.
  b Find the angle that AF makes with the face ABCD.
  c i Calculate the length of AC.
    ii Calculate the length of FC.
    iii Find the angle between the lines AF and FC.

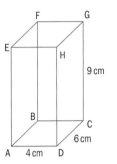

4 The diagram shows the rectangular-based right pyramid ABCDE with AD = 4 cm, CD = 3 cm and EO = 7 cm.
  a Find the length of AC.
  b Find the length of AE.
  c Find angle AEC.
  d Find the angle that AE makes with the base of the pyramid.
  e Find the angle that the base of the pyramid makes with EM, where M is the midpoint of CD.

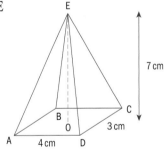

5 The diagram shows a cone with base center O and radius 3 cm. A is 5 cm vertically above O. T and P are on the circumference of the base and O is the midpoint of PT.
  a Find AT, the slant height of the cone.
  b Find the angle that AT makes with the base of the cone.
  c Find angle PAT.

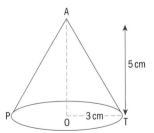

6 A beach tent has the shape of a right cone. The center of the base is O and the base area is $5\,\text{m}^2$. The tent is $2\,\text{m}$ high. It is attached to the sand at points P and T, and O is the midpoint of PT.

 a Find the radius of the base.

 b Find angle PAT.

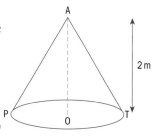

10.4 Surface areas of three-dimensional solids

→ The **surface area** of a solid is the sum of the areas of all its faces. Surface area is measured in square units, e.g. cm^2, m^2.

To calculate surface areas, first sketch the solid.

There are two types of solid:

- Solids with all their faces plane
  e.g. prisms (except cylinders), pyramids (except cones), or combinations of these
- Solids with at least one curved face
  e.g. cylinders, spheres, hemispheres, cones, or combinations of these

Surface areas of solids with all faces plane

Example 8

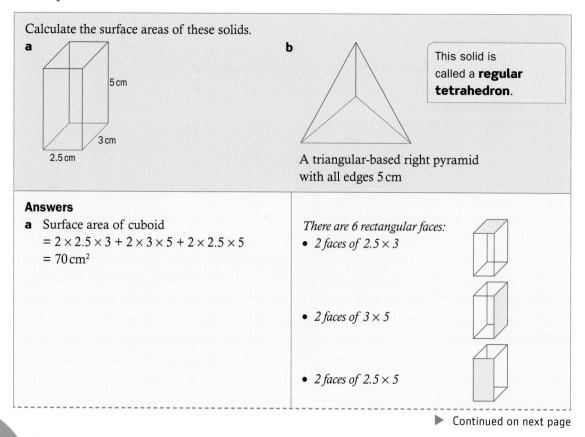

Calculate the surface areas of these solids.

a

5 cm

3 cm

2.5 cm

b

This solid is called a **regular tetrahedron**.

A triangular-based right pyramid with all edges 5 cm

Answers

a Surface area of cuboid
$$= 2 \times 2.5 \times 3 + 2 \times 3 \times 5 + 2 \times 2.5 \times 5$$
$$= 70\,\text{cm}^2$$

There are 6 rectangular faces:
- *2 faces of 2.5×3*

- *2 faces of 3×5*

- *2 faces of 2.5×5*

▶ Continued on next page

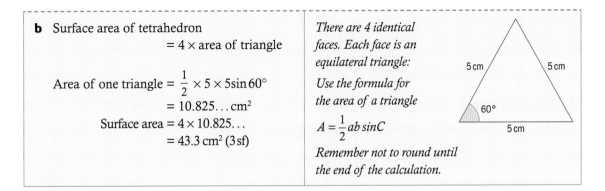

b Surface area of tetrahedron
$$= 4 \times \text{area of triangle}$$

$$\text{Area of one triangle} = \frac{1}{2} \times 5 \times 5\sin 60°$$
$$= 10.825\ldots \text{cm}^2$$
$$\text{Surface area} = 4 \times 10.825\ldots$$
$$= 43.3 \text{cm}^2 \text{ (3 sf)}$$

There are 4 identical faces. Each face is an equilateral triangle:

Use the formula for the area of a triangle

$$A = \frac{1}{2} ab \sin C$$

Remember not to round until the end of the calculation.

Exercise 10E

1 Calculate the surface areas of these solids.

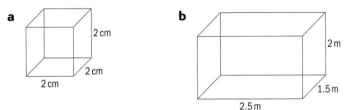

a 2 cm, 2 cm, 2 cm

b 2 m, 1.5 m, 2.5 m

c ABCDEF is a prism. CF is 5 cm and triangle ABC is equilateral with sides 4 cm.

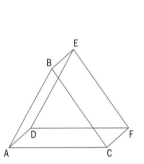

EXAM-STYLE QUESTIONS

2 ABCDEF is a right prism. BE is 4 cm, and triangle ABC is isosceles with AC = CB = 3 cm and angle BCA = 120°.
   a Find the area of triangle ABC.
   b Find the length of the edge AB.
   c Find the surface area of the prism.

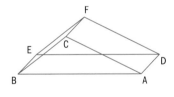

3 ABCDE is a square-based right pyramid and O is the middle of the base. The side length of the base is 5 cm. The height of the pyramid is 6 cm.
   a Calculate the length of EM, where M is the midpoint of BC.
   b Calculate the area of triangle CDE.
   c Calculate the surface area of the pyramid.

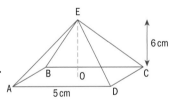

4 The surface area of a cube is 600 m². Calculate its side length. Give your answer in cm.

EXAM-STYLE QUESTION

5 Each edge of a cube is 5.4 m.
   a Calculate the surface area of the cube.
   b Give your answer to part **a** in the form $a \times 10^k$ where $1 \le a < 10$, $k \in \mathbb{Z}$.

6 The diagram represents Jamal's room, which is in the shape of a cuboid. He is planning to paint all the surfaces except the floor, the door and the window.

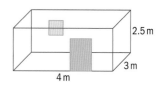

The door is 2 metres high and 1.3 metres wide, and the window is a square with a side length of 1 metre.

a Calculate the surface area that Jamal intends to paint.

Jamal needs 1.2 litres of paint to cover $1\,m^2$.

b Calculate the number of litres of paint that Jamal needs. Round **up** your answer to the next whole litre.

One litre of paint costs US$4.60.

c Calculate how much Jamal will spend on paint. Give your answer correct to 2 decimal places.

> Round up your answer to the next whole number as you buy paint in litres.

Surface areas of solids with at least one curved face

● Cylinder

A cylinder has three faces: one curved and two plane. If you cut the curved face and open it out, you get a rectangle. The length of the rectangle is the circumference of the base of the cylinder.

If h is the height and r is the radius of the base

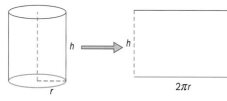

> $C = 2\pi r$
>
>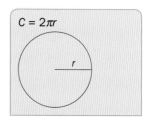

> This formula is in the Formula booklet.

→ Area of curved surface of a cylinder = $2\pi rh$

Area of a circle = πr^2

The cylinder has two equal circular faces

Area of two circles = $2\pi r^2$

Therefore

> Was π invented or discovered? When was it first used? Why is it denoted with a Greek letter?

→ Total surface area of a cylinder = $2\pi rh + 2\pi r^2$

- **Sphere**

A sphere has one curved face.

Let r be the radius of the sphere, then

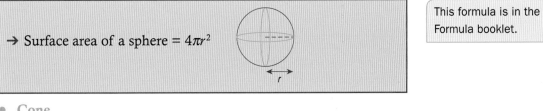

→ Surface area of a sphere = $4\pi r^2$

This formula is in the Formula booklet.

- **Cone**

A cone has two faces: one plane and one curved.

Let r be the radius and l the slant height of the cone, then

→ Area of curved surface of a cone = $\pi r l$

This formula is in the Formula booklet.

The base of a cone is a circle, therefore

→ Total surface area of a cone = $\pi r l + \pi r^2$

Example 9

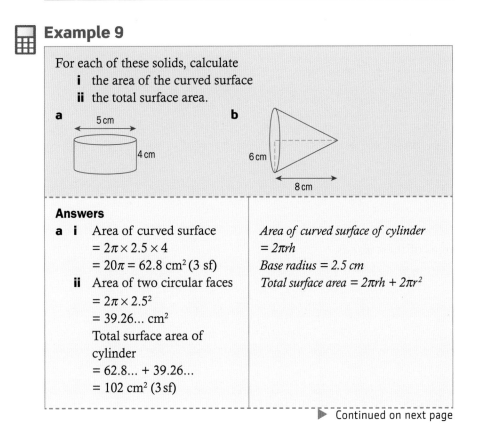

For each of these solids, calculate
  i the area of the curved surface
  ii the total surface area.

a 5 cm   4 cm

b 6 cm   8 cm

Answers

a i Area of curved surface
    $= 2\pi \times 2.5 \times 4$
    $= 20\pi = 62.8$ cm^2 (3 sf)

  ii Area of two circular faces
    $= 2\pi \times 2.5^2$
    $= 39.26...$ cm^2
    Total surface area of
    cylinder
    $= 62.8... + 39.26...$
    $= 102$ cm^2 (3 sf)

Area of curved surface of cylinder
= $2\pi r h$
Base radius = 2.5 cm
Total surface area = $2\pi r h + 2\pi r^2$

▶ Continued on next page

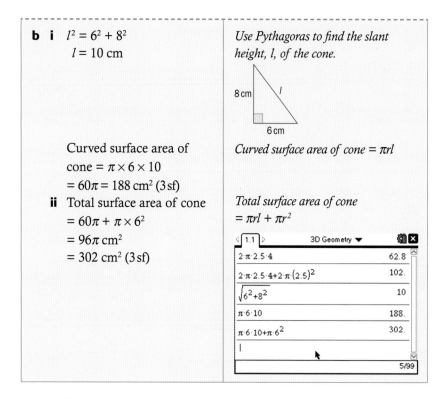

b i $l^2 = 6^2 + 8^2$
    $l = 10$ cm

Use Pythagoras to find the slant height, l, of the cone.

8 cm l
    6 cm

Curved surface area of
cone $= \pi \times 6 \times 10$
$= 60\pi = 188$ cm² (3 sf)

Curved surface area of cone $= \pi r l$

ii Total surface area of cone
    $= 60\pi + \pi \times 6^2$
    $= 96\pi$ cm²
    $= 302$ cm² (3 sf)

*Total surface area of cone
$= \pi r l + \pi r^2$*

1.1 ▷	3D Geometry ▼	
$2 \cdot \pi \cdot 2.5 \cdot 4$		62.8
$2 \cdot \pi \cdot 2.5 \cdot 4 + 2 \cdot \pi \cdot (2.5)^2$		102.
$\sqrt{6^2 + 8^2}$		10
$\pi \cdot 6 \cdot 10$		188.
$\pi \cdot 6 \cdot 10 + \pi \cdot 6^2$		302.
		5/99

GDC help on CD: *Alternative demonstrations for the TI-84 Plus and Casio FX-9860GII GDCs are on the CD.*

Exercise 10F

EXAM-STYLE QUESTIONS

1 Calculate the surface area of each solid.

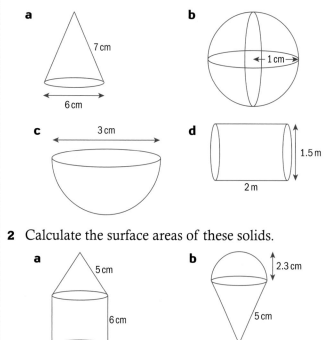

a   7 cm   6 cm

b   1 cm

c   3 cm

d   1.5 m   2 m

2 Calculate the surface areas of these solids.

> Split each solid into two solids.

a   5 cm   6 cm   2 cm

b   2.3 cm   5 cm

3 The surface area of a sphere is 1000 cm². Find its radius.

4 The first diagram shows a cylindrical pencil container made of leather. The base diameter is 8 cm and the height is 13 cm.

a Calculate the area of leather needed to make this pencil container.

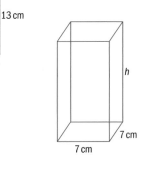

Another container is made in the shape of a cuboid as shown. The square base has sides of 7 cm. This container uses the same area of leather as the cylindrical one.

b Find the height, h, of the cuboid. Give your answer correct to 2 significant figures.

10.5 Volumes of three-dimensional solids

→ The volume of a solid is the amount of space it occupies and is measured in cubic units, e.g. cm^3, m^3, etc.

Remember that one cubic centimetre is the space occupied by a cube with edge length of 1 cm.

Volume of a prism

To calculate the volume of a prism you need to know

- the area of the **cross-section** of the prism (the end face)
- the **height** (distance between the two end faces).

In each of the prisms shown one end face is shaded and the height is labeled.

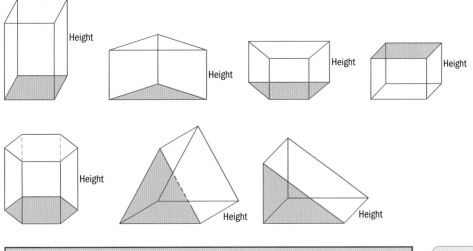

→ Volume of a prism is
V = area of cross-section × height

This formula is in the Formula booklet.

A cuboid is a prism with cross-section a rectangle.

Volume of a cuboid = area of cross-section × height

Area of cross-section = length × width

so Volume of a cuboid = length × width × height

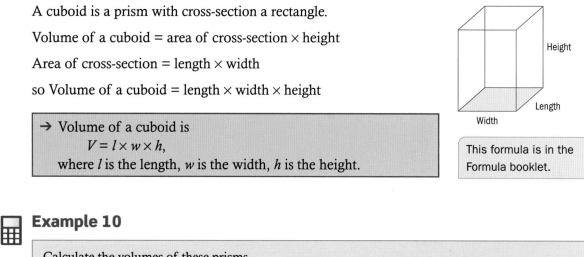

→ Volume of a cuboid is
$$V = l \times w \times h,$$
where l is the length, w is the width, h is the height.

This formula is in the Formula booklet.

Example 10

Calculate the volumes of these prisms.

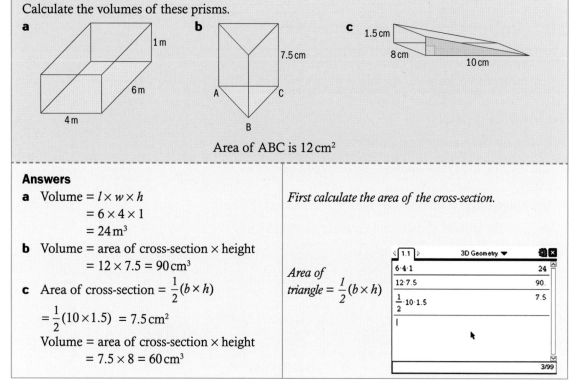

Area of ABC is 12 cm²

Answers

a Volume $= l \times w \times h$
$\qquad = 6 \times 4 \times 1$
$\qquad = 24\,\text{m}^3$

b Volume = area of cross-section × height
$\qquad = 12 \times 7.5 = 90\,\text{cm}^3$

c Area of cross-section $= \dfrac{1}{2}(b \times h)$

$\qquad = \dfrac{1}{2}(10 \times 1.5) = 7.5\,\text{cm}^2$

Volume = area of cross-section × height
$\qquad = 7.5 \times 8 = 60\,\text{cm}^3$

First calculate the area of the cross-section.

Area of triangle $= \dfrac{1}{2}(b \times h)$

1.1 ▷	3D Geometry ▼	
6·4·1		24
12·7.5		90.
$\frac{1}{2}$·10·1.5		7.5
I		

3/99

Exercise 10G

1 Calculate the volume of each prism.

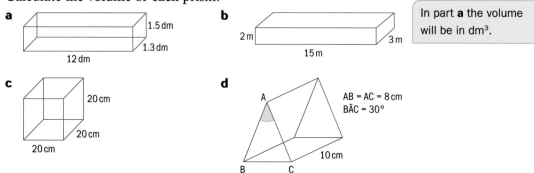

In part **a** the volume will be in dm³.

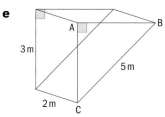

e

3 m, A, 5 m, 2 m, C, B

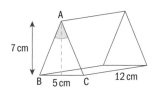

f

7 cm, A, B, 5 cm, C, 12 cm

2 The diagram shows a triangular prism.
Angle CAB = 90°.
  a Calculate the length of AB.
  b Calculate the area of triangle ABC.
  c Calculate the volume of the prism.

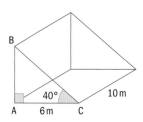

3 The diagram shows a prism with ABCDEF
a regular hexagon. Each side of the hexagon
is 5 cm and the height of the prism is 13.5 cm.
  a What size is angle COB?
  b Find the area of triangle COB.
  c Find the area of the regular hexagon
     ABCDEF.
  d Find the volume of the prism.

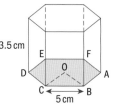

What type of triangle
is OCB?

4 Find expressions for the volume, V, of each of these prisms.
Give answers in their simplest form.
(All the dimensions are in cm.)

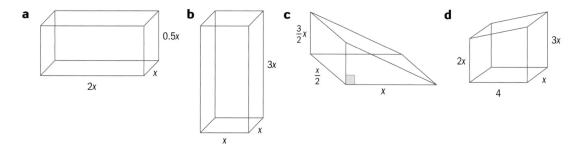

a 0.5x, x, 2x

b 3x, x, x

c $\frac{3}{2}x$, $\frac{x}{2}$, x

d 3x, 2x, x, 4

5 A box with a square base has a volume of 11 025 cm³
and a height of 25 cm. Each side of the base is x cm.
  a Write down an expression in terms of x for the volume of
     the box.
  b Hence write down an equation in x.
  c Find the value of x.

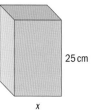

25 cm, x

6 An open box is cubical in shape. It has no lid.
The volume of the box is 9261 cm³.
  a Find the length of one edge of the box.
  b Find the total external surface area of the box.

Extension material on CD:
Worksheet 10 - Volume of a truncated cone

Volume of a cylinder

A cylinder is a prism with a circular cross-section.

Volume of cylinder = area of cross-section × height

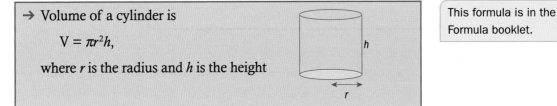

→ Volume of a cylinder is

$$V = \pi r^2 h,$$

where r is the radius and h is the height

This formula is in the Formula booklet.

Volume of a pyramid

→ Volume of a pyramid = $\frac{1}{3}$(area of base × vertical height)

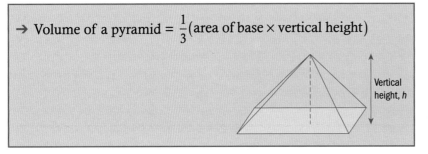

Vertical height, h

This formula is in the Formula booklet.

Volume of a cone

→ Volume of a cone = $\frac{1}{3}\pi r^2 h,$

where r is the radius and h is the vertical height

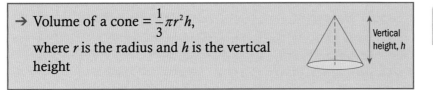

Vertical height, h

This formula is in the Formula booklet.

Volume of a sphere

→ Volume of a sphere = $\frac{4}{3}\pi r^3,$

where r is the radius

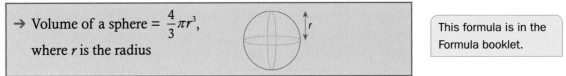

This formula is in the Formula booklet.

Investigation – relationships between volumes

Write an expression for the volume of each solid.
What do you notice?

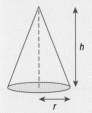
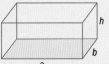

What can you say about the volume of a cuboid and the volume of a pyramid with the same base and height as that cuboid?

What is the relationship between the volumes of these two solids?

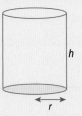

Take a cone and cylinder of the same height and radius. Fill the cone with rice.
Pour it into the cylinder. How many times do you have to do this to fill the cylinder?

Example 11

Calculate the volume of each solid.

a 2.6 m 3.7 m 5 m

b 6 cm 10 cm

c 30.5 cm 13.7 cm

Answers

a Volume of pyramid

$$= \frac{1}{3}(3.7 \times 5 \times 2.6)$$

$$= 16.0 \text{ m}^3 \text{ (3 sf)}$$

b Volume of cone

$$= \frac{1}{3}\pi \times 3^2 \times 10 = 30\pi$$

$$= 94.2 \text{ cm}^3 \text{ (3 sf)}$$

c Volume of cylinder

$$= \pi \times 13.7^2 \times 30.5$$

$$= 18\,000 \text{ cm}^3 \text{ (3 sf)}$$

Volume of pyramid

$$= \frac{1}{3}(\text{area of base} \times \text{vertical height})$$

Area of base = 3.7×5

Volume of cone $= \frac{1}{3}\pi r^2 h$

Volume of cylinder $= \pi r^2 h$

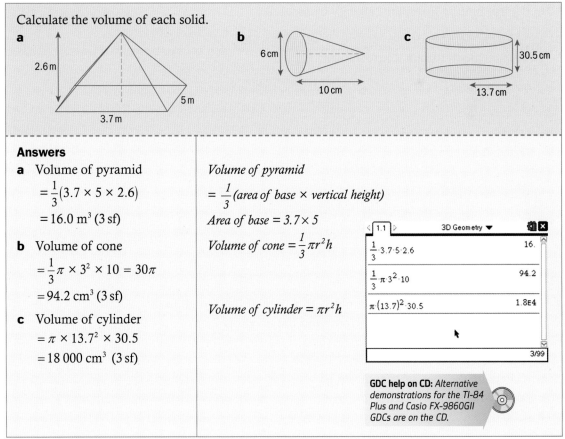

GDC help on CD: *Alternative demonstrations for the TI-84 Plus and Casio FX-9860GII GDCs are on the CD.*

Exercise 10H

1 Calculate the volume of each solid.

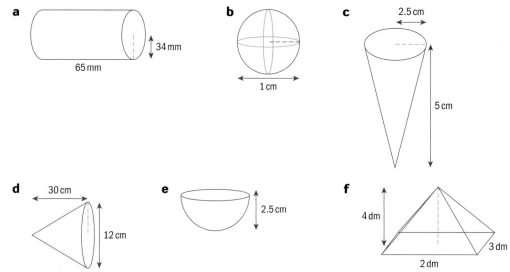

a 65 mm, 34 mm

b 1 cm

c 2.5 cm, 5 cm

d 30 cm, 12 cm

e 2.5 cm

f 4 dm, 2 dm, 3 dm

EXAM-STYLE QUESTION

2 A cylindrical water tank has height 3 m and base radius 1.2 m.
  a Calculate the volume of the tank in m³.
  b Give your answer to part **a** in dm³.
  c Hence find, in litres, the capacity of the tank.

> Capacity is the amount of liquid a container can hold when it is full.

3 Find an expression for the volume, V, of each solid.
Give each answer in its simplest form.

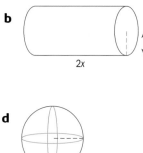

a h, x, x

b x, $2x$

c x, $6x$

d $3x$

EXAM-STYLE QUESTION

4 The diagram shows a right pyramid with a regular hexagonal base.
The volume of the pyramid is 84 cm³ and the height is 7 cm.
O is the center of the base.
  a Calculate the area of the base.
  b Calculate the area of triangle AOB.
  c What size is angle AOB?
  d Calculate the length of AB.

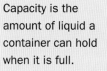

7 cm, O, A, B

5 A spherical ball has a volume of 200 cm³.
  a Find the radius of the ball.
  b Give your answer to part **a** correct to the nearest mm.

EXAM-STYLE QUESTIONS

6 A cylindrical container has base radius 15 cm and height 30 cm.
  It is full of sand.
  a Calculate the volume of sand in the container.
  The sand is poured into a second container in the shape of a
  cuboid. The length of the cuboid is 60 cm, the width is 20 cm,
  and the height is 17 cm.
  b Is the second container big enough for all the sand?
     Justify your decision.

7 A cylindrical pencil is 13.5 cm long with diameter 0.7 cm.
  It is sharpened to a cone as shown in the diagram.

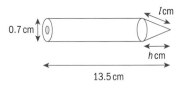

The length of the cylindrical part is now 12.3 cm.
  The height of the cone is h cm and its slant height is l cm.
  a i Write down the value of h.
     ii Find the value of l.
  b Hence find
     i the total surface area of the pencil
     ii the volume of the pencil.
     Give your answers to 3 sf.

> Sketch the cone.
> Use Pythagoras to
> find l.

The pencils are packed in boxes. The boxes are cuboids of
  width 5.6 cm, height 1.4 cm and length 13.5 cm.
  c Show that the maximum number of pencils that will fit in the
     box is 16.
  d Find the space in a full box that is **not** occupied by the pencils.
  e Write your answer to part **d** as a percentage of the volume of
     the box. Give your answer correct to 2 sf.

> Sketch the box.

Review exercise

Paper 1 style questions

1 The cuboid ABCDEFGH is shown in the diagram.
AB = 20 cm, BC = 42 cm and AE = 34 cm.

   a Calculate the surface area of the cuboid.

   b Calculate the volume of the cuboid, giving your answer in dm³.

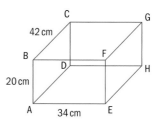

2 The cuboid ABCDEFGH is shown in the diagram.
AB = 5 cm, BC = 4 cm and AE = 10 cm.

   a Calculate the length of AH.

   b Calculate the angle that AG makes with the face ADHE.

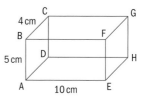

3 The diagram shows a rectangular-based right pyramid ABCDE.
The height of the pyramid is 8 cm. The base is 5 cm long and 4 cm wide.

   Calculate

   a the length of AC

   b the length of EC

   c the angle AEC.

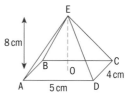

4 The diagram shows a square-based right pyramid ABCDE.
The height of the pyramid is 9 cm. Each edge of the base is 6 cm.

   Calculate

   a the distance between the midpoint of DC and E

   b the area of triangle DCE

   c the surface area of the pyramid.

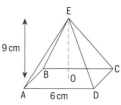

5 The diagram shows a hollow cube ABCDEFGH. Its volume is 512 cm³.

   a Write down the length of any edge of the cube.

   b Find the distance AC.

   Rosaura puts a pencil in the cube. The pencil is 13.5 cm long.

   c Does the pencil fit in the cube? Justify your decision.

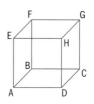

6 A cone has the dimensions shown in the diagram.
Point B is on the circumference of the base, point O is the center of the base and point A is the apex of the cone.

   a Calculate the size of the angle that AB makes with the base of the cone.

   b **i** Calculate the height of the cone.

      ii Calculate the volume of the cone.

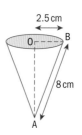

7 The diagram represents a tent in the shape of a prism.
The front of the tent, ABC, is an isosceles triangle with
AB = BC = 2.4 m and $A\hat{B}C = 110°$. The tent is 3.5 m long.

a Calculate the area of the front face of the tent ABC.

b Calculate the space inside the tent.

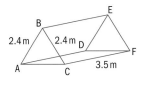

Paper 2 style questions

1 An office tower is shown in the diagram. It consists of a cuboid
with a square base and a square-based right pyramid.

a Calculate the distance from O to M, the midpoint of HG.

b Calculate the height of the tower.

c Find the angle that OM makes with the plane EFGH.

A cleaning services company charges US$ 78 per m² to clean
the outside of a building.

d Calculate the cost of cleaning the tower, giving your answer
correct to the nearest US$.

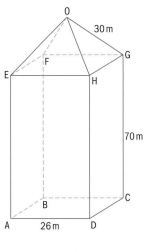

2 A solid sculpture consist of a hemisphere of radius 3 cm and
a right circular cone of slant height *l* as shown in the diagram.

a Show that the volume of the hemisphere is $18\pi\,cm^3$.

The volume of the hemisphere is two-thirds that of the cone.

b Find the vertical height of the cone.

c Calculate the slant height of the cone.

d Calculate the angle between the slanting side of the cone
and the flat face of the hemisphere.

The sculpture is made of a material that weighs 10.8 g per cm³.

e Calculate the weight of the sculpture, giving your answer in kg.

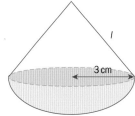

3 ABCDE is a solid glass right pyramid.
The base of the pyramid is a square of side 5 cm and center O.
The vertical height is 7 cm.

a Calculate the volume of the pyramid.

The glass weighs 8.7 grams per cm³.

b Calculate the weight of the pyramid, giving your answer
correct to the nearest g.

c Find the length of a sloping edge of the pyramid, giving
your answer correct to 4 significant figures.

d Calculate the angle made between the edge ED and the base
of the pyramid.

e Calculate the size of the angle AED.

f Calculate the total surface area of the pyramid.

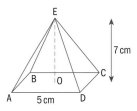

4 The diagram shows a square-based right pyramid ABCDV.
The midpoint of DC is M and VM is inclined at 65° to the base.
The sides of the base are 8 cm and O is the center of the base.

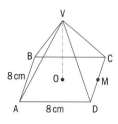

  a Find the height of the pyramid, giving your answer correct
  to 3 significant figures.

  b Calculate

    i the length of VM

    ii the size of angle DVC.

  c Find the total surface area of the pyramid.

  d Find the volume of the pyramid, giving your answer correct
  to the nearest cm^3.

CHAPTER 10 SUMMARY

Geometry of three-dimensional solids

- In a **right prism** the end faces are the same shape and size
  and are parallel. All the other faces are rectangles that are
  perpendicular to the end faces.
- If you cut parallel to the end face of a right prism the
  cross-section will always be the same shape and size.
- The base of a **pyramid** is a polygon. The other faces are
  triangles that meet at a Point called the **apex**. In a **right
  pyramid** the apex is vertically above the center of the base.

Angles between two lines, or between a line and a plane

- When two faces of a solid, X and Y, are perpendicular, any
  line in face X is perpendicular to any line in face Y.

Surface areas of three-dimensional solids

- The **surface area** of a solid is the sum of the areas of all
  its faces. Surface area is measured in square units, e.g. cm^2, m^2.

- Area of curved surface of a cylinder = $2\pi rh$
  Total surface area of a cylinder = $2\pi rh + 2\pi r^2$

- Surface area of a sphere = $4\pi r^2$

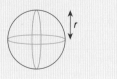

Continued on next page

- Area of curved surface of a cone = $\pi r l$
- Total surface area of a cone = $\pi r l + \pi r^2$

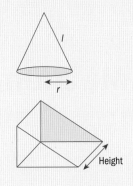

Volumes of three-dimensional solids

- The volume of a solid is the amount of space it occupies and is measured in cubic units, e.g. cm³, m³, etc.
- Volume of a prism is
$$V = \text{area of cross-section} \times \text{height}$$

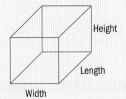

- Volume of a cuboid is
$$V = l \times w \times h,$$
where l is the length, w is the width, h is the height

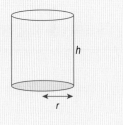

- Volume of a cylinder is
$$V = \pi r^2 h,$$
where r is the radius, h is the height

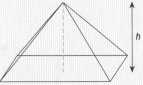

- Volume of a pyramid =
$$\frac{1}{3}(\text{area of base} \times \text{vertical height})$$

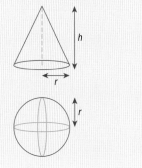

- Volume of a cone $= \dfrac{1}{3}\pi r^2 h$,

where r is the radius and h is the vertical height

- Volume of a sphere $= \dfrac{4}{3}\pi r^3$,
where r is the radius

Mathematical proof

Grass is green.

What about Black Mondo Grass?

In mathematics, you can't just say a statement is true – you have to prove it.

A mathematical proof has to be rigorous. This means, it has to be true in all cases. In fact, one way of proving a statement isn't true is to find a counter example – just one example where it isn't true.

▲ Black Mondo Grass (*Ophiopogon planiscarpus nigrescens*) is native to Korea.

■ Find a counter example to disprove the statement:
All prime numbers are odd.

Searching for the truth

To produce a proof the mathematician starts with basic, self-evident assumptions called axioms. He or she then uses the rules of logic and deductive reasoning to reason toward proving a new theorem.

A new theorem then provides a base for further reasoning.

■ How does mathematical proof differ from proof using 'good reasons' and 'sufficient evidence' in other areas of knowledge?

The Greek mathematician Euclid (c.300 B.C.E.) introduced the axiomatic method of proof. In the west, his book Elements was one of the standard geometry texts for students until the middle of the 20th century and forms the basis of what you learn in geometry today.

In the medieval period Islamic mathematicians developed further the ideas of arithmetic and algebra. These became the basis of more general proofs. In the 10th century C.E., the Iraqi mathematician Al-Hashimi provided general proofs for numbers and proved the existence of irrational numbers.

"A mathematician is a device for turning coffee into theorems."

Attributed variously to both the Hungarian mathematicians Paul Erdos (1913–86) and Alfréd Rényi (1921–70)

Proof problem

Complete a table of values like this for the equation $y = x^5 - 10x^4 + 35x^3 - 50x^2 + 5x$.

x	0	1	2	3	4
y					

- Predict the value of y when $x = 5$.
- Now work out the value when $x = 5$. Was your prediction correct?

A proof using mathematical induction shows that if one particular case is true, then so is the next one. It also shows that one particular base case is true.

- How did you use inductive reasoning to predict the value of y when $x = 5$?
- What were the problems with using inductive reasoning?

A mathematical proof

Prove the theorem:

The sum of any three consecutive even numbers is divisible by 6

Proof: Write the three consecutive even numbers as $2m$, $2m + 2$, $2m + 4$ where m is a whole number.

- What axioms have we used here?

Find their sum, S by adding them together:

$S = 2m + 2m + 2 + m + 4$

$\quad = 6m + 6$

$\quad = 6(m + 1)$

So S is a multiple of 6, and S is **always** divisible by 6.

- Use a similar method to show that the product of three consecutive even numbers is always divisible by 8.

Elegant and economical proof

Here are two solutions to the problem...

Prove that $(x + y + z)(x - y - z) = x^2 - (y + z)^2$

Solution 1

$(x + y + z)(x - y - z)$

$= x^2 - xy - xz + xy - y^2 - yz + xz - yz - z^2$

$= x^2 - y^2 - 2yz - z^2$

$= x^2 - (y^2 + 2yz + z^2)$

$= x^2 - (y + z)^2$

Solution 2

$(x + y + z)(x - y - z)$

$= (x + (y + z))(x - (y + z))$

$= x^2 - (y + z)^2$

> "The mathematics are distinguished by a particular privilege, that is, in the course of ages, they may always advance and can never recede."
>
> Edward Gibbon, Decline and Fall of the Roman Empire

- Which is the better solution?

Solution 1 and solution 2 both give the same answer, so neither is better than the other. However, solution 2 is more elegant and insightful.

A mathematician will seek to find a proof that is
* Economical – as short as possible.
* Elegant – with a surprise or moment of insight.

11 Project

CHAPTER OBJECTIVES:

As part of your Mathematical Studies course, you need to write a project that is assessed and counts towards your final grade.

This chapter gives you advice on planning your project, as well as hints and tips to help you get a good grade by making sure your project satisfies the assessment criteria, suggestions for project topics, and a useful checklist to help you ensure your final project is complete.

11.1 The project

The project is an opportunity for you to show that you can apply mathematics to an area that interests you.

The project is worth 20% of your final grade so it is worth spending time on it.

You should aim to spend:

> You can't receive a grade for Mathematical Studies SL if you don't submit a project.

25 hours of class time	25 hours of your own time
<ul><li>Discussing the project and the assessment criteria</li><li>Looking at and 'marking' previous projects</li><li>Discussing suitable titles</li><li>Discussing methods of data collection and sampling</li><li>Discussing your progress with your teacher</li></ul>	<ul><li>Planning your project</li><li>Collecting and organizing data</li><li>Applying mathematical processes</li><li>Discussing results and validity</li><li>Making sure your project is well structured and reads well</li><li>Checking that your mathematics, notation and terminology are correct</li></ul>

Your school will set you deadlines for submitting drafts and the final piece of work.

If you do not submit a project then you will receive a grade of "N" for Mathematical Studies SL, which means you automatically fail your Diploma.

> Every candidate taking Mathematical Studies SL needs to submit a project. Make sure that you know your school's deadlines and keep to them.

11.2 Internal assessment criteria

Your project will be moderated by your teacher against given criteria. It will then be externally moderated by the IB using the same assessment criteria.

The final mark for each project is the sum of the scores for each criterion.

The maximum possible final mark is 20.

This is 20% of your final mark for Mathematical Studies SL.

The criteria are split into seven areas, A to G:

Criterion A	Introduction
Criterion B	Information/measurement
Criterion C	Mathematical processes
Criterion D	Interpretation of results
Criterion E	Validity
Criterion F	Structure and communication
Criterion G	Notation and terminology

> Make sure you understand these criteria and consult them frequently when writing your project. Marking someone else's project against the criteria will help you to understand them. There are some projects to mark on the CD.

Criterion A: Introduction

In this context,

- "task" means "what the student is going to do"
- "plan" means "how the student is going to do it".

> A good project should be clear and easily understood by a non-mathematician, and self-explanatory all the way through.

> Every project should start with a clear statement of the task and have a clear title.

Achievement level	Descriptor
0	The project does not contain a clear statement of the task. *There is no evidence in the project of any statement of what the student is going to do or has done.*
1	The project contains a clear statement of the task. *For this level to be achieved the task should be stated explicitly.*
2	The project contains a title, a clear statement of the task and a description of the plan. *The plan need not be highly detailed, but must describe how the task will be performed. If the project does not have a title this achievement level cannot be awarded.*
3	The project contains a title, a clear statement of the task and a detailed plan that is followed. *The plan should specify what techniques are to be used at each stage and the purpose behind them, thus lending focus to the task.*

Criterion B: Information/measurement

In this context,

- "information or generated measurements" means information or measurements generated by computer, by observation, by investigation, by prediction from a mathematical model or by experiment
- "mathematical information" includes geometrical figures and data collected empirically or assembled from outside sources. It does not just mean data for statistical analysis. If you use a questionnaire or survey, make sure you include a copy of this along with the raw data.

Achievement level	Descriptor
0	The project does not contain any relevant information collected or relevant measurements generated. *No attempt has been made to collect any relevant information or generate any relevant measurements.*
1	The project contains relevant information collected or relevant generated measurements. *This achievement level can be awarded even if a fundamental flaw exists in the instrument used to collect the information, for example, a faulty questionnaire or an interview conducted in an invalid way.*
2	The relevant information collected, or set of measurements generated, is organized in a form appropriate for analysis or is sufficient in both quality and quantity. *A satisfactory attempt has been made to structure the information/ measurements ready for the process of analysis, or the information/ measurements collection process has been thoroughly described and the quantity of information justified. The raw data must be included for this achievement level to be awarded.*

▶ Continued on next page

Achievement level	Descriptor
3	The relevant information collected, or set of measurements generated, is organized in a form appropriate for analysis and is sufficient in both quality and quantity.
	The information/measurements have been properly structured ready for analysis and the information/measurements collection process has been thoroughly described and the quantity of information justified. If the information/measurements are too sparse or too simple this achievement level cannot be awarded.
	If the information/measurements are from a secondary source then there must be evidence of sampling if appropriate. All sampling processes should be completely described.

Your project

To get a good mark for Criterion B: Information/measurement

✓ Gather your information/measurements from a survey, a questionnaire, calculation, the internet, etc.

✓ Make sure you collect sufficient information/measurements to perform the mathematical processes you mentioned in Criterion A.

✓ Include all your raw information/measurements in the project – you can put this in an appendix if you wish.

✓ Make sure that the information/measurements you collect are relevant and organized ready for use.

✓ Reorganize the information/measurements each time to suit the calculations you do.

✓ Remember to include a copy of your survey or questionnaire, if you used one to collect your information/measurements.

✓ If your information/measurements are taken from a **secondary** source then you need to describe any sampling process that you used.

✓ You can also use mathematical processes that are not in the Mathematical Studies SL syllabus.

Criterion C: Mathematical processes

When presenting data in diagrams:

- Use a ruler and don't just sketch. A freehand sketch would not be considered a correct mathematical process.
- If you use technology to create your diagram, you need to show a clear understanding of the mathematical processes used.
- All graphs must contain all relevant information.

Achievement level	Descriptor
0	The project does not contain any mathematical processes. *For example where the processes have been copied from a book with no attempt being made to use any collected/generated information.* *Projects consisting of only historical accounts will achieve this level.*
1	At least two simple mathematical processes have been carried out. *Simple processes are considered to be those that a mathematical studies student could carry out easily; for example, percentages, areas of plane shapes, graphs, trigonometry, bar charts, pie charts, mean and standard deviation, substitution into formulae and **any** calculations/graphs using technology only.*
2	At least two simple mathematical processes have been carried out correctly. *A small number of isolated mistakes should not disqualify a student from achieving this level. If there is incorrect use of formulae, or consistent mistakes in using data, this level cannot be awarded.*
3	At least two simple mathematical processes have been carried out correctly. All processes used are relevant. *The simple mathematical processes must be relevant to the stated aim of the project.*
4	The simple relevant mathematical processes have been carried out correctly. In addition at least one relevant further process has been carried out. *Examples of further processes are differential calculus, mathematical modeling, optimisation, analysis of exponential functions, statistical tests and distributions, compound probability. For this level to be achieved it is not required that the calculations of the further process be without error. At least one further process must be calculated showing full working.*
5	The simple relevant mathematical processes have been carried out correctly. In addition at least one relevant further process has been carried out. All processes, both simple and further, that have been carried out are without error. *If the measurements, information or data are limited in scope then this achievement level cannot be awarded.*

If a project contains no simple mathematical processes then the first two further processes are assessed as simple.

The teacher is responsible for determining the accuracy of the mathematics used and must indicate any errors on the final project.

Your project

To get a good mark for Criterion C: Mathematical processes

✓ Always include at least two **relevant** simple mathematical processes.

✓ Always put scales and labels on your graphs.

✓ State which process you are going to use and why.

✓ Discuss the validity of these processes.

✓ Check to make sure that your results are accurate.

✓ Check that your results are sensible.

✓ Comment on your results.

✓ Introduce at least one **relevant** further mathematical process.

✓ State why you are using this further process and make sure that it is relevant and valid.

✓ For both simple and further processes, make sure that you do one calculation of each process by hand. You can use your GDC to perform similar calculations.

✓ If you find the standard deviation then comment on it.

✓ For the chi-squared test to be valid the entries must be frequencies – not raw data or percentages, and, if the degree of freedom is 1, then Yate's continuity correction should be applied. No expected values should be less than 5.

✓ For linear correlation there is no point finding the equation of the regression line if the correlation coefficient is weak or if you can see from the scatter diagram that there is no correlation.

Criterion D: Interpretation of results

Achievement level	Descriptor
0	The project does not contain any interpretations or conclusions. *For the student to be awarded this level there must be no evidence of interpretation or conclusions anywhere in the project, or a completely false interpretation is given without reference to any of the results obtained.*
1	The project contains at least one interpretation or conclusion. *Only minimal evidence of interpretation or conclusions is required for this level. This level can be achieved by recognizing the need to interpret the results and attempting to do so, but reaching only false or contradictory conclusions.*
2	The project contains interpretations and/or conclusions that are consistent with the mathematical processes used. *A "follow through" procedure should be used and, consequently, it is irrelevant here whether the processes are either correct or appropriate; the only requirement is consistency.*

▶ Continued on next page

3	The project contains a meaningful discussion of interpretations and conclusions that are consistent with the mathematical processes used.
	To achieve this level the student would be expected to produce a discussion of the results obtained and the conclusions drawn based on the level of understanding reasonably to be expected from a student of Mathematical Studies SL. This may lead to a discussion of underlying reasons for results obtained.
	If the project is a very simple one, with few opportunities for substantial interpretation, this achievement level cannot be awarded.

Your project

To get a good mark for Criterion D: Interpretation of results

✓ After every graph or calculation make a comment – are your results what you expected? Are they meaningful?

✓ Always give a thorough and detailed analysis of **all** your results.

✓ Make sure that you "follow through" with the results of your mathematical processes. Even if your mathematics contains errors, as long as your interpretation or conclusion follows on from that wrong answer then you will be awarded the marks.

✓ Make sure that your project is not a "simple one" with only a few simple mathematical processes. If there are only a few processes then there is very little to comment on. This is also the case when the project is very short.

Criterion E: Validity

This criterion looks at whether appropriate techniques were used to collect information, whether appropriate mathematics was used to deal with this information and whether the mathematics used has any limitations in its applicability within the project. Any limitations or qualifications of the conclusions and interpretations are also judged within this criterion.

Achievement level	Descriptor
0	There is no awareness shown that validity plays a part in the project.
1	There is an indication with reasons if and where validity plays a part in the project.
	There is discussion of the validity of the techniques used or recognition of any limitations that might apply. A simple statement such as
	"I should have used more information/measurements" is not sufficient to achieve this level. If the student considers that validity is not an issue, this must be fully justified.

Criterion F: Structure and communication

In this context "structure" means how you have organized your information, calculations and interpretations. The project should present a logical sequence of thought and activities – starting with the task and the plan, and finishing with the conclusions and limitations.

Avoid large numbers of repetitive procedures.

Make sure all graphs are fully labelled and have an appropriate scale.

Projects that do not reflect a significant time commitment will not score highly on this assessment criterion.

> It is not expected that spelling, grammar and syntax are perfect. Your teacher will encourage you to correct any language errors.

Achievement level	Descriptor
0	No attempt has been made to structure the project. *It is not expected that many students will be awarded this level.*
1	Some attempt has been made to structure the project. *Partially complete and very simple projects would only achieve this level.*
2	The project has been structured in a logical manner so that it is easily followed. *There must be a logical development to the project. The project must reflect the appropriate commitment for this achievement level to be awarded.*
3	The project has been well structured in accordance with the stated plan and is communicated in a coherent manner. *To achieve this level the project would be expected to read well, and contain footnotes and a bibliography, as appropriate. The project must be focused and contain only relevant discussions.*

Criterion G: Notation and terminology

You need to use correct mathematical terminology and mathematical notation. Calculator or spreadsheet notation is not acceptable.

Achievement level	Descriptor
0	The project does not contain correct mathematical notation or terminology. *It is not expected that many students will be awarded this level.*
1	The project contains some correct mathematical notation or terminology.
2	The project contains correct mathematical notation and terminology throughout. *Variables should be explicitly defined. An isolated slip in notation need not preclude a student from achieving this level. If it is a simple project requiring little or no notation/terminology this achievement level cannot be awarded.*

11.3 Moderating the project

Once you have submitted the final version of your project, your teacher will moderate it. The teacher looks at each criterion in turn, starting from the lowest grade. As soon as your project fails to meet one of the grade descriptors, then the mark for that criterion is set.

The teacher submits these marks to the International Baccalaureate, via a special website. A sample of your school's projects is selected automatically from the marks that are entered and this sample is sent to an external moderator to be checked. This person moderates the projects according to the assessment criteria and checks that your teacher has moderated the projects accurately.

If your teacher has moderated the projects too severely then all your project marks may be increased.

If your teacher has moderated the projects too leniently then all your project marks may be decreased.

11.4 Academic Honesty

This is extremely important in all your work. Make sure that you have read and are familiar with the IB Academic Honesty document.

> Your teacher or IB Diploma Programme coordinator will be able to give you this document.

Academic Honesty means:

- that your work is authentic
- that your work is your own intellectual property
- that you conduct yourself properly in written examinations
- that any work taken from another source is properly cited.

Authentic work:

- is work based on your own original ideas
- can draw on the work and ideas of others, but this must be fully acknowledged (e.g. in footnotes and bibliography)
- must use your own language and expression – for both written and oral assignments.
- must acknowledge all sources fully and appropriately (e.g. in a bibliography).

Malpractice

The IB defines **malpractice** as 'behavior that results in, or may result in, the candidate or any other candidate gaining an unfair advantage in one or more assessment components'.

Malpractice includes:

- plagiarism – copying from others' work, published or otherwise
- collusion – working secretly with at least one other person in order to gain an undue advantage. This includes having someone else write your exploration, and passing it off as your own
- duplication of work
- any other behavior that gains an unfair advantage.

'Plagiarism' is a word derived from Latin, meaning 'to kidnap'.

Advice to schools:

- A school-wide policy must be in place to promote Academic Honesty
- All candidates must clearly understand this policy
- All subject areas must promote the policy
- Candidates must be clearly aware of the penalties for academic dishonesty
- Schools must enforce penalties, if incurred.

Acknowledging sources

Remember to acknowledge all your sources. Both teachers and moderators can usually tell when a project has been plagiarised. Many schools use computer software to check for plagiarism. If you are found guilty of plagiarism then you will not receive your Diploma. It is not worth taking the risk.

You will find a definition of plagiarism in the Academic Honesty document.

11.5 Record keeping

Make notes of any books or websites you use, as you go along, so you can include them in your bibliography.

- There are different ways of referencing books, websites, etc. Make sure that you use the style advised by your school and **be consistent**.
- Keep a record of your actions so that you can show your teacher how much time you are spending on your project.
- Remember to follow your teacher's advice and meet the school's deadlines.
- The teacher is there to help you – so do not be afraid to ask for guidance.

11.6 Choosing a topic

You need to choose a topic that you are interested in, because then you will put more effort into the project. Discuss the topic with your teacher to make sure that you can generate sufficient data to perform both simple and further mathematical processes.

Many candidates choose a project based on statistics. Other topics are also suitable, such as optimisation in calculus, modeling equations, quadratics or exponential equations, trigonometry, set theory, probability and finance.

If you cannot think of a topic yourself then ask your teacher to show you the list of topics in the Teacher Support Manual or a list of topics from the Online Curriculum Center. Below are some ideas – perhaps you will find something that interests you here.

PROJECT SUGGESTIONS

- Comparing heights from sports data
- Rollerblading: the mathematics behind it
- Traffic study of Schiphol International Airport
- Is there a connection between the time it takes to get to school and the distance a student lives from school?
- Does gender influence someone's choice of favorite animal?
- The effect of sport on GPA
- Does eating breakfast affect your grade?
- Is there a relation between BMI and hours playing sports?
- The effect of blood alcohol content law on the number of traffic collisions
- M.C. Escher: symmetry and infinity in art
- A statistical investigation of leaves
- Olympic games: track and field times
- Analysis of basic US stocks
- Modeling the decrease in swimming times since the Olympics began
- Examining the relation between lung capacity, smoking and sports
- Relations between international and bilingual students re: jobs, pocket money and spending behavior
- Investing in a hotel in Costa Rica
- Statistical comparison of the number of words in a sentence in different languages

- Teenage drinking and its effect on GPA
- Relationship between unemployment and criminality in Sweden
- How many peas are there in a 500 g box?
- Correlation between women's participation in higher education and politics from 1955–2000
- Investigating eating trends of today's youth
- Correlation study of TV versus sleep times
- Which type of movies do males and females prefer?
- Power to weight ratio
- The Ferris wheel
- The effect that different temperatures have on the level of growth of bacteria in water from a garden pool
- Music and the brain
- Blood pressure and stress levels
- Sunspot cycles
- Public transportation costs and car usage: a personal comparison
- The geometry involved in billiards
- Investigation into different brands of batteries
- Costs of products bought online compared to local grocery stores
- Investigating the most economical packaging for one-litre drink cartons
- Modeling the temperature each week for various cities in the world
- Modeling the trajectory of an arrow fired from various angles
- Modeling the cooling rate of hot drinks placed in different locations
- Investigation of how to get from A to B in New York
- Testing if the weights of 1-kilogram bags of sugar are normally distributed

Mathematical Studies SL Project Check List

	Check ✓
Does your project have a front cover with the title of your project?	
Is your candidate name and number and the examination session on the front cover?	
Have you stated clearly what you are going to do?	
Have you explained how you are going to do it?	
Have you explained what mathematical processes you will use and why?	
Did you do everything that you said you would do?	
Have you collected data or generated measurements or information?	
Is your raw data included in the project or in an appendix?	
Is your data relevant?	
Is your data sufficient in quantity?	
Do you have quality data?	
Is your data set up for use?	
Have you described the sampling process clearly?	
Have you performed at least two simple mathematical processes?	
Are these simple processes correct?	
Are the simple processes relevant?	
Have you performed a sophisticated mathematical process?	
Is this sophisticated process relevant?	
Is the sophisticated process correct?	
Have you commented on your results?	
Are your comments consistent with your analysis?	
Have you commented thoroughly on everything that you have done?	
Have you commented on validity?	
Does the project contain only correct notation?	
Does the project contain only correct terminology?	
Is your project laid out in a logical manner?	
Do you have an appendix if one is needed?	
Do you have a bibliography?	

12

CHAPTER OBJECTIVES:

This chapter shows you how to use your graphic display calculator (GDC) to solve the different types of problems that you will meet in your Mathematical Studies course. You should not work through the whole of the chapter – it is simply here for reference purposes. When you are working on problems in the mathematical chapters, you can refer to this chapter for extra help with your GDC if you need it.

GDC instructions on CD: *The instructions in this chapter are for the TI-Nspire model. Instructions for the same techniques using the TI-84 Plus and the Casio FX-9860GII are available on the CD.*

Chapter contents

Use this list to help you to find the topic you need.

Before you start

You should know:

- Important keys on the keyboard: On, menu, esc, tab, ctrl, shift, enter, del
- The home screen
- Opening new documents, adding new pages, changing settings
- Moving between pages in a document
- Panning and grabbing axes to change a window in a Graphs page
- Changing window settings in a Graphs page
- Using zoom tools in a Graphs page
- Using trace in a Graphs page

> For a reminder of how to perform the basic operations, have a look at your GDC manual.

1 Number and algebra 1

1.1 Solving simultaneous linear equations

When solving simultaneous equations in an examination, you do not need to show any method of solution. You should simply write out the equations in the correct form and then give the solutions. The GDC will do all the working for you.

> You do not need the equations to be written in any particular format to use the linear equation solver, as long as they are both *linear*, that is, neither equation contains x^2 or higher order terms.

Example 1

Solve the equations:
$2x + y = 10$
$x - y = 2$

Open a new document and add a Calculator page. Press menu 3:Algebra | 2:Solve Systems of Linear Equations… Press enter You will see this dialog box, showing two equations and two variables, x and y. **Note:** This is how you will use the linear equation solver in your examinations. In your project, you might want to solve a more complicated system with more equations and more variables.	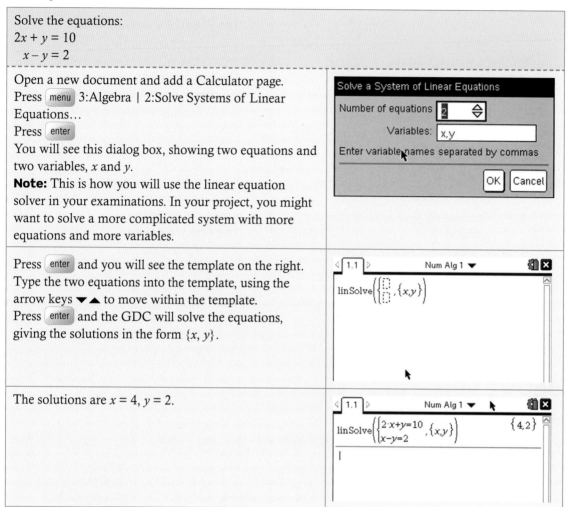
Press enter and you will see the template on the right. Type the two equations into the template, using the arrow keys ▼▲ to move within the template. Press enter and the GDC will solve the equations, giving the solutions in the form $\{x, y\}$.	
The solutions are $x = 4$, $y = 2$.	

1.2 Solving quadratic equations

When solving quadratic equations in an examination, you do not need to show any method of solution. You should simply write out the equations in the correct form and then give the solutions. The GDC will do all the working for you.

Example 2

Solve $3x^2 - 4x - 2 = 0$	
Press menu 3:Algebra | 3:Polynomial Tools | 1:Find Roots of a Polynomial… Press enter You will see this dialog box, showing a polynomial of degree 2 (a quadratic equation) with real roots. You do not need to change anything. Press enter	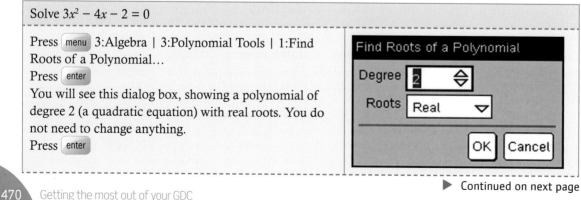

▶ Continued on next page

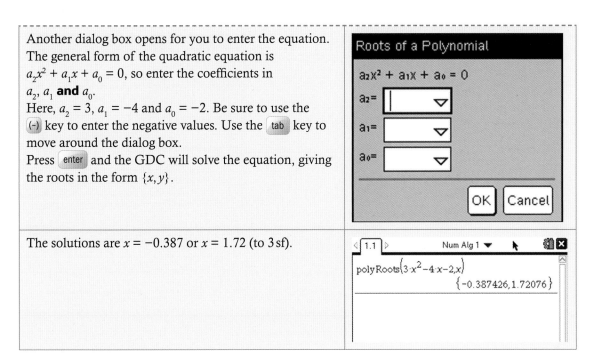

Another dialog box opens for you to enter the equation. The general form of the quadratic equation is $a_2x^2 + a_1x + a_0 = 0$, so enter the coefficients in a_2, a_1 **and** a_0.

Here, $a_2 = 3$, $a_1 = -4$ and $a_0 = -2$. Be sure to use the (−) key to enter the negative values. Use the tab key to move around the dialog box.

Press enter and the GDC will solve the equation, giving the roots in the form $\{x, y\}$.

The solutions are $x = -0.387$ or $x = 1.72$ (to 3 sf).

1.3 Standard form

Numbers written in standard form are in the form $a \times 10^n$, where $1 \le a < 10$ and $n \in \mathbb{Z}$.

There are three different ways of entering a number in standard form.

For example, to enter 2.4×10^4 press the keys

2 . 4 × 1 0 ^ 4   or

2 . 4 × 10ˣ 4   or

2 . 4 EE 4

The GDC changes the appearance of the number as you enter it.

Example 3

Given that $x = 2.4 \times 10^4$ and $y = 3.6 \times 10^3$, find the value of
a $2x + 3y$ **b** xy^2

Enter the values using one of the three methods shown above.

In normal mode, the GDC will display the result either as a normal number or, if it is a large number, in standard form.

Write your answer in standard form.

For 3.11E11, write 3.11×10^{11}.

▶ Continued on next page

To change the answer to standard form, press
🏠 On | 5:Settings & Status 2:Settings | 1:General
From the drop-down menu, choose 'Scientific' for the
Exponential Format.
Press 4:Current to return to the calculation page.
Note: Remember to return the settings back to normal
when you have finished.

All results are now given in standard form:
a 5.88×10^4
b 3.11×10^{11}

1.4 Significant figures

GDC limitations: rounding

You can use your GDC to round numbers to 3 sf, but you need to remember that (i) the GDC will leave off zeros at the end of a number when they come after the decimal point and (ii) the GDC will not round numbers with more than three digits before the decimal point.

Example 4

Do these calculations. Give each of your answers correct to 3 significant figures (3 sf).
a $4 \times \pi$　　**b** 3.629×2.76　　**c** 123×12

To change an answer to 3 sf, press 🏠 On | 5:Settings &
Status 2:Settings | 1:General
From the drop-down menu, choose 'Float 3' for Display
Digits.
Press 4:Current to return to the calculation page.
Note: Remember to return the settings back to normal
when you have finished.

For Mathematical Studies, answers to questions should
be rounded to 3 significant figures (3 sf), unless you are
told otherwise.

All results are now given in rounded form:
a 12.6 is correct to 3 sf.
b 10 is 10.0 to 3 sf. The GDC has left off the final 0
after the decimal point.
c 1476 is 1480 to 3 sf. The digits before the decimal
point are all included and have not been rounded.

2 Descriptive statistics

You can use your GDC to draw charts to represent data and to calculate basic statistics such as mean, median, etc. Before you can do this, you need to enter the data into a list or spreadsheet. This is done in a Lists & Spreadsheet page in your document.

Entering data

There are two ways of entering data: as a list or as a frequency table.

2.1 Entering lists of data

Example 5

Enter the data in the list
1, 1, 3, 9, 2

Open a new document and add a Lists & Spreadsheet page.
Type 'data' in the first cell.
Type the numbers from the list in the first column.
Press enter or ▼ after each number to move down to the next cell.
Note: The word 'data' is a label that will be used later when you want to create a chart or do some calculations with the data. You can use any letter or name to label the list.

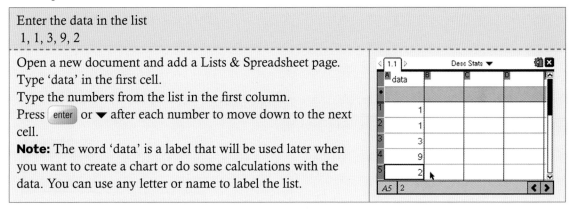

2.2 Entering data from a frequency table

Example 6

Enter the data in a table

Number	1	2	3	4	5
Frequency	3	4	6	5	2

Add a new Lists & Spreadsheet page to your document.
To label the columns, type 'number' in the first cell and 'freq' in the cell to its right.
Enter the numbers in the first column and the frequencies in the second.
Use the ▼ ▲ ◀▶ keys to navigate around the spreadsheet.

Drawing charts

You can draw charts from a list or from a frequency table.

2.3 Drawing a frequency histogram from a list

Example 7

Draw a frequency histogram for these data: 1, 1, 3, 9, 2	
Enter the data in a list called 'data' (see Example 5). Add a new Data & Statistics page to your document. **Note:** You do not need to worry about what this screen shows.	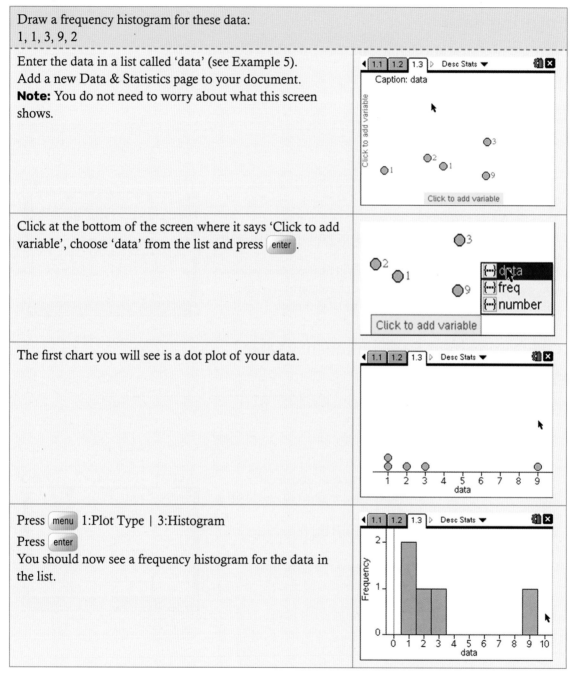
Click at the bottom of the screen where it says 'Click to add variable', choose 'data' from the list and press ⌷enter⌷.	
The first chart you will see is a dot plot of your data.	
Press ⌷menu⌷ 1:Plot Type | 3:Histogram Press ⌷enter⌷ You should now see a frequency histogram for the data in the list.	

2.4 Drawing a frequency histogram from a frequency table

Example 8

Draw a frequency histogram for these data:

Number	1	2	3	4	5
Frequency	3	4	6	5	2

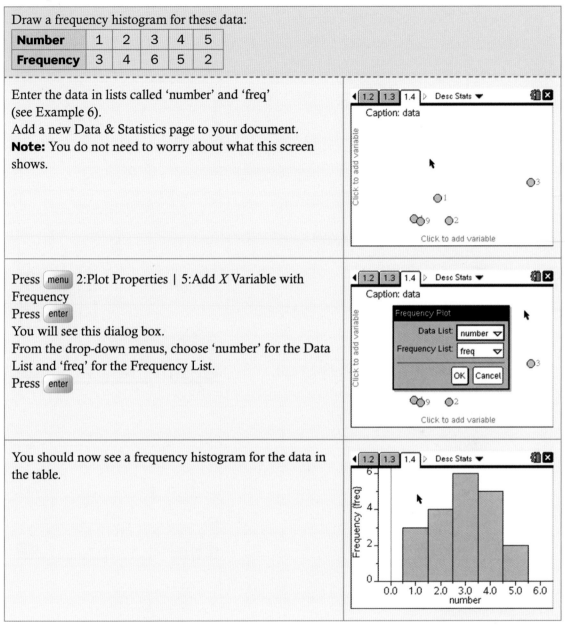

Enter the data in lists called 'number' and 'freq' (see Example 6).
Add a new Data & Statistics page to your document.
Note: You do not need to worry about what this screen shows.

Press menu 2:Plot Properties | 5:Add X Variable with Frequency
Press enter
You will see this dialog box.
From the drop-down menus, choose 'number' for the Data List and 'freq' for the Frequency List.
Press enter

You should now see a frequency histogram for the data in the table.

2.5 Drawing a box and whisker graph from a list

Example 9

Draw a box and whisker graph for these data:
1, 1, 3, 9, 2

Enter the data in a list called 'data' (see Example 5). Add a new Data & Statistics page to your document. **Note:** You do not need to worry about what this screen shows.	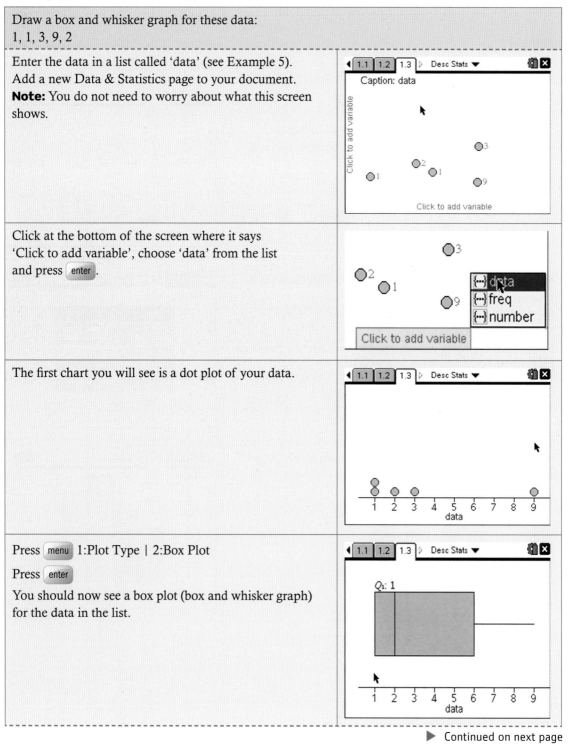
Click at the bottom of the screen where it says 'Click to add variable', choose 'data' from the list and press `enter`.	
The first chart you will see is a dot plot of your data.	
Press `menu` 1:Plot Type | 2:Box Plot Press `enter` You should now see a box plot (box and whisker graph) for the data in the list.	

▶ Continued on next page

Move the cursor over the plot and you will see the quartiles, Q_1 and Q_3, the median, and the maximum and minimum values.

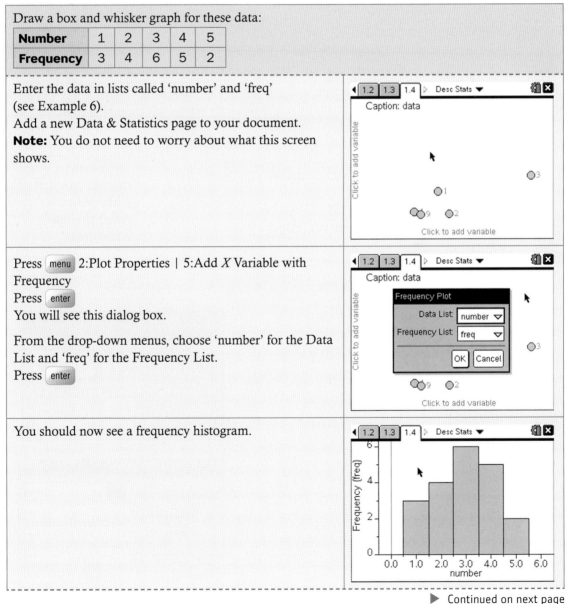

2.6 Drawing a box and whisker graph from a frequency table

Example 10

Draw a box and whisker graph for these data:

Number	1	2	3	4	5
Frequency	3	4	6	5	2

Enter the data in lists called 'number' and 'freq' (see Example 6).
Add a new Data & Statistics page to your document.
Note: You do not need to worry about what this screen shows.

Press menu 2:Plot Properties | 5:Add X Variable with Frequency
Press enter
You will see this dialog box.

From the drop-down menus, choose 'number' for the Data List and 'freq' for the Frequency List.
Press enter

You should now see a frequency histogram.

▶ Continued on next page

Press menu 1:Plot Type | 2:Box Plot Press enter You should now see a box plot (box and whisker graph) for the data in the table.	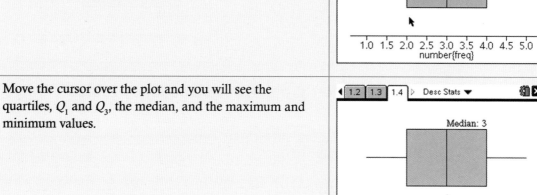
Move the cursor over the plot and you will see the quartiles, Q_1 and Q_3, the median, and the maximum and minimum values.	

Calculating statistics

You can calculate statistics such as mean, median, etc. from a list, or from a frequency table.

> Mean, median, range, quartiles, standard deviation, etc. are called **summary statistics**.

2.7 Calculating statistics from a list

Example 11

Calculate the summary statistics for these data: 1, 1, 3, 9, 2

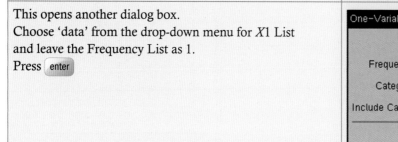

Enter the data in a list called 'data' (see Example 5).
Add a new Calculator page to your document.
Press menu 6:Statistics \| 1:Stat Calculations \| 1:One-Var Statistics...
Press enter
This opens a dialog box.
Leave the number of lists as 1 and press enter .

This opens another dialog box.
Choose 'data' from the drop-down menu for $X1$ List and leave the Frequency List as 1.
Press enter

▶ Continued on next page

The information shown will not fit on a single screen. You can scroll up and down to see it all. The statistics calculated for the data are:

mean	$\bar{x}$
sum	Σx
sum of squares	Σx^2
sample standard deviation	s_x
population standard deviation	σ_x
number	n
minimum value	$\mathrm{Min}X$
lower quartile	$Q_1 X$
median	$\mathrm{Median}X$
upper quartile	$Q_3 X$
maximum value	$\mathrm{Max}X$
sum of squared deviations from the mean	$\mathrm{SS}X$

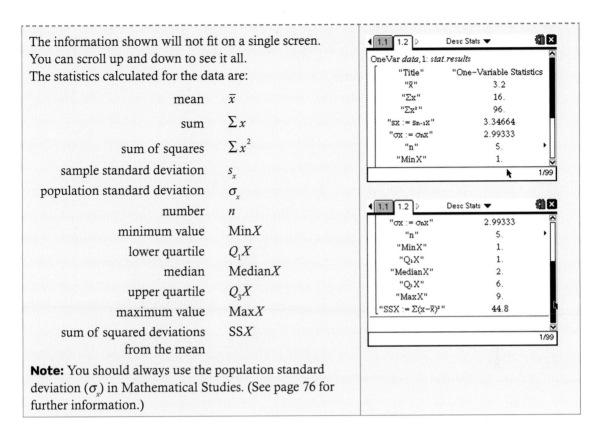

Note: You should always use the population standard deviation (σ_x) in Mathematical Studies. (See page 76 for further information.)

2.8 Calculating statistics from a frequency table

Example 12

Calculate the summary statistics for these data:

Number	1	2	3	4	5
Frequency	3	4	6	5	2

Enter the data in lists called 'number' and 'freq' (see Example 6).
Add a new Calculator page to your document.
Press menu 6:Statistics | 1:Stat Calculations | 1:One-Var Statistics…
Press enter
This opens a dialog box.
Leave the number of lists as 1 and press enter .

This opens another dialog box.
From the drop-down menus, choose 'number' for X1 List and 'freq' for the Frequency List.
Press enter

▶ Continued on next page

The information shown will not fit on a single screen. You can scroll up and down to see it all. The statistics calculated for the data are:

mean	$\bar{x}$
sum	Σx
sum of squares	Σx^2
sample standard deviation	s_x
population standard deviation	σ_x
number	n
minimum value	$\mathrm{Min}X$
lower quartile	Q_1X
median	$\mathrm{Median}X$
upper quartile	Q_3X
maximum value	$\mathrm{Max}X$
sum of squared deviations from the mean	$\mathrm{SS}X$

Note: You should always use the population standard deviation (σ_x) in Mathematical Studies. (See page 76 for further information.)

2.9 Calculating the interquartile range

Example 13

The interquartile range is the difference between the upper and lower quartiles ($Q_3 - Q_1$).

Calculate the interquartile range for these data:

Number	1	2	3	4	5
Frequency	3	4	6	5	2

First calculate the summary statistics for these data (see Example 12).
Add a new Calculator page to your document.

The values of the summary statistics are stored after One-Variable Statistics have been calculated and remain stored until the next time they are calculated.

Press [var]
A dialog box will appear with the names of the statistical variables.
Scroll down to stat.q_3x using the touchpad, or the ▼ ▲ keys, and then press [enter].

▶ Continued on next page

Type $(-)$ and press `var` again. Scroll down to stat.q_1x using the touchpad, or the ▼ ▲ keys, and then press `enter`.	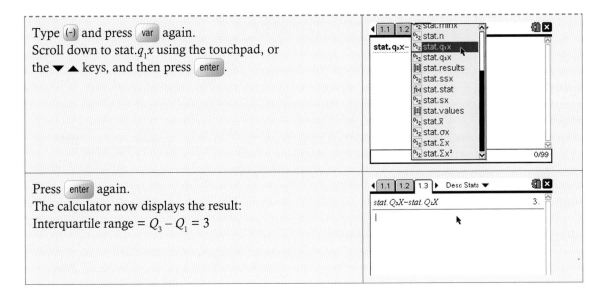
Press `enter` again. The calculator now displays the result: Interquartile range = $Q_3 - Q_1 = 3$	

2.10 Using statistics

Example 14

Calculate $\bar{x} + \sigma_x$ for these data: 	Number	1	2	3	4	5
---	---	---	---	---	---
Frequency	3	4	6	5	2		The calculator stores the values you calculate in One-Variable Statistics so that you can access them in other calculations. The values are stored until you do another One-Variable Statistics calculation.

First calculate the summary statistics for these data (see Example 12). Add a new Calculator page to your document. Press `var` A dialog box will appear with the names of the statistical variables. Scroll down to stat. $\bar{x}$ using the touchpad, or the ▼ ▲ keys, and then press `enter`.	
Type $(+)$ and press `var` again. Scroll down to stat.σx using the touchpad, or the ▼ ▲ keys, and then press `enter`.	
Press `enter` again. The calculator now displays the result: $\bar{x} + \sigma_x = 4.15$ (to 3 sf)	

3 Geometry and trigonometry 1

3.1 Graphing linear functions

Example 15

Draw the graph of the function $y = 2x + 1$

Open a new document and add a Graphs page.
The entry line is displayed at the bottom of the work area.
The default graph type is Function,
so the form '$f1(x)=$' is displayed.
The default axes are $-10 \leq x \leq 10$ and
$-6.67 \leq y \leq 6.67$.
Type $2x + 1$ and press enter .

The graph of $y = 2x + 1$ is now displayed and labeled
on the screen.

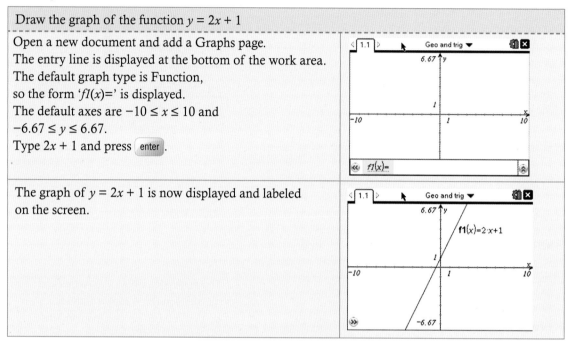

Finding information about the graph

Your GDC can give you a lot of information about the graph of a
function, such as the coordinates of points of interest and the
gradient (slope).

3.2 Finding a zero

The x-intercept is known as a **zero** of the function.

> At the x-intercept,
> $y = 0$.

Example 16

Find the zero of $y = 2x + 1$

First draw the graph of $y = 2x + 1$ (see Example 15).

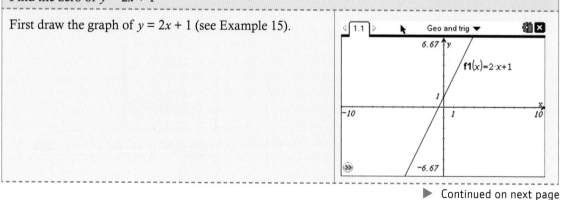

▶ Continued on next page

Press menu 6:Analyze Graph | 1:Zero Press enter To find the zero, you need to give the lower and upper bounds of a region that includes the zero. The GDC shows a line and asks you to set the lower bound. Move the line using the touchpad and choose a position to the left of the zero. Click the touchpad.	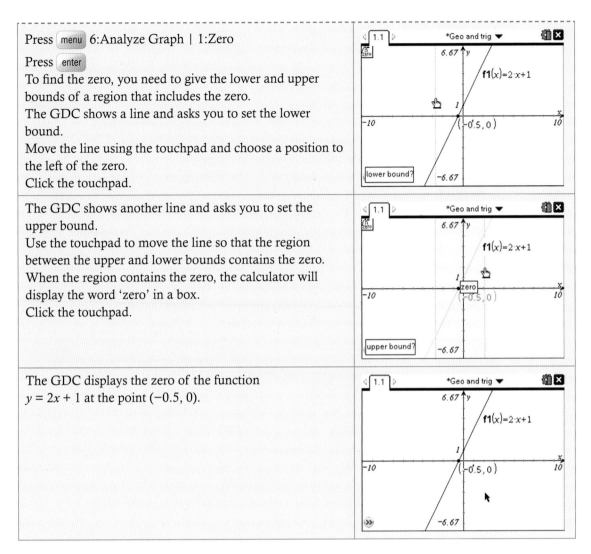
The GDC shows another line and asks you to set the upper bound. Use the touchpad to move the line so that the region between the upper and lower bounds contains the zero. When the region contains the zero, the calculator will display the word 'zero' in a box. Click the touchpad.	
The GDC displays the zero of the function $y = 2x + 1$ at the point $(-0.5, 0)$.	

3.3 Finding the gradient (slope) of a line

The correct mathematical notation for gradient (slope) is $\dfrac{dy}{dx}$, and this is how the GDC denotes gradient.

Example 17

Find the gradient of $y = 2x + 1$	
First draw the graph of $y = 2x + 1$ (see Example 15).	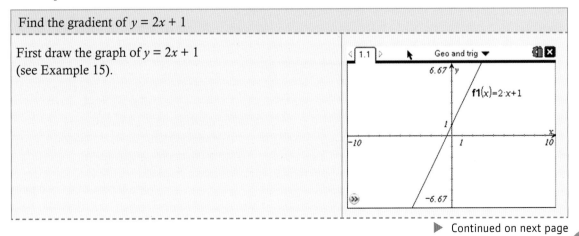

▶ Continued on next page

Press [menu] 6:Analyze Graph | 5:dy/dx Press [enter] Use the touchpad to select a point on the line. Click the touchpad.	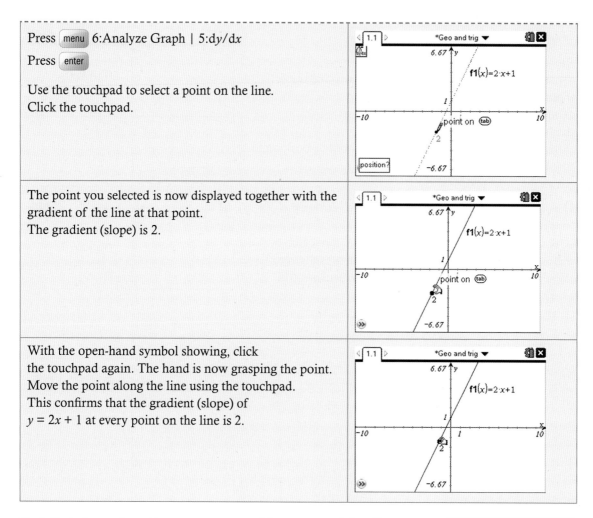
The point you selected is now displayed together with the gradient of the line at that point. The gradient (slope) is 2.	
With the open-hand symbol showing, click the touchpad again. The hand is now grasping the point. Move the point along the line using the touchpad. This confirms that the gradient (slope) of $y = 2x + 1$ at every point on the line is 2.	

3.4 Solving simultaneous equations graphically

To solve simultaneous equations graphically you draw the straight lines and then find their point of intersection. The coordinates of the point of intersection give you the solutions x and y.

> For solving simultaneous equations using a non-graphical method, see Section 1.1 of this chapter.

Example 18

Use a graphical method to solve the simultaneous equations
$2x + y = 10$
  $x - y = 2$

First rewrite both equations in the form '$y =$'.

$2x + y = 10$ $\qquad\qquad$ $x - y = 2$

$\quad y = 10 - 2x$ $\qquad\qquad$ $-y = 2 - x$

$\qquad\qquad\qquad\qquad\qquad$ $y = x - 2$

> The GDC will only draw the graphs of functions that are expressed explicitly as y = a function of x. If the equations are in a different form, rearrange them first.

▶ Continued on next page

To draw the graphs $y = 10 - 2x$ and $y = x - 2$:
Open a new document and add a Graphs page.
The entry line is displayed at the bottom of the work area.
The default graph type is Function, so the form '$f1(x)=$' is displayed.
The default axes are $-10 \le x \le 10$ and $-6.67 \le y \le 6.67$.
Type $10 - 2x$ and press enter.

The GDC displays the first straight-line graph:
$f1(x) = 10 - 2x$

Use the touchpad to click on the arrows in the bottom left-hand corner of the screen.
This will open the entry line again. This time '$f2(x)=$' is displayed.
Type $x - 2$ and press enter.

The GDC now displays both straight-line graphs:
$f1(x) = 10 - 2x$
$f2(x) = x - 2$

Press menu 6:Analyze Graph | 4:Intersection Point(s)

Press enter

To find the intersection you need to give the lower and upper bounds of a region that includes the intersection.
The GDC shows a line and asks you to set the lower bound. Move the line using the touchpad and choose a position to the left of the intersection.
Click the touchpad.

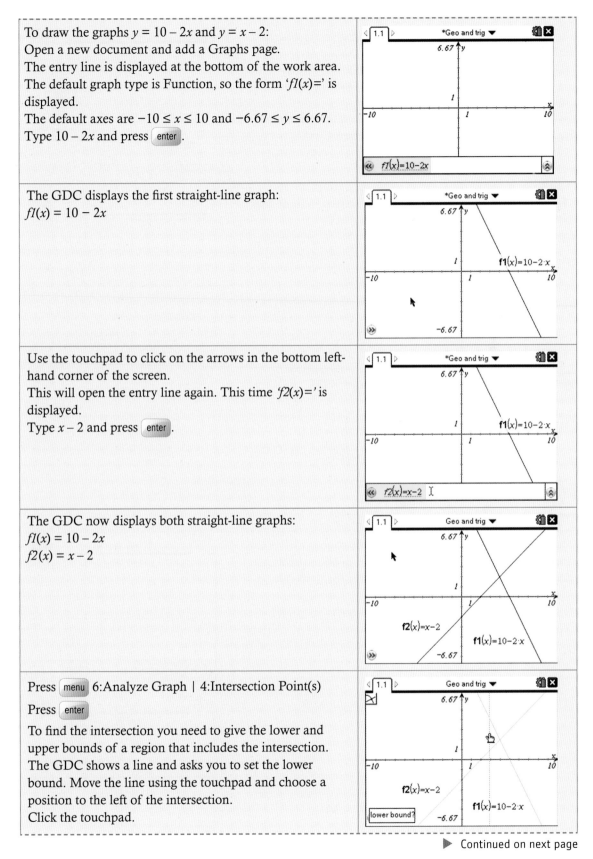

▶ Continued on next page

The GDC shows another line and asks you to set the upper bound.

Use the touchpad to move the line so that the region between the upper and lower bounds contains the intersection.

When the region contains the intersection, the calculator will display the word 'intersection' in a box.

Click the touchpad.

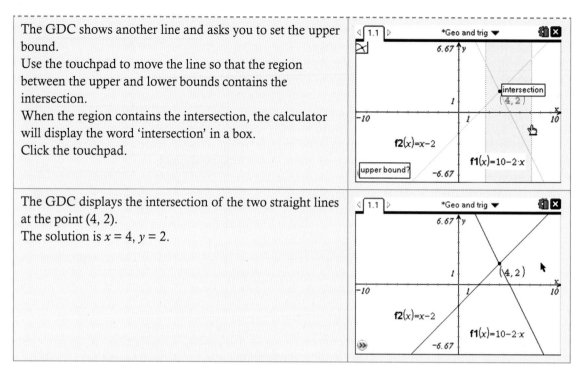

The GDC displays the intersection of the two straight lines at the point (4, 2).

The solution is $x = 4$, $y = 2$.

4 Mathematical models

Quadratic functions

4.1 Drawing a quadratic graph

Example 19

Draw the graph of $y = x^2 - 2x + 3$ and display using suitable axes.

Open a new document and add a Graphs page.

The entry line is displayed at the bottom of the work area.

The default graph type is Function, so the form '$f1(x) =$' is displayed.

The default axes are $-10 \le x \le 10$ and $-6.67 \le y \le 6.67$.

Type $x^2 - 2x + 3$ and press ⟨enter⟩.

The GDC displays the curve with the default axes.

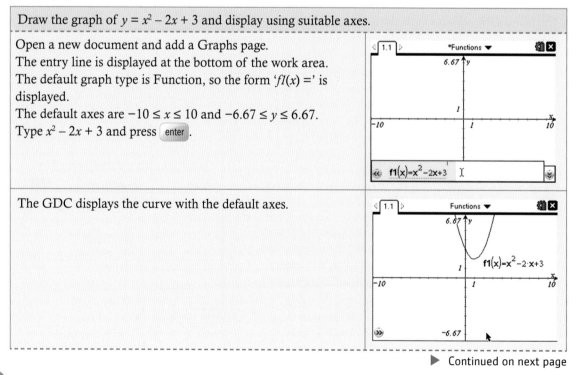

▶ Continued on next page

Pan the axes to get a better view of the curve.

> For help with panning,
> see your GDC manual.

Grab the x-axis and change it to make the quadratic curve fit the screen better.

> For help with changing axes, see your GDC manual.

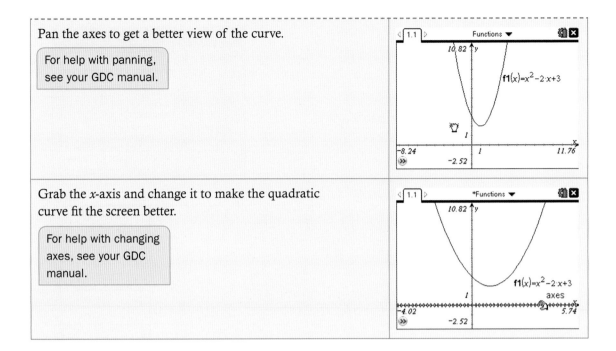

4.2 Finding a local minimum or maximum point

Example 20

Find the minimum point on the graph of $y = x^2 - 2x + 3$

First draw the graph of $y = x^2 - 2x + 3$ (see Example 19).

Method 1: Using a table
You can look at the graph **and** a table of the values by using a split screen.
Press menu 2:View | 9:Show Table
(or simply press ctrl T).
The minimum value shown in the table is 2 when $x = 1$.

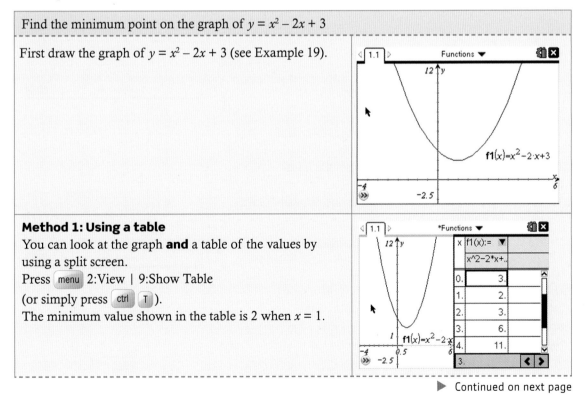

▶ Continued on next page

Look more closely at the values of the function around $x = 1$. Change the settings in the table. Choose any cell and press menu 5:Table | 5:Edit Table Settings… Set Table Start to 0.98 and Table Step to 0.01. Press enter	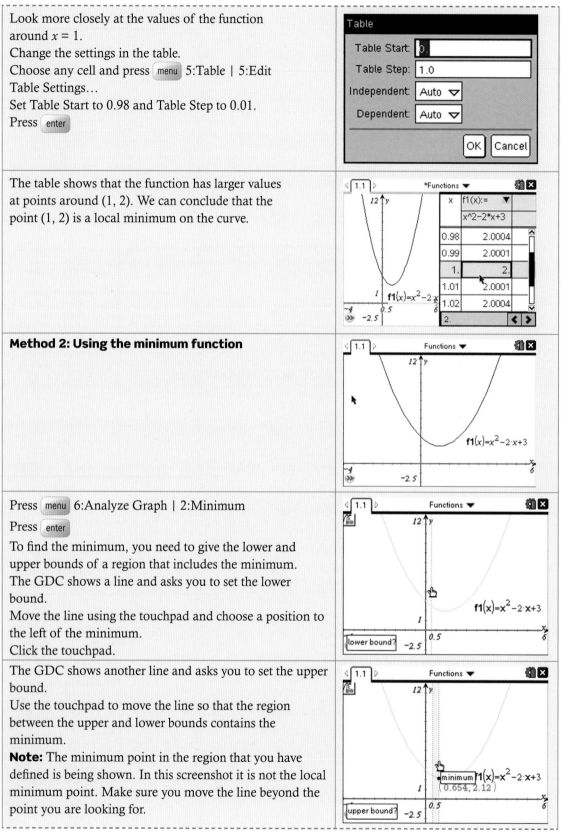
The table shows that the function has larger values at points around (1, 2). We can conclude that the point (1, 2) is a local minimum on the curve.	
Method 2: Using the minimum function	
Press menu 6:Analyze Graph | 2:Minimum Press enter To find the minimum, you need to give the lower and upper bounds of a region that includes the minimum. The GDC shows a line and asks you to set the lower bound. Move the line using the touchpad and choose a position to the left of the minimum. Click the touchpad.	
The GDC shows another line and asks you to set the upper bound. Use the touchpad to move the line so that the region between the upper and lower bounds contains the minimum. **Note:** The minimum point in the region that you have defined is being shown. In this screenshot it is not the local minimum point. Make sure you move the line beyond the point you are looking for.	

▶ Continued on next page

| When the region contains the minimum, the GDC will display the word 'minimum' in a box and a point that lies between the lower and upper bounds. The point displayed is clearly between the upper and lower bounds.
Click the touchpad. | 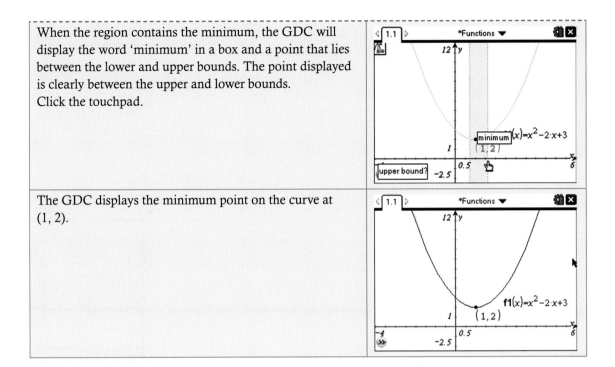 |
| The GDC displays the minimum point on the curve at $(1, 2)$. | |

Example 21

Find the maximum point on the graph of $y = -x^2 + 3x - 4$	
First draw the graph of $y = -x^2 + 3x - 4$: Open a new document and add a Graphs page. The entry line is displayed at the bottom of the work area. The default graph type is Function, so the form '$f1(x)=$' is displayed. The default axes are $-10 \le x \le 10$ and $-6.67 \le y \le 6.67$. Type $-x^2 + 3x - 4$ and press [enter].	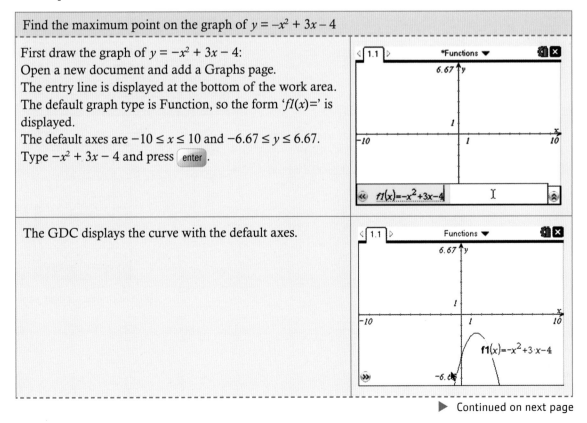
The GDC displays the curve with the default axes.	

▶ Continued on next page

Pan the axes to get a better view of the curve.
Grab the x-axis and change it to make the quadratic curve fit the screen better.

> For help with panning or changing axes, see your GDC manual.

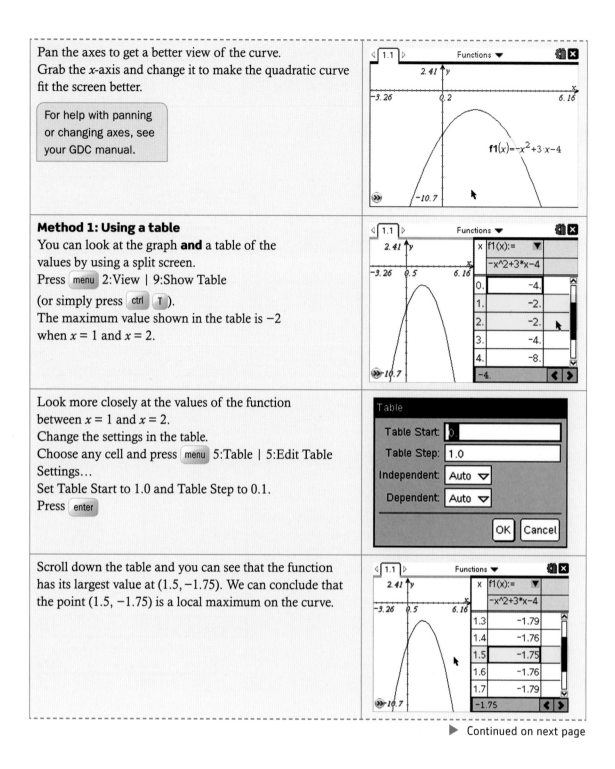

Method 1: Using a table

You can look at the graph **and** a table of the values by using a split screen.
Press menu 2:View | 9:Show Table
(or simply press ctrl T).
The maximum value shown in the table is -2 when $x = 1$ and $x = 2$.

Look more closely at the values of the function between $x = 1$ and $x = 2$.
Change the settings in the table.
Choose any cell and press menu 5:Table | 5:Edit Table Settings…
Set Table Start to 1.0 and Table Step to 0.1.
Press enter

Scroll down the table and you can see that the function has its largest value at $(1.5, -1.75)$. We can conclude that the point $(1.5, -1.75)$ is a local maximum on the curve.

▶ Continued on next page

Method 2: Using the maximum function

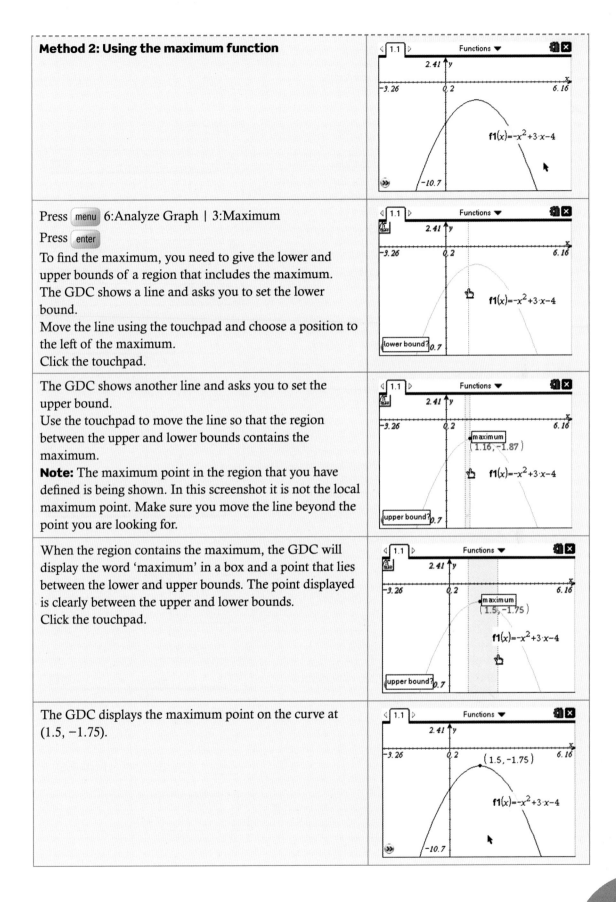

Press menu 6:Analyze Graph | 3:Maximum

Press enter

To find the maximum, you need to give the lower and upper bounds of a region that includes the maximum. The GDC shows a line and asks you to set the lower bound.

Move the line using the touchpad and choose a position to the left of the maximum.

Click the touchpad.

The GDC shows another line and asks you to set the upper bound.

Use the touchpad to move the line so that the region between the upper and lower bounds contains the maximum.

Note: The maximum point in the region that you have defined is being shown. In this screenshot it is not the local maximum point. Make sure you move the line beyond the point you are looking for.

When the region contains the maximum, the GDC will display the word 'maximum' in a box and a point that lies between the lower and upper bounds. The point displayed is clearly between the upper and lower bounds.

Click the touchpad.

The GDC displays the maximum point on the curve at $(1.5, -1.75)$.

Exponential functions

4.3 Drawing an exponential graph

Example 22

Draw the graph of $y = 3^x + 2$

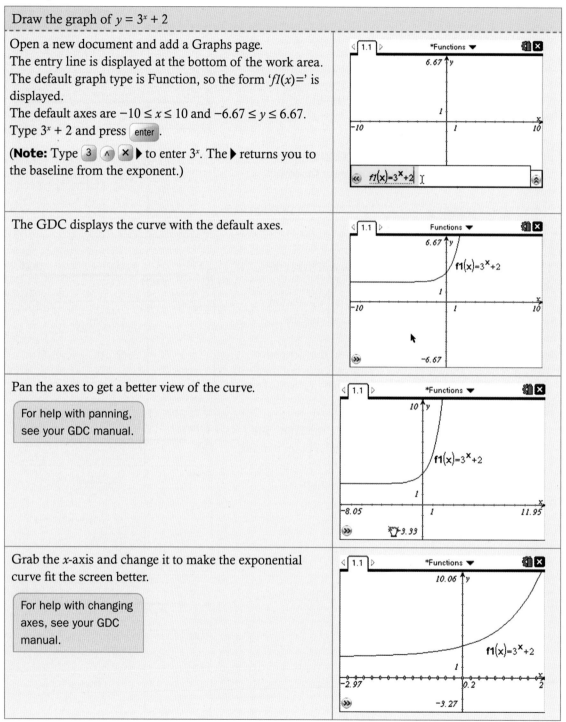

Open a new document and add a Graphs page. The entry line is displayed at the bottom of the work area. The default graph type is Function, so the form '$f1(x)=$' is displayed. The default axes are $-10 \leq x \leq 10$ and $-6.67 \leq y \leq 6.67$. Type $3^x + 2$ and press **enter**. (**Note:** Type ③ ^ ✕ ▶ to enter 3^x. The ▶ returns you to the baseline from the exponent.)	
The GDC displays the curve with the default axes.	
Pan the axes to get a better view of the curve. For help with panning, see your GDC manual.	
Grab the x-axis and change it to make the exponential curve fit the screen better. For help with changing axes, see your GDC manual.	

4.4 Finding a horizontal asymptote

Example 23

Find the horizontal asymptote to the graph of $y = 3^x + 2$

First draw the graph of $y = 3^x + 2$ (see Example 22).

You can look at the graph **and** a table of the values by using a split screen.

Press menu 2:View | 9:Show Table

(or simply press ctrl T).

The values of the function are clearly decreasing as $x \to 0$.

Press and hold ▲ to scroll up the table.
The table shows that, as the values of x get smaller, $f1(x)$ approaches 2.

Eventually, the value of $f1(x)$ reaches 2. On closer inspection, you can see, at the bottom of the screen, that the actual value of $f1(x)$ is $2.000\,001\,881\,6...$
We can say that $f1(x) \to 2$ as $x \to -\infty$.
The line $x = 2$ is a horizontal asymptote to the curve $y = 3^x + 2$.

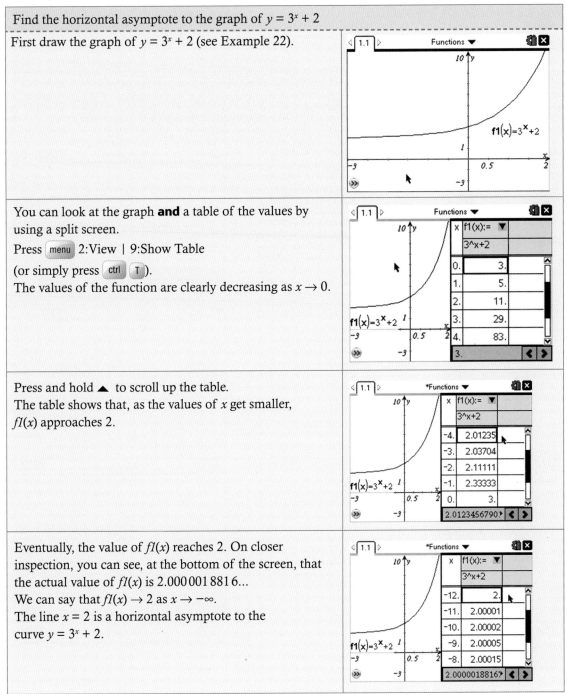

More complicated functions

4.5 Solving a combined quadratic and exponential equation

> Follow the same GDC procedure here as you followed when solving simultaneous equations graphically (see Example 18).

Example 24

Solve the equation $x^2 - 2x + 3 = 3 \times 2^{-x} + 4$

> To solve the equation, find the point of intersection of the quadratic function $f1(x) = x^2 - 2x + 3$ with the exponential function $f2(x) = 3 \times 2^{-x} + 4$.

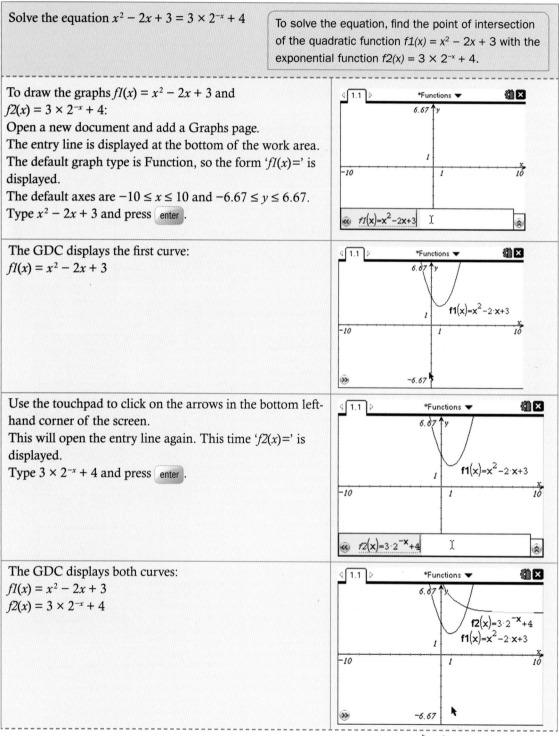

To draw the graphs $f1(x) = x^2 - 2x + 3$ and $f2(x) = 3 \times 2^{-x} + 4$:
Open a new document and add a Graphs page.
The entry line is displayed at the bottom of the work area.
The default graph type is Function, so the form '$f1(x)=$' is displayed.
The default axes are $-10 \leq x \leq 10$ and $-6.67 \leq y \leq 6.67$.
Type $x^2 - 2x + 3$ and press [enter].

The GDC displays the first curve:
$f1(x) = x^2 - 2x + 3$

Use the touchpad to click on the arrows in the bottom left-hand corner of the screen.
This will open the entry line again. This time '$f2(x)=$' is displayed.
Type $3 \times 2^{-x} + 4$ and press [enter].

The GDC displays both curves:
$f1(x) = x^2 - 2x + 3$
$f2(x) = 3 \times 2^{-x} + 4$

► Continued on next page

Pan the axes to get a better view of the curves. For help with panning, see your GDC manual.	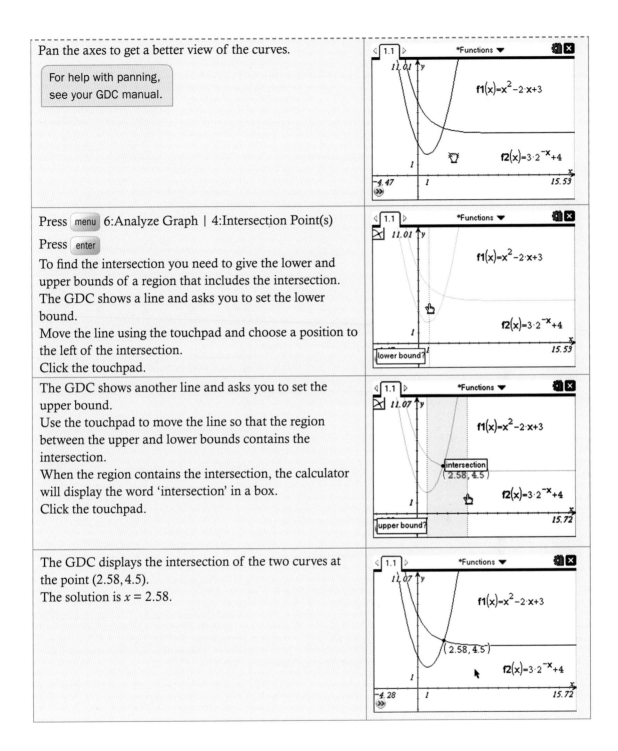
Press menu 6:Analyze Graph | 4:Intersection Point(s) Press enter To find the intersection you need to give the lower and upper bounds of a region that includes the intersection. The GDC shows a line and asks you to set the lower bound. Move the line using the touchpad and choose a position to the left of the intersection. Click the touchpad.	
The GDC shows another line and asks you to set the upper bound. Use the touchpad to move the line so that the region between the upper and lower bounds contains the intersection. When the region contains the intersection, the calculator will display the word 'intersection' in a box. Click the touchpad.	
The GDC displays the intersection of the two curves at the point $(2.58, 4.5)$. The solution is $x = 2.58$.	

Fitting a model to data

You can find a function to model data by transforming a curve or by using sliders.

You can also model a linear function by finding the equation of the least squares regression line (see Section 5.3 of this chapter).

4.6 Using transformations to model a quadratic function

Example 25

These data are approximately connected by a quadratic function.

x	−2	−1	0	1	2	3	4
y	9.1	0.2	−4.8	−5.9	−3.1	4.0	15.0

Find a function that fits the data.

Transform a basic quadratic curve to find an equation to fit some quadratic data.

Open a new document and add a Lists & Spreadsheet page.
Enter the data in two lists:
Type '*x*' in the first cell and '*y*' in the cell to its right.
Enter the *x*-values in the first column and the *y*-values in the second. Remember to use ⁽⁻⁾ to enter a negative number.
Use the ▼ ▲ ◀▶ keys to navigate around the spreadsheet.

Add a Graphs page to your document.
Press menu 3:Graph Type | 4:Scatter Plot
Press enter
The entry line is displayed at the bottom of the work area.
Scatter plot type is displayed.
Enter the names of the lists, *x* and *y*, into the scatter plot function.
Use the tab key to move from *x* to *y*.
Press enter

Press menu A:Zoom – Fit from the Window/Zoom menu
This is a quick way to choose an appropriate scale to show all the points.
You should recognize that the points are in the shape of a quadratic function.

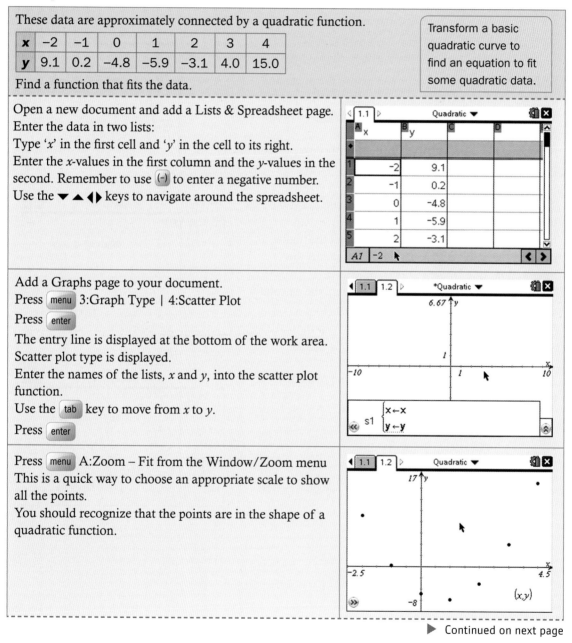

▶ Continued on next page

	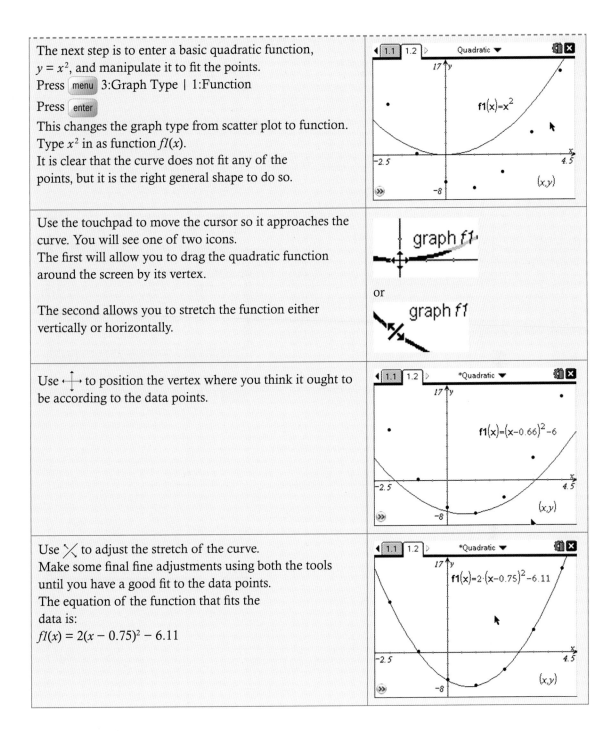
The next step is to enter a basic quadratic function, $y = x^2$, and manipulate it to fit the points. Press menu 3:Graph Type | 1:Function Press enter This changes the graph type from scatter plot to function. Type x^2 in as function $f1(x)$. It is clear that the curve does not fit any of the points, but it is the right general shape to do so.	
Use the touchpad to move the cursor so it approaches the curve. You will see one of two icons. The first will allow you to drag the quadratic function around the screen by its vertex. The second allows you to stretch the function either vertically or horizontally.	
Use ⟵⟶ to position the vertex where you think it ought to be according to the data points.	
Use ✕ to adjust the stretch of the curve. Make some final fine adjustments using both the tools until you have a good fit to the data points. The equation of the function that fits the data is: $f1(x) = 2(x - 0.75)^2 - 6.11$	

4.7 Using sliders to model an exponential function

Example 26

In general, an exponential function has the form $y = ka^x + c$.
For these data, it is known that the value of a is 1.5, so $y = k(1.5)^x + c$.

x	−3	−2	−1	0	1	2	3	4	5	6	7	8
y	3.1	3.2	3.3	3.5	3.8	4.1	4.7	5.5	6.8	8.7	11.5	15.8

Find the values of the constants k and c.

Open a new document and add a Lists & Spreadsheet page. Enter the data in two lists: Type 'x' in the first cell and 'y' in the cell to its right. Enter the x-values in the first column and the y-values in the second. Remember to use (-) to enter a negative number. Use the ▼ ▲ ◀▶ keys to navigate around the spreadsheet.	
Add a Graphs page to your document. Press menu 3:Graph Type | 4:Scatter Plot Press enter The entry line is displayed at the bottom of the work area. Scatter plot type is displayed. Enter the names of the lists, x and y, into the scatter plot function. Use the tab key to move from x to y. Press enter	
Adjust the window settings to fit the data and to display the axes clearly.	

▶ Continued on next page

Press menu I:Actions | A:Insert Slider Position the slider somewhere where it is not in the way and change the name of the constant to k. Repeat and add a second slider for c. For help with sliders, see your GDC manual.	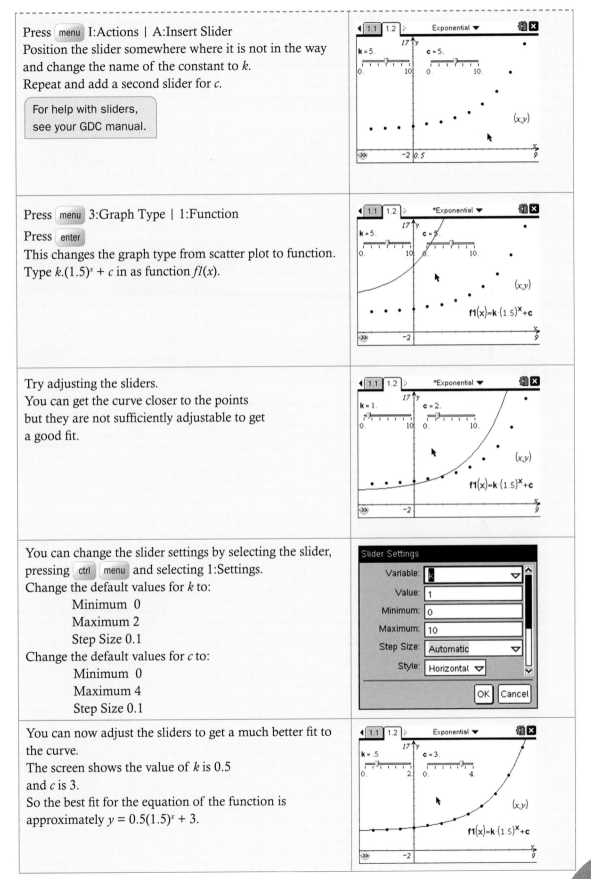
Press menu 3:Graph Type | 1:Function Press enter This changes the graph type from scatter plot to function. Type $k.(1.5)^x + c$ in as function $f1(x)$.	
Try adjusting the sliders. You can get the curve closer to the points but they are not sufficiently adjustable to get a good fit.	

You can change the slider settings by selecting the slider,
pressing ctrl menu and selecting 1:Settings.
Change the default values for k to:
    Minimum 0
    Maximum 2
    Step Size 0.1
Change the default values for c to:
    Minimum 0
    Maximum 4
    Step Size 0.1	

You can now adjust the sliders to get a much better fit to the curve. The screen shows the value of k is 0.5 and c is 3. So the best fit for the equation of the function is approximately $y = 0.5(1.5)^x + 3$.	

5 Statistical applications

Calculating normal probabilities

5.1 Calculating normal probabilities from X-values

Example 27

A random variable X is normally distributed with a mean of 195 and a standard deviation of 20, or $X \sim N(195, 20^2)$. Calculate
a   the probability that X is less than 190
b   the probability that X is greater than 194
c   the probability that X lies between 187 and 196.

Open a new document and add a Calculator page.
Press menu 5:Probability | 5:Distributions | 2:Normal Cdf

Press enter

You need to enter the values Lower Bound, Upper Bound, μ and σ in the dialog box.
For the Lower Bound, enter -9×10^{999} as $-9\text{E}999$. This is the smallest number that can be entered in the GDC, so it is used in place of $-\infty$. To enter the E, you need to press the key marked E E.

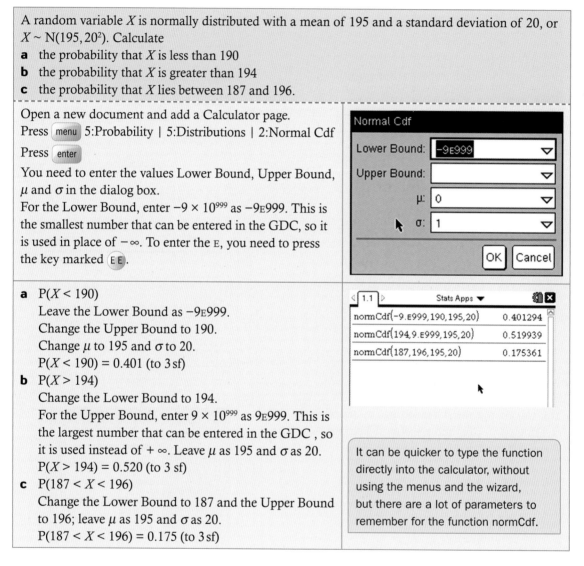

a $P(X < 190)$
    Leave the Lower Bound as $-9\text{E}999$.
    Change the Upper Bound to 190.
    Change μ to 195 and σ to 20.
    $P(X < 190) = 0.401$ (to 3 sf)
b $P(X > 194)$
    Change the Lower Bound to 194.
    For the Upper Bound, enter 9×10^{999} as $9\text{E}999$. This is the largest number that can be entered in the GDC , so it is used instead of $+\infty$. Leave μ as 195 and σ as 20.
    $P(X > 194) = 0.520$ (to 3 sf)
c $P(187 < X < 196)$
    Change the Lower Bound to 187 and the Upper Bound to 196; leave μ as 195 and σ as 20.
    $P(187 < X < 196) = 0.175$ (to 3 sf)

It can be quicker to type the function directly into the calculator, without using the menus and the wizard, but there are a lot of parameters to remember for the function normCdf.

5.2 Calculating *X*-values from normal probabilities

In some problems you are given the probabilities and have to calculate the associated values of *X*. To do this, use the invNorm function.

> When using the inverse normal function (invNorm), make sure that you find the probability on the correct side of the normal curve. The areas are always the lower tail, that is, they are of the form P(*X* < *x*) (see Example 28).

> If you are given the upper tail, P(*X* > *x*), you must first subtract the probability from 1 to before you can use invNorm (see Example 29).

Example 28

A random variable *X* is normally distributed with a mean of 75 and a standard deviation of 12, or $X \sim N(75, 12^2)$. If P(*X* < *x*) = 0.4, find the value of *x*.

> You are given a **lower**-tail probability, so you can find P(*X* < *x*) directly.

Open a new document and add a Calculator page.
Press menu 5:Probability | 5:Distributions | 3:Inverse Normal...
Press enter
Enter the probability (area = 0.4), mean (μ = 75) and standard deviation (σ = 12) in the dialog box.

> It can be quicker to type the function directly into the calculator, without using the menus and the wizard, but there are a lot of parameters to remember for the function invNorm.

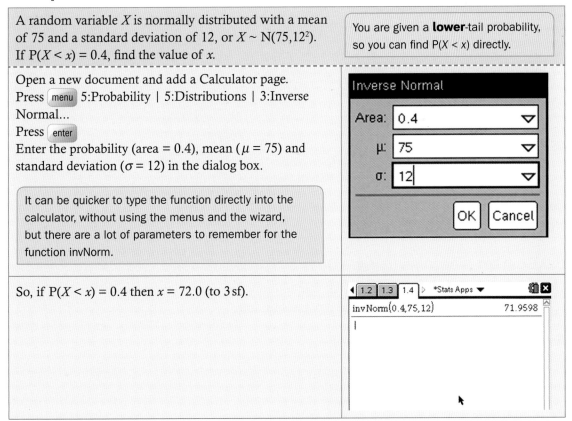

So, if P(*X* < *x*) = 0.4 then *x* = 72.0 (to 3 sf).

Example 29

A random variable X is normally distributed with a mean of 75 and a standard deviation of 12, or $X \sim N(75, 12^2)$.
If $P(X > x) = 0.2$, find the value of x.

You are given an **upper**-tail probability, so you must first find $P(X < x) = 1 - 0.2 = 0.8$. You can now use the invNorm function as before.

Open a new document and add a Calculator page.
Press menu 5:Probability | 5:Distributions | 3:Inverse Normal...
Press enter
Enter the probability (area = 0.8), mean ($\mu = 75$) and standard deviation ($\sigma = 12$) in the dialog box.

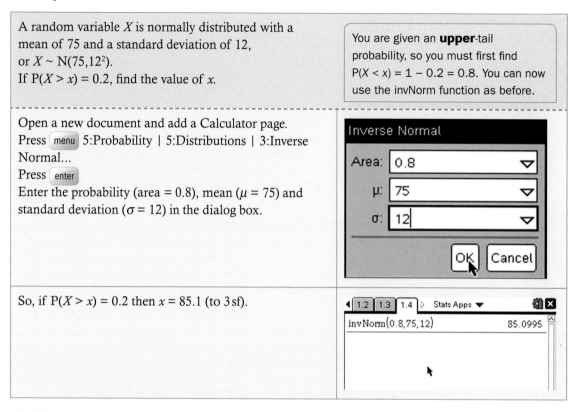

So, if $P(X > x) = 0.2$ then $x = 85.1$ (to 3 sf).

This sketch of a normal distribution curve shows the value of x and the probabilities for Example 29.

Scatter diagrams, linear regression and the correlation coefficient

For Pearson's product–moment correlation coefficient, see Section 5.4 of this chapter.

5.3 Scatter diagrams using a Data & Statistics page

Using a Data & Statistics page is a quick way to draw scatter graphs and find the equation of a regression line.

Example 30

These data are approximately connected by a linear function.

x	1.0	2.1	2.4	3.7	5.0
y	4.0	5.6	9.8	10.6	14.7

Find the equation of the least squares regression line for y on x.
Use the equation to predict the value of y when $x = 3.0$.

▶ Continued on next page

Open a new document and add a Lists & Spreadsheet page. Enter the data in two lists: Type '*x*' in the first cell and '*y*' in the cell to its right. Enter the *x*-values in the first column and the *y*-values in the second. Use the ▼ ▲ ◀▶ keys to navigate around the spreadsheet.	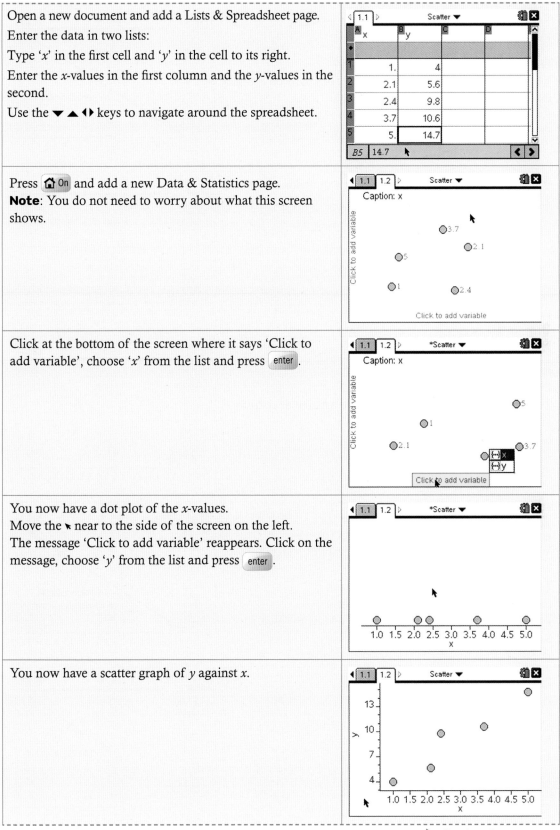
Press ⌂ On and add a new Data & Statistics page. **Note**: You do not need to worry about what this screen shows.	
Click at the bottom of the screen where it says 'Click to add variable', choose '*x*' from the list and press enter.	
You now have a dot plot of the *x*-values. Move the ⭦ near to the side of the screen on the left. The message 'Click to add variable' reappears. Click on the message, choose '*y*' from the list and press enter.	
You now have a scatter graph of *y* against *x*.	

▶ Continued on next page

Press <kbd>menu</kbd> 4:Analyze | 6:Regression | 1:Show
Linear ($mx + b$)
Press <kbd>enter</kbd>
You will see the least squares regression line for y on x and
its equation:
$y = 2.6282x + 1.47591$

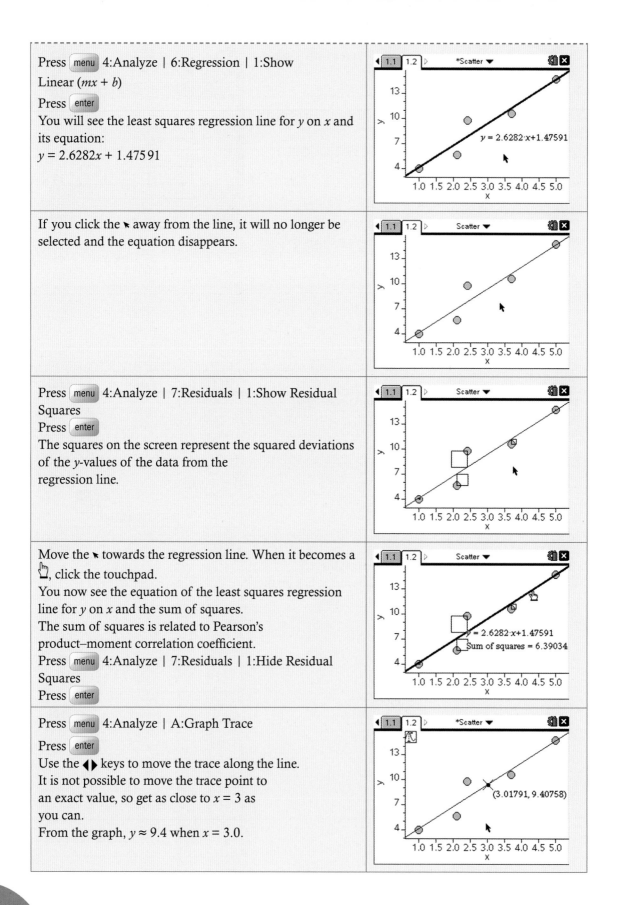

If you click the ☚ away from the line, it will no longer be
selected and the equation disappears.

Press <kbd>menu</kbd> 4:Analyze | 7:Residuals | 1:Show Residual
Squares
Press <kbd>enter</kbd>
The squares on the screen represent the squared deviations
of the y-values of the data from the
regression line.

Move the ☚ towards the regression line. When it becomes a
☝, click the touchpad.
You now see the equation of the least squares regression
line for y on x and the sum of squares.
The sum of squares is related to Pearson's
product–moment correlation coefficient.
Press <kbd>menu</kbd> 4:Analyze | 7:Residuals | 1:Hide Residual
Squares
Press <kbd>enter</kbd>

Press <kbd>menu</kbd> 4:Analyze | A:Graph Trace
Press <kbd>enter</kbd>
Use the ◀▶ keys to move the trace along the line.
It is not possible to move the trace point to
an exact value, so get as close to $x = 3$ as
you can.
From the graph, $y \approx 9.4$ when $x = 3.0$.

5.4 Scatter diagrams using a Graphs page

Using a Graphs page takes a little longer than the Data & Statistics page, but you will get more detailed information about the data such as Pearson's product–moment correlation coefficient.

Example 31

These data are approximately connected by a linear function.

x	1.0	2.1	2.4	3.7	5.0
y	4.0	5.6	9.8	10.6	14.7

These are the same data as in Example 30.

a Find the equation of the least squares regression line for y on x.
b Find Pearson's product–moment correlation coefficient.
c Predict the value of y when $x = 3.0$.

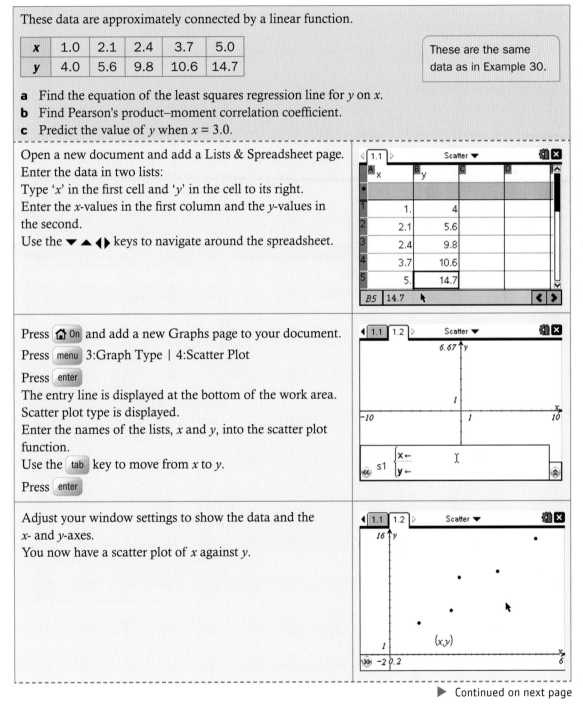

Open a new document and add a Lists & Spreadsheet page.
Enter the data in two lists:
Type 'x' in the first cell and 'y' in the cell to its right.
Enter the x-values in the first column and the y-values in the second.
Use the ▼ ▲ ◀▶ keys to navigate around the spreadsheet.

Press 🏠 On and add a new Graphs page to your document.
Press menu 3:Graph Type | 4:Scatter Plot
Press enter
The entry line is displayed at the bottom of the work area.
Scatter plot type is displayed.
Enter the names of the lists, x and y, into the scatter plot function.
Use the tab key to move from x to y.
Press enter

Adjust your window settings to show the data and the x- and y-axes.
You now have a scatter plot of x against y.

▶ Continued on next page

Press `ctrl` `◀` to return to the Lists & Spreadsheet page.

Press `menu` 4:Statistics | 1:Stat Calculations | 3:Linear Regression (*mx* + *b*)

Press `enter`

From the drop-down menus, choose '*x*' for X List and '*y*' for Y List. You should press `tab` to move between the fields.

Press `enter`

On the screen, you will see the result of the linear regression in lists next to the lists for x and y.

The values of m (2.6282) and b (1.47591) are shown separately.

a The equation of the least squares regression line for y on x is $y = 2.6282x + 1.47591$.

Scroll down the table to see the value of Pearson's product–moment correlation coefficient, given by r.

b Pearson's product–moment correlation coefficient, $r = 0.954741$.

Press `ctrl` `▶` to return to the Graphs page.

Using the touchpad, click on the entry line at the bottom of the work area.

You will see that the equation of the regression line has been pasted into $f1(x)$.

Press `enter`

The regression line is now shown on the graph.

Use the trace function `menu` 5:Trace | 1:Graph Trace to find the point where x is 3.0.

Using the `▶` `◀` keys, move the trace point close, then edit the x-coordinate and change it to exactly 3.0.

c When $x = 3.0$, $y = 9.36$.

The χ^2 test for independence

5.5 Using contingency tables

Example 32

A survey of the favorite color for a mobile phone produced the following data.

	Black	Red	Blue	Silver
Male	48	35	33	54
Female	35	66	42	27

Perform a χ^2 test, at the 5% significance level, to determine whether the choice of color is independent of gender.

> You need to enter the data from a contingency table into a matrix. The GDC then calculates the expected frequencies, the χ^2 value, the number of degrees of freedom and the p-value.

Open a new document and add a Calculator page.
Press menu 7:Matrices & Vectors | 1:Create | 1:Matrix
Press enter
Using tab to move around the dialog box, enter 2 rows and 4 columns.

Enter the data from the table in the matrix.
Use tab to move from one cell to the next.
When you have entered all the figures press ctrl sto→.
Enter a name for the matrix, e.g. 'observe'.
Press enter
This matrix is now the 'observed frequencies' for the χ^2 calculations.

Press menu 6:Statistics | 7:Statistical Tests... | 8: χ^2 2-way Test...
Press enter
From the drop-down menu, choose 'observe' for the Observed Matrix.
Press enter

The results screen shows:
$\chi^2 = 21.631$
p-value (PVal) = 0.000078
number of degrees of freedom (df) = 3

Since $0.000078 < 0.05$ (p-value < significance level), we would reject the null hypothesis. That is, there is enough evidence to conclude that choice of color is dependent on gender.

Continued on next page

Chapter 12   507

When you have finished, you should always check the table of expected frequencies to ensure that at least 80% of the values are greater than 5. These values are in the ExpMatrix variable but this is not displayed directly.
Press `var`
Using the ▼ key, scroll down to stat.expmatrix and press `enter`.

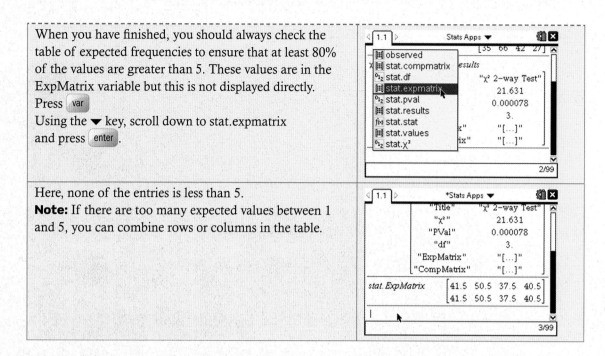

Here, none of the entries is less than 5.
Note: If there are too many expected values between 1 and 5, you can combine rows or columns in the table.

6 Introductory differential calculus

Finding gradients, tangents and maximum and minimum points

6.1 Finding the gradient at a point

Example 33

Find the gradient of the cubic function $y = x^3 - 2x^2 - 6x + 5$

Open a new document and add a Graphs page.
The entry line is displayed at the bottom of the work area. The default graph type is Function, so the form '$f1(x)=$' is displayed.
The default axes are $-10 \leq x \leq 10$ and $-6.67 \leq y \leq 6.67$.
Type $x^3 - 2x^2 - 6x + 5$ and press `enter`.
(**Note:** Type `x` `∧` `3` ▶ to enter x^3. The ▶ returns you to the baseline from the exponent.)

The GDC displays the curve.
Pan the axes to get a better view of the curve and then grab the x- and y-axes to fit the curve to the window.

> For help with panning and changing axes, see your GDC manual.

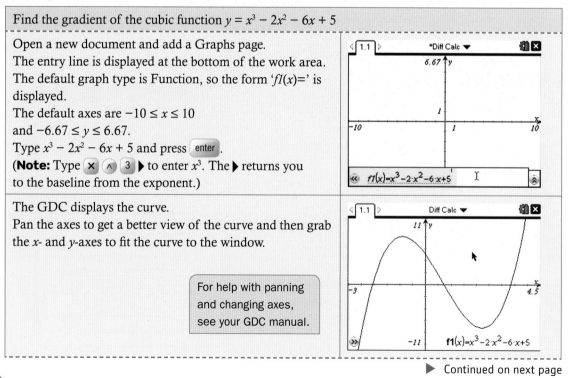

▶ Continued on next page

Press menu 6:Analyze Graph | 5: dy/dx

Press enter

Using the touchpad, move the 🖑 towards the curve. As

it approaches the curve, it turns to 🖉 and displays the
numerical value of the gradient.

Press enter to attach a point on the curve.

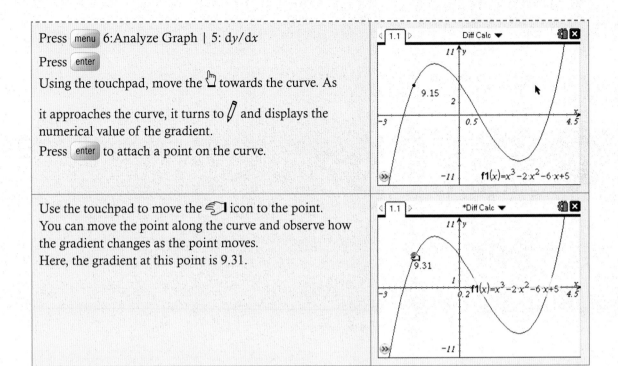

Use the touchpad to move the 🖐 icon to the point.
You can move the point along the curve and observe how
the gradient changes as the point moves.
Here, the gradient at this point is 9.31.

6.2 Drawing a tangent to a curve

Example 34

Draw a tangent to the curve $y = x^3 - 2x^2 - 6x + 5$

First draw the graph of $y = x^3 - 2x^2 - 6x + 5$
(see Example 33).

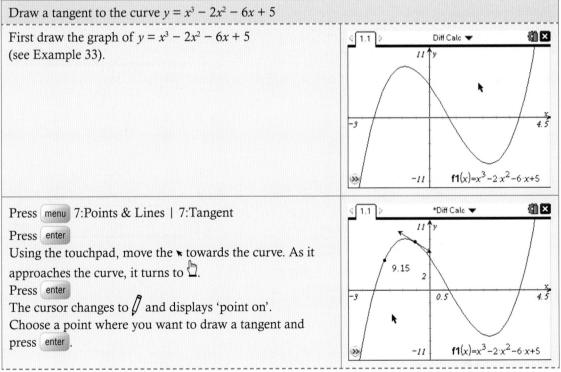

Press menu 7:Points & Lines | 7:Tangent

Press enter

Using the touchpad, move the ↖ towards the curve. As it
approaches the curve, it turns to 🖑.

Press enter

The cursor changes to 🖉 and displays 'point on'.

Choose a point where you want to draw a tangent and
press enter .

▶ Continued on next page

You can move the point that the tangent line is attached to with the touchpad.	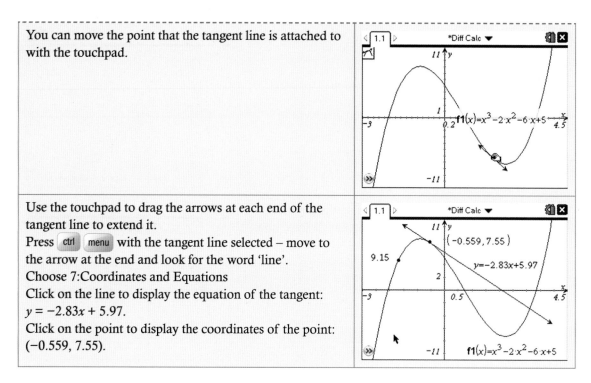
Use the touchpad to drag the arrows at each end of the tangent line to extend it. Press ctrl menu with the tangent line selected – move to the arrow at the end and look for the word 'line'. Choose 7:Coordinates and Equations Click on the line to display the equation of the tangent: $y = -2.83x + 5.97$. Click on the point to display the coordinates of the point: $(-0.559, 7.55)$.	

6.3 Finding maximum and minimum points

Example 35

Find the local maximum and local minimum points on the cubic curve:
$y = x^3 - 2x^2 - 6x + 5$

First draw the graph of $y = x^3 - 2x^2 - 6x + 5$ (see Example 33).	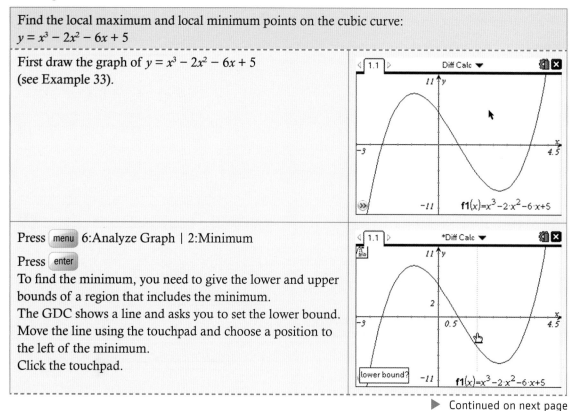
Press menu 6:Analyze Graph | 2:Minimum Press enter To find the minimum, you need to give the lower and upper bounds of a region that includes the minimum. The GDC shows a line and asks you to set the lower bound. Move the line using the touchpad and choose a position to the left of the minimum. Click the touchpad.	

▶ Continued on next page

The GDC shows another line and asks you to set the upper bound.

Use the touchpad to move the line so that the region between the upper and lower bounds contains the minimum.

Note: The minimum point in the region that you have defined is being shown. In this screenshot it is not the local minimum point. Make sure you move the line beyond the point you are looking for.

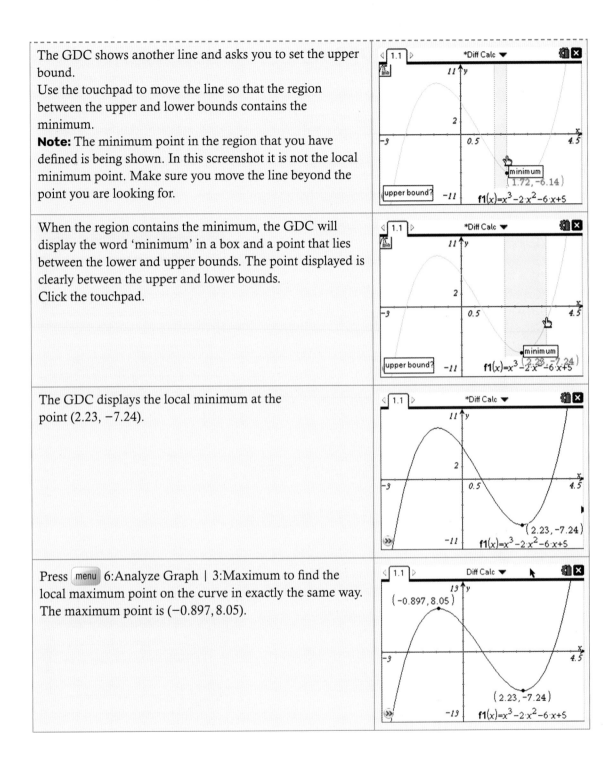

When the region contains the minimum, the GDC will display the word 'minimum' in a box and a point that lies between the lower and upper bounds. The point displayed is clearly between the upper and lower bounds.

Click the touchpad.

The GDC displays the local minimum at the point $(2.23, -7.24)$.

Press menu 6:Analyze Graph | 3:Maximum to find the local maximum point on the curve in exactly the same way. The maximum point is $(-0.897, 8.05)$.

7 Number and algebra 2

The Finance Solver

The Finance Solver will solve problems involving simple loans, mortgages, and investments.

> In general, in financial problems, a negative monetary amount indicates an amount you give to the bank and a positive monetary amount indicates an amount you receive from the bank. This can be a little confusing.

Example 36

Entering data in the Finance Solver

Open a new document and add a Calculator page.
Press menu Finance | Finance Solver

Press enter

You will see this dialog box, where

N:	the total number of payments
I(%):	the annual interest rate as a percent
PV:	the present value, which is negative for investments
Pmt:	the payment or regular deposit, which is negative for investments
FV:	the future value
PpY:	the payments per year
CpY:	the interest calculations period per year
PmtAt:	payments made at the end or beginning of each period

7.1 Finding the total value of an investment

Example 37

$1500 is invested at a rate of 5.25% per annum. The interest is compounded two times per year. How much will the investment be worth at the end of 6 years?

Open a new document and add a Calculator page.
Press menu Finance | Finance Solver

Press enter

Using the tab key to move around the dialog box, enter:

N:	6
I(%):	5.25
PV:	−1500
Pmt:	0
FV:	Leave blank: to be calculated
PpY:	1
CpY:	2
PmtAt:	END

> The present value (PV) is negative as the investment is paid to the bank.

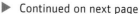

▶ Continued on next page

Select FV and press enter . The final amount is $2047.05.	**Finance Solver** PV: -1500 Pmt: 0 **FV:** 2047.0540063921 PpY: 1 CpY: 2 PmtAt: END Finance Solver info stored into tvm.n, tvm.i, tvm.pv, tvm.pmt, …
As described on page 315, you can also calculate this using the compound interest formula: $1500\left(1+\dfrac{5.25}{200}\right)^{12}$	1.1 Num alg 2 ▼ $1500\cdot\left(1+\dfrac{5.25}{200}\right)^{12}$ 2047.05

7.2 Calculating payments for a loan

Example 38

> Calculate the monthly payment required to repay a 4-year loan of $12 000 that is compounded monthly at an annual interest rate of 4.25%. Payments are made at the end of each month.

Open a new document and add a Calculator page. Press menu Finance | Finance Solver Press enter Using the tab key to move around the dialog box, enter: N: 48 I(%): 4.25 PV: 12 000 Pmt: Leave blank: to be calculated FV: 0 PpY: 12 CpY: 12 PmtAt: END	The repayments are made each month so the total number of payments, N, is 4 years × 12 = 48. **Finance Solver** N: 48 I(%): 4.25 PV: 12000 Pmt: 0 FV: 0. PpY: 12 Press ENTER to calculate Number of Payments, N
Select Pmt and press enter . The monthly payments will be $272.29.	**Finance Solver** PV: 12000 **Pmt:** -272.29317885746 FV: 0. PpY: 12 CpY: 12 PmtAt: END Finance Solver info stored into tvm.n, tvm.i, tvm.pv, tvm.pmt, …

The answer, Pmt, is negative because it is a payment made to the bank.

13 Prior learning

CHAPTER OBJECTIVES:

This chapter contains short explanations, examples and practice exercises on topics that you should know before starting the course. You do not need to work through the whole of this chapter in one go. You can dip into it for help when you need it.

For example, before you start Chapter 2 Descriptive Statistics, work through Section 4 Statistics in this chapter.

Chapter contents

1 Number

1.1 Calculation

There are several versions of the rules for the order of operations. They all amount to the same thing:

- Brackets or parentheses are calculated first.
- Next come exponents, indices or orders.
- Then multiplication and division, in order from left to right.
- Additions and subtractions, in order from left to right.

A fraction line or the line above a square root counts as a bracket too.

Your GDC follows the rules, so if you enter a calculation correctly you should get the correct answers.

BEDMAS:	Brackets, exponents, division, multiplication, addition, subtraction.
BIDMAS:	Brackets, indices, division, multiplication, addition, subtraction.
BEMDAS:	Brackets, exponents, multiplication, division, addition, subtraction.
BODMAS:	Brackets, orders, division, multiplication, addition, subtraction.
BOMDAS:	Brackets, orders, multiplication, division, addition, subtraction.
PEMDAS:	Parentheses, exponents, multiplication, division, addition, subtraction.

1.1 Prior Learning	
$2 \cdot 3^2$	18
$\dfrac{18}{2 \cdot 3}$	3
$\sqrt{16+9}$	5
	3/99

GDC help on CD: *Alternative demonstrations for the TI-84 Plus and Casio FX-9860GII GDCs are on the CD.*

Simple calculators, like the ones on phones, do not always follow the calculation rules.

The GDC shows divisions as fractions, which makes the order of operations clearer.

Example 1

a Evaluate $\dfrac{11 + (-1)^2}{4 - (3 - 5)}$

$$= \frac{11 + 1}{4 - (-2)}$$ *Brackets first*

$$= \frac{12}{6}$$ *Simplify numerator and denominator.*

$$= 2$$

b Evaluate $\dfrac{-3 + \sqrt{9 - 8}}{4}$

$$= \frac{-3 + \sqrt{1}}{4}$$ *Simplify the terms inside the square root.*
    Evaluate the root.

$$= \frac{-3 + 1}{4}$$

$$= \frac{-2}{4}$$ *Simplify the numerator and denominator.*

$$= -\frac{1}{2}$$

▶ Continued on next page

On your GDC you can either use templates for the fractions and roots or you can use brackets.

1.1 ▷	Presumed K...dge ▼
$\dfrac{6\cdot2}{4}$	3
$4-2+6$	8
$3\cdot2+4\cdot5$	26
$5-\dfrac{8}{4}$	3

4/99

GDC help on CD: *Alternative demonstrations for the TI-84 Plus and Casio FX-9860GII GDCs are on the CD.*

Exercise 1A

Do the questions by hand first, then check your answers with your GDC.

1 Calculate

a $12 - 5 + 4$ **b** $6 \div 3 \times 5$ **c** $4 + 2 \times 3 - 2$

d $8 - 6 \div 3 \times 2$ **e** $4 + (3 - 2)$ **f** $(7 + 2) \div 3$

g $(1 + 4) \times (8 - 4)$ **h** $1 - 3 + 5 \times (2 - 1)$

2 Find

a $\dfrac{6 + 9}{4 - 1}$ **b** $\dfrac{2 \times 9}{3 \times 4}$ **c** $\dfrac{2 - (3 + 4)}{4 \times (2 - 3)}$ **d** $\dfrac{6 \times 5 \times 4}{3 \times 2 - 1}$

3 Determine

a $3 \times (-2)^2$ **b** $2^2 \times 3^3 \times 5$ **c** $4 \times (5 - 3)^2$ **d** $(-3)^2 - 2^2$

4 Calculate

a $\sqrt{3^2 + 4^2}$ **b** $\left(\sqrt{4}\right)^3$ **c** $\sqrt{4^3}$ **d** $\sqrt{2 + \sqrt{2 + 2}}$

5 Find

a $\sqrt{\dfrac{13^2 - \left(3^2 + 4^2\right)}{2 \times 18}}$ **b** $2\sqrt{\dfrac{3 + 5^2}{7}}$ **c** $2(3^2 - 4(-2)) - (2 - \sqrt{7 - 3})$

1.2 Primes, factors and multiples

A **prime** number is an integer, greater than 1, that is not a multiple of any other number apart from 1 and itself.

Example 2

List all the factors of 42.	
Answer	
$42 = 1 \times 42$, $42 = 2 \times 21$, $42 = 3 \times 14$, $42 = 6 \times 7$	*Write 42 as a product of two numbers every way you can.*
The factors of 42 are 1, 2, 3, 6, 7, 14, 21 and 42.	

In 2009, the largest known prime was a 12 978 189-digit number. Prime numbers have become big business because they are used in cryptography.

Example 3

Write the number 24 as a product of prime factors.	
Answer $\begin{array}{r} 2\overline{)24} \\ 2\overline{)12} \\ 2\overline{)6} \\ 3\overline{)3} \\ 1 \end{array}$ $\begin{aligned} 24 &= 2 \times 2 \times 2 \times 3 \\ &= 2^3 \times 3 \end{aligned}$	*Begin dividing by the smallest prime number. Repeat until you reach an answer of 1.*

Example 4

Find the **lowest common multiple** (LCM) of 12 and 15.	
Answer The multiples of 12 are 12, 24, 36, 48, 60, 72, 84, 96, 108, 120, 132, 144, ...	
The multiples of 15 are 15, 30, 45, 60, 75, 90, 105, 120, 135, ...	
The common multiples are 60, 120, ...	*List all the multiples until you find some in both lists. The LCM is the smallest number in each of the lists.*
The LCM is 60.	

Example 5

Find the **highest common factor** (HCF) of 36 and 54.
Answer $\begin{array}{r} 2\overline{)36} \\ 2\overline{)18} \\ 3\overline{)9} \\ 3\overline{)3} \\ 1 \end{array}$ $36 = 2 \times 2 \times 3 \times 3$ $\qquad$ $\begin{array}{r} 2\overline{)54} \\ 3\overline{)27} \\ 3\overline{)9} \\ 3\overline{)3} \\ 1 \end{array}$ $54 = 2 \times 3 \times 3 \times 3$
The HCF of 36 and 54 is $2 \times 3 \times 3 = 18$.

Write each number as a product of prime factors. Find the product of all the factors that are common to both numbers.

Exercise 1B

1 List all the factors of
   a 18   **b** 27   **c** 30   **d** 28   **e** 78

2 Write as products of prime factors.
   a 36   **b** 60   **c** 54   **d** 32   **e** 112

3 Find the LCM of
   a 8 and 20   **b** 6, 10 and 16

4 Find the HCF of
   a 56 and 48   **b** 36, 54 and 90

1.3 Fractions and decimals

There are two types of fraction:

- **common** fractions (often just called 'fractions') like

$$\frac{4}{5} \quad \text{numerator} \atop \text{denominator}$$

- **decimal** fractions (often just called 'decimals') like 0.125.

Fractions can be:

- **proper** like $\frac{2}{3}$ where the numerator is less than the denominator

- **improper** like $\frac{4}{3}$ where the numerator is greater than the denominator

- **mixed numbers** like $6\frac{7}{8}$.

Fractions where the numerator and denominator have no common factor are in their **lowest terms**.

$\frac{1}{3}$ and $\frac{4}{12}$ are **equivalent** fractions.

0.675 is a **terminating** decimal.

0.32... or $0.\overline{32}$ or $0.\dot{3}\dot{2}$ are different ways of writing the **recurring** decimal 0.323 232 323 2...

Non-terminating, non-recurring decimals are **irrational** numbers, like π or $\sqrt{2}$.

Using a GDC, you can either enter a fraction using the fraction template $\frac{\square}{\square}$ or by using the divide key ÷. Take care – you will sometimes need to use brackets.

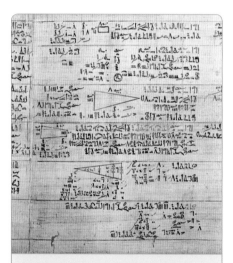

The Rhind Papyrus from ancient Egypt in around 1600 BCE shows calculations using fractions. Egyptians used **unit** fractions, so for $\frac{4}{5}$ they would write $\frac{1}{2} + \frac{1}{4} + \frac{1}{20}$. This is not generally regarded as a very helpful way of writing fractions.

$\pi \approx$ 3.14159265358979323846264 33832795028841971693 99375...
$\sqrt{2} \approx$ 1.41421356237309504880 16887242096980785696 7187537...
They do not terminate and there are no repeating patterns in the digits.

Example 6

a Evaluate

$$\frac{1}{2}+\frac{3}{8}\times\frac{4}{9}$$

$\times$ *before* $+$.

$$=\frac{1}{2}+\frac{1}{6}$$

$$=\frac{4}{6}$$

Simplify.

$$=\frac{2}{3}$$

b Evaluate

$$\frac{\frac{1}{2}+\frac{1}{3}}{\frac{1}{2}\times\frac{1}{3}}$$

Evaluate the numerator and denominator first.

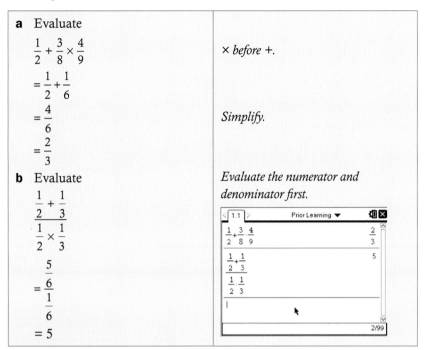

$$=\frac{\frac{5}{6}}{\frac{1}{6}}$$

$$=5$$

GDC help on CD: *Alternative demonstrations for the TI-84 Plus and Casio FX-9860GII GDCs are on the CD.*

Example 7

a Convert $\dfrac{7}{16}$ to a decimal. **b** Write $3\dfrac{7}{8}$ as an improper fraction.

Answers

a $\dfrac{7}{16}=0.4375$

b $3\dfrac{7}{8}=\dfrac{24}{8}+\dfrac{7}{8}$

$$=\dfrac{31}{8}$$

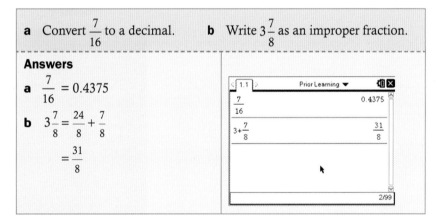

GDC help on CD: *Alternative demonstrations for the TI-84 Plus and Casio FX-9860GII GDCs are on the CD.*

 ### *Exercise 1C*

1 Calculate

a $\dfrac{1}{2}+\dfrac{3}{4}\times\dfrac{5}{9}$ **b** $\dfrac{2}{3}\div\dfrac{5}{6}\times1\dfrac{1}{3}$

c $\sqrt{\left(\dfrac{3}{5}\right)^2+\left(\dfrac{4}{5}\right)^2}$ **d** $\dfrac{1-\left(\dfrac{2}{3}\right)^5}{1-\dfrac{2}{3}}$

2 Write the following fractions in their lowest terms.

a $\dfrac{16}{36}$ **b** $\dfrac{35}{100}$ **c** $\dfrac{34}{51}$ **d** $\dfrac{125}{200}$

3 Write these mixed numbers as improper fractions.

a $3\dfrac{3}{5}$ **b** $3\dfrac{1}{7}$ **c** $23\dfrac{1}{4}$ **d** $2\dfrac{23}{72}$

4 Write these improper fractions as mixed numbers.

a $\dfrac{32}{7}$ **b** $\dfrac{100}{3}$ **c** $\dfrac{17}{4}$ **d** $\dfrac{162}{11}$

5 Convert to decimals.

a $\dfrac{8}{25}$ **b** $\dfrac{5}{7}$ **c** $3\dfrac{4}{5}$ **d** $\dfrac{45}{17}$

> There are some useful tools for working with fractions. Look in menu 2:Number.

> To convert a fraction to a decimal, divide the numerator by the denominator. Pressing ctrl ≈ will give the result as a decimal instead of a fraction.

1.4 Percentages

A percentage is a way of expressing a fraction or a ratio as part of a hundred.

For example 25% means 25 parts out of 100.

As a fraction, $25\% = \dfrac{25}{100} = \dfrac{1}{4}$.

As a decimal, $25\% = 0.25$.

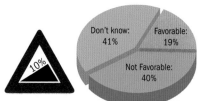

Example 8

Lara's mark in her Mathematics test was 25 out of 40. What was her mark as a percentage?	
Answer	
$\dfrac{25}{40} \times 100 = 62.5\%$	*Write the mark as a fraction.* *Multiply by 100.* *Use your GDC.*

Example 9

There are 80 students taking the IB in a school. 15% take Mathematical Studies. How many students is this?	
Answer	
Method 1	
$\dfrac{15}{100} \times 80 = 12$	*Write the percentage as a fraction out of a hundred and then multiply by 80.*
Method 2	
$15\% = 0.15$	*Write the percentage as a decimal.*
$0.15 \times 80 = 12$	*Multiply by 80.*

Exchange Rates

International currencies

Questions in the Mathematical Studies examination may use international currencies. For example: Swiss franc (CHF); US dollar (USD); British pound sterling (GBP); euro (EUR); Japanese yen (JPY) and Australian dollar (AUD).

Exercise 1D

1 Write as percentages
  a 13 students from a class of 25
  b 14 marks out of 20

2 Find the value of

  | 7% = 0.07 |

  a 7% of 32 CHF
  b $4\frac{1}{2}$% of 12.00 GBP
  c 25% of 750.28 EUR
  d 130% of 8000 JPY

Percentage increase and decrease

Consider an increase of 35%.

The new value after the increase will be 135% of the original value.

So, to increase an amount by 35%, find 135% of the amount.

Multiply by $\frac{135}{100}$ or 1.35.

Now consider a decrease of 15%.

After a 15% decrease, the new value will be 85% of the original. So to decrease an amount by 15% find 85%. Multiply by $\frac{85}{100}$ or by 0.85.

Example 10

a The manager of a shop increases the prices of CDs by 12%.
A CD originally cost 11.60 CHF.
What will it cost after the increase?

b The cost of a plane ticket is decreased by 8%.
The original price was 880 GBP. What is the new price?

c The rent for an apartment has increased from 2700 EUR to 3645 EUR per month.
What is the percentage increase?

> After a 12% increase, the amount will be 112% of its original value.

> After an 8% decrease, the amount will be 92% of its original value.

Answer

a 11.60 × 1.12 = 12.99 CHF
(to the nearest 0.01 CHF)

b 880 × 0.92 = 809.60 GBP

c **Method 1**

The increase is 3645 − 2700
= 945 EUR
Find the increase.

The percentage increase is $\frac{945}{2700} \times 100 = 35\%$
Work out the increase as a percentage of the original amount.

Method 2

$\frac{3645}{2700} = 1.35 = 135\%$
Percentage increase is 35%.
Calculate the new price as a percentage of the old price.

> Percentage increase = $\frac{\text{actual increase}}{\text{original amount}} \times 100\%$

Example 11

In a shop, an item's price is given as 44 AUD, **including** tax.
The tax rate is 10%.
What was the price without the tax?

Answer	
Call the original price x.	
After tax has been added, the price will be $1.10x$.	$110\% = 1.10$
Hence $1.10x = 44$	*Solve for x.*
$\qquad x = 44 \div 1.10$	*Divide both sides by 1.10.*
$\qquad\quad = 40$	
The price without tax is 40 AUD.	

Exercise 1E

1   In the UK, prices of some goods include a government tax called VAT, which is at 20%.
    A TV costs 480 GBP before VAT. How much will it cost including VAT?

2   In a sale in a shop in Tokyo, a dress that was priced at 17 000 JPY is reduced by 12.5%. What is the sale price?

3   The cost of a weekly train ticket goes up from 120 GBP to 128.40 GBP. What is the percentage increase?

4   Between 2004 and 2005, oil production in Australia fell from 731 000 to 537 500 barrels per day. What was the percentage decrease in the production?

5   Between 2005 and 2009 the population of Venezuela increased by 7%. The population was 28 400 000 in 2009. What was it in 2005 (to the nearest 100 000)?

6   An item appears in a sale marked as 15% off with a price tag of 27.20 USD. What was the original price before discount?

7   The rate of GST (goods and service tax) that is charged on items sold in shops was increased from 17% to 20%. What would the price increase be on an item that costs 20 GBP before tax?

8   A waiter mistakenly adds a 10% service tax onto the cost of a meal which was 50.00 AUD. He then reduces the price by 10%. Is the price now the same as originally? If not, what is the percentage change from the original price?

1.5 Ratio and proportion

The **ratio** of two numbers r and s is $r : s$. It is equivalent to the fraction $\frac{r}{s}$. Like the fraction, it can be written in its lowest terms.

For example, $6 : 12$ is equivalent to $1 : 2$ (dividing both numbers in the ratio by 6).

> When you write a ratio in its lowest terms, both numbers in the ratio should be positive whole numbers.

In a **unitary ratio**, one of the terms is 1.

For example $1 : 4.5$ or $25 : 1$.

If two quantities a and b are in **proportion**, then the ratio $a : b$ is fixed.

> When you write a unitary ratio, you can use decimals.

We also write $a \propto b$ (a is proportional to b).

Example 12

200 tickets were sold for a school dance. 75 were bought by boys and the rest by girls. Write down the ratio of boys to girls at the dance, in its lowest terms.

Answer
The number of girls is $200 - 75 = 125$
The ratio of boys to girls is $75 : 125 = 3 : 5$

> Always give the ratio in its lowest terms.

Map scales are often written as a ratio. A scale of $1 : 50\,000$ means that 1 cm on the map represents $50\,000$ cm $= 0.5$ km on the earth.

Example 13

An old English map was made to the scale of 1 inch to a mile. Write this scale as a ratio.

Answer

1 mile $= 1760 \times 3 \times 12$ $\quad = 63\,360$ inches The ratio of the map is $1 : 63\,360$.	*Always make sure that the units in ratios match each other.*

> 12 inches = 1 foot
> 3 feet = 1 yard
> 1760 yards = 1 mile

Example 14

Three children, aged 8, 12 and 15, win a prize of 140 USD. They decide to share the prize money in the ratio of their ages. How much does each receive?

Answer

140 USD is divided in the ratio $8 : 12 : 15$. This is a total of $8 + 12 + 15$ $= 35$ parts. $140 \div 35 = 4$ USD $8 \times 4 = 32$, $12 \times 4 = 48$ and $15 \times 4 = 60$ The children receive 32 USD, 48 USD and 60 USD.	*Divide the money into 35 parts.* *One part is 4 USD.*

Exercise 1F

1 Aspect ratio is the ratio of an image's width to its height. A photograph is 17.5 cm wide by 14 cm high. What is its aspect ratio, in its lowest terms?

2 Gender ratio is expressed as the ratio of men to women in the form $n : 100$. Based on the figures for 2008, the gender ratio of the world was $102 : 100$. In Japan, there were 62 million men and 65.2 million women in 2008. What was the gender ratio in Japan?

3 Ryoka was absent for a total of 21 days during a school year of 32 weeks. What is the ratio of the number of days that she was absent to the number of possible days she could have spent at the school during the year, in its simplest terms? (A school week is 5 days.)

4 A model airplane has a wingspan of 15.6 cm. The model is built to a scale of $1 : 72$. What is the wingspan of a full-sized airplane (in metres)?

5 On a map, a road measures 1.5 cm. The actual road is 3 km long. What is the scale of the map? How long would a footpath that is 800 m long be on the map?

6 A joint collection is made for two charities and it is agreed that the proceeds should be split in the ratio $5 : 3$ between an animal charity and one for sick children. 72 USD is collected. How much is donated to the two charities?

7 For a bake sale, a group of students decide to make brownies, chocolate chip cookies and flapjacks in the ratio $5 : 3 : 2$. They plan to make 150 items all together. How many of each will they need to make?

Leonardo da Vinci drew this famous drawing of Vitruvian Man around 1487. The drawing is based on ideal human proportions described by the ancient Roman architect Vitruvius.

1.6 The unitary method

In the unitary method, you begin by finding the value of **one** part or item.

Example 15

A wheelbarrow full of concrete is made by mixing together 6 spades of gravel, 4 spades of sand, 2 spades of cement and water as required. When there are only 3 spades of sand left, what quantities of the other ingredients will be required to make concrete?

▶ Continued on next page

Since the value you want to change is the sand, make sand equal to 1 by dividing through by 4. Then multiply through by 3 to make the quantity of sand equal to 3.

Exercise 1G

1 Josh, Jarrod and Se Jung invested 5000 USD, 7000 USD and 4000 USD to start up a company. In the first year, they make a profit of 24 000 USD which they share in the ratio of the money they invested. How much do they each receive?

2 Amy is taking a Mathematics test. She notices that there are three questions worth 12, 18 and 20 marks. The test lasts one hour and fifteen minutes. She decides to allocate the time she spends on each question in the ratio of the marks. How long does she spend on each question?

2 Algebra

The word **algebra** comes from the title of a book *Hisab al-jabr w'al-muqabala* written by Abu Ja'far Muhammad ibn Musa Al-Khwarizmi in Baghdad around 800 CE. It is regarded as the first book to be written about algebra.

2.1 Expanding brackets and factorization

The **distributive law** is used to expand brackets and factorize expressions.

$a(b + c) = ab + ac$

Two other laws used in algebra are the **commutative law** $ab = ba$ and the **associative law** $(ab)c = a(bc)$.

Example 16

Expand $2y(3x + 5y - z)$

Answer

$2y(3x + 5y - z) = 2y(3x) + 2y(5y) + 2y(-z)$
$\qquad\qquad\qquad = 6xy + 10y^2 - 2yz$

Example 17

Factorize $6x^2y - 9xy + 12xz^2$

Answer

$6x^2y - 9xy + 12xz^2 = 3x(2xy - 3y + 4z^2)$

> Look for a common factor. Write this outside the bracket. Find the terms inside the bracket by dividing each term by the common factor.

Exercise 2A

1 Expand

   a $3x(x - 2)$ **b** $\dfrac{x}{y}(x^2y - y^2 + x)$ **c** $a(b - 2c) + b(2a + b)$

2 Factorize

   a $3pq - 6p^2q^3r$ **b** $12ac^2 + 15bc - 3c^2$ **c** $2a^2bc + 3ab^2c - 5abc^2$

2.2 Formulae

Rearranging formulae

Example 18

The formula for the area of a circle is $A = \pi r^2$, where A is the area and r is the radius.
The subject of the formula is A.
Rearrange the formula to make r the subject.

> The subject of a formula is the letter on its own on one side of the = sign.

Answer

$A = \pi r^2$

$r^2 = \dfrac{A}{\pi}$

$r = \sqrt{\dfrac{A}{\pi}}$

Use the same techniques as for solving equations. Whatever you do to one side of the formula, you must do to the other.
Divide both sides by π.
Take the square root of both sides.

> You can use this formula to work out the radius of a circle when you know its area.

Exercise 2B

Rearrange these formulae to make the quantity shown in brackets the subject.

1 $v = u - gt$ (t) 2 $a = \sqrt{b^2 + c^2}$ (c) 3 $c = 2\pi r$ (r)

4 $\dfrac{\sin A}{a} = \dfrac{\sin B}{b}$ (b) 5 $a^2 = b^2 + c^2 - 2bc \cos A$ $(\cos A)$

Substituting into formulae

You can always use your GDC in Mathematical Studies.

When using formulae, let the calculator do the calculation for you. You should still show your working.

1. Find the formula you are going to use (from the Formula booklet, from the question or from memory) and write it down.
2. Identify the values that you are going to substitute into the formula.
3. Write out the formula with the values substituted for the letters.
4. Enter the formula into your calculator. Use templates to make the formula look the same on your GDC as it is on paper.
5. If you think it is necessary, use brackets. It is better to have too many brackets than too few!
6. Write down, with units if necessary, the result from your calculator (to the required accuracy).

Example 19

x and y are related by the formula $y = \dfrac{x^2 + 1}{2\sqrt{x+1}}$.
Find y when x is 3.1.

Answer	*Write the formula with 3.1 instead*
$y = \dfrac{3.1^2 + 1}{2\sqrt{3.1+1}}$	*of x.*
$= 2.62$ (to 3 sf)	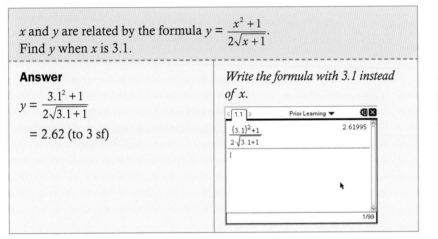

GDC help on CD: *Alternative demonstrations for the TI-84 Plus and Casio FX-9860GII GDCs are on the CD.*

Exercise 2C

1. If $a = 2.3$, $b = 4.1$ and $c = 1.7$, find d where
$$d = \frac{3a^2 + 2\sqrt{b}}{ac + b}$$

2. If $b = 8.2$, $c = 7.5$ and $A = 27°$, find a where
$$a = \sqrt{b^2 + c^2 - 2bc \cos A}$$

3. If $u_1 = 10.2$, $r = 0.75$ and $n = 14$, find the value of S, where
$$S = u_1 \frac{1 - r^n}{1 - r}$$

2.3 Solving linear equations

'Solve an equation' means 'find the value of the unknown variable' (the letter).

Rearrange the equation so that the unknown variable x becomes the subject of the equation. To keep the equation 'balanced' always do the same to both sides.

Example 20

Solve the equation $3x + 5 = 17$

Answer	
$3x + 5 = 17$	
$3x + 5 - 5 = 17 - 5$	*Subtract 5.*
$3x = 12$	
$\dfrac{3x}{3} = \dfrac{12}{3}$	*Divide by 3.*
$x = 4$	

Add, subtract, multiply or divide on both sides of the equation until the x is by itself on one side. (This can be either the left or the right-hand side.)

Example 21

Solve the equation $4(x - 5) = 8$

Answer	
$4(x - 5) = 8$	
$\dfrac{4(x - 5)}{4} = \dfrac{8}{4}$	*Divide by 4.*
$x - 5 = 2$	
$x - 5 + 5 = 2 + 5$	*Add 5.*
$x = 7$	

Always take care with − signs.

Example 22

Solve the equation $7 - 3x = 1$

Answer	
$7 - 3x = 1$	
$7 - 3x - 7 = 1 - 7$	*Subtract 7.*
$-3x = -6$	
$\dfrac{-3x}{-3} = \dfrac{-6}{-3}$	*Divide by −3.*
$x = 2$	

An alternative method for this equation would be to start by **adding** 3x. Then the x would be positive, but on the right-hand side.

Example 23

Solve the equation $3(2 + 3x) = 5(4 - x)$

Answer	
$3(2 + 3x) = 5(4 - x)$	
$6 + 9x = 20 - 5x$	
$6 + 9x + 5x = 20 - 5x + 5x$	*Add 5x.*
$6 + 14x = 20$	
$6 + 14x - 6 = 20 - 6$	*Subtract 6.*
$14x = 14$	
$\dfrac{14x}{14} = \dfrac{14}{14}$	*Divide by 14.*
$x = 1$	

Compare this method with the one in Example 21. Sometimes it can be quicker to **divide** first rather than expanding the brackets.

Exercise 2D

Solve these equations.

1 $3x - 10 = 2$

2 $\dfrac{x}{2} + 5 = 7$

3 $5x + 4 = -11$

4 $3(x + 3) = 18$

5 $4(2x - 5) = 20$

6 $\dfrac{2}{5}(3x - 7) = 8$

7 $21 - 6x = 9$

8 $12 = 2 - 5x$

9 $2(11 - 3x) = 4$

10 $4(3 + x) = 3(9 - 2x)$

11 $2(10 - 2x) = 4(3x + 1)$

12 $\dfrac{5x + 2}{3} = \dfrac{3x + 10}{4}$

2.4 Simultaneous linear equations

Simultaneous equations involve two variables.
There are two methods which you can use to solve them,
called substitution and elimination. You can also sometimes solve
them graphically.

Example 24

Solve the equations $3x + 4y = 17$ and $2x + 5y = 16$.

Answer
Graphical method

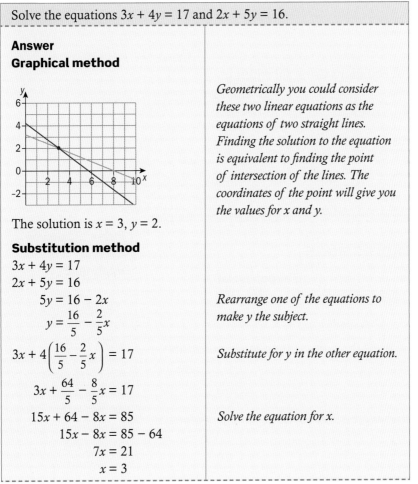

The solution is $x = 3$, $y = 2$.

Substitution method

$3x + 4y = 17$
$2x + 5y = 16$

$\qquad 5y = 16 - 2x$

$\qquad y = \dfrac{16}{5} - \dfrac{2}{5}x$

$3x + 4\left(\dfrac{16}{5} - \dfrac{2}{5}x\right) = 17$

$\qquad 3x + \dfrac{64}{5} - \dfrac{8}{5}x = 17$

$\qquad 15x + 64 - 8x = 85$

$\qquad\qquad 15x - 8x = 85 - 64$

$\qquad\qquad\qquad 7x = 21$

$\qquad\qquad\qquad x = 3$

Geometrically you could consider
these two linear equations as the
equations of two straight lines.
Finding the solution to the equation
is equivalent to finding the point
of intersection of the lines. The
coordinates of the point will give you
the values for x and y.

Rearrange one of the equations to
make y the subject.

Substitute for y in the other equation.

Solve the equation for x.

▶ Continued on next page

$3(3) + 4y = 17$ $9 + 4y = 17$ $4y = 8$ $y = 2$ The solution is $x = 3$, $y = 2$.	*Substitute for x in one of the original equations and solve for y.*
Elimination method $3x + 4y = 17$ (1) $2x + 5y = 16$ (2) Multiply equation (1) by 2 and equation (2) by 3.	*This is to make the coefficients of x equal.*
$6x + 8y = 34$ (3) $6x + 15y = 48$ (4) Subtract the equations. $[(4) - (3)]$ $7y = 14$ $y = 2$	*Subtracting now eliminates x from the equations.*
$3x + 4(2) = 17$ $3x + 8 = 17$ $3x = 17 - 8$ $3x = 9$ $x = 3$ The solution is $x = 3$, $y = 2$.	*Substitute for y in one of the original equations and solve for x.*

Exercise 2E

1 Solve these simultaneous equations using substitution.
    a $y = 3x - 2$ and $2x + 3y = 5$ **b** $4x - 3y = 10$ and $2y + 5 = x$
    c $2x + 5y = 14$ and $3x + 4y = 7$

2 Solve these simultaneous equations using elimination.
    a $2x - 3y = 15$ and $2x + 5y = 7$ **b** $3x + y = 5$ and $4x - y = 9$
    c $x + 4y = 6$ and $3x + 2y = -2$ **d** $3x + 2y = 8$ and $2x + 3y = 7$
    e $4x - 5y = 17$ and $3x + 2y = 7$

2.5 Exponential expressions

Repeated multiplication can be written as an **exponential** expression. For example, squaring a number:

> $3 \times 3 = 3^2$ or $5.42 \times 5.42 = 5.42^2$

If we multiply a number by itself three times then the exponential expression is a cube. For example

> $4.6 \times 4.6 \times 4.6 = 4.6^3$

You can also use exponential expressions for larger integer values. So, for example,

> $3^7 = 3 \times 3 \times 3 \times 3 \times 3 \times 3 \times 3$

Index and **power** are other names for **exponent**.

You use squares in Pythagoras' theorem $c^2 = a^2 + b^2$ or in the formula for the area of a circle $A = \pi r^2$.
You use a cube in the formula for the volume of a sphere $V = \dfrac{4}{3} \pi r^3$.

Where the exponent is not a positive integer, these rules apply:

$a^0 = 1, a \neq 0$

$a^{-n} = \dfrac{1}{a^n}$

Example 25

Write down the values of 10^2, 10^3, 10^1, 10^0, 10^{-2}, 10^{-3}.

Answer

$10^2 = 10 \times 10 = 100$

$10^3 = 10 \times 10 \times 10 = 1000$

$10^1 = 10$

$10^0 = 1$

$10^{-2} = \dfrac{1}{10^2} = \dfrac{1}{100} = 0.01$

$10^{-3} = \dfrac{1}{10^3} = \dfrac{1}{1000} = 0.001$

To evaluate an exponential function with the GDC use either the ∧ key or the template key and the exponent template.

GDC help on CD: *Alternative demonstrations for the TI-84 Plus and Casio FX-9860GII GDCs are on the CD.*

Exercise 2F

Evaluate these expressions.

1 a $2^3 + 3^2$ **b** $4^2 \times 3^2$ **c** 2^6

2 a 5^0 **b** 3^{-2} **c** 2^{-4}

3 a 3.5^5 **b** 0.495^{-2} **c** $2\dfrac{(1-0.02)^{10}}{1-0.02}$

2.6 Solving inequalities

Inequalities can be solved in a similar way to equations.

Example 26

Solve the inequalities   **a** $2x + 5 < 7$ **b** $3(x - 2) \geq 4$

Answers

a $2x + 5 < 7$

$2x < 2$

$x < 1$

b $3(x - 2) \geq 4$

$x - 2 \geq 1\dfrac{1}{3}$

$x \geq 3\dfrac{1}{3}$

Add, subtract, multiply or divide on both sides until the x is by itself on one side.

Multiplying or dividing by a negative number reverses the inequality.

If you multiply or divide an inequality by a negative value, change the signs on both sides of the inequality and reverse the inequality sign. For example, $4 > 2$, but $-4 < -2$.

Example 27

Solve the inequality $7 - 2x \le 5$	
Answer $\begin{aligned} 7 - 2x &\le 5 \\ -2x &\le -2 \\ x &\ge 1 \end{aligned}$	*Subtract 7.* *Divide by −2.* *Change $\le$ to $\ge$.*

Example 28

Solve the inequality $19 - 2x > 3 + 6x$	
Answer $\begin{aligned} 19 - 2x &> 3 + 6x \\ 19 &> 3 + 8x \\ 16 &> 8x \\ 2 &> x \\ x &< 2 \end{aligned}$	*Reverse the inequality.*

Sometimes the x ends up on the right-hand side of the inequality. In this case reverse the inequality as in this example.

Exercise 2G

1 Solve the inequality for x and represent it on the number line.

    a $3x + 4 \le 13$ **b** $5(x - 5) > 15$ **c** $2x + 3 < x + 5$

2 Solve for x.

    a $2(x - 2) \ge 3(x - 3)$ **b** $4 < 2x + 7$ **c** $7 - 4x \le 11$

Properties of inequalities

→ When you add or subtract a real number from both sides of an inequality the direction of the inequality is unchanged.

For example:

- $4 > 6 \Rightarrow 4 + 2 > 6 + 2$
- $15 \le 20 \Rightarrow 15 - 6 \le 20 - 6$
- $x - 7 \ge 8 \Rightarrow x - 7 + 7 \ge 8 + 7$
- $x + 5 < 12 \Rightarrow x + 5 - 5 < 12 - 5$

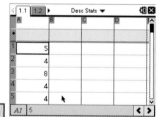

→ When you multiply or divide both sides of an inequality by a positive real number the direction of the inequality is unchanged. When you multiply or divide both sides of an inequality by a negative real number the direction of the inequality is reversed.

For example:

- $4 > 5 \Rightarrow 2(4) > 2(5)$
- $6 \le 10 \Rightarrow -2(6) \ge -2(10)$
- $10 \le 30 \Rightarrow \dfrac{10}{5} \le \dfrac{30}{5}$
- $18 > 24 \Rightarrow \dfrac{18}{-3} < \dfrac{24}{-3}$
- $-12 > -20 \Rightarrow \dfrac{-12}{4} > \dfrac{-20}{4}$

2.7 Absolute value

The absolute value (or modulus) of a number $|x|$ is the numerical part of the number without its sign. It can be written as $|x| = \begin{cases} -x, & \text{if } x < 0 \\ x, & \text{if } x \geq 0 \end{cases}$

Example 29

Write down $|a|$ where $a = -4.5$ and $a = 2.6$

Answer
If $a = -4.5$ then $|a| = 4.5$
If $a = 2.6$ then $|a| = 2.6$

Example 30

Write the value of $|p - q|$ where $p = 3$ and $q = 6$.

Answer
$|p - q| = |3 - 6| = |-3| = 3$

Exercise 2H

1 Write the value of $|a|$ when a is
 a 3.25 **b** −6.18 **c** 0

2 Write the value of $|5 - x|$ when $x = 3$ and when $x = 8$.

3 If $x = 6$ and $y = 4$, write the values of
 a $|x - y|$ **b** $|x - 2y|$ **c** $|y - x|$

3 Geometry

Euclid's *Elements*, written around 300 BCE, was one of the first mathematics textbooks and remained a required text until the 20th century. Euclid began his first book with postulates: self-evident truths.

A point is that which has no part.

A line is breadthless length.

A plane is a surface which lies evenly with the straight lines on itself. An angle is the inclination to one another of two lines in a plane which meet one another and do not lie in a straight line.

3.1 Pythagoras' theorem

→ In a right-angled triangle ABC with sides a, b and c, where a is the *hypotenuse*:

$c^2 = a^2 + b^2$

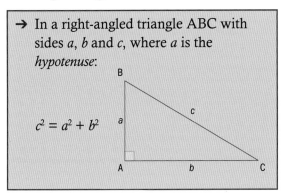

Although the theorem is named after the Greek mathematician Pythagoras, it was known several hundred years earlier to the Indians in their Sulba Sutras and thousands of years before to the Chinese as the Gougu Theorem.

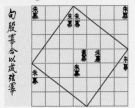

Example 31

Find the length labeled a.

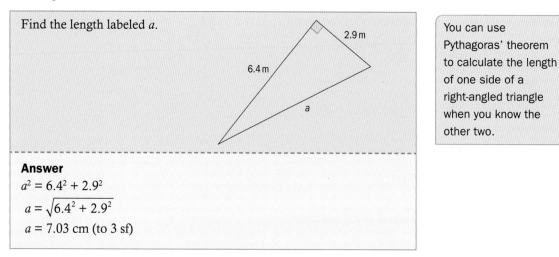

You can use Pythagoras' theorem to calculate the length of one side of a right-angled triangle when you know the other two.

Answer

$a^2 = 6.4^2 + 2.9^2$

$a = \sqrt{6.4^2 + 2.9^2}$

$a = 7.03$ cm (to 3 sf)

Sometimes you have to find a shorter side.

Example 32

Find the length labeled b.

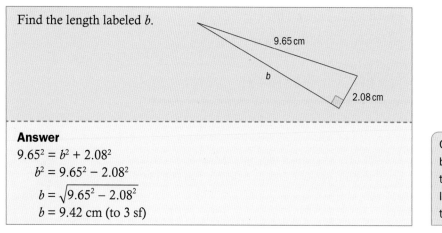

Answer

$9.65^2 = b^2 + 2.08^2$

$b^2 = 9.65^2 - 2.08^2$

$b = \sqrt{9.65^2 - 2.08^2}$

$b = 9.42$ cm (to 3 sf)

Check your answer by making sure that the hypotenuse is the longest side of the triangle.

Exercise 3A

In each diagram, find the length of the side marked x.
Give your answer to 3 signifigant figures.

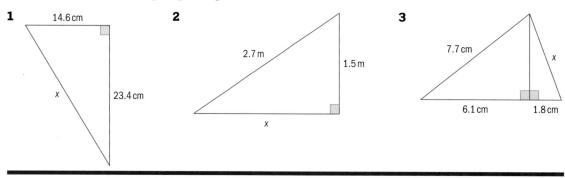

3.2 Points, lines, planes and angles

The most basic ideas of geometry are points, lines and planes. A **straight line** is the shortest distance between two points. Planes can be **finite**, like the surface of a desk or a wall, or can be **infinite**, continuing in every direction.

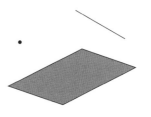

We say that a point has zero dimensions, a line has one dimension and a plane has two dimensions.

Angles are measured in degrees.

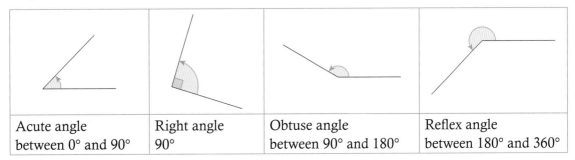

Acute angle between 0° and 90°	Right angle 90°	Obtuse angle between 90° and 180°	Reflex angle between 180° and 360°

Exercise 3B

1 Draw a sketch of:
  a a reflex angle   **b** an acute angle
  c a right angle   **d** an obtuse angle.

2 State whether the following angles are acute, obtuse or reflex.

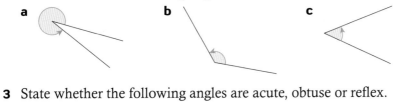

3 State whether the following angles are acute, obtuse or reflex.
  a 173°   **b** 44°   **c** 272°
  d 82°   **e** 308°   **f** 196°

3.3 Two-dimensional shapes

Triangles

Scalene triangle	Isosceles triangle	Equilateral triangle	Right-angled triangle

> The small lines on these diagrams show equal lines and the arrows show parallel lines.

Quadrilaterals

Irregular	Rectangle	Parallelogram	Rhombus
Square	Trapezium	Kite	Arrowhead

Polygons

Pentagon	Hexagon	Octagon	Decagon

Exercise 3C

1 Sketch the quadrilaterals from the table above, with their diagonals. Copy and complete the following table.

Diagonals	Irregular	Rectangle	Parallelogram	Rhombus	Square	Trapezium	Kite
Perpendicular					☐		
Equal					☐		
Bisect					☐		
Bisect angles					☐		

For example, the diagonals of a square are perpendicular to each other, equal in length, bisect each other and bisect the angles of the square.

2 List the names of all the shapes that are contained in each of these figures.

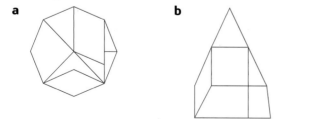

a

b

3.4 Perimeter

The **perimeter** of a figure is defined as the length of its boundary. The perimeter of a polygon is found by adding together the sum of the lengths of its sides.

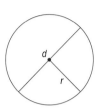

The perimeter of a circle is called its **circumference**.

In the circle on the left, r is the radius and d is the diameter. If C is the circumference.

$$C = 2\pi r \quad \text{or} \quad C = \pi d$$

$\pi = 3.141592653589793238462...$
Many mathematical enthusiasts around the world celebrate Pi day on March 14 (3/14). The use of the symbol π was popularized by the Swiss mathematician Leonhard Euler (1707–1783).

Example 33

Find the perimeter of this shape.

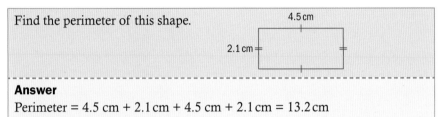

Answer
Perimeter = 4.5 cm + 2.1 cm + 4.5 cm + 2.1 cm = 13.2 cm

Example 34

Find the perimeter of this shape.

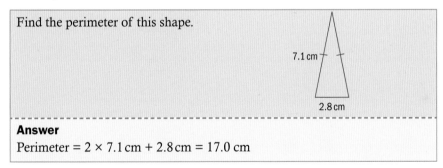

Answer
Perimeter = 2 × 7.1 cm + 2.8 cm = 17.0 cm

Exercise 3D

Find the perimeters of these shapes.

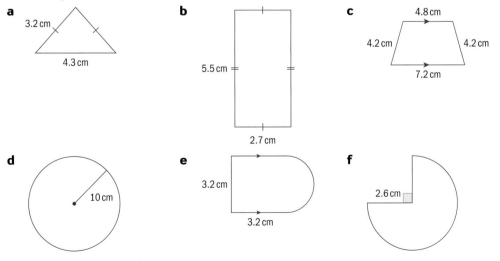

3.5 Area

These are the formulae for the areas of a number of plane shapes.

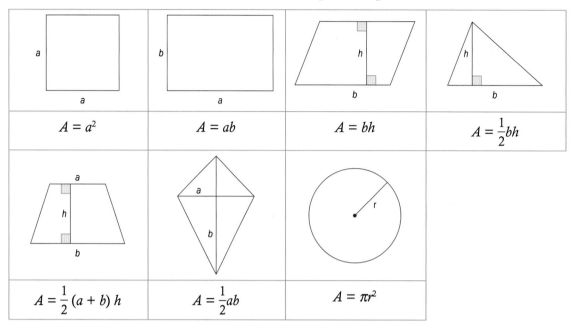

$A = a^2$	$A = ab$	$A = bh$	$A = \dfrac{1}{2}bh$

$A = \dfrac{1}{2}(a + b)\,h$	$A = \dfrac{1}{2}ab$	$A = \pi r^2$

Example 35

Find the area of this shape.

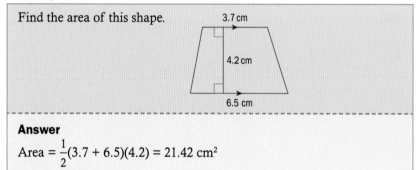

3.7 cm

4.2 cm

6.5 cm

Answer

Area $= \dfrac{1}{2}(3.7 + 6.5)(4.2) = 21.42$ cm^2

Example 36

Find the area of this shape, giving your answer to 3 significant figures.

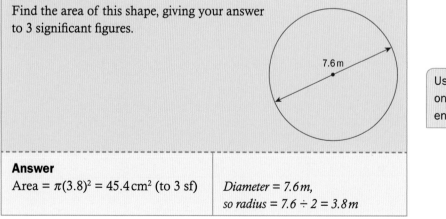

7.6 m

Use the π button on your calculator to enter π.

Answer

Area $= \pi(3.8)^2 = 45.4$ cm^2 (to 3 sf)

*Diameter = 7.6 m,
so radius = 7.6 ÷ 2 = 3.8 m*

Exercise 3E

Find the areas of these shapes. Give your answer to 3 signifigant figures.

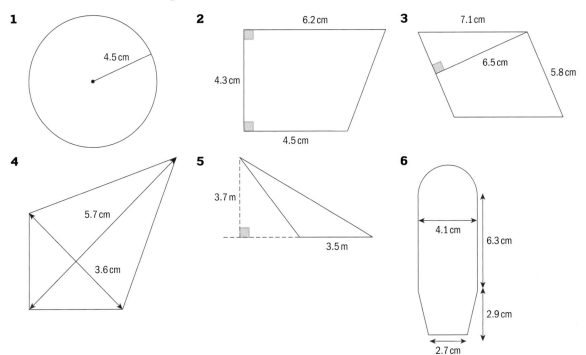

1 4.5 cm

2 6.2 cm, 4.3 cm, 4.5 cm

3 7.1 cm, 6.5 cm, 5.8 cm

4 5.7 cm, 3.6 cm

5 3.7 m, 3.5 m

6 4.1 cm, 6.3 cm, 2.9 cm, 2.7 cm

3.6 Coordinate geometry

Coordinates

Coordinates describe the position of points in the plane. Horizontal positions are shown on the *x*-axis and vertical positions on the *y*-axis.

René Descartes introduced the use of coordinates in a treatise in 1637. You may see axes and coordinates described as Cartesian axes and Cartesian coordinates.

Example 37

Draw axes for and $-10 \le x \le 10$ and $-10 \le y \le 10$.
Plot the points with coordinates: (4, 7), (3, −6), (−5, −2) and (−8, 4).

Answer

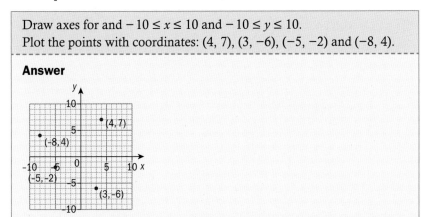

Exercise 3F

1 Draw axes for $-8 \le x \le 8$ and $-5 \le y \le 10$.
Plot the points with coordinates:
$(5, 0), (2, -2), (-7, -4)$ and $(-1, 9)$.

2 Write down the coordinates of the points shown in this diagram.

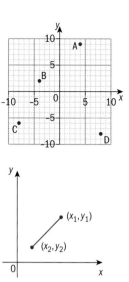

Midpoints

The midpoint of the line joining the points with

coordinates (x_1, y_1) and (x_2, y_2) is given by $\left(\dfrac{x_1 + x_2}{2}, \dfrac{y_1 + y_2}{2} \right)$.

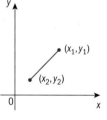

Example 38

Find the midpoint of the line joining the points with coordinates
$(1, 7)$ and $(-3, 3)$.

Answer

The midpoint is $= \left(\dfrac{1 + (-3)}{2}, \dfrac{7 + 3}{2} \right) = (-1, 5)$

Exercise 3G

Calculate the midpoints of the lines joining these pairs of points.

1 $(2, 7)$ and $(8, 3)$ **2** $(-6, 5)$ and $(4, -7)$ **3** $(-2, -1)$ and $(5, 6)$.

Distance between two points

The distance between points with coordinates

(x_1, y_1) and (x_2, y_2) is given by $\sqrt{(x_2 - x_1)^2 + (y_2 - y_1)^2}$.

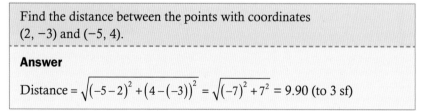

Example 39

Find the distance between the points with coordinates
$(2, -3)$ and $(-5, 4)$.

Answer

Distance $= \sqrt{(-5 - 2)^2 + (4 - (-3))^2} = \sqrt{(-7)^2 + 7^2} = 9.90$ (to 3 sf)

Exercise 3H

Calculate the distance between the following pairs of points. Give your answer to 3 signifigant figures where appropriate.

1 (1, 2) and (4, 6)

2 (−2, 5) and (3, −3)

3 (−6, −6) and (1, 7)

4 Statistics

4.1 Statistical graphs

In a statistical investigation we collect information, known as **data**. To represent the data in a clear way we can use graphs. Three types of statistical graph are bar charts, pie charts and pictograms.

Bar charts

A **bar chart** is a graph made from rectangles, or bars, of equal width whose length is proportional to the quantity they represent, or frequency. Sometimes we leave a small gap between the bars.

Example 40

Juliene collected some data about the ways in which her class travel to school.

Type of transport	Bus	Car	Taxi	Bike	Walk
Frequency	7	6	4	1	2

Represent this information in a bar chart.

Answer

Example 41

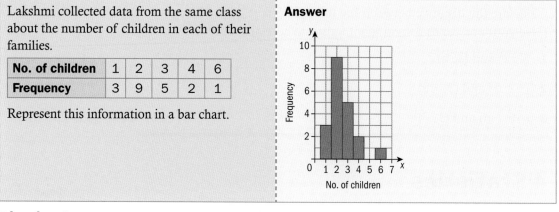

Lakshmi collected data from the same class about the number of children in each of their families.

No. of children	1	2	3	4	6
Frequency	3	9	5	2	1

Represent this information in a bar chart.

Answer

Pie charts

A **pie chart** is a circle divided into sectors, like slices from a pie.
The sector angles are proportional to the quantities they represent.

Example 42

Use Juliene's data from Example 40 to construct a pie chart.

Answer

Type of transport	Frequency		Sector angle
Bus	7	$\frac{7}{20} \times 360°$	126°
Car	6	$\frac{6}{20} \times 360°$	108°
Taxi	4	$\frac{4}{20} \times 360°$	72°
Bike	1	$\frac{1}{20} \times 360°$	18°
Walk	2	$\frac{2}{20} \times 360°$	36°

The total of the frequencies is 20. The total angle for the whole circle is 360°.

■ Bike
■ Walk
■ Bus
■ Car
■ Taxi

Start by drawing a radius and then measure, with your protractor, each angle in turn. The total of the sector angles should be 360°.

Pictograms

Pictograms are similar to bar charts, except that pictures are used. The number of pictures is proportional to the quantity they represent. The pictures can be relevant to the items they show or just a simple character such as an asterisk.

Example 43

Use Juliene's data from Example 40 to construct a pictogram.

Answer

Key: = 1   = 1   = 1   = 1   = 1

> In this pictogram, different symbols are used for each category but the symbols describe the category as well.

Example 44

Use these data on the number of children in a sample of families to construct a pictogram.

Number of children	1	2	3	4	6
Frequency	4	9	6	2	1

Answer

No. of children

1	△△△△
2	△△△△△△△△△
3	△△△△△△
4	△△
6	△

Key: △ = 1

Exercise 4A

1   Adam carried out a survey of the cars passing by his window on the road outside. He noted the colors of the cars that passed by for 10 minutes and collected the following data.

Color	Black	Red	Blue	Green	Silver	White
Frequency	12	6	10	7	14	11

Draw a bar chart, a pie chart and a pictogram to represent the data.

2   Ida asked the members of her class how many times they had visited the cinema in the past month. She collected the following data.

Number of times visited	1	2	3	4	8	12
Number of students	4	7	4	3	1	1

Draw a bar chart, a pie chart and a pictogram to represent the data.

Time allowed: 1 hour 30 minutes
- Answer all the questions
- Unless otherwise stated in the question, all numerical answers must be given exactly or correct to three significant figures.

Maximum marks will be given for correct answers. Where an answer is wrong, some marks may be given for correct method, provided this is shown by written working. Working may be continued below the box, if necessary. Solutions found from a graphic display calculator should be supported by suitable working, e.g. if graphs are used to find a solution, you should sketch these as part of your answer.

Practice papers on CD: *IB examination papers include spaces for you to write your answers. There is a version of this practice paper with space for you to write your answers on the CD. You can also find an additional set of papers for further practice.*

Worked solutions on CD: *Detailed worked solutions for this practice paper are given as a PowerPoint presentation on the CD.*

1 a Find the exact value of $\dfrac{\sqrt{b^2 - ac}}{3125}$, given that $a = 6.4$, $b = 7$ and $c = -5$. *[2 marks]*

 b Write your answer to **a**
  i correct to 3 decimal places;
  ii correct to 2 significant figures;
  iii in the form $a \times 10^k$, where $1 \le a < 10$, $k \in \mathbb{Z}$. *[4 marks]*

2 The table below shows the number of children in the families of a class in a school.

Number of children	1	2	3	4	5
Frequency	3	8	7	4	2

 a Write down the number of families in the class. *[1 mark]*
 b Calculate the mean number of children per family. *[2 marks]*
 c Calculate the standard deviation of the number of children per family. *[1 mark]*
 d Find the median number of children per family. *[2 marks]*

3 The diagram shows the straight line L_1.

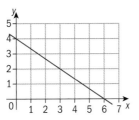

a Calculate the gradient of L_1. *[2 marks]*

b Write down the equation of L_1. *[1 mark]*

A second line L_2 is perpendicular to L_1 and passes through the point $(3, 2)$.
The equation of L_2 is $y = mx + c$.

c Find the value of m and of c. *[3 marks]*

4 a Complete the next two columns of the truth table. *[2 marks]*

p	q	¬ p	¬ p ⇒ q	Inverse
T	T			
T	F			
F	T			
F	F			

b Write down the inverse of the statement $\neg p \Rightarrow q$ *[2 marks]*

c Complete the final column of the truth table with the
truth values for **b**. *[1 mark]*

d The statement $\neg p \Rightarrow q$ and its inverse are **not** equivalent.
State the reason why not. *[1 mark]*

5 The second term, u_2, of a geometric sequence is 162. The fifth
term, u_5, of the same sequence is -6.

a Find the common ratio of the sequence *[4 marks]*

b Find u_1, the first term in the sequence. *[2 marks]*

6 A room is in the shape of a cuboid. Its floor measures 3 m by 4 m
and its height is 2.5 m.

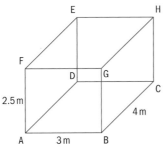

a Calculate the length of BD, the diagonal of the floor of the room. *[2 marks]*

b Calculate the length of BE, the diagonal of the room. *[2 marks]*

c Calculate the angle of depression of B from E. *[2 marks]*

7 The quadratic function $f(x) = 5 + 6x - 2x^2$ intersects the y-axis at point A and has its vertex at point B.

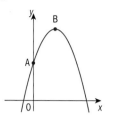

  a Write down the coordinates of A.   *[1 mark]*

  b Find the coordinates of B.   *[2 marks]*

  Point C has the same y coordinate as A.

  c Label point C on the diagram.   *[1 mark]*

  d Write down the coordinates of C.   *[2 marks]*

8 The probability that it rains today is 0.8. If it rains today, the probability that it will rain tomorrow is 0.7. If it does not rain today, the probability that it will rain tomorrow is 0.9.

  a Complete the tree diagram below.   *[3 marks]*

  b Calculate the probability that it does not rain tomorrow.   *[3 marks]*

9 The graph shows the function $f(x) = 10 - (8)a^{-x}$. It intersects the y-axis at point A and has the line L as a horizontal asymptote.

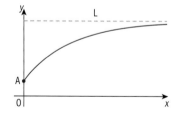

  a Find the y coordinate of A.   *[2 marks]*

  b Write down the equation of L.   *[2 marks]*

  $f(x)$ passes through the point (1, 8).

  c Calculate the value of a.   *[2 marks]*

10 The weights in kg of 40 adult females were collected and summarized in the box and whisker plot shown below.

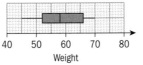
Weight

a Write down the median weight of the females. *[1 mark]*

b Calculate the interquartile range. *[2 marks]*

Two females are chosen at random.

c Find the probability that both females chosen weighed more than 66 kg. *[3 marks]*

11 Jing Yi invests 4000 euros in an account which pays a nominal annual interest rate of 3%, **compounded monthly**.
Give all answers correct to two decimal places.
Find:

a the value of the investment after 5 years *[3 marks]*

b the difference in the final value of the investment if the interest was compounded quarterly at the same nominal rate. *[3 marks]*

12 Given the sequence: 437, 422, 407, 392, …

a Write down the common difference of the sequence. *[1 mark]*

b Calculate the sum of the first 50 terms of the sequence. *[2 marks]*

u_k is the first term in the sequence which is negative.

c Find the value of k. *[3 marks]*

13 a Express in set notation the shaded region on the Venn diagram below. *[2 marks]*

b Given that $x \in (A \cap B' \cap C')$, place x in its correct position on the Venn diagram. *[2 marks]*

c Shade carefully on the above Venn diagram the region which represents $(A \cup B)' \cap C$. *[2 marks]*

14 Consider $f(x) = x^2 - kx$.

a Find $f'(x)$. *[2 marks]*

The graph of $y = f(x)$ has a minimum point with coordinates $(3, p)$.

b Find the value of k. *[2 marks]*

c Find the value of p. *[2 marks]*

15 Consider the statement p:

"If a quadrilateral is a rhombus then the four sides of the quadrilateral are equal".

    a Write down the contrapositive of statement p in words.   *[2 marks]*

    b Write down the converse of statement p in words.   *[2 marks]*

    c Determine whether the converse of statement p is always true. Give an example to justify your answer.   *[2 marks]*

> Use the mark scheme in the Answer section at the back of this book to mark your answers to this practice paper.

Practice paper 2

Time allowed: 1 hour 30 minutes
- Answer all the questions
- Unless otherwise stated in the question, all numerical answers must be given exactly or correct to three significant figures.

Maximum marks will be given for correct answers. Where an answer is wrong, some marks may be given for correct method, provided this is shown by written working. Working may be continued below the box, if necessary. Solutions found from a graphic display calculator should be supported by suitable working, e.g. if graphs are used to find a solution, you should sketch these as part of your answer.

1 The 350 students of an international school play three sports: hockey (H), football (F) and tennis (T).

150 play hockey
220 play football
35 play tennis
80 play hockey and football but not tennis
10 play football and tennis but not hockey
8 play tennis and hockey but not football
5 play all three sports.

   a Draw a Venn diagram that illustrates the above information.   *[4 marks]*
   b Find the number of students that play tennis only.   *[1 mark]*
   c Find the number of students that do not play any sport.   *[2 marks]*

One student is chosen at random. Calculate the probability that this student
   d plays hockey or football but not both,   *[2 marks]*
   e plays hockey given that the student plays tennis.   *[2 marks]*

Two students are chosen at random.
   f Calculate the probability that these two students play both football and tennis.   *[3 marks]*

2 University students were given a number of additional Physics lessons before they took the Physics exam. The following table shows the results (y) in this exam of 10 of these students with the number of additional lessons they took (x).

Number of additional lessons (x)	2	3	4	5	7	8	9	10	12	14
Result (y)	70	72	75	76	79	80	79	82	87	91

a **i** Use your graphic display calculator to find r, the correlation coefficient between x and y.

    ii Use your answer for r to describe the correlation between x and y. *[4 marks]*

b Write down the equation of the regression line y on x. *[2 marks]*

c Use your equation in **b** to estimate the score for a student who took 6 additional lessons. *[2 marks]*

Peter believes that the time when the students took the additional lessons (morning or afternoon) influenced their result in the Physics exam. He records the number of students attending these lessons in the table below and performs a chi-squared test at the 5% significance level to determine whether he is correct.

	Students' result on the Physics exam (y)		
	$y \leq 40$	$40 < y < 60$	$60 \leq y \leq 100$
Morning	35	22	14
Afternoon	48	18	9

d Write down the null hypothesis, H_0. *[1 mark]*

e Write down the number of degrees of freedom. *[1 mark]*

f Show that the expected number of students that took the additional lessons during the **morning** and had a result in the Physics exam **between 40 and 60** is 19 correct to the nearest integer. *[2 marks]*

g Use your graphic display calculator to find the chi-test statistic. *[2 marks]*

The χ^2 value at the 5% is 5.991.

h Peter accepts H_0. Give a reason for his decision. *[1 mark]*

3 The following is the graph of the function $f(x) = 2x^2 - tx$ where t is a constant.

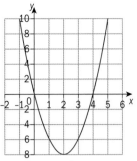

a i Factorize the expression $f(x) = 2x^2 - tx$.

  ii Using the graph of $f(x)$ write down the solutions to the equation $f(x) = 0$.

  iii Hence or otherwise find the value of t.   *[6 marks]*

The function $f(x)$ is increasing for $x > a$.

b Write down the value of a.   *[1 mark]*

The graph of the function $g(x) = mx + c$ intersects the graph of $f(x)$ at the points A and B, where $x = 1$ and $x = 5$ respectively.

c Write down the y-coordinate of

  i A

  ii B.   *[2 marks]*

d Hence write down two equations in m and c.   *[2 marks]*

e Find the value of m and of c.   *[2 marks]*

f Find the x coordinate of the point at which $g(x)$ intersects the x-axis.   *[2 marks]*

g Write down the interval of values of x for which $g(x) > f(x)$.   *[2 marks]*

4 Three cities, Pemberley (P), Vimy (V) and Ridge (R) are the vertices of a triangle; the distance between Pemberley and Vimy is 45 km, the distance between Vimy and Ridge is 60 km. The angle PVR is 75°. This information is given in the diagram.

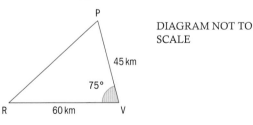

DIAGRAM NOT TO SCALE

a Calculate the area of triangle PVR. Give your answer correct to the nearest km².   *[4 marks]*

b Find the length of PR.   *[3 marks]*

c Find the angle RPV.   *[3 marks]*

A road is constructed from R and meets PV at T such that RT is perpendicular to PV. A company wants to build a water reservoir for the three cities at M, the midpoint of RT.

d Show that the distance MR is $29\,\text{km}$ correct to the nearest km. *[4 marks]*

This water reservoir will be in the shape of a square of side $150\,\text{m}$ and have a depth of $2.85\,\text{m}$

e Calculate the volume of the reservoir. *[2 marks]*

To construct the reservoir, the company will pay a fee of 1.25 Swiss Francs (CHF) per m^3 of its volume.

f Calculate this fee in CHF. Give your answer correct to **two decimal places**. *[2 marks]*

One third of the capacity of the reservoir will be used by Pemberley. $1\,\text{m}^3$ is equal to 1000 litres.

g Calculate the number of litres of water that Pemberley will use. *[2 marks]*

h Give your answer to **g** in the form $a \times 10^k$, where $1 \le a < 10$, and $k \in \mathbb{Z}$. *[2 marks]*

5 Consider the function $f(x) = x^2 + \dfrac{2}{x}$, $x \ne 0$

a Sketch the graph of $f(x)$ for $-3 \le x \le 3$ and $-10 \le y \le 10$. Indicate clearly any asymptotes to the graph. *[4 marks]*

b Write down the x-intercept of the graph of $f(x)$. *[1 mark]*

c Find $f'(x)$. *[3 marks]*

The graph of $f(x)$ has a local minimum at point P.

d Use your answer to **c** to show that the x coordinate of P is 1. *[3 marks]*

e Write down the y coordinate of P. *[1 mark]*

f Describe the behavior of the graph of $f(x)$ in the interval $x > 1$. *[2 marks]*

Let T be the tangent to the graph of $f(x)$ at $x = -2$.

g i Find the gradient of the graph of $f(x)$ at $x = -2$.

ii Write down the equation of T. Give your answer in the form $ax + by + d = 0$. *[5 marks]*

h Find the distance between P and the point of intersection of T with the y-axis. *[3 marks]*

> Use the mark scheme in the Answer section at the back of this book to mark your answers to this practice paper.

Answers

Chapter 1

Skills check

1 a -0.033 b -12.1
   c 0.88

2 a $x = 7$ b $x = 8$
   c $x = 1$ d $x = 4, x = -4$

3 a 96 b 0.234

4 a $x \geq 9$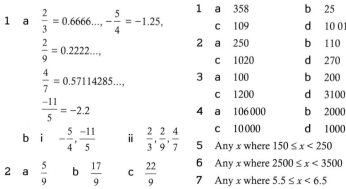
         9

   b $x > 6$
         6

   c $x \leq 0$
         0

5 a 5 b $\dfrac{1}{2}$
   c 2 d 50

Exercise 1A

a   i 8 ii 12
   iii -12 iv 4

b   i Natural   ii Natural
   iii Not natural   iv Natural

Investigation – natural numbers

a T   b T
c F eg. $3 - 8 = -5$. Negative numbers are not natural.

Exercise 1B

1 a $x = -\dfrac{1}{2}$ b Not an integer

2 a $x = 2; x = -2$
   b Both are integers

3 a i -3 ii 9.75
   b i Integer   ii Not an integer

Investigation – integers

a T   b T
c F eg. $\dfrac{1}{2} = 0.5$ d T

Exercise 1C

1 a $\dfrac{2}{3} = 0.6666...,\ -\dfrac{5}{4} = -1.25,$
     $\dfrac{2}{9} = 0.2222...,$
     $\dfrac{4}{7} = 0.57114285...,$
     $\dfrac{-11}{5} = -2.2$

   b i $-\dfrac{5}{4}, \dfrac{-11}{5}$ ii $\dfrac{2}{3}, \dfrac{2}{9}, \dfrac{4}{7}$

2 a $\dfrac{5}{9}$ b $\dfrac{17}{9}$ c $\dfrac{22}{9}$

---

3 a For example 0.8
   b For example $0.1\dot{2}$
   c For example $3.45\dot{7}\dot{8}$

Exercise 1D

1 For example $2.1, 2.2, 2.23$

2 a 2.5 b It is rational

3 a For example $1.81; 1.82; 1.83$
   b i For example $-2.14; -2.12; -2.1$
     ii infinite

Investigation – rational numbers

a T   b T
c T   b F eg. $\sqrt{2}$

Exercise 1E

1 a 2.5 cm   b rational

2 a (25π) cm²   b irrational

Exercise 1F

1 a i $1 < x \leq 3$ ii $x \leq 2$
   b i
      1   3
     ii
       2

   c i Both are solutions
     ii q is solution and t isn't.

2 a i $x > -1$ ii $3 \leq x \leq 7$
    iii $x < 3$
   b i
      -1
    ii
      3   7
    iii
      3

Inequality p	$2x+1>-1$	$4 \leq x+1 \leq 8$	$2-x>-1$
$-\dfrac{2}{3}$	✓		✓
$\sqrt{10}$	✓	✓	
2π	✓	✓	

Exercise 1G

1 a 358 b 25
   c 109 d $10\,016$

2 a 250 b 110
   c 1020 d 270

3 a 100 b 200
   c 1200 d 3100

4 a $106\,000$ b 2000
   c $10\,000$ d 1000

5 Any x where $150 \leq x < 250$

6 Any x where $2500 \leq x < 3500$

7 Any x where $5.5 \leq x < 6.5$

Exercise 1H

1 a 45.7 b 301.1
   c 2.4 d 0.1

2 a 0.00 b 201.31
   c 9.62 d 28.08

3 a 10.049 b 3.900
   c 201.781 d 0.008

4 a 3025.0 b 3024.98
   c 3024.984 d 3000
   e 3000

5 a 15.60 b 15.603
   c 16 d 20

6 Any x where $2.365 \leq x < 2.375$

7 Any x where $4.05 \leq x < 4.15$

Exercise 1I

1 a 3 b $1, 2$ or 3
   c 1 d 3 or 4
   e 4

2 a 300 b 0.07
   c 400 d 0.001

3 a 360 b 0.080
   c 1.1 d 1600

4 a 2970 b 0.326
   c $10\,400$ d 0.501

5 a 400 b 426
   c 425.9 d 425.88

6 a 3 b 3.14
   c 3.1 d 3.142

7 a 200 b 4610
   c 2.70

8 a 0.3703703704
   b i 0.37
     ii 0.370
     iii 0.3704

Exercise 1J

1 a 1.828 cm   b 11 cm

2 a 2.288 b 20.9
   c 4.5 cm²

Exercise 1K

1 a 3000 b 16
   c 15 d 10

2 4000 pipes

3 300 people per km²

4 20 reams

5 15 km h⁻¹

6 $20\,000\,000$ visitors per year

7 Peter is not correct.
An estimate of the area is $10\,000$ m².

Exercise 1L

1 a 119.423   b 17.2% (3 sf)
2 a 8.17 (3 sf)   b 8
   c 2.04% (3 sf)
3 a $18.5832\,m^2$
   b 5.3 m and 3.5 m
   c 0.179% (3 sf)
4 a 5.323 m   b 33.4 m (3 sf)
   c 10% (2 sf)

Exercise 1M

1 2.5×10^{-3}, 10^{10}
2 a 1.356×10^5 b 2.45×10^{-3}
   c 1.6×10^{10} d 1.08×10^{-4}
   e 2.3×10^2
3 3.4×10^5, 0.21×10^7,
   215×10^4, 2.3×10^6
4 3.621×10^4, 0.3621×10^4,
   3.261×10^3, 31.62×10^2

Exercise 1N

1 a 1.764×10^{17} b 2.25×10^{-4}
   c 1.5×10^{-2}
2 a 2.99×10^6
   b 3 000 000 or 3×10^6
3 a 2.205×10^9 b 700
   c 7×10^2
4 a 2.25×10^{10}
   b True because
      $x^2 = 5.06 \times 10^{20} > 10^{20}$
   c i 150 000   ii 1.5×10^5

Investigation – SI units

a Many different names
   eg. mm, cm, km
b 10^6 = M (mega), $10^{-6} = \mu$ (micro)
c length eg. Millimetres or
   centimetres weight eg. Kilograms

Exercise 1O

1 a $km\,h^{-2}$ b $kg\,m^{-3}$
   c $m\,s^{-1}$
2 a decagram
   b centisecond
   c millimetre
   d decimetre
3 a 32 000 m   b 0.087 dam
   c 1.28 m
4 a 0.5 kg   b 35 700 dag
   c 1.080 hg
5 a 80 ms   b 120 das
   c 800 ds
6 a 68 kg   b 36 km
   c 6.54×10^2 mg

Exercise 1P

1 a $23\,600\,cm^2$
   b $0.000\,15\,dam^2$
   c $54\,cm^2$
   d $60\,000\,mm^2$
   e $80\,hm^2$
   f $0.035\,km^2$
2 a $5 \times 10^6\,cm^3$
   b $10^2\,m^3$
   c $3.5 \times 10^0\,dm^3$
   d $2.55 \times 10^{11}\,mm^3$
   e $1.2 \times 10^1\,dam^3$
   f $7.802 \times 10^2\,dam^3$
3 a $169\,cm^2$ b $0.0169\,m^2$
4 a $0.614\,125\,m^3$
      or $0.614\,m^3$ (3 sf)
   b $614\,125\,cm^3$ or
      $614\,000\,cm^3$ (3 sf)
5 $7560\,cm^2$, $0.8\,m^2$, $82\,dm^2$,
   $8\,000\,000\,mm^2$, $0.081\,dam^2$
6 $1200\,dm^3$, $0.01\,dam^3$, $10\,900\,000\,cm^3$,
   $11\,020\,000\,000\,mm^3$, $11.2\,m^3$

Exercise 1Q

1 a 94 980 s   b 95 000 s
2 a 173 100 s
   b 1.731×10^5 s or 1.73×10^5 s
3 a 5000 ml
   b 0.000 005 6 hl
   c 4 500 000 cl
4 a $5 \times 10^5\,cm^3$
   b $1.458 \times 10^1\,dm^3$
   c $8 \times 10^5\,cm^3$
5 a $13\,\ell$ b 4 hl
   c 81 cl
6 a 75 min   b 4500 s
7 a $3.375\,m^3$ b $3375\,dm^3$
   c No, only $3375\,\ell$ can be poured.
8 a $0.176\,\ell$ b 8 tea cups
9 a 8.625 h   b $696.5\,km\,h^{-1}$
   c 10:08 p.m. (nearest minute,
      Buenos Aires time)

Exercise 1R

1 a 6.9 °C   b 26.7 °C
2 a 70 °F   b 36 °F
3 a 16.85 °C   b 62.33 °F
4 a $t_K = t_C + 273.15$
   b $t_C = \dfrac{5}{9}(t_F - 32)$

Review exercise

Paper 1 style questions

1

	5	$\dfrac{\pi}{2}$	−3	$\dfrac{5}{4}$	2.3
N	✓				
Z	✓		✓		
Q	✓		✓	✓	✓
ℝ	✓	✓	✓	✓	✓

2 a $\sqrt{2}$ b 1.4142
   c $0.001\,39 \times 10^2$, 14.1×10^{-1}, $\sqrt{2}$,
      1414×10^{-2}, 1.4×10^2
3 a 2.69×10^3 kg
   b i 2700 kg   ii 0.372% (3 sf)
4 a $300\,000\,000\,m\,s^{-1}$
   b 300 000 km
   c $1.08 \times 10^9\,km\,h^{-1}$
5 a 0.58 kg   b 0.6 kg
   c 33.3%
6 a $1.56\,dm^3$ b $1.17\,\ell$
   c i 21   ii $0.43\,\ell$
7 a 31.25   b 31.3
   c 3.13×10^1
8 a $A = x^2$
   b i 1600 m   ii 6400 m
9 a 80.33 °F   b 311 K
10 a $x > 1$
   b
   c $\sqrt{3}$; $2.0\dot{6}$; $\dfrac{101}{100}$
11 a $62\,370\,mm^2$
   b $0.062\,370\,m^2$
   c 4.68 g   d 2.34 kg

Paper 2 style questions

1 a 7.52 km   b $2463.85
   c 1.06%   d $3.15\,km^2$
2 a 2857 m   b 4 laps
   c 0.150 h   d 44.842 minutes
   e 1.88%
3 a $8.18\,cm^3$ b $73.63\,cm^3$
   c $\dfrac{15}{2.5} = 6$ d $24.5\,cm^3$
   e $24\,500\,mm^3$ f $2.45 \times 10^4\,mm^3$
      or
      $2.45 \times 10^1\,cm^3$

Chapter 2

Skill check

1 (Example)

Which age group do you belong to?

‡ under 16 ‡ over 16

Are you

‡ male? ‡ female?

2 a

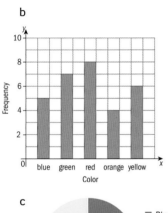

```
blue    * * * * * * * *
green   * * * * * * * *
red     * * * * * * * *
orange  * * * * * * * *
yellow  * * * * * * * *
```

Key: * = 1 sweet

b

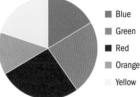

c

![Pie chart with legend: Blue, Green, Red, Orange, Yellow]

3 Axes drawn to scale of 1 cm to 2 units on the *x*-axis, 1 cm to 10 units on the *y*-axis

Investigation – population distribution

Tower Hamlets has a greater proportion of younger people compared to the UK population, whereas Christchurch has a greater proportion of older people compared to the UK population.

Tower Hamlets is in the city of London where there are many jobs and schools and therefore younger people are more likely to live there.

Exercise 2A

1 a Discrete   b Continuous
   c Discrete   d Discrete

e Continuous   f Discrete
g Continuous   h Continuous
i Continuous   j Discrete
k Continuous   l Discrete

2 a Biased   b Random
   c Biased   d Random
   e Biased

Exercise 2B

1

Number of goals	Frequency
0	4
1	7
2	7
3	4
4	1
5	2

2

Number of heads	Frequency
0	1
1	1
2	4
3	4
4	3
5	7
6	9
7	4
8	5
9	2
10	4
11	3
12	3

3

Age	Frequency
9	4
10	9
11	8
12	7
13	4
14	1
15	4
16	3

4

Number of crisps	Frequency
88	3
89	6
90	16
91	3
92	2

5

Number	Frequency
1	7
2	9
3	11
4	6
5	7
6	10

6 $m = 6$, $n = 3$

Exercise 2C

1 Answers will depend on width of class intervals chosen. Example:

a

Number	Frequency
$0 \le x < 5$	1
$5 \le x < 10$	7
$10 \le x < 15$	3
$15 \le x < 20$	4
$20 \le x < 25$	6
$25 \le x < 30$	1
$30 \le x < 35$	5
$35 \le x < 40$	0
$40 \le x < 45$	2
$45 \le x < 50$	1

b

Number	Frequency
$10 \le x < 20$	7
$20 \le x < 30$	5
$30 \le x < 40$	7
$40 \le x < 50$	5
$50 \le x < 60$	7
$60 \le x < 70$	5
$70 \le x < 80$	5
$80 \le x < 90$	2
$90 \le x < 100$	2

c

Number	Frequency
$1 \le x < 3$	3
$3 \le x < 5$	7
$5 \le x < 7$	4
$7 \le x < 9$	3
$9 \le x < 11$	6
$11 \le x < 13$	3
$13 \le x < 15$	4
$15 \le x < 17$	3
$17 \le x < 19$	1
$19 \le x < 21$	1

Exercise 2D

1 a

Class	Lower boundary	Upper boundary
9–12	8.5	12.5
13–16	12.5	16.5
17–20	16.5	20.5
21–24	20.5	24.5

b

Time (*t* seconds)	Lower boundary	Upper boundary
$2.0 \le t < 2.2$	2.0	2.2
$2.2 \le t < 2.4$	2.2	2.4
$2.4 \le t < 2.6$	2.4	2.6

Exercise 2E

1

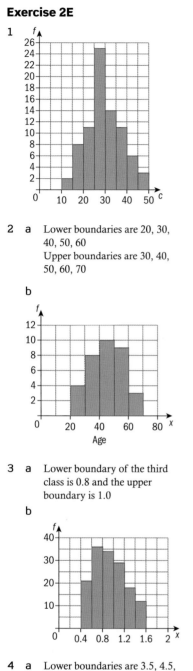

2 a Lower boundaries are 20, 30, 40, 50, 60
Upper boundaries are 30, 40, 50, 60, 70

b

3 a Lower boundary of the third class is 0.8 and the upper boundary is 1.0

b

4 a Lower boundaries are 3.5, 4.5, 5.5, 6.5, 7.5, 8.5, 9.5
Upper boundaries are 4.5, 5.5, 6.5, 7.5, 8.5, 9.5, 10.5

b

5 a

Number	Frequency
$0 \leq x < 10$	8
$10 \leq x < 20$	10
$20 \leq x < 30$	7
$30 \leq x < 40$	6
$40 \leq x < 50$	3
$50 \leq x < 60$	6
$60 \leq x < 70$	5
$70 \leq x < 80$	4
$80 \leq x < 90$	1

b

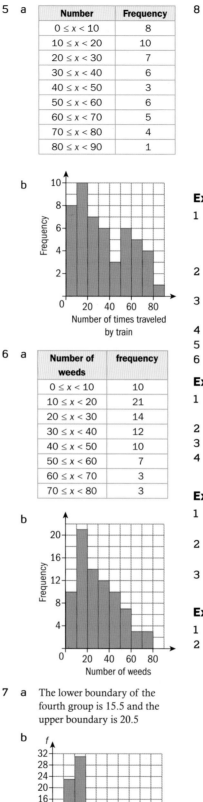

6 a

Number of weeds	frequency
$0 \leq x < 10$	10
$10 \leq x < 20$	21
$20 \leq x < 30$	14
$30 \leq x < 40$	12
$40 \leq x < 50$	10
$50 \leq x < 60$	7
$60 \leq x < 70$	3
$70 \leq x < 80$	3

b

7 a The lower boundary of the fourth group is 15.5 and the upper boundary is 20.5

b

8

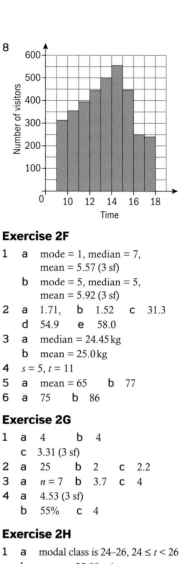

Exercise 2F

1 a mode = 1, median = 7, mean = 5.57 (3 sf)
b mode = 5, median = 5, mean = 5.92 (3 sf)
2 a 1.71, **b** 1.52 **c** 31.3
d 54.9 **e** 58.0
3 a median = 24.45 kg
b mean = 25.0 kg
4 $s = 5, t = 11$
5 a mean = 65 **b** 77
6 a 75 **b** 86

Exercise 2G

1 a 4 **b** 4
c 3.31 (3 sf)
2 a 25 **b** 2 **c** 2.2
3 a $n = 7$ **b** 3.7 **c** 4
4 a 4.53 (3 sf)
b 55% **c** 4

Exercise 2H

1 a modal class is 24–26, $24 \leq t < 26$
b mean = 25.88 min
2 a $70 \leq s < 80$
b mean = 88.3 km h⁻¹ (3 sf)
3 a modal class is 40–50
b mean = 51.8

Exercise 2I

1 a 50 **b** $a = 8, b = 14, c = 38$
2 a $s = 13, t = 122$
b

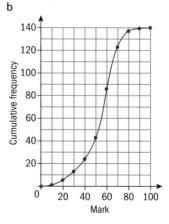

c **i** Median mark is about 60

ii Lower quartile approximately 46

iii 60% of 140 = 84 therefore pass mark is about 60

3 **a**

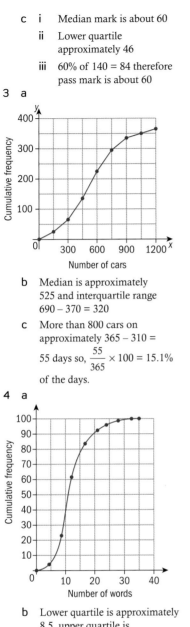

b Median is approximately 525 and interquartile range 690 − 370 = 320

c More than 800 cars on approximately 365 − 310 = 55 days so, $\frac{55}{365} \times 100 = 15.1\%$ of the days.

4 **a**

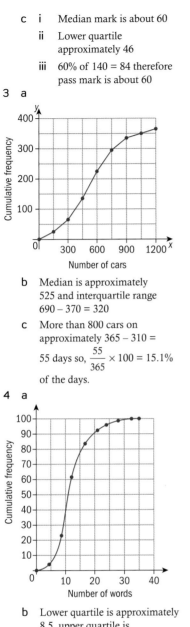

b Lower quartile is approximately 8.5, upper quartile is approximately 14.5, median is approximately 11.5

5 **a**

Length (x cm)	Cumulative frequency
≤ 28	3
≤ 31	7
≤ 34	18
≤ 37	41
≤ 40	69
≤ 43	84
≤ 46	96
≤ 49	100

b

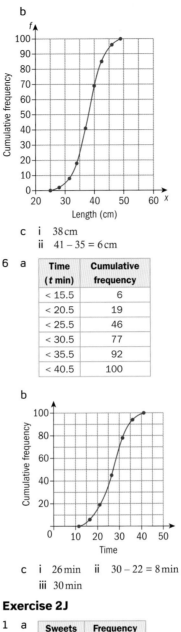

c **i** 38 cm
ii 41 − 35 = 6 cm

6 **a**

Time (t min)	Cumulative frequency
< 15.5	6
< 20.5	19
< 25.5	46
< 30.5	77
< 35.5	92
< 40.5	100

b

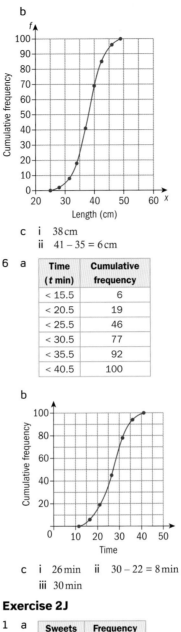

c **i** 26 min **ii** 30 − 22 = 8 min
iii 30 min

Exercise 2J

1 **a**

Sweets	Frequency
30	1
31	3
32	10
33	16
34	9
35	4
36	2

b Median = 33, lower quartile = 32, upper quartile = 34

c

Number of sweets

2 **a** Median = 3, lower quartile = 2, upper quartile = 4

b

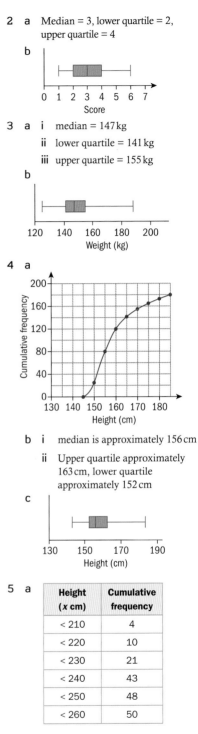

3 **a** **i** median = 147 kg

ii lower quartile = 141 kg

iii upper quartile = 155 kg

b

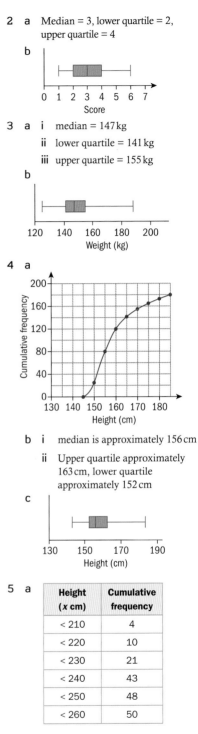

4 **a**

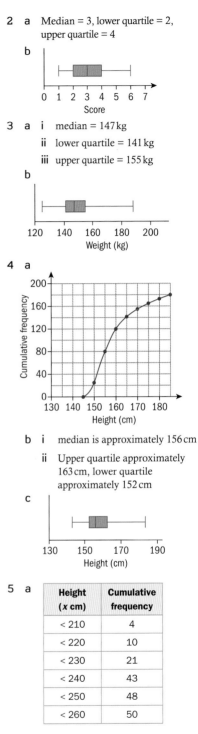

b **i** median is approximately 156 cm

ii Upper quartile approximately 163 cm, lower quartile approximately 152 cm

c

5 **a**

Height (x cm)	Cumulative frequency
< 210	4
< 220	10
< 230	21
< 240	43
< 250	48
< 260	50

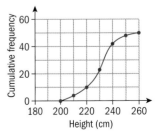

b Median is about 232 cm

c Lower quartile approximately 223 cm, upper quartile approximately 237 cm

d

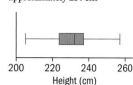

Exercise 2K

1 a median for boys = 55
median for girls = 55

b IQR boys = 64 − 40 = 24
IQR girls = 68 − 45 = 23

c 50%   **d** 25%

2 a 0   **b** 12

c 14   **d** 28   **e** 25%

3 a 22   **b** 44

c 53   **d** 22

Exercise 2L

1 a i 19   **ii** 6

b i 13   **ii** 4

c i 7   **ii** 4.5

Exercise 2M

1 a standard deviation = 3.17

b standard deviation = 2.29

2 mean = 8.8
standard deviation = 5.44

3 a range = 5   **b** IQR = 2

c mean = 3.26
standard deviation = 1.28

4 a range = 6   **b** IQR = 2

c mean = 7.32
standard deviation = 1.41

5 a mean = 67.2

b standard deviation = 4.94

c range = 18   **d** IQR = 6

6 a $x = 45$

b standard deviation = 15.6

c range = 46

d IQR = 27

7 a $m = 9$ **b** mean = 12.7

c standard deviation = 1.49

d IQR = 2

8 a range = 7, IQR = 3

b mean = 7.92
standard deviation = 1.87

9 mean = 32 min
standard deviation = 7.57 min

10 a girls' mean = 55.4 and
standard deviation = 11.5

boys' mean = 51.8 and
standard deviation = 23.1

---

b There is a big difference in the standard deviation implying that the boys' marks are much more widespread than the girls' marks.

Review exercise

Paper 1 style questions

1 a 9   **b** 5.5

2 a 1   **b** 5.5

3 a 6.62   **b** 6   **c** 6

4 a i 6.54 m   **ii** 3.08 m

b 6.1 m

5 a discrete   **b** 1.93

c 1.25

6 a 46.2

b

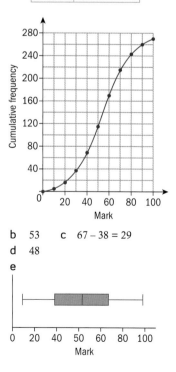

7 a 41   **b** 31

c 49   **d** 18

8

Number of horses

Paper 2 style questions

1 a i mean = 98
ii mode = 96

b i

Number	Frequency
94	1
96	4
97	3
98	3
99	3
100	3
101	2

ii median = 98, IQR = 4

2 a i

Number	Cumulative frequency
< 4.5	18
< 9.5	43
< 14.5	75
< 19.5	89
< 24.5	96
< 29.5	100

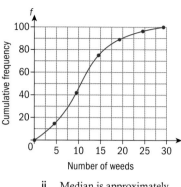

Number of weeds

ii Median is approximately 10.6

iii 11%

b i Mean is approximately 10.95

ii Standard deviation is approximately 6.53

iii Total number of weeds is approximately 87 60000

3 a

Mark	Cumulative frequency
< 10	3
< 20	17
< 30	38
< 40	73
< 50	115
< 60	170
< 70	213
< 80	245
< 90	260
< 100	270

Mark

b 53   **c** 67 − 38 = 29

d 48

e

Mark

4 a Median is approximately 13 000, IQR is approximately 6200

b

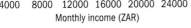
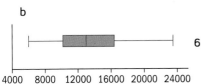

Monthly income (ZAR)

c

Monthly income (ZAR)	Frequency
$6000 \leq x < 8000$	10
$8000 \leq x < 10000$	19
$10000 \leq x < 12000$	30
$12000 \leq x < 14000$	29
$14000 \leq x < 16000$	20
$16000 \leq x < 18000$	15
$18000 \leq x < 20000$	11
$20000 \leq x < 22000$	9
$22000 \leq x < 24000$	7

d Mean is approximately 13 747 and standard deviation 4237

5 a Modal group is 60–65, $60 \leq w < 65$

b Mean is approximately 63.2 and standard deviation 6.62

c

Weight (kg)	Cumulative frequency
< 50	4
< 55	20
< 60	65
< 65	123
< 70	166
< 75	194
< 80	200

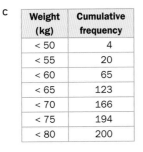

Weight (kg)

d median = 63, lower quartile = 59, upper quartile = 68

e

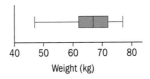

Weight (kg)

6 a Mean is approximately 26.9 and standard deviation 4.40

b Modal class is 24–28

c

Age (years)	Cumulative frequency
≤ 20	3
≤ 24	15
≤ 28	37
≤ 32	52
≤ 36	59
≤ 40	60

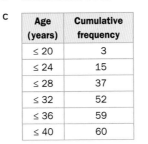

Age (years)

d median = 27, IQR = 5.5

e

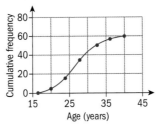

Age (years)

7 a Modal class is 30–40

b Estimate of mean is 34.3 and standard deviation is 16.6

c

Time (s)	Cumulative frequency
< 10	5
< 20	24
< 30	42
< 40	64
< 50	80
< 60	92
< 70	100

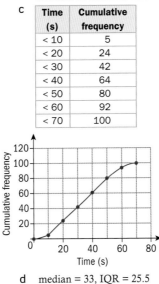

Time (s)

d median = 33, IQR = 25.5

8 a median = 4750, lower quartile = 4570, upper quartile = 5000

b

Number of visitors

c

Visitors	Frequency
$4000 \leq x < 4200$	1
$4200 \leq x < 4400$	3
$4400 \leq x < 4600$	5
$4600 \leq x < 4800$	9
$4800 \leq x < 5000$	6
$5000 \leq x < 5200$	3
$5200 \leq x < 5400$	2
$5400 \leq x < 5600$	1
$5600 \leq x < 5800$	1

d Modal class is 4600–4800

e Estimate of mean is 4784 and standard deviation is 355

9 a

Weight(x kg)	Frequency
$120 \leq x < 130$	10
$130 \leq x < 140$	35
$140 \leq x < 150$	75
$150 \leq x < 160$	50
$160 \leq x < 170$	15
$170 \leq x < 180$	10
$180 \leq x < 190$	5

b Modal class is $140 \leq x < 150$

c Estimate of mean is 149

Chapter 3

Skills check

1 a $h = 20$ cm

b $\sqrt{50}$ cm $= 7.07$ (3 sf)

2 a i $(0, 6)$ **ii** $\sqrt{40} = 6.32$ (3 sf)

b $q = 3, p = 6$

Exercise 3A

1 a −1 **b** 8 **c** −8 **d** 1

2 a i A(1, 5), B(0, 1) **ii** 4

b i A(−1, 5), B(0, 1), **ii** −4

c i A(0, 3), B(3, 2), **ii** $-\frac{1}{3}$

d i A(0, −1), B(1, 0), **ii** 1

e i $A(-1,-2), B(2,0)$ ii $\dfrac{2}{3}$

f i $A(2,4), B(4,1)$ ii $-\dfrac{3}{2}$

Exercise 3B

1

a

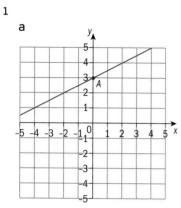

b

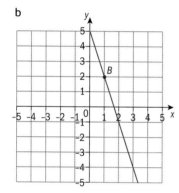

c

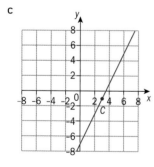

2 a i 2 ii $p = 9$

   b i 4 ii $t = 10$

   c i -5 ii $q = -10$

   d i 1 ii $s = 3$

   e i -3 ii $r = -2$

3 a $\dfrac{5}{a+1}$ b $a = \dfrac{1}{4}$

4 a 0.5 b $\dfrac{t-6}{-5}$ c $t = 3.5$

Exercise 3C

1 a 4.5

   b & c

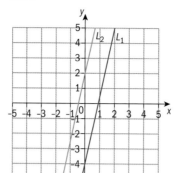

2 a Parallel to x-axis
   b Parallel to y-axis c Neither

3 a x
   b y c zero

4 $a = 3$ 5 $m = -5$

Exercise 3D

1 a, b, d
2 b, d
3 a $\dfrac{1}{3}$ b -1.5
   c 4 d -1 e 1

4 a $\dfrac{3}{7}$ b $-\dfrac{5}{12}$

5 a i -3 ii $\dfrac{1}{2}$

   iii

b i $\dfrac{1}{2}$ ii -2

iii

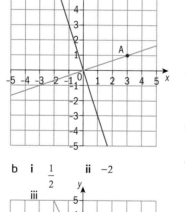

c i $\dfrac{4}{3}$ ii $-\dfrac{3}{4}$

iii

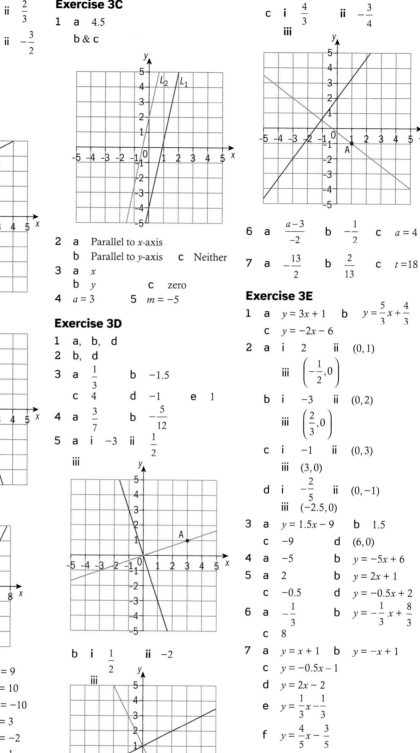

6 a $\dfrac{a-3}{-2}$ b $-\dfrac{1}{2}$ c $a = 4$

7 a $-\dfrac{13}{2}$ b $\dfrac{2}{13}$ c $t = 18$

Exercise 3E

1 a $y = 3x + 1$ b $y = \dfrac{5}{3}x + \dfrac{4}{3}$
   c $y = -2x - 6$

2 a i 2 ii $(0,1)$

      iii $\left(-\dfrac{1}{2},0\right)$

   b i -3 ii $(0,2)$

      iii $\left(\dfrac{2}{3},0\right)$

   c i -1 ii $(0,3)$

      iii $(3,0)$

   d i $-\dfrac{2}{5}$ ii $(0,-1)$

      iii $(-2.5,0)$

3 a $y = 1.5x - 9$ b 1.5
   c -9 d $(6,0)$

4 a -5 b $y = -5x + 6$

5 a 2 b $y = 2x + 1$
   c -0.5 d $y = -0.5x + 2$

6 a $-\dfrac{1}{3}$ b $y = -\dfrac{1}{3}x + \dfrac{8}{3}$
   c 8

7 a $y = x + 1$ b $y = -x + 1$
   c $y = -0.5x - 1$
   d $y = 2x - 2$
   e $y = \dfrac{1}{3}x - \dfrac{1}{3}$
   f $y = \dfrac{4}{5}x - \dfrac{3}{5}$

Exercise 3F

1 a $4x + y - 20 = 0$
   b $x - 2y + 4 = 0$
   c $-5x - 4y + 7 = 0$
   d $x - y + 5 = 0$

2 **a** $y = -3x$ **b** $y = -x - 1$
 c $y = -2x + 1$ **d** $y = 0.5x$
 e $y = -2x + 3$

3 **a** $y = 0.5x + 1$
 b $x = -2$ **c** $y = 1$

4 **a** A, C, D, F
 b $a = 6.5$ **c** $t = 8$

5 **a** A, B, E
 b $a = \dfrac{2}{3}$ **c** $t = 31$

6

Line	Conditions
A	H
B	G
C	F
D	E

7 **a** 2   **b** 6   **c** $c = -2.25$
 d $t = 16$ **e** 2
 f $y = 2x + 4$ ($2x - y + 4 = 0$)

8 **a** $y = -2x + 4$
   ($2x + y - 4 = 0$)
 b Yes, A, B and C are collinear.
   The coordinates of A, B and C
   all satisfy the equation of L_1.

Investigation—vertical and horizontal lines

1 **a** $(-3, -1)$, $(-3, 0)$, $(-3, 1)$, $(-3, 2)$
   and $(-3, 3)$
 b All the coordinates have the
   x-coordinate as -3.
 c To lie on L1, the x-coordinate
   must be -3. ie. $x = -3$

2 **a** $(2, -1)$, $(2, 0)$, $(2, 1)$, $(2, 2)$
   and $(2, 3)$
 b All the coordinates have the
   x-coordinate as 2
 c To lie on L3, the x-coordinate
   must be 2. ie. $x = 2$

3 $x = 1$

4 **a** $(-1, 1)$, $(0, 1)$, $(1, 1)$, $(2, 1)$
   and $(3, 1)$
 b All the coordinates have the
   y-coordinate as 1
 c To lie on L3, the y-coordinate
   must be 1. ie. $y = 1$

5 **a** $(-1, -2)$, $(0, -2)$, $(1, -2)$,
   $(2, -2)$ and $(3, -2)$
 b All the coordinates have the
   y-coordinate as -2
 c To lie on L4, the y-coordinate
   must be -2. ie. $y = -2$

6 $y = -3$

Exercise 3G

1 **a** $x = 3$ **b** $y = 1$

2 **a** $(2, 0)$ **b** $(5, 1)$ **c** $(-7, 3)$
 d $(-2, 1)$ **e** $(4, -1)$ **f** $(0, 4)$

3 $L_1: y = 5x - 1$
   $L_2: y = 5x + 2$
   L_1 and L_2 have same gradient but
   different y-intercepts.

4 **a** no point
 b an infinite number of points
 c only one point
 d infinite number of points

5 **a** $y = 5x - 5$ ($5x - y - 5 = 0$)
 b $y = -\dfrac{1}{5}x + \dfrac{1}{5}$ ($x + 5y - 1 = 0$)

Investigation - right-angled triangles

1 Angles are identical

2 $\dfrac{AE}{AC} = 1.5$

3 $\dfrac{AD}{AB} = 1.5$

4 $\dfrac{DE}{BC} = 1.5$
   All the ratios are identical.

Exercise 3H

1

H	Opp	Adj
XY	YZ	XZ
CB	AB	AC
RQ	PQ	PR

2 **a** $\cos\delta = \dfrac{AC}{AB}$
   $\sin\delta = \dfrac{BC}{AB}$
   $\tan\delta = \dfrac{BC}{AC}$
 b $\cos\delta = \dfrac{QR}{PQ}$
   $\sin\delta = \dfrac{PR}{PQ}$
   $\tan\delta = \dfrac{PR}{QR}$
 c $\cos\delta = \dfrac{EF}{DF}$
   $\sin\delta = \dfrac{ED}{DF}$
   $\tan\delta = \dfrac{ED}{EF}$

3 **a** **i** $\sin\alpha = \dfrac{4}{\sqrt{41}}$
   ii $\cos\alpha = \dfrac{5}{\sqrt{41}}$
   iii $\tan\alpha = \dfrac{4}{5}$

 b **i** $\sin\alpha = \dfrac{\sqrt{28}}{8}$
   ii $\cos\alpha = \dfrac{6}{8}$
   iii $\tan\alpha = \dfrac{\sqrt{28}}{6}$
 c **i** $\sin\alpha = \dfrac{10}{14}$
   ii $\cos\alpha = \dfrac{\sqrt{96}}{14}$
   iii $\tan\alpha = \dfrac{10}{\sqrt{96}}$

4 **a** $\sin\beta = \dfrac{x}{10}$
 b $\cos\beta = \dfrac{x}{5}$
 c $\tan\beta = \dfrac{x}{12}$
 d $\tan\beta = \dfrac{7}{x}$
 e $\sin\beta = \dfrac{14}{x}$
 f $\cos\beta = \dfrac{3}{x}$

Exercise 3I

1 $h = 3.11\,\text{cm}$ **2** $x = 6.41\,\text{cm}$

3 $m = 4.88\,\text{cm}$ **4** $y = 13.94\,\text{cm}$

5 $t = 386.37\,\text{m}$ **6** $s = 86.60\,\text{m}$

Exercise 3J

1 **a**

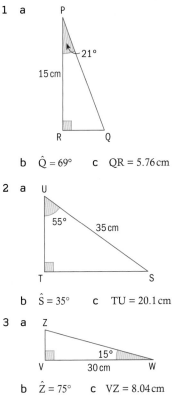

 b $\hat{Q} = 69°$ **c** $QR = 5.76\,\text{cm}$

2 **a** U

 b $\hat{S} = 35°$ **c** $TU = 20.1\,\text{cm}$

3 **a** Z

 b $\hat{Z} = 75°$ **c** $VZ = 8.04\,\text{cm}$

4 **a**

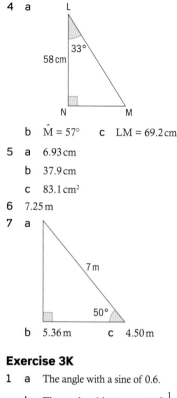

b $\hat{M} = 57°$ **c** $LM = 69.2\,cm$

5 **a** $6.93\,cm$
  b $37.9\,cm$
  c $83.1\,cm^2$

6 $7.25\,m$

7 **a**

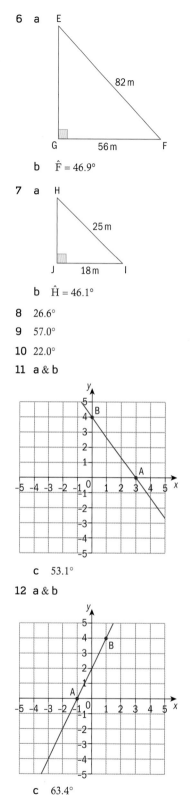

b $5.36\,m$ **c** $4.50\,m$

Exercise 3K

1 **a** The angle with a sine of 0.6.

  b The angle with a tangent of $\frac{1}{2}$.

  c The angle with a cosine of $\frac{2}{3}$.

2 **a** $36.9°$ **b** $26.6°$ **c** $48.2°$

3 **a** $11.5°$ **b** $48.2°$ **c** $45°$

4 **a** $\hat{A} = 53.6°, \hat{C} = 36.4°$

  b $\hat{R} = 41.4°, \hat{Q} = 48.6°$

  c $\hat{M} = 36.9°, \hat{Q} = 53.1°$

  d $\hat{Y} = 41.4°, \hat{Z} = 48.6°$

  e $\hat{J} = 70.1°, \hat{I} = 19.9°$

  f $\hat{D} = 25.9°, \hat{F} = 64.1°$

5 **a**

54 cm
D   **42 cm**   **C**

b $\hat{C} = 52.1°$

6 **a**

82 m
G   **56 m**   **F**

b $\hat{F} = 46.9°$

7 **a** H

25 m
J   **18 m**   **I**

b $\hat{H} = 46.1°$

8 $26.6°$

9 $57.0°$

10 $22.0°$

11 **a & b**

c $53.1°$

12 **a & b**

c $63.4°$

Investigation – 2-D shapes

1 The diagonals of a rhombus are perpendicular to each other.

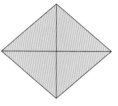

There are 4 right-angled triangles in the rhombus. They are congruent because all the sides are the same length.

2 The diagonals of a kite are perpendicular to each other.

There are 4 right-angled triangles in the kite. They are not congruent because the diagonals are not of equal length.

3

Cutting the parallelogram as shown gives 3 shapes with 2 right-angled triangles. This explains why the area of a parallelogram is the same as the area of a rectangle, length x width.

4

The height of the triangle is perpendicular to the base. The two right-angled triangles formed would be congruent if the triangle was equilateral or isosceles.

5 Triangles are congruent if $AD = BC$ on the trapezium.

6 ABO is an isosceles triangle since the two radii are the same length.

The two triangles formed are congruent because all the sides are of equal length.

Investigation - rhombus

1

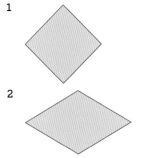

2

3 There are infinitely many different rhombuses with a side length of 6 cm. They differ in that their diagonals are not of equal length.

Exercise 3L

1 a

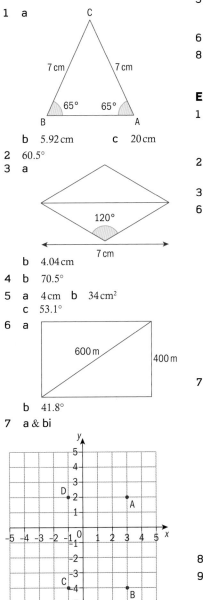

In triangle: C at top, 7 cm on left (to B), 7 cm on right (to A), 65° at B, 65° at A.

b 5.92 cm **c** 20 cm

2 60.5°

3 a

Triangle with 120° angle, 7 cm base.

b 4.04 cm

4 b 70.5°

5 a 4 cm **b** 34 cm²
c 53.1°

6 a

Rectangle with 600 m diagonal, 400 m side.

b 41.8°

7 a & bi

Coordinate grid with points D (-4, 2), A (3, 2), C (-3, -4), B (3, -4).

---

b ii B(3, −4)

c i 6 **ii** 4 **d** 56.3°

Exercise 3M

1 27.5° **2** 52.1°
3 125 m **4** 6.89°
5 21.5 m **6** 32.3 m
7 a 12° **b** 425 m

Exercise 3N

1 a $y = 13.7$ km
  b $r = 3.47$ cm
  c $c = 11.0$ km
2 8.34 cm **3** 2.65 cm
4 7.32 km
5 a $\hat{C} = 37.9°$ **b** $\hat{R} = 58.6°$
  c $\hat{Y} = 27.6°$
6 $\hat{C} = 42.9°$ **7** $\hat{R} = 46.3°$
8 a 150° **b** 5.08 m
  c 2.54 m

Exercise 3O

1 a $y = 13.5$ km
  b $p = 9.74$ cm
  c $c = 6.84$ m
2 a $x = 38.0°$ **b** $y = 59.4°$
  c $a = 50.1°$
3 193 m **4** 7.14 cm **5** 55.8°
6 a

Triangle with Y at top, 25 km to Z, 30 km to X, 18 km from X to Z.

b $\hat{Z} = 86.9°$
7 a

Figure with A, S, 8 m, 110°, 12 m, J.

b JS = 16.5 m **c** 2.74 m
8 113°
9 a PR = 15.9 m
  b PR̂Q = 30.5° **c** 2.54 m

Exercise 3P

1 a 41.6 km² **b** 1890 m²
2 a 100° **b** 49.2 cm²
3 a 80° **b** 4.43 m²
4 60.8 km²
5 a 50.5° **b** 1930 m²
6 a $A = 0.25x^2$ or equivalent
  b $x = 4$
7 a $\sqrt{61}$ cm or 7.81 cm (3 sf)
  b 15.4 cm **c** 56.5 cm²
  d 71.5 cm²

Review exercise

Paper 1 style questions

1 a $-\dfrac{1}{2}$

  b $y = -\dfrac{1}{2}x + 4$ or $x + 2y - 8 = 0$

2 a −1 **b** 1
  c $y = x$ or $y - x = 0$
3 a i (−1.5, 0) **ii** (0, 3)
  b

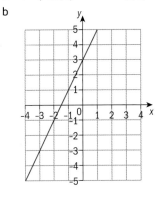

Coordinate grid with a line.

  c 63.4°
4 a $a = 1$
  b $b = -19$
  c (1, 4)
5 a

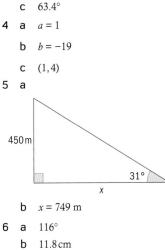

Right triangle with 450 m, 31°, x.

  b $x = 749$ m
6 a 116°
  b 11.8 cm
  c 62.5 cm²
7 a 9 m
  b 38.9°
  c 14.1 m²

8 a 97.2°

  b 12.4 cm²

  c 66.1 cm²

Paper 2 style questions

1 a & b i

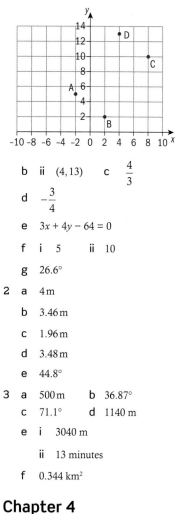

  b ii (4, 13)   **c** $\dfrac{4}{3}$

  d $-\dfrac{3}{4}$

  e $3x + 4y - 64 = 0$

  f i 5   **ii** 10

  g 26.6°

2 a 4 m

  b 3.46 m

  c 1.96 m

  d 3.48 m

  e 44.8°

3 a 500 m   **b** 36.87°

  c 71.1°   **d** 1140 m

  e i 3040 m

    ii 13 minutes

  f 0.344 km²

Chapter 4

Skills check

1 a 18.5   **b** 2   **c** $-\dfrac{31}{4}$

2 a 1.30, –2.30   **b** –0.781, 1.28

  c $x = -19, y = -11$

3 a $-\dfrac{3}{4}$ **b** $\dfrac{5}{2}$

Exercise 4A

1 a function since each student is in only one mathematics class

  b not a function since each teacher teaches two of the students

---

2 a function since each element of A is related to one and only one element of B

  b not a function since one element of B (16) is not related to any element of A

  c function since each element of C is related to one and only one element of A

3 a i not a function since one element of A (4) is not related to any element of B

    ii not a function since one element of A (4) is not related to any element of C

    iii not a function since one element of C (1) is not related to any element of A

    iv function since each element of B is related to one and only one element of C

    v not a function since one element of C (6) is not related to any element of A

  b iv

B	C
	1
2	2
4	4
6	6

4 a $y = 2x$ **b** $y = \dfrac{x}{2}$

  c $y = \sqrt[3]{x}$ **d** $y = \dfrac{x^3}{2}$

5 a function   **b** function

  c not a function since negative elements in the first set are not related to any element in the second set

  d function

Exercise 4B

1 a i

x	$-\dfrac{1}{2}$	0	1	3.5	6
y = 2x	–1	0	2	7	12

    ii the set of all real numbers

    iii yes, $y = 0$ is the image of $x = 0$

  b i

x	–3	0	2	$\dfrac{1}{4}$	–2	×
y = x² + 1	10	1	5	$\dfrac{17}{16}$	5	5

    ii the set of all real numbers

---

    iii no, there is no solution to the equation $0 = x^2 + 1$

  c i

x	–2	–1	0	$\dfrac{1}{2}$	3	5
$y = \dfrac{1}{x+1}$	–1	×	1	$\dfrac{2}{3}$	$\dfrac{1}{4}$	$\dfrac{1}{6}$

    ii the set of all real numbers except $x = -1$

    iii no, there is no solution to the equation $0 = \dfrac{1}{x+1}$

  d i

x	–3	0	$\dfrac{1}{4}$	1	9	100
$y = \sqrt{x}$	×	0	$\dfrac{1}{2}$	1	3	10

    ii the set of all non-negative real numbers

    iii yes, $y = 0$ is the image of $x = 0$

2 a false, there is no solution to the equation $0 = \dfrac{2}{x}$

  b true, $y = x^2 \geq 0$ for all values of x

  c true, $y = x^2 = 3 \geq 3$ for all values of x

  d true, $y = 3$ when $x = \pm 2$

  e true, $\dfrac{-3}{3} - 1 = -2$

  f false, the image of $x = -1$ is $y = 4$

Exercise 4C

1 a

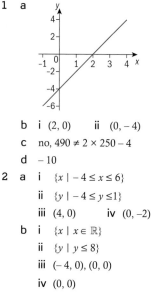

  b i (2, 0)   **ii** (0, –4)

  c no, $490 \neq 2 \times 250 - 4$

  d –10

2 a i $\{x \mid -4 \leq x \leq 6\}$

    ii $\{y \mid -4 \leq y \leq 1\}$

    iii (4, 0)   **iv** (0, –2)

  b i $\{x \mid x \in \mathbb{R}\}$

    ii $\{y \mid y \leq 8\}$

    iii (–4, 0), (0, 0)

    iv (0, 0)

c i $\{x \mid -1 \le x \le 1\}$

ii $\{y \mid 0 \le y \le 1\}$

iii $(-1, 0), (1, 0)$

iv $(0, 1)$

d i $\{x \mid x \ge -1\}$

ii $\{y \mid y \ge 4\}$

iii no points

iv $(0, 8)$

3 a i false   ii false   iii true

b i false   ii true   iii false

c i false   ii true   iii false

d i false   ii false   iii true

4 a

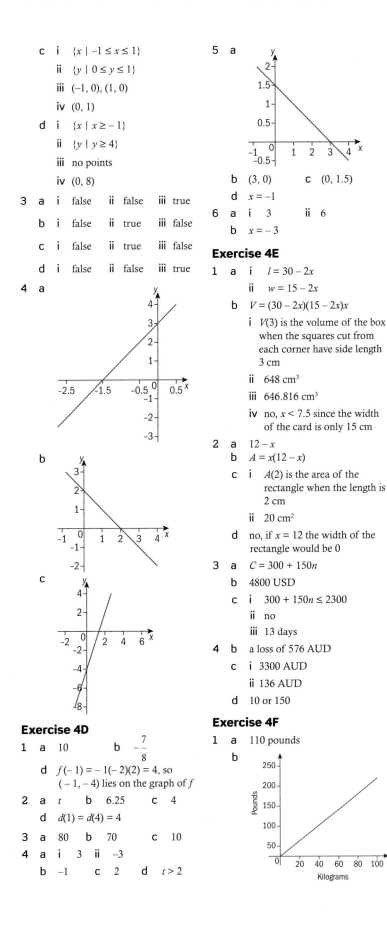

b

c

Exercise 4D

1 a 10   b $-\dfrac{7}{8}$

d $f(-1) = -1(-2)(2) = 4$, so $(-1, -4)$ lies on the graph of f

2 a t b 6.25   c 4

d $d(1) = d(4) = 4$

3 a 80   b 70   c 10

4 a i 3   ii −3

b −1   c 2   d $t > 2$

5 a

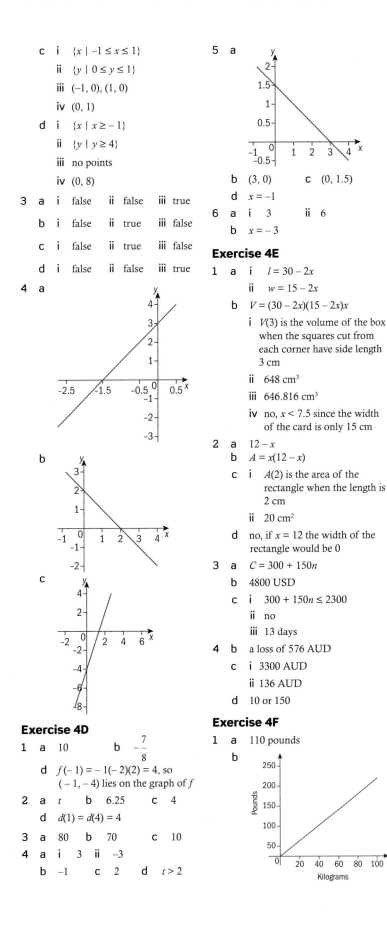

b $(3, 0)$ c $(0, 1.5)$

d $x = -1$

6 a i 3   ii 6

b $x = -3$

Exercise 4E

1 a i $l = 30 - 2x$

ii $w = 15 - 2x$

b $V = (30 - 2x)(15 - 2x)x$

i $V(3)$ is the volume of the box when the squares cut from each corner have side length 3 cm

ii 648 cm³

iii 646.816 cm³

iv no, $x < 7.5$ since the width of the card is only 15 cm

2 a $12 - x$

b $A = x(12 - x)$

c i $A(2)$ is the area of the rectangle when the length is 2 cm

ii 20 cm²

d no, if $x = 12$ the width of the rectangle would be 0

3 a $C = 300 + 150n$

b 4800 USD

c i $300 + 150n \le 2300$

ii no

iii 13 days

4 b a loss of 576 AUD

c i 3300 AUD

ii 136 AUD

d 10 or 150

Exercise 4F

1 a 110 pounds

b

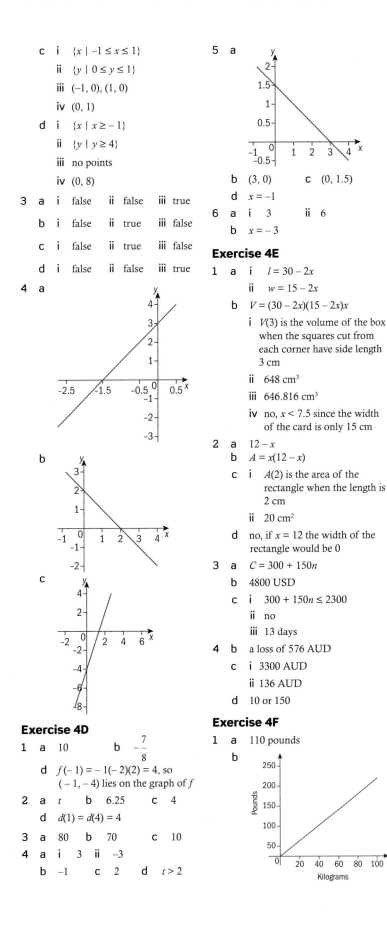

c gradient = 2.2 $p(x) = 2.2x$

d $p(75) = 165$ $p(125) = 275$

e $k(x) = \dfrac{x}{2.2}$

f $k(75) = 34.1$ $k(100) = 45.5$

2 a 102.5 SGD

b

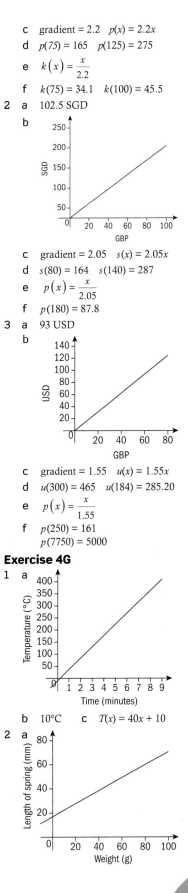

c gradient = 2.05 $s(x) = 2.05x$

d $s(80) = 164$ $s(140) = 287$

e $p(x) = \dfrac{x}{2.05}$

f $p(180) = 87.8$

3 a 93 USD

b

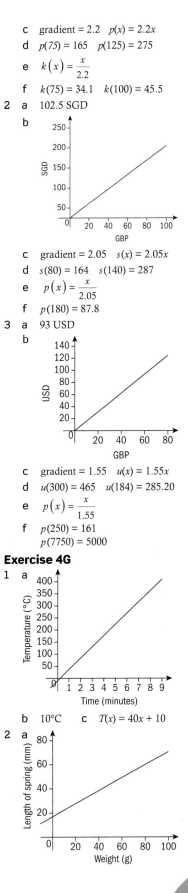

c gradient = 1.55 $u(x) = 1.55x$

d $u(300) = 465$ $u(184) = 285.20$

e $p(x) = \dfrac{x}{1.55}$

f $p(250) = 161$
$p(7750) = 5000$

Exercise 4G

1 a

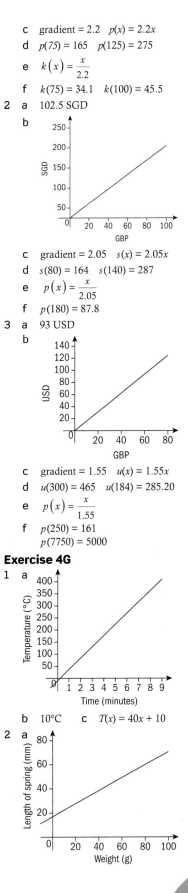

b 10°C   c $T(x) = 40x + 10$

2 a

b   18 mm

c   20 mm

d   0.5 mm

e $L(x) = 0.5x + 18$

3   b $T(x) = \dfrac{2}{3}x + 10$

    c   66.7°C

4   b   20 cm

    c   20 cm

    d   350 g

    e $L(x) = 0.08x + 20$

Exercise 4H

1   a   flour = $80s + 60f$

    b   fat = $50s + 90f$

    c   5 sponge cakes, 7 fruit cakes

2   3 tables, 9 chairs

3   7 vans, 8 cars

4   4 passenger planes, 6 transport planes

5   16 volume 1, 8 volume 2

Investigation - the curve $y = ax^2$

1   The curves are related by being a reflection of each other in the x-axis.

2   a   The curve is still a parabola. Positive numbers give a U-shaped parabola. Negative numbers give a ∧-shaped graph.

    b   There is a vertical line of symmetry in each graph at $x = 0$.

    c   The vertex is (0,0). It is a minimum point for the U-shaped graphs and a maximum point for the ∧- shaped graph.

4   'a' represents the steepness of the graph. Higher values of a give a steeper curve. Smaller values for a give a shallow graph. Negative values for a give a ∧- shaped graph.

Investigation - the curve $x^2 + c$

Changing the value of c translates the graph of $y = x^2$ vertically up and down. The value of c is the y-intercept of the graph.

Exercise 4I

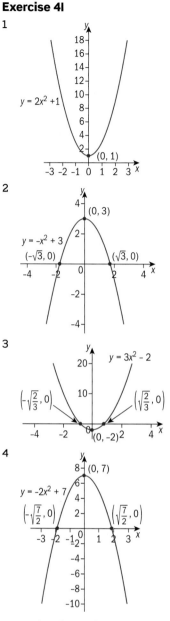

1

2

3

4

Investigation - the curves $y = (x + p)^2$ and $y = (x + p)^2 + q$

1   Changing the value of p shifts the graph of $y = x^2$ to the left if p is positive and to the right if p is negative.

2   The axis of symmetry is $x = -p$
The coordinates of the vertex is $(-p, q)$

Exercise 4J

1   $(-3, -2)$ $x = -3$

2   $(-5, 4)$ $x = -5$

3   $(4, -1)$ $x = 4$

4   $(5, 7)$ $x = 5$

5   $(-3, 4)$ $x = -3$

Investigation - the curves $y = kx - x^2$ and $y = x^2 - kx$

Part A:

1   The axis of symmetry is $x = 2$

The coordinates of the vertex is $(2,4)$

X-axis intercepts are $(0, 0)$ and $(4, 0)$.

2   Graphs of required curves.

3   Varying the value of k changes where the graph intersects the x-axis.

The axis of symmetry is $x = \dfrac{1}{2}k$

X-axis intercepts are $(0, 0)$ and $(k, 0)$

Part B:

Graphs of required curves.

4   Varying the value of k changes where the graph intersects the x-axis. This time the shape of the graph is U-shaped.

The axis of symmetry is $x = \dfrac{1}{2}k$

X-axis intercepts are $(0, 0)$ and $(k, 0)$.

Investigation - curves of the form $y = (x - k)(x - l)$

1   The curve intersects the x-axis at $(1, 0)$ and $(3, 0)$.

The axis of symmetry is at $x = 2$
The co-ordinate of the vertex is $(2, -1)$

2   The curve intersects the x-axis at $(k, 0)$ and $(1, 0)$.

The axis of symmetry is at $x = \dfrac{(k + 1)}{2}$

Exercise 4K

1   a $x = 2$ b $(0, 0), (4, 0)$

    c $(2, -4)$

2   a $x = -3$ b $(0, 0), (-6, 0)$

    c $(-3, -9)$

3   a $x = 4$ b $(0, 0), (8, 0)$

    c $(4, 16)$

4   a $x = \dfrac{3}{2}$ b $(0, 0), (3, 0)$

    c $\left(\dfrac{3}{2}, \dfrac{9}{4}\right)$

5   a $x = 1$ b $(0, 0), (2, 0)$

    c $(1, -1)$

6   a $x = \dfrac{1}{2}$ b $(0, 0), (1, 0)$

    c $\left(\dfrac{1}{2}, -\dfrac{1}{4}\right)$

7 a $x = -2$ **b** $(0, 0), (-4, 0)$
 c $(-2, -4)$

8 a $x = -\dfrac{1}{2}$ **b** $(0, 0), (-1, 0)$

 c $\left(-\dfrac{1}{2}, -\dfrac{1}{4}\right)$

9 a $x = 1$ **b** $(-1, 0), (3, 0)$
 c $(1, -4)$

10 a $x = 1$ **b** $(5, 0), (-3, 0)$
 c $(1, -16)$

11 a $x = 4$ **b** $(2, 0), (6, 0)$
 c $(4, -4)$

12 a $x = 1$ **b** $(4, 0), (-2, 0)$
 c $(1, -9)$

Investigation – the general quadratic curve $y = ax^2 + bx + c$

Part A

1 The curve intersects the x-axis at $(1, 0)$ and $(3, 0)$.

The axis of symmetry is at $x = 2$
The co-ordinate of the vertex is $(2, -1)$

2 In the case $a = 1$:

The curve intersects the x-axis at

$$\left(\dfrac{-b - \sqrt{b^2 - 4ac}}{2a}, 0\right) \text{ and }$$

$$\left(\dfrac{-b + \sqrt{b^2 - 4ac}}{2a}, 0\right).$$

The axis of symmetry is at $x = \dfrac{-b}{2}$
The co-ordinate of the vertex is
$\left(\dfrac{-b}{2}, c - \dfrac{-b}{4}\right)$

Part B

1 The curve does not intercept the x-axis

The axis of symmetry is at $x = 1$
The co-ordinate of the vertex is $(1, 1)$

Exercise 4L

1 a $x = 1$ **b** no points
 c $(1, 2)$

2 a $x = -2$ **b** $(1, 0), (-5, 0)$
 c $(-2, -9)$

3 a $x = -3$
 b $(-0.764, 0), (-5.24, 0)$
 c $(-3, -5)$

4 a $x = 1$ **b** $(0.423, 0), (1.58, 0)$
 c $(1, -1)$

5 a $x = 2$
 b $(-0.121, 0), (4.12, 0)$
 c $(2, -9)$

6 a $x = -\dfrac{3}{2}$
 b $(0.898, 0), (-3.90, 0)$

 c $\left(-\dfrac{3}{2}, -\dfrac{23}{2}\right)$

7 a $x = 1$
 b no points

 c $\left(1, \dfrac{3}{2}\right)$

8 a $x = -3$
 b $(1.12, 0), (-7.12, 0)$

 c $\left(-3, -\dfrac{17}{2}\right)$

Exercise 4M

1 a i $(0, -3)$ **ii** $x = -1$
 iii $(-1, -4)$ **iv** $(-3, 0), (1, 0)$
 v $y \geq -4$

b

$y = x^2 + 2x - 3$, with point $(-1, -4)$

2 a i $(0, 7)$ **ii** $x = -4$
 iii $(-4, -9)$ **iv** $(-7, 0), (-1, 0)$
 v $y \geq -9$

b

$y = x^2 + 8x + 7$, with point $(-4, -9)$

3 a i $(0, -7)$ **ii** $x = 3$
 iii $(3, -16)$
 iv $(7, 0), (-1, 0)$
 v $y \geq -16$

b

$y = x^2 - 6x - 7$, with point $(3, -16)$

4 a i $(0, -4)$ **ii** $x = \dfrac{3}{2}$

 iii $\left(\dfrac{3}{2}, -\dfrac{25}{4}\right)$

 iv $(4, 0), (-1, 0)$

 v $y \geq -\dfrac{25}{4}$

b

$y = x^2 - 3x - 4$, with point $\left(\dfrac{3}{2}, -\dfrac{25}{4}\right)$

5 a i $(0, -10)$ **ii** $x = \dfrac{3}{2}$

 iii $\left(\dfrac{3}{2}, -\dfrac{49}{4}\right)$

 iv $(5, 0), (-2, 0)$

 v $y \geq -\dfrac{49}{4}$

b

$y = x^2 - 3x - 10$, with point $\left(\dfrac{3}{2}, -\dfrac{49}{4}\right)$

6 a i $(0, -3)$

ii $x = -\dfrac{1}{4}$ **iii** $\left(-\dfrac{1}{4}, -\dfrac{25}{8}\right)$

iv $(1, 0), \left(-\dfrac{3}{2}, 0\right)$

v $y \geq -\dfrac{25}{8}$

b

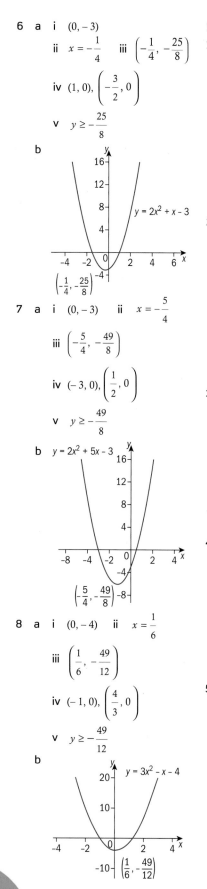

$y = 2x^2 + x - 3$

$\left(-\dfrac{1}{4}, -\dfrac{25}{8}\right)$

7 a i $(0, -3)$ **ii** $x = -\dfrac{5}{4}$

iii $\left(-\dfrac{5}{4}, -\dfrac{49}{8}\right)$

iv $(-3, 0), \left(\dfrac{1}{2}, 0\right)$

v $y \geq -\dfrac{49}{8}$

b $y = 2x^2 + 5x - 3$

$\left(-\dfrac{5}{4}, -\dfrac{49}{8}\right)$

8 a i $(0, -4)$ **ii** $x = \dfrac{1}{6}$

iii $\left(\dfrac{1}{6}, -\dfrac{49}{12}\right)$

iv $(-1, 0), \left(\dfrac{4}{3}, 0\right)$

v $y \geq -\dfrac{49}{12}$

b

$y = 3x^2 - x - 4$

$\left(\dfrac{1}{6}, -\dfrac{49}{12}\right)$

Exercise 4N

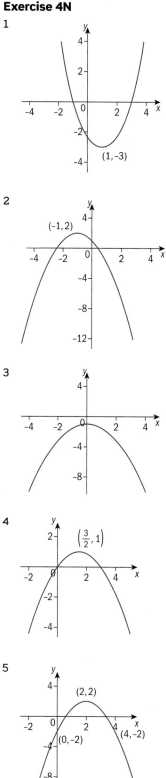

1

$(1, -3)$

2

$(-1, 2)$

3

4

$\left(\dfrac{3}{2}, 1\right)$

5

$(2, 2)$

$(0, -2)$ $(4, -2)$

6

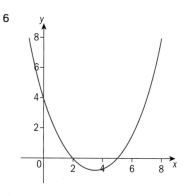

Exercise 4O

1 a $(-3, -5), (1, -1)$

b yes: $x = -3$ or $x = 1$

c $(-2, -7), (1, -1)$

2 $(0, -5), (-4, -1)$

3 a $(4, 1), (-1, 1)$

b $(2, 7), (-1, 1)$

4 b $f: \{y \mid -3.125 \leq y \leq 18\}$
$g: \{y \mid -2 \leq y \leq 4\}$

c $x = -1$ or $x = 2$

e $x = -1$ or $x = \dfrac{5}{2}$

f $(-2, 7), (2, 3)$

5 a $(2.12, 1.5), (-2.12, 1.5)$

b $-2.12 < x < 2.12$

Exercise 4P

1 $f(x) = x^2 + 4x - 1$
$g(x) = x^2 + 2x - 2$

2 $f(x) = x^2 - 4x + 5$
$g(x) = x^2 - 2x + 3$

3 $f(x) = -x^2 + 4x + 5$
$g(x) = -x^2 + 2x + 3$

4 $f(x) = -3x^2 - 6x + 2$
$g(x) = -2x^2 - 8x - 3$

5 $f(x) = 2x^2 + 2x$
$g(x) = -x^2 + 3$

Exercise 4Q

1 a length = 42.5 m,
width = 42.5 m

b length = 31.25 m,
width = 31.25 m

2 a 13 531.25 riyals

b 3000 riyals

c 69 or 1369 units

3 a 270 m

b 342.25 m

c 37 s

Investigation - exponential graphs

1

Weeks	Number of water lilies
1	4
2	8
3	16
4	32
5	64
6	128
7	256
8	512
9	1024
10	2048
11	4096
12	8192

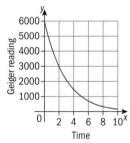

2

Time	Geiger reading
0	6000
2	3000
4	1500
6	750
8	375
10	187.5

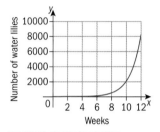

Exercise 4R

For all questions: y-intercept is $(0, 1)$, horizontal asymptote is $y = 0$

Investigation - graphs of $f(x) = ka^x$

1 Graph of $y = 2(3)^x$
   a $k = 2$ **b** $(0, 2)$
   c $y = 0$

2 Graph of $y = 3(\frac{1}{2})^x$
   a $k = 3$ **b** $(0, 3)$ **c** $y = 0$

3 Graph of $y = -3(2)^x$
   a $k = -3$ **b** $(0, -3)$ **c** $y = 0$

The y-intercept is given by $(0, k)$ and all graphs have a horizontal asymptote of $y = 0$.

Investigation - graphs of $f(x) = ka^x + c$

1 Graph of $y = 2^x + 3$
   a $k = 1, c = 3$
   b $(0, 4)$ **c** $y = 3$

2 Graph of $y = 3(\frac{1}{2})^x - 4$
   a $k = 3, c = -4$
   b $(0, -1)$ **c** $y = -4$

3 Graph of $y = -2(3)^x + 5$
   a $k = -2, c = 5$
   b $(0, 3)$ **c** $y = 5$

The y-intercept is given by $(0, k + c)$ and the horizontal asymptote is $y = c$.

Exercise 4S

1 a $(0, 1)$ **b** $y = 0$

2 a $(0, 1)$ **b** $y = 0$

3 a $(0, 1)$ **b** $y = 0$

4 a $(0, 1)$ **b** $y = 0$

5 a $(0, 7)$ **b** $y = 4$

6 a $(0, -3)$ **b** $y = -1$

7 a $(0, 2)$ **b** $y = 3$

8 a $(0, 2)$ **b** $y = -2$

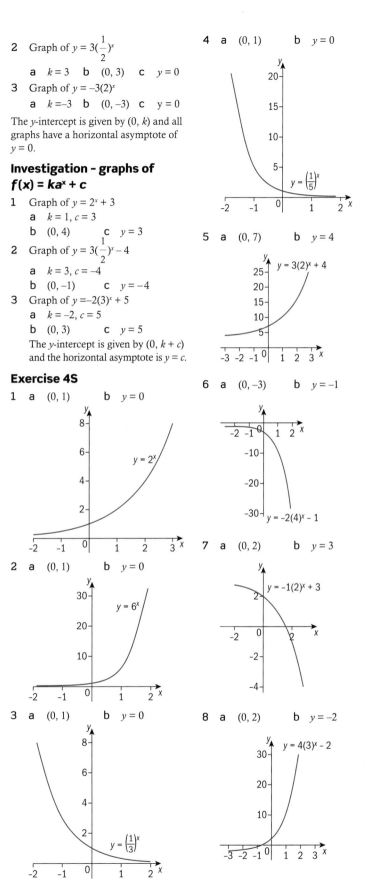

9 a $(0, 3.5)$ **b** $y = 3$

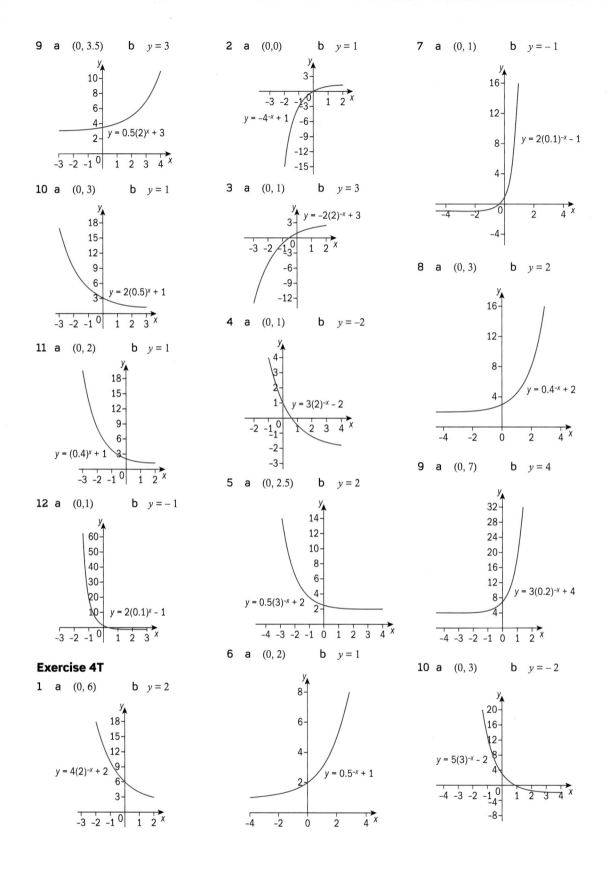

10 a $(0, 3)$ **b** $y = 1$

11 a $(0, 2)$ **b** $y = 1$

12 a $(0,1)$ **b** $y = -1$

Exercise 4T

1 a $(0, 6)$ **b** $y = 2$

2 a $(0,0)$ **b** $y = 1$

3 a $(0, 1)$ **b** $y = 3$

4 a $(0, 1)$ **b** $y = -2$

5 a $(0, 2.5)$ **b** $y = 2$

6 a $(0, 2)$ **b** $y = 1$

7 a $(0, 1)$ **b** $y = -1$

8 a $(0, 3)$ **b** $y = 2$

9 a $(0, 7)$ **b** $y = 4$

10 a $(0, 3)$ **b** $y = -2$

Exercise 4U

1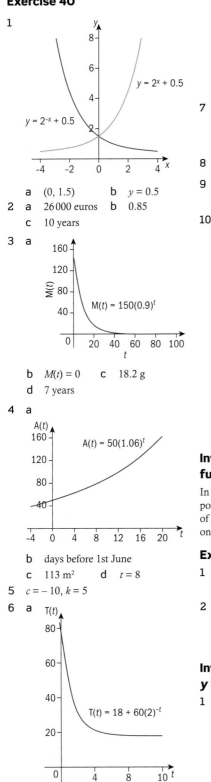

The graph shows $y = 2^x + 0.5$ and $y = 2^{-x} + 0.5$

  a (0, 1.5)   b $y = 0.5$

2 a 26 000 euros b 0.85

  c 10 years

3 a

M(t) graph with $M(t) = 150(0.9)^t$

  b $M(t) = 0$ c 18.2 g

  d 7 years

4 a

A(t) graph with $A(t) = 50(1.06)^t$

  b days before 1st June

  c 113 m²   d $t = 8$

5 $c = -10, k = 5$

6 a

T(t) graph with $T(t) = 18 + 60(2)^{-t}$

  b 78°C

  c 19.875°C

  d 1.45 minutes

  e 18°C, $T = 18$ is an asymptote, as t increases T gets closer to 18 °C

7 a 18 000 USD

  b 10 628.82 USD

  c 7 years

8 $a = 5, b = 0.2$

9 a $a = 4, b = 5$

  b $y = 3$

10 a $a = 1.667, b = 19$

  b

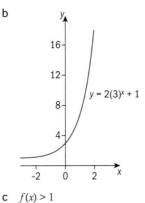

Graph with $y = 2(3)^x + 1$

  c $f(x) > 1$

Investigation - quartic functions

In general, a quartic graph has 3 turning points and intercepts the x-axis a total of 4 times. It intercepts the y-axis just once.

Exercise 4V

1 b 8.77 hours

  c 1.80 hours, 17.4 hours

2 a 6

  b 2

  c $f(x) \geq 6$

Investigation - graphs of $y = ax^{-n}$

1 The graphs of $y = x^{-3}$ and $y = x^{-1}$ (the odd powers) have very similar shapes to each other.
The graphs of $y = x^{-4}$ and $y = x^{-2}$ (even powers) also have very similar shapes to each other.

2 The graphs with odd powers all have similar shapes to each other but the graph of $y = 2x^{-3}$ is further out than the graph of $y = x^{-3}$, which is further out than the graph of $y = x^{-1}$.
The graphs with even powers all have similar shapes to each other but the graph of $y = 3x^{-4}$ is further out than the graph of $y = x^{-4}$, which is further out than the graph of $y = x^{-2}$.

Exercise 4W

1 b 28.9 °C

  c 2.72 minutes

  d $x = 0$

  e $y = 21$

  f 21°C

2 b 90 °C

  c 1.43 minutes

  d 100 °C

3 b ± 0.791

  c $x = 0, y = 0$

  d $f(x) > 0$

4 b 3.75

  c 3

  d $x = 0, y = 3$

  e $\{y \mid y \in \mathbb{R}, y \neq 3\}$

Exercise 4X

1 b minimum value = 17.5 (when $x = 1.71$)

  c 75.3 m s⁻¹

  d 0.403 s, 4.79 s

2 a $V = 2x^2 y$

  b $A = 2x^2 + \dfrac{900}{x}$

  d length = 6.08, breadth = 12.2 cm, height = 4.05 cm

3 a $V = \dfrac{1}{3}x^2 h$

  c $A = x^2 + 2x\sqrt{h^2 + \left(\dfrac{x}{2}\right)^2}$

  d $A = x^2 + 2x\sqrt{\dfrac{4500^2}{x^4} + \dfrac{x^2}{4}}$

  f side length = 14.7 m, height = 20.8 m

4 2670 cm²

Exercise 4Y

1 a $\{x \mid x \in \mathbb{R}, x \neq 0\}$

   b

x	−10	−5	−4	−2	−1	−0.5	−0.2	0	0.2	0.5	1	2	4	5	10
f(x)	0.8	0.6	0.5	0	−1	−3	−9	×	11	5	3	2	1.5	1.4	1.2

   c

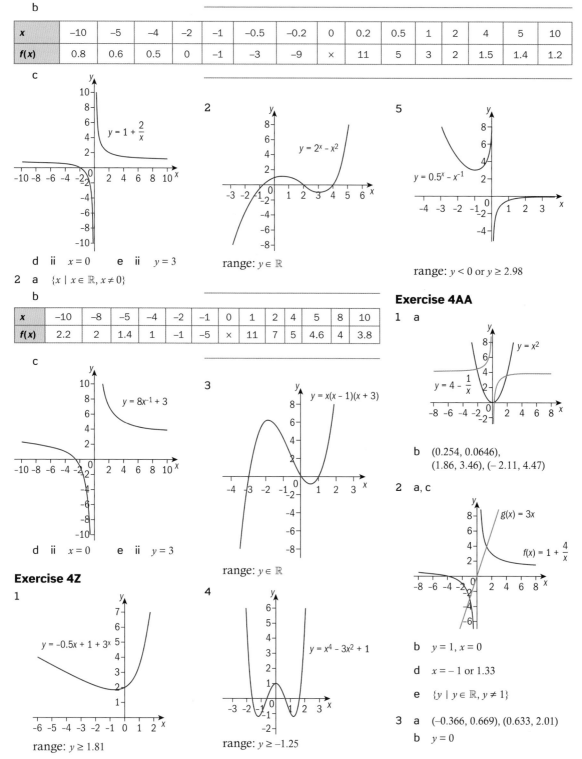

   d ii $x = 0$ e ii $y = 3$

2 a $\{x \mid x \in \mathbb{R}, x \neq 0\}$

   b

x	−10	−8	−5	−4	−2	−1	0	1	2	4	5	8	10
f(x)	2.2	2	1.4	1	−1	−5	×	11	7	5	4.6	4	3.8

   c

   d ii $x = 0$ e ii $y = 3$

Exercise 4Z

1

range: $y \geq 1.81$

2

range: $y \in \mathbb{R}$

3

range: $y \in \mathbb{R}$

4

range: $y \geq -1.25$

5

range: $y < 0$ or $y \geq 2.98$

Exercise 4AA

1 a

   b (0.254, 0.0646),
      (1.86, 3.46), (− 2.11, 4.47)

2 a, c

   b $y = 1, x = 0$

   d $x = -1$ or 1.33

   e $\{y \mid y \in \mathbb{R}, y \neq 1\}$

3 a (−0.366, 0.669), (0.633, 2.01)

   b $y = 0$

4 a

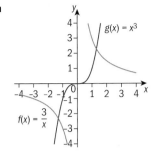

b two solutions

c 1.32 or –1.32

5

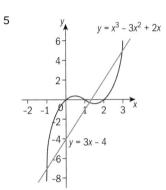

(–1.11, –7.34), (1.25, –0.238), (2.86, 4.58)

6

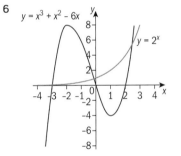

(–2.99, 0.126), (–0.147, 0.903), (2.41, 5.31)

7

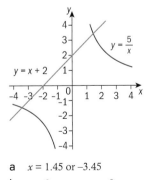

a $x = 1.45$ or -3.45

b $y = 0$ **c** $x = 0$

Exercise 4AB

1 a time in hours, water consumption in litres

b 07:00–20:00

c 07:00– 12:00, 14:00–16:00

d 12:00–14:00, 16:00–20:00

e 12:00 (local maximum at 16:00)

f 07:00, 20:00 (local minimum at 14:00)

2 a time in minutes, temperature in °C

b 100 °C   **c** 35 °C

d $\frac{1}{2}$ minute   **e** no

f approximately 22 °C

3 a

t	0	5	10	15	20
N	1	2	4	8	16

b 13 s   **c** 4096

4 a 45 m   **b** 1.5 s and 5.5 s

c 0–3.5 s   **d** 3.5–7 s

e 90 m, 3.5 s

f ball returns to ground level

5 a i 3.8 m   **ii** 2.2 m

iii 02:00 and 06:00

b $2 < t < 6$

6 a twice   **b** 04:00–09:00

c 16:00   **d** 5 °C

e 11:00–16:00

f 13:00 and 19:30

g no, the temperature at the start of the following day is 1 °C whereas it was 3 °C at the start of this day

7 a $y = \dfrac{16}{x^2}$

b

x	0.5	1	2	4	8	10
y = f(x)	64	16	4	1	0.25	0.16

c

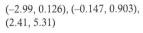

d height tends to 0

8 a 3000 cm³   **b** $y = \dfrac{3000}{x^2}$

c $A = x^2 + \dfrac{12\,000}{x}$

d

x (m)	5	10	15	20	25	30	35
A(x) (cm²)	2400	1300	1000	1000	1100	1300	1600

e

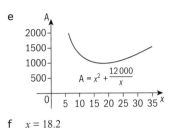

f $x = 18.2$

Review exercise

Paper 1 style questions

1 a 00:00–06:00

b 11:30–17:00

c 13 °C

2 a 4500 SGD

b 8000 SGD

3 a $x(x + 5)$

b

4 a 40 m

b 45 m

c 4 s

5 a $m = 5$

b $n = \dfrac{1}{5}$ $f(2) = \dfrac{4}{5}$

6 a $(x - 5)(x + 3)$

b i (–3, 0)   **ii** (1, –16)

7 a ii   b i   c iii   d iv

8 a i (–1.68, 1.19)

ii (2.41, – 1.81)

b $-1.68 < x < 2.41$

c $y = -2$

9 a $2.2 - x$

b $A = x(2.2 - x)$

c $x = 1.1$ m

10 a

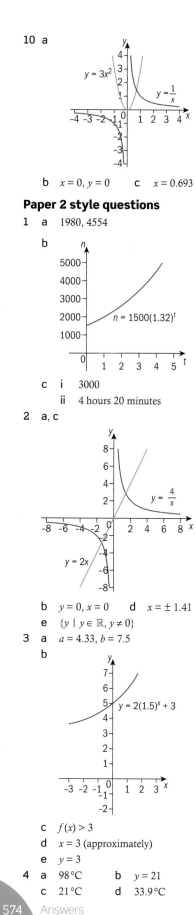

b $x = 0, y = 0$ c $x = 0.693$

Paper 2 style questions

1 a 1980, 4554

b

c i 3000
   ii 4 hours 20 minutes

2 a, c

b $y = 0, x = 0$ d $x = \pm 1.41$
e $\{y \mid y \in \mathbb{R}, y \neq 0\}$

3 a $a = 4.33, b = 7.5$

b

c $f(x) > 3$
d $x = 3$ (approximately)
e $y = 3$

4 a 98 °C   b $y = 21$
c 21 °C   d 33.9 °C

5 a

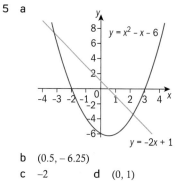

b $(0.5, -6.25)$
c -2 d $(0, 1)$
e $(2.19, -3.39), (-3.19, 7.39)$
f $x = 2.19, -3.19$

6 a, c

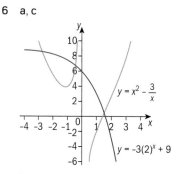

b $x = 0$ d $y = 9$
e $(-2.73, 8.55), (-0.454, 6.81),$
   $(1.53, 0.362)$

7 a

x	0	10	20	30	40	50	60	70	80	90
P	−60	30	100	150	180	190	180	150	100	30

b

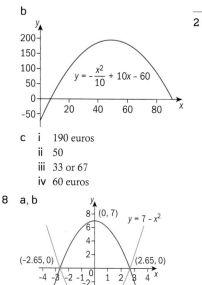

c i 190 euros
   ii 50
   iii 33 or 67
   iv 60 euros

8 a, b

c $x = \pm 2.65$
d $c = 1, 2, 3, 4, 5$

9 a $(0, 0), (2, 2)$
b $x = 2$ c $k = 2$
d

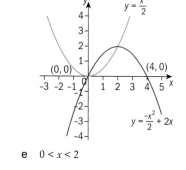

e $0 < x < 2$

Chapter 5

Skills check

1 a mean = 3.61 (3 sf)
   standard deviation = 1.21 (3 sf)
   The standard deviation implies
   that the data are close to the
   mean.

b mean = 4
   standard deviation = 0.643(3 sf)
   Mean = median, as the
   frequencies are symmetrical
   about the mean. The standard
   deviation implies that the data
   are very close to the mean.

2 a

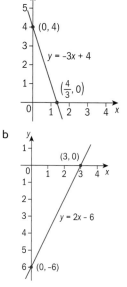

b

Investigation – related data?

There is a positive correlation between
height and shoe size. Inevitably the data
will not fall on a straight line but as a

general trend, the taller the height of a person, the greater their shoe size.

Exercise 5A

1 a

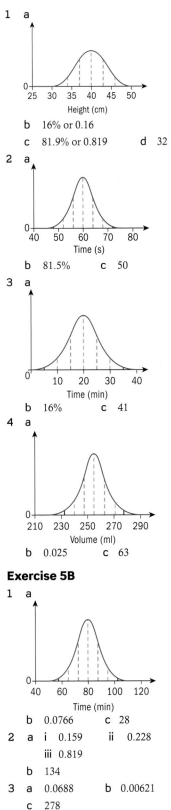

  b 16% or 0.16
  c 81.9% or 0.819   d 32

2 a

  b 81.5%   c 50

3 a

  b 16%   c 41

4 a

  b 0.025   c 63

Exercise 5B

1 a

  b 0.0766   c 28
2 a i 0.159   ii 0.228
    iii 0.819
  b 134
3 a 0.0688   b 0.00621
  c 278

4 a

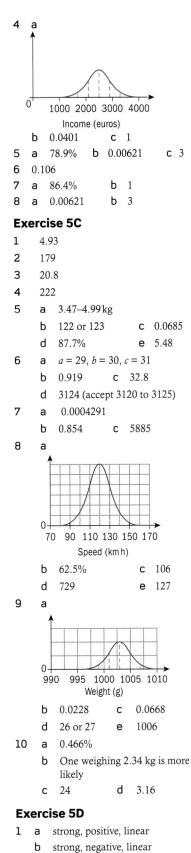

  b 0.0401   c 1
5 a 78.9%   b 0.00621   c 3
6 0.106
7 a 86.4%   b 1
8 a 0.00621   b 3

Exercise 5C

1 4.93
2 179
3 20.8
4 222
5 a 3.47–4.99 kg
  b 122 or 123   c 0.0685
  d 87.7%   e 5.48
6 a $a = 29$, $b = 30$, $c = 31$
  b 0.919   c 32.8
  d 3124 (accept 3120 to 3125)
7 a 0.0004291
  b 0.854   c 5885
8 a

  b 62.5%   c 106
  d 729   e 127
9 a

  b 0.0228   c 0.0668
  d 26 or 27   e 1006
10 a 0.466%
  b One weighing 2.34 kg is more likely
  c 24   d 3.16

Exercise 5D

1 a strong, positive, linear
  b strong, negative, linear
  c moderate, positive, linear
  d weak, positive, linear

  e none
  f perfect, negative, linear
  g non-linear
  h moderate, negative
2 a moderate, positive correlation

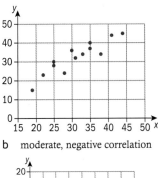

  b moderate, negative correlation

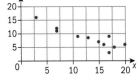

Exercise 5E

1 a i, iii very strong, positive, linear correlation

    ii 13 and 25.75
  b i, iii strong, negative, linear correlation

    ii 16.5 and 20.2
2 a, c moderate, positive, linear correlation

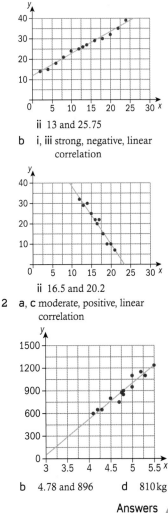

  b 4.78 and 896   d 810 kg

3 **a, c** moderate, positive, linear correlation

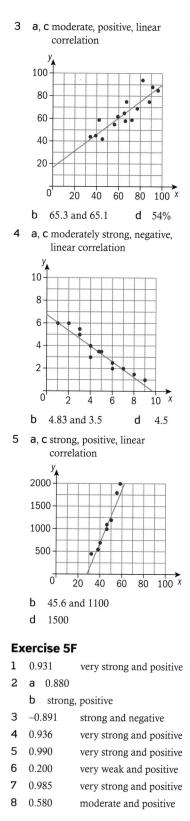

b 65.3 and 65.1 **d** 54%

4 **a, c** moderately strong, negative, linear correlation

b 4.83 and 3.5 **d** 4.5

5 **a, c** strong, positive, linear correlation

b 45.6 and 1100
d 1500

Exercise 5F

1 0.931 very strong and positive
2 **a** 0.880
 b strong, positive
3 −0.891 strong and negative
4 0.936 very strong and positive
5 0.990 very strong and positive
6 0.200 very weak and positive
7 0.985 very strong and positive
8 0.580 moderate and positive

Exercise 5G

1 **a** 0.994 very strong and positive
 b $y = 1.47x + 116$
 c 1586 rupees
2 **a** 0.974
 b $y = 0.483x + 15.6$
 c 19.5 cm
3 **a** mean of $x = 68.6$ and standard deviation of $x = 6.55$
  mean of $y = 137.7$ and standard deviation of $y = 5.97$
 b −0.860
 c strong and negative
 d $y = -0.784x + 191.5$
 e 137 s
4 **a** 0.792
 b $y = 0.193x + 1.22$ **c** 4
5 **a** $y = 0.0127x + 0.688$
 b 1.58 AUD
6 **a** $y = 0.751x + 11.6$ **b** 49
7 **a** $y = 1.04x - 2.53$ **b** 60
8 **a** $y = 0.279x + 2.20$
 b 13.4 hours

Exercise 5H

1 **a** H_0: Genre of books is independent of age
  H_1: Genre of books is not independent of age
 b $130 \times \dfrac{97}{300} = 42.0$
 c 4 **d** 26.9
 e 26.9 > 9.488 so reject null hypothesis
2 **a** H_0: Hair color is independent of eye color
  H_1: Hair color is not independent of eye color
 b $85 \times \dfrac{90}{227} = 33.7$
 c 4 **d** 44.3
 e 44.3 > 7.779 so reject the null hypothesis
3 **a** H_0: Favorite flavor is independent of breed
  H_1: Favorite flavor is not independent of breed
 b $35 \times \dfrac{44}{140} = 11$
 c $(3 - 1)(4 - 1) = 6$

d 0.675
e 0.675 < 12.59 so do not reject the null hypothesis
4 **a** H_0: Film genre is independent of gender
  H_1: Film genre is not independent of gender
 b $39 \times \dfrac{21}{80} = 10.2$
 c 3 **d** 19.1
 e 19.1 > 11.345 so reject the null hypothesis
5 **a** H_0: Grade is independent of number of hours spent playing computer games
  H_1: Grade is not independent of number of hours spent playing computer games
 b $90 \times \dfrac{96}{220} = 39.27 \approx 39.3$
 c $(3 - 1)(3 - 1) = 4$ **d** 42.1
 e 42.1 > 9.488 so reject the null hypothesis
6 **a** H_0: Employment grade is independent of gender
  H_1: Employment grade is not independent of gender
 b

11.5	71.5	539
20.5	127.5	960

 c 2 **d** 180
 e 180 > 4.605 so reject the null hypothesis
7 **a** H_0: Amount of sushi is independent of day of the week
  H_1: Amount of sushi is not independent of day of the week
 b $170 \times \dfrac{145}{470} = 52.4$
 c 4 **d** 0.840
 e 0.840 < 9.488 so do not reject the null hypothesis.
8 **a** H_0: Puppy's weight is independent of its parent's weight
  H_1: Puppy's weight is not independent of its parent's weight
 b $46 \times \dfrac{41}{141} = 13.38 \approx 13.4$
 c 4
 d 13.7
 e 13.7 > 13.277 so reject the null hypothesis

Review exercise

Paper 1 style questions

1 a
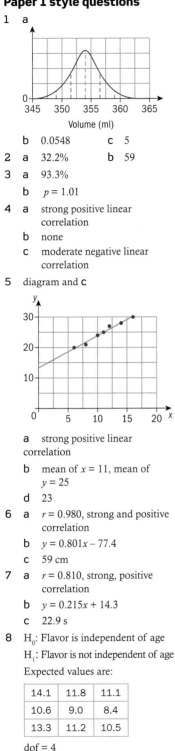

   b 0.0548   c 5
2 a 32.2%   b 59
3 a 93.3%
   b $p = 1.01$
4 a strong positive linear correlation
   b none
   c moderate negative linear correlation
5 diagram and c

   a strong positive linear correlation
   b mean of $x = 11$, mean of $y = 25$
   d 23
6 a $r = 0.980$, strong and positive correlation
   b $y = 0.801x - 77.4$
   c 59 cm
7 a $r = 0.810$, strong, positive correlation
   b $y = 0.215x + 14.3$
   c 22.9 s
8 H_0: Flavor is independent of age
   H_1: Flavor is not independent of age
   Expected values are:

14.1	11.8	11.1
10.6	9.0	8.4
13.3	11.2	10.5

   dof = 4
   $\chi^2 = 0.604$
   p-value = 0.963
   0.963 > 0.05 so do not reject H_0

9 a H_0: The number of pins knocked down is independent of the hand used
   b dof = 2
   c $\dfrac{20 \times 60}{120} = 10$
   d 0.422 > 0.10 so do not reject H_0
10 a H_0: The time to prepare for a test is independent of the outcome
   b 2
   c 0.069 > 0.05 so do not reject H_0

Paper 2 style questions

1 a
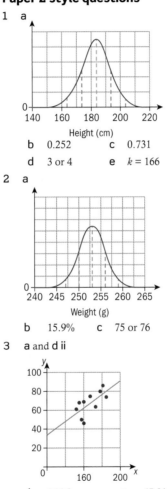

   b 0.252   c 0.731
   d 3 or 4   e $k = 166$
2 a

   b 15.9%   c 75 or 76
3 a and d ii

   b 166.9 cm   c 67.3 kg
   d i $y = 0.719x - 52.8$
   e 69.4 kg
4 a $r = 0.823$
   b strong, positive correlation
   c $y = 0.219x + 3.85$
   d 12 hours
5 a 0.9
   b strong, positive correlation
   c $y = 0.0666x - 2.36$

6 a $r = 0.89$
   b strong, positive correlation
   c $y = 0.0151x + 0.229$
   d 1.44 euros
7 a $y = 0.163x - 15.0$
   b 12.7   c 0.741
   d moderate, positive correlation
8 H_0: Game chosen is independent of gender
   H_1: Game chosen is not independent of gender
   dof = 2
   expected values

39.4	14.8	26.8
29.6	11.2	20.2

   $\chi^2 = 0.667$
   p-value = 0.717
   0.717 > 0.05, so do not reject H_0
9 a $p = 21.6$, $q = 14.4$, $r = 13.6$
   b i H_0: Extra-curricular activity is independent of gender
      ii $(2 - 1)(3 - 1) = 2$
   c $\chi^2 = 4.613$
   d 4.163 > 4.605 so reject H_0
10 a i $\dfrac{180 \times 300}{500} = 108$
      ii $b = 72$, $c = 132$, $d = 88$
   b H_0: Position is independent of gender
      H_1: Position is not independent of gender
   c i $\chi^2 = 59.7$
      ii dof = 2
      iii $\chi^2 >$ critical value so reject H_0
11 a H_1: Choice of candidate is not independent of where the voter lives
   b $\dfrac{3680 \times 3720}{8000} = 1711.2$
      ≈ 1711
   c i $\chi^2 = 58.4$ ii 2
   d i reject H_0
      ii 58.4 > 9.21
12 a $\dfrac{90 \times 110}{200} = 49.5$
   b i H_0: Grade is independent of gender
      ii 2
      iii $\chi^2 = 0.400$
   c 0.400 < 5.991 so do not reject H_0

Chapter 6

Skills check

1 a $f(5) = 3 - 2(5) = -7$

$f(-5) = 3 - 2(-5) = 13$

b $f(2) = 3(2) + 5 = 11$

$f(-3) = 3(-3) + 5 = -4$

c $g(5) = 5^2 = 25$

$g\left(\frac{1}{2}\right) = \left(\frac{1}{2}\right)^2 = \frac{1}{4}$

d $g(2) = \frac{3}{2} = 1.5$

$g(15) = \frac{3}{15} = 0.2$

e $f(4) = \frac{4^2}{(4+1)} = 3.2$

$f(-3) = \frac{(-3)^2}{(-3+1)} = -4.5$

2 a $\frac{C}{2\pi} = r$ b $\sqrt{\frac{A}{\pi}} = r$

c $\sqrt{\frac{A}{4\pi}} = r$ d $\sqrt{\frac{3V}{\pi h}} = r$

e $\sqrt[3]{\frac{3V}{2\pi}} = r$ f $r = \frac{2A}{C}$

3 a 16 b $\frac{1}{8}$ c $\frac{1}{16}$

4 a x^{-1} b x^{-4} c x^2

d x^{-3} e x

5 a $y = 2x - 13$

b $y = -3x + 14$

Investigation - tangent and the gradient function

5

x-coordinate	-3	-2	-1	0	1	2	3	4	x
Gradient of Tangent	-6	-4	-2	0	2	4	6	8	2x

8

x-coordinate	-3	-2	-1	0	1	2	3	4	x
Gradient of Tangent	-12	-8	-4	0	4	8	12	16	4x

Investigation - GDC and the gradient function

3 a −3.5

b 2

c 0

d −1

e 0

f −0.5

4

Curve	Gradient function
$y = x^2$	2x
$y = 2x^2$	4x
$y = 3x^2$	6x

5

Curve	$y = 4x$	$y = -3.5x$	$y = 2x + 4$	$y = 5$	$y = 3 - x$	$y = -3.5$	$y = 2 - \frac{1}{2}x$
Gradient function	4	-3.5	2	0	-1	0	$-\frac{1}{2}$

Curve	$y = x^2$	$y = 2x^2$	$y = 3x^2$	$y = 4x^2$	$y = -x^2$	$y = -2x^2$	$y = \frac{1}{2}x^2$
Gradient function	2x	4x	6x	8x	-2x	-4x	x

6

x-coordinate	-3	-2	-1	0	1	2	3	4
Gradient of Tangent	-3	-1	1	3	5	7	9	11

The algebraic rule: gradient of the tangent = $2x + 3$

a $2x + 3$

b $2x - 5$

c $4x - 3$

d $6x - 1$

e $5 - 4x$

f $2 - 2x$

g $2x$

h $2x$

i $-2x$

j $2x + 1$

k $4x - 1$

l $3 - 2x$

For the general curve: $ax^2 + bx + c$, gradient = $2ax + b$

1 $10x + 7$

2 $5 + 14x$

3 $x - 6$

4 $-3x + 8$

Curve	$y = x^3 + 3x^2 + 2$	$y = x^3 + 4x^2 + 3x$	$y = x^3 + 5x^2 - 4x + 1$	$y = x^3 - x^2 - 5x - 4$
Gradient function	$3x^2 + 6x$	$3x^2 + 8x + 3$	$3x^2 + 10x - 4$	$3x^2 - 2x - 5$

Function	Formula	Gradient Function
Constant	$y = a$	0
Linear	$y = ax + b$	a
Quadratic	$y = ax^2 + bx + c$	$2ax + b$
Cubic	$y = ax^3 + bx^2 + cx + d$	$3ax^2 + 2bx + c$

Investigation - the gradient function of any curve

1 $4x^3$

2 $5x^4$

3 nx^{n-1}

Function	Gradient function
$y = ax^n$	nax^{n-1}

Exercise 6A

1 a $8x$ b $18x^2$ c $28x^3$

d $15x^2$ e $4x^3$ f 5

g 1 h 12 i $18x$

j $\frac{3x^2}{2}$ k x l $3x^3$

2 a 0 b $-9x^2$ c $-x^3$

d $-2x^2$ e -1 f 0

g $30x^5$ h $-63x^8$ i $4x^7$

j $9x^{11}$ k $-6x^8$ l 0

Investigation - the gradient function of a cubic curve

Curve	$y = x^3$	$y = 2x^3$	$y = 3x^3$	$y = 4x^3$	$y = -x^3$	$y = -2x^3$	$y = \frac{1}{2}x^3$
Gradient function	$3x^2$	$6x^2$	$9x^2$	$12x^2$	$-3x^2$	$-6x^2$	$\frac{3}{2}x^2$

Curve	$y = x^3 - 4$	$y = 2x^3 - 3$	$y = x^3 + 5x$	$y = x^3 - 2x$
Gradient function	$3x^2$	$6x^2$	$3x^2 + 5$	$3x^2 - 2$

Curve	$y = x^3 + 2x^2$	$y = 2x^3 + \frac{1}{2}x^2$
Gradient function	$3x^2 + 4x$	$6x^2 + x$

3 **a** $6x + 15x^2$ **b** $20x^3 - 4$
   c $9 - 33x^2$ **d** $4x^3 + 3$

4 **a** $24x^5 - 5$ **b** $18x - 5$
   c $7 + 20x^4$ **d** $4x + 3$

Exercise 6B

1 **a** $36 - 12t^2$ **b** 12
   c $3t^2 - 10t$ **d** $4t + 1$
   e $7 - 4t$ **f** $36t - 9$
   g $3t^2 - 2t + 3$ **h** $6t - 3$

2 **a** $2r$ **b** $2r + 6$
   c $8r - 12$ **d** $8r - 20$
   e $6r + 30$ **f** $f'(r) = 70 - 10r$

Exercise 6C

1 $\dfrac{dy}{dx} = -\dfrac{6}{x^3}$ **2** $f'(x) = -\dfrac{8}{x^5}$

3 $\dfrac{dy}{dx} = -\dfrac{7}{x^2}$ **4** $f'(x) = -\dfrac{16}{x^9}$

5 $\dfrac{dy}{dx} = -\dfrac{35}{x^8}$ **6** $\dfrac{dy}{dx} = -\dfrac{2}{x^2}$

7 $f'(x) = 14x - \dfrac{20}{x^6}$

8 $\dfrac{dy}{dx} = -4 - \dfrac{5}{x^3}$

9 $g'(x) = 3x^2 - \dfrac{6}{x^3}$

10 $\dfrac{dy}{dx} = 4 + \dfrac{3}{x^2}$

11 $g'(x) = 15x^2 + \dfrac{4}{x^5}$

12 $\dfrac{dy}{dx} = 2x^3 + \dfrac{6}{x^9}$

13 $\dfrac{dy}{dx} = \dfrac{x^3}{2} + 6x - \dfrac{10}{3x^5}$

14 $g'(x) = 6x^2 - 2x + \dfrac{3}{x^3}$

15 $A'(x) = 2x + \dfrac{5}{2x^2} - \dfrac{3}{2x^3}$

Exercise 6D

1 $\dfrac{dy}{dx} = 2x - 3 \Rightarrow \dfrac{dy}{dx} = 5$

2 $\dfrac{dy}{dx} = 6 - 3x^2 \Rightarrow \dfrac{dy}{dx} = 6$

3 $\dfrac{dy}{dx} = -8x^3 - 9x^2 \Rightarrow \dfrac{dy}{dx} = 135$

4 $\dfrac{dy}{dx} = 20x + 8 \Rightarrow \dfrac{dy}{dx} = -12$

5 $\dfrac{dy}{dx} = 3x^2 - 5 \Rightarrow \dfrac{dy}{dx} = 103$

6 $\dfrac{dy}{dx} = -2x^3 \Rightarrow \dfrac{dy}{dx} = 16$

7 $\dfrac{dy}{dx} = 21 - 36x^2 \Rightarrow \dfrac{dy}{dx} = -15$

8 $\dfrac{dy}{dx} = 6x - 5 \Rightarrow \dfrac{dy}{dx} = -17$

9 $\dfrac{ds}{dt} = 40 - 10t \Rightarrow \dfrac{ds}{dt} = 10$

10 $\dfrac{ds}{dt} = 35 + 12t \Rightarrow \dfrac{ds}{dt} = 35$

11 $\dfrac{dv}{dt} = 80 \Rightarrow \dfrac{dv}{dt} = 80$

12 $\dfrac{dv}{dt} = 0.7 \Rightarrow \dfrac{dv}{dt} = 0.7$

13 $\dfrac{dA}{dh} = 42h^2 \Rightarrow \dfrac{dA}{dh} = 18\dfrac{2}{3}$

14 $\dfrac{dW}{dp} = 21.75p^2$

   $\Rightarrow \dfrac{dW}{dp} = 87$

15 $\dfrac{dV}{dr} = 8r - \dfrac{18}{r^2}, \dfrac{dV}{dr} = 22$

16 $\dfrac{dA}{dr} = 5 - \dfrac{16}{r^3}, \dfrac{dA}{dr} = 4.75$

17 $\dfrac{dV}{dr} = 21r^2 + \dfrac{8}{r^2}, \dfrac{dV}{dr} = 86$

18 $\dfrac{dA}{dr} = 2\pi r + \dfrac{2\pi}{r^2}, \dfrac{dA}{dr} = 4\pi$

19 $\dfrac{dV}{dr} = 6 - \dfrac{15}{2r^2}, \dfrac{dV}{dr} = 5.7$

20 $\dfrac{dC}{dr} = 45 - \dfrac{36}{r^4}, \dfrac{dC}{dr} = 9$

Exercise 6E

1 **a** $\dfrac{dy}{dx} = 2x + 3$ **b** $x = 2$
   c $y = 6$

2 **a** $\dfrac{dy}{dx} = 4x - 1$ **b** $x = -2$
   c $y = 11$

3 **a** $\dfrac{dy}{dx} = 3 - 2x$ **b** $a = 3, b = 4$

4 $\dfrac{dy}{dx} = 2x - 6, a = 6, b = 0$

5 $\left(\dfrac{1}{2}, -3\dfrac{3}{4} \right)$

6 $(-1, -10)$

7 $(-1, 0), (1, 8)$

8 $(-1, 6), (1, -4); y = -5x + 1$

9 $(2, -11), (-2, 21); y + 8x - 5 = 0$

10 **a** $b = -2$ **b** $\dfrac{dy}{dx} = 2x - 4$
   c at $x = 1$, $\dfrac{dy}{dx} = -2 = b$
   d $c = 1, d = -2$

11 **a** $b = 7$ **b** $\dfrac{dy}{dx} = 2x - 3$
   c at $x = 5$, $\dfrac{dy}{dx} = 2(5) - 3 = 7 = b$
   d at Q, $-3 = 2x - 3 \Rightarrow c = 0$
     $d = 0^2 - 3(0) - 3 = -3$

12 **a** $f'(x) = 4 - 2x$
   b $f(5) = -6$, $f'(5) = -6$
   c $(1, 2)$

13 **a** $f'(x) = 4x - 1$
   b $f'(2) = 7$, $f(2) = 7$
   c $(0.5, 1)$

14 **a** $f'(x) = 3 - 2x$
   b $f'(1) = 3 - 2(1) = 1$
     $f'(1) = 3(1) - 1^2 - 1 = 1$
   c $(4, -5)$

15 **a** $f'(x) = 4x - 1$ **b** $(0, -1),$
                   $\left(\dfrac{5}{2}, 9 \right)$

16 **a** $f'(x) = 2x + 5$ **b** $(-5, -5),$
                   $(2, 9)$

17 $(-1, 2)$

Exercise 6F

1 **a** $\dfrac{dy}{dx} = 2x$ $y = 6x - 9$

   b $\dfrac{dy}{dx} = 6x^2$ $y = 6x - 4$

   c $\dfrac{dy}{dx} = 6 - 2x$ $y = 2x + 4$

   d $\dfrac{dy}{dx} = 6x$ $y = 6x - 13$

   e $\dfrac{dy}{dx} = 4x - 5$ $y = 7x - 14$

   f $\dfrac{dy}{dx} = 10 - 3x^2$ $y = -2x + 21$

   g $\dfrac{dy}{dx} = -4x$ $y = -12x + 29$

   h $\dfrac{dy}{dx} = 6 - 2x$ $y = 2x + 9$

   i $\dfrac{dy}{dx} = 8x - 3x^2$ $y = -16x + 24$

   j $\dfrac{dy}{dx} = 5 - 6x$ $y = 11x + 3$

   k $\dfrac{dy}{dx} = 12x - 6x^2$ $y = 8$

   l $\dfrac{dy}{dx} = 60 - 10x^2$ $y = 40x + 27$

   m $\dfrac{dy}{dx} = 2x^3$ $y = 128x - 391$

   n $\dfrac{dy}{dx} = 10x - 3$ $y = -3x + 17$

o $\dfrac{dy}{dx} = 10 - 4x$ $y = 10x$

p $\dfrac{dy}{dx} = \dfrac{3x^2}{4} - 4x$ $y = -x - 4$

q $\dfrac{dy}{dx} = \dfrac{3}{2}x$ $y = -3x$

r $\dfrac{dy}{dx} = 2x^2$ $y = 2x + \dfrac{5}{3}$

s $\dfrac{dy}{dx} = \dfrac{3}{4}x^2 - 14x$ $y = 31x + 37$

2 a $\dfrac{dy}{dx} = \dfrac{-24}{x^3}$ $3x + y - 9 = 0$

 b $\dfrac{dy}{dx} = \dfrac{-18}{x^4}$ $18x + y - 29 = 0$

 c $\dfrac{dy}{dx} = 6 + \dfrac{16}{x^3}$ $4x - y - 6 = 0$

 d $\dfrac{dy}{dx} = 3x^2 - \dfrac{12}{x^3}$ $15x - y + 20 = 0$

 e $\dfrac{dy}{dx} = 5 + \dfrac{8}{x^2}$ $11x - 2y - 8 = 0$

Exercise 6G

1 $\dfrac{dy}{dx} = 4x$ $x + 4y - 9 = 0$

2 $\dfrac{dy}{dx} = 12x^2$ $x + 3y - 11 = 0$

3 $\dfrac{dy}{dx} = 0.5 - 2x$ $2x - 7y - 25 = 0$

4 $\dfrac{dy}{dx} = 3x + 1$ $x - 5y + 22 = 0$

5 $\dfrac{dy}{dx} = 3 - 2x$ $x + 3y - 30 = 0$

6 $\dfrac{dy}{dx} = 2x + 4$ $x + 4y - 16 = 0$

7 $\dfrac{dy}{dx} = -\dfrac{4}{x^2}$ $x - y = 0$

8 $\dfrac{dy}{dx} = \dfrac{-12}{x^3}$ $x + 12y - 71 = 0$

9 $\dfrac{dy}{dx} = 6 - \dfrac{8}{x^2}$ $x - 2y + 27 = 0$

10 $\dfrac{dy}{dx} = 4x^3 + \dfrac{9}{x^4}$ $x + 5y - 19 = 0$

11 $\dfrac{dy}{dx} = -2 + \dfrac{1}{x^2}$ $2x + 4y - 5 = 0$

12 $\dfrac{dy}{dx} = 5 + \dfrac{9}{2x^2}$ $4x + 22y - 309 = 0$

Exercise 6H

1 $\dfrac{dy}{dx} = 2x - 8$ (5, 1)
 $y = 2x - 9$

2 $\dfrac{dy}{dx} = 3x^2 - 3$ (−2, −2)
 $y = 9x + 16$

3 $\dfrac{dy}{dx} = 1 - \dfrac{6}{x^2}$ (4, 5.5)

 $m(\text{normal}) = -\dfrac{8}{5}$

 $16x + 10y - 119 = 0$

4 $\dfrac{dy}{dx} = 2x + \dfrac{2}{x^3}$ (−1, 0)

 $m(\text{normal}) = \dfrac{1}{4}$

 $x - 4y + 1 = 0$

5   (2, 8) $y = 10x - 12$

 $\left(\dfrac{-4}{3}, 8\right)$ $10x + y + 5\dfrac{1}{3} = 0$

6   (5, 20) $y = -14x + 50$

 (−2, 20) $y = 14x + 8$

7   $11y - x + 1 = 0$

8   $y = \dfrac{-x}{6} - \dfrac{37}{6}$

9 a $x = \dfrac{3}{4}$ b $y = 0$

10 a $x = 2$ b $y = 12$

11 a $x = 4$ b $y = 5x - 11$

12 a $x = 0$ b $y = 3x - 3$

 c $y = -\dfrac{1}{3}x - 3$ or

     $x + 3y + 9 = 0$

13 a $x = -1$ b $y = 16x + 15$

 c $x + 16y + 17 = 0$

14 At $x = 2$, $y = 9$ tangent is
 $y = 36x - 63$

 At $x = -5$, $y = 100$ tangent is
 $y = 36x + 280$

15 $k = 1$, $b = 12$

16 $k = 5$, $b = -6$

17 $k = \dfrac{1}{2}$, $b = 3$

18 $k = 7$, $b = -2$

19 $p = 2.25$, $q = -2$

20 $p = -4$, $q = -18$

Exercise 6I

1 a $V(0) = 100 \, \text{cm}^3$
 b $V(3) = 133 \, \text{cm}^3$
 c   the rate of change of the
     volume of water in the
     container
 d $\dfrac{dV}{dt} = 2 + 3t^2$. At $t = 3$,

     $\dfrac{dV}{dt} = 2 + 3(3)^2 = 29 \, \text{cm}^3\text{s}^{-1}$

 e   There is 133 cm³ of water
     in the container and, at this
     time, water is flowing into the
     container at 29 cm³s⁻¹

2 a $A(0) = 0$ b $A(5) = 45 \, \text{cm}^2$

 c   the rate of change of the
     area of the pool of water

 d $\dfrac{dA}{dt} = 4 + 2t$. At $t = 5$,

     $\dfrac{dA}{dt} = 14 \, \text{cm}^2\text{s}^{-1}$

 e   The area of the pool is
     45 cm² and, at this time, the
     area is increasing at 14 cm²s⁻¹

3 a $W(1) = 685$ tonnes

 b $\dfrac{dW}{dt} = 10t - \dfrac{640}{t^2}$

 c   i $\dfrac{dW}{dt}(3) = -41\dfrac{1}{9}$ tonnes⁻¹

      ii $\dfrac{dW}{dt}(5) = 24.4$ tonnes⁻¹

 d   At $t = 3$, oil is flowing from the
     tank, but at $t = 5$, oil is flowing
     into the tank.

 e $t = 4$
 f   At $t = 4$, the weight of oil in the
     tank is at its maximum value.
     (This is 280 tonnes.)

4 a   At $t = 1$, $\dfrac{dV}{dt} = 8 \, \text{m}^3\text{min}^{-1}$

 b $65 = 10 + 6t + t^2$
     $\Rightarrow t^2 + 6t - 55 = 0$
     $\Rightarrow t = 5 (>0)$
     At $t = 5$

     $\dfrac{dV}{dt} = 16 \, \text{m}^3\text{min}^{-1}$

5 a $\dfrac{dy}{dt}(2) = -16 \, \text{cm}\,\text{s}^{-1}$

     $\dfrac{dy}{dt}(3) = -31 \, \text{cm}\,\text{s}^{-1}$

 b   Use the GDC to solve
     $500 - 4t - t^3 = 0$, $7.77s$

6 a $3.5 \, \text{cm}^2\text{s}^{-1}$
 b   GDC gives $t = 6$.
     Hence, 9.5 cm²s⁻¹

7 a   −23.75 tonnes/hour
 b $t = 3$ hours

8 a   44 degrees per second

 b $t = \dfrac{1}{6}$ seconds

9 a   −15 and −215; these are losses
     of USD 15 000 and 215 000.

 b $\dfrac{dP}{dx} = -30x^2 + 80x + 10$

 c   i $P(2) = 85$, $\dfrac{dP}{dx}(2) = 50$

      ii $P(3) = -10(3)^3 + 40(3)^2 +$
        $10(3) - 15 = 105$,

      $\dfrac{dP}{dx}(3) = -30(3)^2 + 80(3) + 10$

      $= -20$

d i A profit is being made and if production is increased, the profit will increase also.

ii A profit is being made but if production is increased, the profit will decrease.

e $\dfrac{dP}{dx} = -30x^2 + 80x + 10 = 0;$

$x = 2.79, p = 107.$

GDC gives the answer. At this point the level of production yields a profit that is a maximum. Maximum profit = $107000 when 2.79 tonnes are made.

Exercise 6J

1 $x = 3$ **2** $x = 3$

3 $x = -5$ **4** $x = -\dfrac{5}{2}$

5 $x = \pm 3$ **6** $x = \pm 2$

7 $x = \pm\dfrac{1}{2}$ **8** $x = \pm\dfrac{1}{4}$

9 $6x^2 - 18x + 12 = 0$ $x = 1, 2$

10 $9 + 12x + 3x^2 = 0$ $x = -1, -3$

11 $3x^2 - 6x - 45 = 0$

$x^2 - 2x - 15 = 0$ $x = 5, -3$

12 $24x + 3x^2 + 36 = 0$

$x^2 + 8x + 12 = 0$ $x = -2, -6$

13 $6x^2 - 12x = 0$ $x = 0, 2$

14 $60x - 15x^2 = 0$ $x = 0, 4$

15 $\dfrac{dy}{dx} = 1 - \dfrac{1}{x^2} = 0$ $1 = \dfrac{1}{x^2}$

$x^2 = 1$ $x = \pm 1$

16 $\dfrac{dy}{dx} = 1 - \dfrac{4}{x^2} = 0$ $1 = \dfrac{4}{x^2}$

$x^2 = 4$ $x = \pm 2$

17 $\dfrac{dy}{dx} = 4 - \dfrac{9}{x^2} = 0$ $4 = \dfrac{9}{x^2}$

$4x^2 = 9$ $x = \pm 1.5$

18 $\dfrac{dy}{dx} = 8 - \dfrac{1}{2x^2} = 0$ $8 = \dfrac{1}{2x^2}$

$x^2 = \dfrac{1}{16}$ $x = \pm\dfrac{1}{4}$

19 $\dfrac{dy}{dx} = 27 - \dfrac{8}{x^3}$ $27 = \dfrac{8}{x^3}$

$x^3 = \dfrac{8}{27}$ $x = \dfrac{2}{3}$

20 $\dfrac{dy}{dx} = 1 - \dfrac{1}{x^3}$ $1 = \dfrac{1}{x^3}$

$x^3 = 1$ $x = 1$

Exercise 6K

1 $3x^2 - 18x + 24 = 0$

$3(x - 4)(x - 2) = 0$

$x = 4$ $-4)$ minimum

$x = 2$ $0)$ maximum

2 $3x^2 + 12x + 9 = 0$

$x = -1$ $(-1, 1)$ minimum

$x = -3$ $(-3, 5)$ maximum

3 $9 + 6x - 3x^2 = 0$

$x = -1$ $(-1, -5)$ minimum

$x = 3$ $(3, 27)$ maximum

4 $3x^2 - 6x = 0$

$x = 2$ $(2, 1)$ minimum

$x = 0$ $(0, 5)$ maximum

5 $27 - 3x^2 = 0$ $(-3, 54)$ minimum

$(3, 54)$ maximum

6 $18x - 3x^2 = 0$ $(0, 0)$ minimum

$(6, 108)$ maximum

7 $\dfrac{dy}{dx} = 1 - \dfrac{1}{x^2}$ $(1, 2)$ minimum

$(-1, -2)$ maximum

8 $\dfrac{dy}{dx} = 1 - \dfrac{9}{x^2}$ $(3, 6)$ minimum

$(-3, -6)$ maximum

9 $\dfrac{dy}{dx} = \dfrac{1}{2} - \dfrac{8}{x^2}$ $(4, 4)$ minimum

$(-4, -4)$ maximum

10 $\dfrac{dy}{dx} = -\dfrac{9}{x^2} + \dfrac{1}{4}$ $(6, 3)$ minimum

$(-6, -3)$ maximum

11 $\dfrac{dy}{dx} = 2x + \dfrac{16}{x^2}$ $(-2, 12)$ minimum

12 $\dfrac{dy}{dx} = 9 - \dfrac{1}{3x^3}$ $\left(\dfrac{1}{3}, 4.5\right)$ minimum

Exercise 6L

1 $(2, 6)$ minimum

2 $(3, 29)$ maximum

3 $\left(-\dfrac{1}{2}, -3\dfrac{1}{4}\right)$ minimum

4 $\left(\dfrac{5}{2}, 1\dfrac{3}{4}\right)$ minimum

5 $\left(\dfrac{3}{2}, 13\dfrac{1}{4}\right)$ maximum

6 $\left(-\dfrac{5}{4}, \dfrac{235}{8}\right)$ maximum

7 $(5, -4)$ minimum

8 $(9, -81)$ minimum

9 $(-2, -4)$ minimum

Exercise 6M

1 a $b = 7 + h$ **b** $A = h(7 + h)$

2 a $x = 10 - t$ **b** $V = 3t(10 - t)$

3 a $y = 5 - 2x$ **b** $p = x^2(5 - 2x)$

4 a $R = \dfrac{1}{2}r^2(r + 25)$

b $R = \dfrac{1}{2}n(n - 25)^2$

5 a $L = 2m(m + 100 - 5m)$

$= 2m(100 - 4m)$

b $L = 2\left(20 - \dfrac{x}{5}\right)\left(20 + \dfrac{4x}{5}\right)$

6 a $V = \pi r^2(17 - 2r)$

b $V = \dfrac{\pi h(17 - h)^2}{4}$

7 a $y = 5x^2 + 6x - \dfrac{3}{2}$

b $\dfrac{dy}{dx} = 10x + 6$

c $x = -0.6, y = -3.3$

d $c = -5.1$

8 a $N = 2n(5 - 1.5 + 1.2n)$

$= 2n(3.5 + 1.2n)$

b $\dfrac{dN}{dn} = 7 + 4.8n$

c $N = -5.10$ (to 3 sf)

d $x = 3.25$

9 $A = \dfrac{1}{2}L\left(\dfrac{3L - 18}{5}\right) = \dfrac{L(3L - 18)}{10}$

$L = 3, A = -2.7, B = -1.8$

10 $C = \pi r\left(\dfrac{30 - r}{3}\right), r = 15,$

$C = 75\pi, f = 5$

11 $a = b + 10$

$X = 2b(b + 10)$

$b = -5, X = -50$

12 $A = t(12 - 2t)$

$t = 3, A = 18,$ max.

13 $A = 2y(30 - 3y)$

$y = 5, A = 150,$ max.

14 $A = 3M(2M - 28)$

$M = +7, A = -294,$ min.

15 $A = g^2 + (8 - g^2); A' = 4g - 16 = 0$

$\Rightarrow g = 4, A = 32,$ min

16 $S = x^2 + (6 - x)^2$

$S' = 4x - 12 \Rightarrow x = 3, y = 3$

17 $V = r^2h = r^2(6 - r)$

$V' = 12r - 3r^2 = 0$

$r = 4, V = 32$

18 $V = m^2(9 - m)$

$V' = 18m - 3m^2 = 0$

$m = 6 \Rightarrow V = 108,$ max.

$m = 0 \Rightarrow V = 0,$ min.

Exercise 6N

1 $w = 10\,\text{m}, l = 20\,\text{m}$ 2 $x = 12$

3 $2x^2 + 6xh = 150$

$$V = 2x^2 h = 2x^2 \left[\frac{150 - 2x^2}{6x}\right]$$

$$\Rightarrow V = \frac{500}{3} \quad w = 5\,\text{cm}, l = 10\,\text{cm}$$

$$h = \frac{10}{3}\,\text{cm}$$

4 $w = 4\,\text{cm}, l = 6\,\text{cm}$

5 $w = 60\,\text{cm}$

6 $V = 268(3\,\text{sf})\,\text{cm}^3$

   $r = 8, h = 4$

7 $V_{max} = 1000\,\text{cm}^3$

8 $2\pi r^2 + 6\pi rh = 600$

   $V = (600 - 2\pi r^2)$

   $V = 300r - \pi r^3$

   $V' = 300 - 3\pi r^2$

   $h = \dfrac{20}{\sqrt{\pi}} \qquad r = \dfrac{10}{\sqrt{\pi}}\,\text{cm}$

9 $V' = 576 - 192x + 12x^2 = 0$

   $\Rightarrow x^2 - 16x + 48 = 0$

   $(x - 4)(x - 12) = 0$

   $x \neq 12, x = 4 \Rightarrow V = 1024\,\text{cm}^3$

10 $V = 160x - 52x^2 + 4x^3$

   $V' = 160 - 104x + 12x^2$

   $40 - 26x + 3x^2 = 0$

   $(20 - 3x)(2 - x) = 0$

   $x = 2$

   $x \neq \dfrac{20}{3}$

   $\Rightarrow V = 144\,\text{cm}^3$

11 a i $\dfrac{14}{\pi} = 4.46\,\text{cm}\ (3\,\text{sf})$

   b i $\dfrac{350}{4\pi} = 27.9\,\text{cm}\ (3\,\text{sf})$

   c i $\pi r^2 h = 350$

   ii $h = \dfrac{350}{\pi r^2}$

   iii $A = 2\pi r^2 + \dfrac{700}{r}$

   iv $r = 3.82\,\text{cm}(3\,\text{sf})$,
       $h = 7.64\,\text{cm}(3\,\text{sf})$

   v $A = 275\,\text{cm}^2(3\,\text{sf})$

12 a $W = 250\,\text{m}$

   b $1150\,\text{m}$

   c $LW = 50000$

   d $W = 182.6\,\text{m}, L = 273.9\,\text{m}$,
      perimeter $= 913\,\text{m}$(all $3\,\text{sf}$)

13 a $\$3950$

   b $LW = 50000$

   c $W = 165\,\text{m}\ (3\,\text{sf})$,
      $L = 303\,\text{m}\ (3\,\text{sf})$,
      cost $= \$3633.18$

14 a $h = 16\,\text{cm}$

      Page area $= 13 \times 22$
      $= 286\,\text{cm}^2$

   b $293\,\text{cm}^2(3\,\text{sf})$ c $A = wh$

   d $P = (w + 4)(h + 6)$

   f Width $9.8\,\text{cm}(3\,\text{sf})$
      Height $14.7\,\text{cm}(3\,\text{sf})$

15 a i   Width $= 50\,\text{cm}$
       iii Frame $= 480\,\text{cm}$

   b $225000 = 2x^2 h$

   d $L = 6x + \dfrac{450000}{x^2}$

   $$\Rightarrow \frac{dL}{dx} = 6 - \frac{900000}{x^3}$$

   Setting this equal to zero

   $\Rightarrow x = \sqrt[3]{150000} = 53.1(3\,\text{sf})$

   Width $53.1\,\text{cm}(3\,\text{sf})$
   Length $106.2\,\text{cm}(3\,\text{sf})$
   Height $39.8\,\text{cm}(3\,\text{sf})$
   Length of frame $478\,\text{cm}(3\,\text{sf})$

Chapter 7

Skills check

1 a $88.0\ (3\,\text{sf})$

   b $s = \dfrac{A - \pi r^2}{\pi r}$

2 a $655.20\,\text{GBP}$

   b 730.24

   c 96 euros

3 $x = 1, y = -5$

Investigation – number sequences

Triangle Numbers:

1	3		3		1
1	4	6	4		1

Natural Numbers:	1	2	3	4	5	6	7	8	9
Cube Numbers:	1	8	27	64	125	216	343	512	729

Investigation – allowances

A: Total allowance $= 10400$ euros

B: Total allowance $= 9693$ euros

Option A the best

Exercise 7A

1 a 31 b 599

2 a $u_1 + 2d = 8$
      $u_1 + 8d = 26$

   b $u_1 = 2, d = 3$

3 3.5

4 a $4n - 1$ b 199

5 a $39, 36$ b 17 c 8

6 a 4 b 53

7 56

8 a 4 b 43 c 21

9 a 2.5 b 35.5 c 73

10 a $19, 26$ b 7 c 187

Exercise 7B

1 a 26 b 246 c 6175

2 a $5k + 2 - (k + 4)$
      $= 10k - 2 - (5k + 2)$
      $4k - 2 = 5k - 4$
      $k = 2$

   b $6, 12, 18$ c 6

   d 150 e 1950

3 a i 6 ii -10 b 28700

4 a $16 - 4n$ b -11680

5 a i -3 ii 5

   b -5010

6 5775 7 127.5

8 a $3k + 4 - (4k - 2)$
      $= 6k - (3k + 4)$
      $-k + 6 = 3k - 4$
      $4k = 10$
      $k = 2.5$

   b $8, 11.5, 15$ c 3.5

   d 57 e 487.5

Exercise 7C

1 a $\$475$ b $\$4725$

2 a $2\,\text{m}\ 50\,\text{s}$ b $32\,\text{m}\ 30\,\text{s}$

3 $p = a = 400$

4 a $\$2400$ b $\$12750$

   c Option 2 has $\$750$ more.

5 a $\$190$ b $\$2550$

6 a 36 b 1050

Exercise 7D

1 a 2 b 2097152

2 a $\dfrac{1}{3}$ b $\dfrac{2}{6561} = 0.000305$

3 a -0.5 b -10

4 a 2 b 320

5 a 1.5 b 205.03125

6 a -8 b -0.125

7 $\dfrac{2}{3}$

8 a 2 b 24576

Investigation – grains of rice

Grains of rice is $2^{64} - 1 = 1.84 \times 10^{19}$

Investigation – becoming a millionaire

After 27 months you would become a millionaire.

Exercise 7E

1 a 4   b 0.25   c 32.0
2 a 4 or –4
  b $r = 4$ then
     sum = 11 184 810
     $r = -4$ then
     sum = –6 710 886
3 a –3   b 29 524
4 a 0.5 or –0.5
  b $r = 0.5$ then sum = 83.9
     $r = -0.5$ then sum = –28.0
5 16 382
6 –64.125

Exercise 7F

1 0.975 m   2 49 431.11 GBP
3 10 230 BGN
4 a 112.57 Dinar
  b 1273.37 Dinar
5 236 221
6 a 142 800   b 157 663
7 a 1.05   b $40 811
8 a Common ratio = $\dfrac{24}{8} = 3$
     and $\dfrac{72}{24} = 3$
  b 648   c 8744

Exercise 7G

1 10 815.82 ringgits
2 a 391.50 euros   b 54.18 GBP
3 a 606.40 euros   b 726.23 CAD
  c 73.77 CAD
4 a 888 euros   b 7338.84 SEK
  c 661.16 SEK
5 a 1992.00 ZAR   b 125.50 BRL
6 a 288.56 euros   b 19.18 GBP
7 a 3297.50 USD   b 939.38 EUR
  c lost 43.98 EUR
8 a 206 yuan   b 174 655 yen
  c 0.85 GBP
9 a 45 euros   b 2518.84 GBP
  c 486.27 euros
10 a 22 475 IDR   b 229 761 CLP
11 a $p = 1.3175$, $q = 107.99$
  b i 176.06 EUR
     ii 146.40 GBP
12 a 1907.10 GBP   b 16.95 GBP

Exercise 7H

1 a 7715.52 JPY   b 11 years
2 a A has 3105.94 euros;
     B has 3090.64 euros and
     C has 3067.47 euros
  b 9.21 or 10 years
  c 16.2 or 17 years
3 a $6110.73   b $r = 3.79$
4 a 23 348.49 EGP
  b 22.4 or 23 years
5 a 61 252.15 SGD
  b 75 070.16 SGD
6 Mr Lin has 11 698.59 CNY and Mr
   Lee has 11 707.24 CNY so Mr Lee
   has the most interest.
7 a 1348.85 GBP
  b 2965 GBP   c 11.6 or 12 years
8 a $a\left(1 + \dfrac{6}{100}\right) + (8000 - a)$

   $\left(1 + \dfrac{5}{100}\right) = 8430$

  b 3000 euros in Bank A and 5000
     euros in Bank B

Exercise 7I

1 3.69 euros   2 3 745 833 MXN
3 8811.63 USD   4 50.77 CAD
5 13.69 KRW   6 28 687.26 GBP
7 60 303.57 USD
8 119 985.99 euros

Review exercise

Paper 1 style questions

1 a 11.8%   b 6.21 or 7 years
2 a USD 256 944
  b 2.32%
3 a GBP 220.10 b 4.49 or 5 years
  c 16.5 or 17 years
4 a 94.13 EUR b 0.99 AUD
5 a £1607   b £8073.70
6 a €35 220
  b €26.4 or 27 months
7 a first term = 6, common
     difference = 3
  b 153   c 3975
8 132
9 a first term = ±15,
     common ratio = ± 2
  b 480   c 3825 or 1275
10 a $\dfrac{1}{3}$ b $\dfrac{2}{27}$
  c 81.0 or $\dfrac{59078}{729}$

11

11 a first term = ±8,
      common ratio = $\pm\dfrac{1}{2}$
  b –0.25   c –15.75 or 5.25
12 a $450   b $1009.11
  c one
13 a 288   b $3r^5 = 96$
  c 2

Paper 2 style questions

1 a i $2750 ii $1920
  b $21 250
  c option two (by $1250)
2 a A: $1800, B: $1767.54,
     C: $1920, D: $1910.06
  b C–largest total amount
  c 6.27%
3 a i $2250, $2500
     ii $6750
     iii $\dfrac{20}{2}(2000 + 6750) = \$87\,500$
  b i $2940
     ii $2800 \times 1.05^4 = 3403.42$
  c $5085 (option 1)
4 a $(6k + 4) - 5k = 5k - (3k + 1)$
     $\Rightarrow k + 4 = 2k - 1 \Rightarrow k = 5$
  b 16, 25, 34   c 9
  d 142   e 2030
5 a 31 496.19 GBP
  b i 18 years
     ii 467.23 GBP
6 a 12   b $\dfrac{1}{5}$ c 2.50

Chapter 8

Skills check

1 a 5 is an integer, real and rational
     $\left(\text{since it can be written as } \dfrac{5}{1}\right)$
  b $1.875 = 1\dfrac{7}{8}$ is not an integer, but
     is both real and rational, since it
     can be written as $\dfrac{15}{8}$
  c $0.333 = \dfrac{333}{1000}$ is not an
     integer, but is both real
     and rational. Note that
     $0.333 \neq \dfrac{1}{3}$
  d 0.303 003 000 3... is real, but not
     rational.
  e $\sqrt{0.5625} = \dfrac{3}{4}$ is both real
     and rational.
  f $\sqrt[3]{2.744} = 1.4 = \dfrac{7}{5}$ is both real
     and rational.
  g π^2 is real, but not rational.

2 For **a–d** : $-2, -1, 0, 1, 2, 3$

3 a i 1, 2, 3, 4, 6, 12

    ii 1, 2, 4, 8 **iii** 1, 17

    iv 1, 5, 25

    v 1, 2, 3, 4, 6, 8, 12, 24

  b i 2, 3 **ii** 2 **iii** 17

    iv 5 **v** 2, 3

  c 17 is prime.

  d Zero has an infinite number of factors. Zero is an integer, it is rational and it is real, but it is not prime.

Investigation – a contradiction?

There is no contradiction, because some pupils study **both** Chemistry and Biology.

The question, 'How many?' cannot be resolved; there is not enough information.

But, **at least 2** study both subjects and there might be as many as 13.

Investigation – intuition

1 a Not fair, and not practical

  b Not fair

  c Fair, but not practical

2 a Fair   **b** Not fair

  c Not fair. Or is it?

  d Not fair **e** Fair

Exercise 8A

1 a & b

$M = \{2, 3, 4\}, n(M) = 3$

$N = \{1, 2, 3, 4, 5\}, n(N) = 5$

$P = \{1, 2, 3, 4, 5\}, n(P) = 5$

$S = \{(1, 4), (2, 3), (3, 2), (4, 1)\}, n(S) = 4$

$T = \{(0, 5), (1, 4), (2, 3), (3, 2), (4, 1), (5, 0)\}, n(T) = 6$

$V = \{ \}$ or $\varnothing, n(V) = 0$

$W = \{1, 2, 4, 5, 10, 20\}, n(W) = 6$

X is an infinite set and so elements cannot be listed, $n(X) = \infty$

2 a $\{4, 5, 6\}$ **b** $\{2, 4, 6\}$

  c $\{7, 9, 11\}$ **d** $\{5, 9, 13, 17, 21\}$

  e $\{(2,2), (4,4), (6,6), (8,8), (10,10)\}$

  f $\{(6, 3), (10, 5)\}$

3 a $\{x \,|\, x = 2y, y \in \mathbb{Z}^+\}$

  b $\{p \,|\, p \text{ is prime}\}$

  c $\{x \,|\, -2 \le x \le 2, x \in \mathbb{Z}\}$

  d $\{x \,|\, 2 \le x \le 8, x \in \mathbb{Z}\}$

  e $\{x \,|\, -2 \le x \le 8, x \text{ is even}\}$

  f $\{x \,|\, x = 3y, 1 \le y \le 6, y \in \mathbb{Z}\}$

Exercise 8B

1 False **2** True **3** False

4 True **5** True **6** True

7 True **8** True

Exercise 8C

1 a False **b** True **c** False

  d True **e** True **f** False

  g False **h** False

2 a i $\varnothing, \{a\}$

    ii $\varnothing, \{a\}, \{b\}, \{a, b\}$

    iii $\varnothing, \{a\}, \{b\}, \{c\}, \{a, b\}, \{a, c\}, \{b, c\}, \{a, b, c\}$

    iv There are 16 of these!

  b 2^n **c** 64 **d** 7

3 a i There are none.

    ii $\{a\}, \{b\}$

    iii $\{a\}, \{b\}, \{c\}, \{a, b\}, \{a, c\}, \{b, c\}$

    iv There are 14 of these!

  b $2^n - 2$ **c** 62 **d** 8

Exercise 8D

1

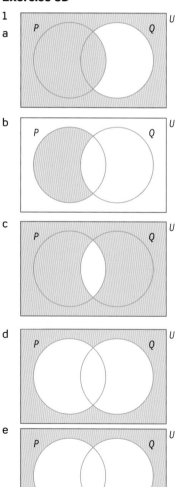

Exercise 8B

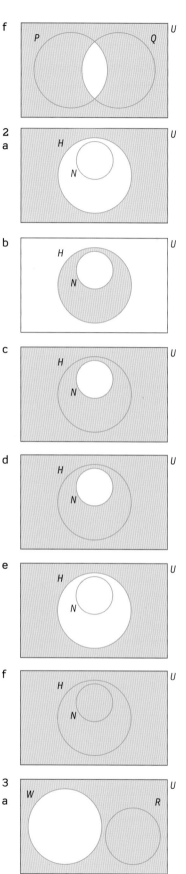

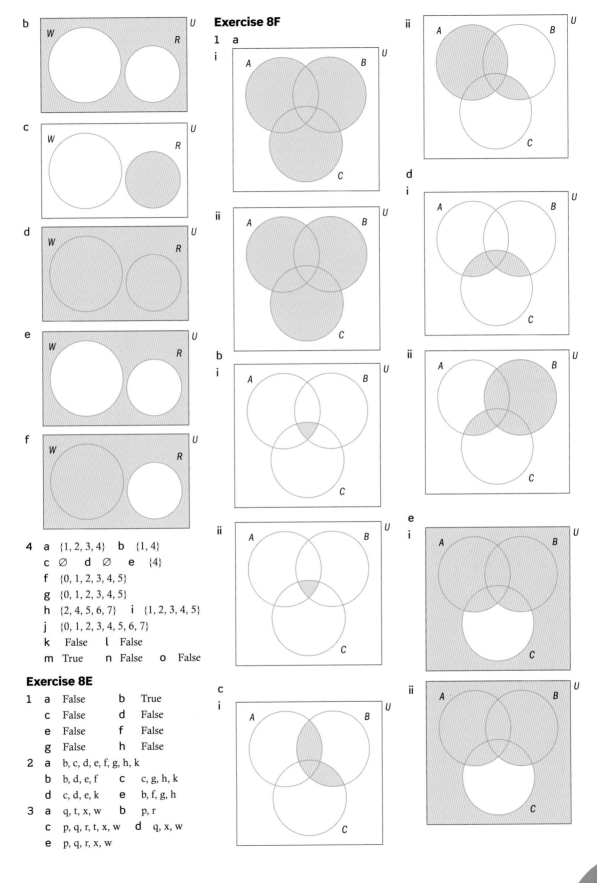

b

c

d

e

f

Exercise 8F

1 a

i

ii

b

i

ii

c

i

ii

d

i

ii

e

i

ii

4 a {1, 2, 3, 4} b {1, 4}
   c ∅ d ∅ e {4}
   f {0, 1, 2, 3, 4, 5}
   g {0, 1, 2, 3, 4, 5}
   h {2, 4, 5, 6, 7} i {1, 2, 3, 4, 5}
   j {0, 1, 2, 3, 4, 5, 6, 7}
   k False l False
   m True n False o False

Exercise 8E

1 a False b True
   c False d False
   e False f False
   g False h False
2 a b, c, d, e, f, g, h, k
   b b, d, e, f c c, g, h, k
   d c, d, e, k e b, f, g, h
3 a q, t, x, w b p, r
   c p, q, r, t, x, w d q, x, w
   e p, q, r, x, w

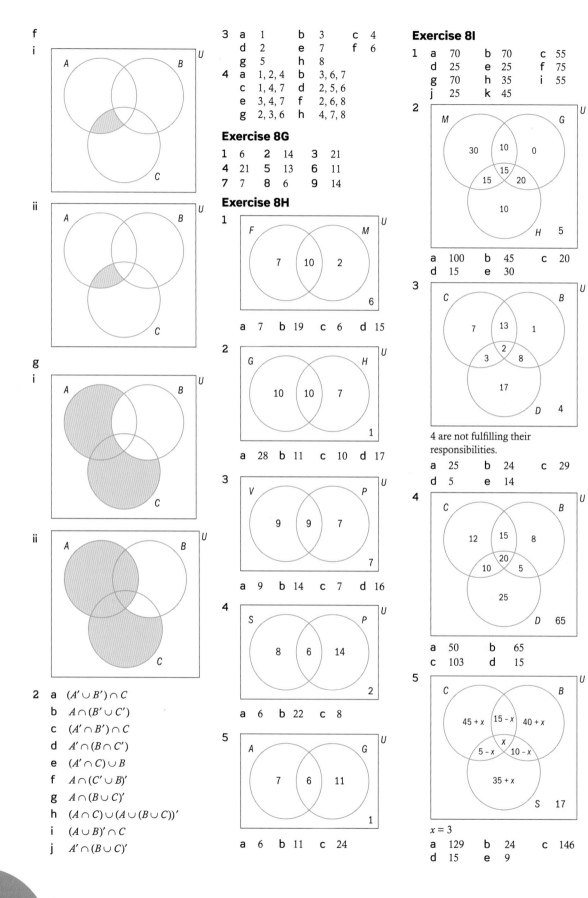

f

i

ii

g

i

ii

2 a $(A' \cup B') \cap C$
  b $A \cap (B' \cup C')$
  c $(A' \cap B') \cap C$
  d $A' \cap (B \cap C')$
  e $(A' \cap C) \cup B$
  f $A \cap (C' \cup B)'$
  g $A \cap (B \cup C)'$
  h $(A \cap C) \cup (A \cup (B \cup C))'$
  i $(A \cup B)' \cap C$
  j $A' \cap (B \cup C)'$

3 a 1   b 3   c 4
  d 2   e 7   f 6
  g 5   h 8
4 a 1, 2, 4   b 3, 6, 7
  c 1, 4, 7   d 2, 5, 6
  e 3, 4, 7   f 2, 6, 8
  g 2, 3, 6   h 4, 7, 8

Exercise 8G

1 6   2 14   3 21
4 21   5 13   6 11
7 7   8 6   9 14

Exercise 8H

1

a 7   b 19   c 6   d 15

2

a 28   b 11   c 10   d 17

3

a 9   b 14   c 7   d 16

4

a 6   b 22   c 8

5

a 6   b 11   c 24

Exercise 8I

1 a 70   b 70   c 55
  d 25   e 25   f 75
  g 70   h 35   i 55
  j 25   k 45

2

a 100   b 45   c 20
d 15   e 30

3

4 are not fulfilling their responsibilities.
a 25   b 24   c 29
d 5   e 14

4

a 50   b 65
c 103   d 15

5

$x = 3$
a 129   b 24   c 146
d 15   e 9

6

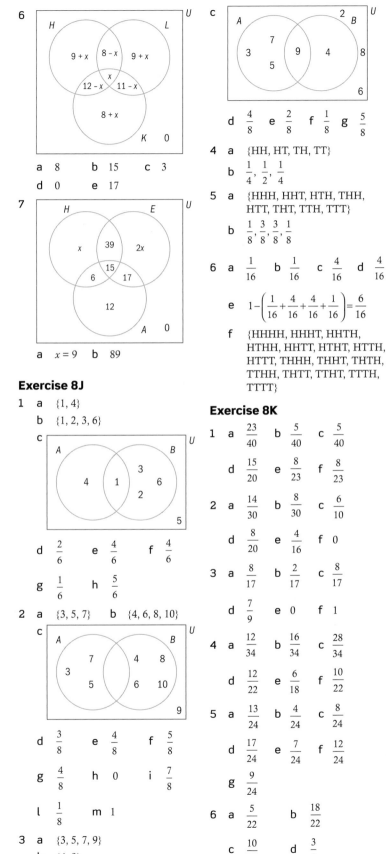

Venn diagram with universal set U, circles H, L, K:
- H only: 9 + x
- H ∩ L only: 8 − x
- L only: 9 + x
- H ∩ K only: 12 − x
- center (H∩L∩K): x
- L ∩ K only: 11 − x
- K only: 8 + x
- outside: 0

a 8   b 15   c 3
d 0   e 17

7

Venn diagram with U, circles H, E, A:
- H only: x
- H ∩ E only: 39
- E only: 2x
- H ∩ A only: 6
- center: 15
- E ∩ A only: 17
- A only: 12
- outside: 0

a x = 9   b 89

Exercise 8J

1 a {1, 4}
b {1, 2, 3, 6}
c
Venn diagram U, circles A and B:
- A only: 4
- A ∩ B: 1
- B: 3, 2, 6
- outside: 5

d $\frac{2}{6}$ e $\frac{4}{6}$ f $\frac{4}{6}$
g $\frac{1}{6}$ h $\frac{5}{6}$

2 a {3, 5, 7}   b {4, 6, 8, 10}
c
Venn diagram U, circles A and B:
- A only: 7, 3, 5
- A ∩ B: 4, 6
- B: 8, 10
- outside: 9

d $\frac{3}{8}$ e $\frac{4}{8}$ f $\frac{5}{8}$
g $\frac{4}{8}$ h 0   i $\frac{7}{8}$
l $\frac{1}{8}$ m 1

3 a {3, 5, 7, 9}
b {4, 9}

c
Venn diagram U, circles A and B:
- A only: 3, 5
- A ∩ B: 7, 9
- B: 4, 8
- outside: 6

d $\frac{4}{8}$ e $\frac{2}{8}$ f $\frac{1}{8}$ g $\frac{5}{8}$

4 a {HH, HT, TH, TT}
b $\frac{1}{4}, \frac{1}{2}, \frac{1}{4}$

5 a {HHH, HHT, HTH, THH, HTT, THT, TTH, TTT}
b $\frac{1}{8}, \frac{3}{8}, \frac{3}{8}, \frac{1}{8}$

6 a $\frac{1}{16}$ b $\frac{1}{16}$ c $\frac{4}{16}$ d $\frac{4}{16}$
e $1-\left(\frac{1}{16}+\frac{4}{16}+\frac{4}{16}+\frac{1}{16}\right)=\frac{6}{16}$
f {HHHH, HHHT, HHTH, HTHH, HHTT, HTHT, HTTH, HTTT, THHH, THHT, THTH, TTHH, THTT, TTHT, TTTH, TTTT}

Exercise 8K

1 a $\frac{23}{40}$ b $\frac{5}{40}$ c $\frac{5}{40}$
d $\frac{15}{20}$ e $\frac{8}{23}$ f $\frac{8}{23}$

2 a $\frac{14}{30}$ b $\frac{8}{30}$ c $\frac{6}{10}$
d $\frac{8}{20}$ e $\frac{4}{16}$ f 0

3 a $\frac{8}{17}$ b $\frac{2}{17}$ c $\frac{8}{17}$
d $\frac{7}{9}$ e 0   f 1

4 a $\frac{12}{34}$ b $\frac{16}{34}$ c $\frac{28}{34}$
d $\frac{12}{22}$ e $\frac{6}{18}$ f $\frac{10}{22}$

5 a $\frac{13}{24}$ b $\frac{4}{24}$ c $\frac{8}{24}$
d $\frac{17}{24}$ e $\frac{7}{24}$ f $\frac{12}{24}$
g $\frac{9}{24}$

6 a $\frac{5}{22}$ b $\frac{18}{22}$
c $\frac{10}{15}$ d $\frac{3}{8}$

c
Venn diagram U, circles A and B:
- A only: 3, 7, 5
- A ∩ B: 9
- B: 4, 8
- outside: 2, 6

d $\frac{4}{8}$ e $\frac{2}{8}$ f $\frac{1}{8}$ g $\frac{5}{8}$

7 a $\frac{12}{28}$ b $\frac{4}{13}$ c $\frac{4}{16}$
d $\frac{3}{28}$ e $\frac{12}{21}$

8 a $\frac{12}{27}$ b $\frac{12}{20}$ c $\frac{7}{19}$
d $\frac{2}{7}$ e $\frac{12}{17}$

Exercise 8L

1 $A \cap B = \{1\}$
2 $A \cap B = \varnothing$, so A and B are mutually exclusive
3 $A \cap B = \{2\}$
4 $A \cap B = \varnothing$, so A and B are mutually exclusive
5 $A \cap B = \{9\}$
6 $A \cap B = \varnothing$, so A and B are mutually exclusive
7 $A \cap B = \{6\}$
8 $A \cap B = \varnothing$, so A and B are mutually exclusive

Exercise 8M

1 Not independent
2 Independent
3 Not independent
4 Independent events
5 Not independent events
6 Not independent events

Exercise 8N

1
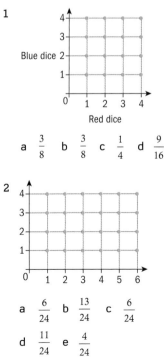

Grid graph, Red dice (x-axis 0–4) vs Blue dice (y-axis 0–4)

a $\frac{3}{8}$ b $\frac{3}{8}$ c $\frac{1}{4}$ d $\frac{9}{16}$

2

Grid graph (x-axis 0–6, y-axis 0–4)

a $\frac{6}{24}$ b $\frac{13}{24}$ c $\frac{6}{24}$
d $\frac{11}{24}$ e $\frac{4}{24}$

3

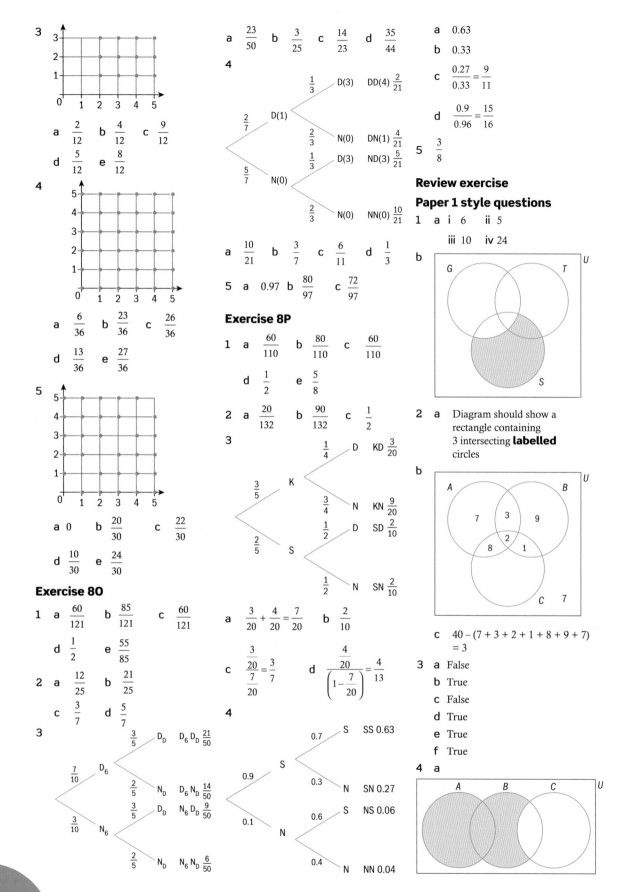

a $\dfrac{2}{12}$ b $\dfrac{4}{12}$ c $\dfrac{9}{12}$

d $\dfrac{5}{12}$ e $\dfrac{8}{12}$

4

a $\dfrac{6}{36}$ b $\dfrac{23}{36}$ c $\dfrac{26}{36}$

d $\dfrac{13}{36}$ e $\dfrac{27}{36}$

5

a 0 b $\dfrac{20}{30}$ c $\dfrac{22}{30}$

d $\dfrac{10}{30}$ e $\dfrac{24}{30}$

Exercise 8O

1 a $\dfrac{60}{121}$ b $\dfrac{85}{121}$ c $\dfrac{60}{121}$

  d $\dfrac{1}{2}$ e $\dfrac{55}{85}$

2 a $\dfrac{12}{25}$ b $\dfrac{21}{25}$

  c $\dfrac{3}{7}$ d $\dfrac{5}{7}$

3

$\dfrac{7}{10}$ D_6 $\dfrac{3}{5}$ D_D $D_6\,D_D$ $\dfrac{21}{50}$

  $\dfrac{2}{5}$ N_D $D_6\,N_D$ $\dfrac{14}{50}$

$\dfrac{3}{10}$ N_6 $\dfrac{3}{5}$ D_D $N_6\,D_D$ $\dfrac{9}{50}$

  $\dfrac{2}{5}$ N_D $N_6\,N_D$ $\dfrac{6}{50}$

a $\dfrac{23}{50}$ b $\dfrac{3}{25}$ c $\dfrac{14}{23}$ d $\dfrac{35}{44}$

4

$\dfrac{2}{7}$ $D(1)$ $\dfrac{1}{3}$ $D(3)$ $DD(4)$ $\dfrac{2}{21}$

  $\dfrac{2}{3}$ $N(0)$ $DN(1)$ $\dfrac{4}{21}$

$\dfrac{5}{7}$ $N(0)$ $\dfrac{1}{3}$ $D(3)$ $ND(3)$ $\dfrac{5}{21}$

  $\dfrac{2}{3}$ $N(0)$ $NN(0)$ $\dfrac{10}{21}$

a $\dfrac{10}{21}$ b $\dfrac{3}{7}$ c $\dfrac{6}{11}$ d $\dfrac{1}{3}$

5 a 0.97 b $\dfrac{80}{97}$ c $\dfrac{72}{97}$

Exercise 8P

1 a $\dfrac{60}{110}$ b $\dfrac{80}{110}$ c $\dfrac{60}{110}$

  d $\dfrac{1}{2}$ e $\dfrac{5}{8}$

2 a $\dfrac{20}{132}$ b $\dfrac{90}{132}$ c $\dfrac{1}{2}$

3

$\dfrac{3}{5}$ K $\dfrac{1}{4}$ D KD $\dfrac{3}{20}$

  $\dfrac{3}{4}$ N KN $\dfrac{9}{20}$

$\dfrac{2}{5}$ S $\dfrac{1}{2}$ D SD $\dfrac{2}{10}$

  $\dfrac{1}{2}$ N SN $\dfrac{2}{10}$

a $\dfrac{3}{20}+\dfrac{4}{20}=\dfrac{7}{20}$ b $\dfrac{2}{10}$

c $\dfrac{\frac{3}{20}}{\frac{7}{20}}=\dfrac{3}{7}$ d $\dfrac{\frac{4}{20}}{\left(1-\frac{7}{20}\right)}=\dfrac{4}{13}$

4

0.9 S 0.7 S SS 0.63

  0.3 N SN 0.27

0.1 N 0.6 S NS 0.06

  0.4 N NN 0.04

a 0.63

b 0.33

c $\dfrac{0.27}{0.33}=\dfrac{9}{11}$

d $\dfrac{0.9}{0.96}=\dfrac{15}{16}$

5 $\dfrac{3}{8}$

Review exercise

Paper 1 style questions

1 a i 6   ii 5

   iii 10   iv 24

  b

2 a Diagram should show a rectangle containing 3 intersecting **labelled** circles

  b

  c $40-(7+3+2+1+8+9+7)$
    $=3$

3 a False
  b True
  c False
  d True
  e True
  f True

4 a

b 15

c **i** 5, 10, 15, 20 **ii** 10, 20, 30

5 a For example 2, –3 etc.

b For example $\dfrac{3}{5}\left(\text{not }\dfrac{6}{1}\right)$

c For example $\dfrac{3}{5}$ or $\dfrac{6}{1}$

d For example $\dfrac{3}{5}, \sqrt{2}, \pi$

e For example $\sqrt{2}, \pi$

f For example $\sqrt{2}, \pi$

6 a $\dfrac{4}{60}$ or 6.67% or 0.0667

b $\dfrac{56}{60}$ or 93.3% or 0.933

c $\dfrac{16}{20}$ or 80% or 0.8

7 a $\dfrac{3}{15}$ or 20% or 0.2

b $\dfrac{3}{14}$ or 21.4% or 0.214

c $\dfrac{4}{15} \times \dfrac{3}{14} = \dfrac{2}{35}$ or 5.71% or 0.0571

8 a 12

b $\dfrac{3}{12} = \dfrac{1}{4}$ or 25%

c $\dfrac{4}{12} = \dfrac{1}{3}$ or 33.3% (3 sf)

9 a $3400 \le w < 3700$

b $\dfrac{5}{50} = \dfrac{1}{10}$ or 10% or 0.1

c $1 - \dfrac{5}{50} = \dfrac{45}{50} = \dfrac{9}{10}$ or 90% or 0.9

d $\dfrac{20}{45} = \dfrac{4}{9}$ or 44.4% or 0.444

Paper 2 style questions

1 a $U = \{8, 9, 10, 11, 12\}$

b

c **i** none **ii** none

d $P \cup Q$: the set of numbers that are either multiples of 4 or factors of 24, or both

e

2 a **i**

ii $48 - x + x + 44 - x = 70$
$\Rightarrow x = 22$

iii Those members who **did not** attend for **both** Drama and Sports (or equivalent)

iv $P(D \text{ or } S) = \left[\dfrac{48 - 22}{70} + \dfrac{44 - 22}{70}\right]$
$= \dfrac{48}{70}$ or $\dfrac{24}{35}$

b **i** $\dfrac{30}{70} = \dfrac{3}{7}$ **ii** $\dfrac{12}{70} = \dfrac{6}{35}$

3 a

b $50 - (2 + 3 + 4 + 15 + 5 + 4 + 12) = 5$

c **i** P(fruit juice) $= \dfrac{35}{50}$

ii $P([P \cup Q] \cap R') = \dfrac{29}{50}$

iii $P(Q' | P) = \dfrac{6}{24}$

d P(both drank all three)
$= \dfrac{3}{50} \times \dfrac{2}{49} = \dfrac{6}{2450}$

4 a **i** $P = \{2, 3, 5, 7, 11\}$
ii $Q = \{1, 2, 3, 6, 9\}$
iii $R = \{3, 6, 9, 12\}$
iv $P \cap Q \cap R = \{3\}$

b

c **i** $\{2, 3, 5, 6, 7, 9, 11\}$
ii $\{1, 4, 8, 10\}$
iii $\{4, 8, 10\}$

d **i** $\dfrac{5}{12}$ **ii** $\dfrac{3}{12}$

iii $\dfrac{4}{12}$ **iv** $\dfrac{2}{5}$

5 a

b **i** P(chocolate, chocolate) $= \dfrac{4}{10} \times \dfrac{4}{10} = 0.16$

ii P(one is plain) = P(chocolate, plain) + P(plain, chocolate)
$= \dfrac{4}{10} \times \dfrac{6}{10} + \dfrac{6}{10} \times \dfrac{4}{10} = 0.48$

c **i** $a = 8, b = 9$ **ii** 0

iii P(at least one plain) = $1 - $ P(two chocolate) = 1

d P(tin, chocolate)
$= \dfrac{1}{2} \times \dfrac{4}{10} + \dfrac{1}{2} \times \dfrac{1}{10} = 0.25$

6 a **i** P(desert) $= \dfrac{13}{60}$

ii P(waterlogged and low growth rate) $= \dfrac{16}{60}$

iii P(not temperate)
$= 1 - \dfrac{18}{60} = \dfrac{42}{60}$

b **i** P (high growth rate or waterlogged environment, but not both)
$= \dfrac{4}{60} + \dfrac{7}{60} + \dfrac{16}{60} = \dfrac{27}{60}$

ii P (low, given desert)

$$= \frac{9}{13}$$

c i $\frac{36}{60} \times \frac{35}{59} = \frac{21}{59}$

ii $\frac{45}{60} \times \frac{44}{59} = \frac{33}{59}$

Chapter 9

Skills check

1 a

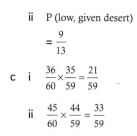

b

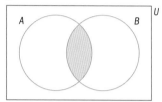

c

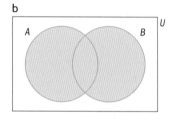

d

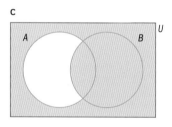

e

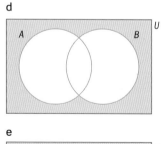

2 a

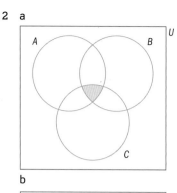

b

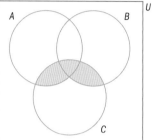

c

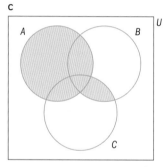

d

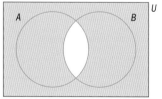

e

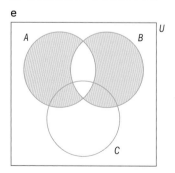

Investigation – logical thinking

1 That food must not be taken into the sports hall by anyone and that drinks must not be taken into the sports hall by anyone either.

2 According to the sign, yes!

3 Again, yes!

4 Unfortunately, it does not.

5 'No food **or** drink allowed in the sports hall.'

'No food **and no** drink allowed in the sports hall.'

Exercise 9A

1	Y	**2**	N
3	N	**4**	Y
5	Y	**6**	Y
7	Y	**8**	N
9	Y	**10**	Y
11	Y	**12**	Y
13	Y	**14**	N

Exercise 9B

1	Exclusive	**2**	Exclusive
3	Inclusive	**4**	Inclusive
5	Inclusive	**6**	Inclusive
7	Exclusive	**8**	Exclusive
9	Exclusive	**10**	Exclusive
11	Exclusive	**12**	Exclusive

Note that there are some cases (notably 7) where there is some ambiguity about the version of 'or'. Such ambiguity **must** be removed.

Exercise 9C

1 a The student is a not council member.

b She does not own a mobile phone.

c n is a composite number.

d ABCD is not a parallelogram.

e Surabaya is not the capital of Indonesia.

2 a This word starts with a consonant.

b There is an odd number of pages in this book.

c This price is exclusive of sales tax.

d This shape is something other than quadrilateral.

e He walked at a variable speed.

3 a i There are other marks in between the highest and the lowest.

ii   There are degrees of difficulty.

iii   She may have scored exactly 50%

iv   Richard may have one foot in the classroom and one foot out

v   Unless the average (mean) was an exact integer, the negation is correctly stated.

b Clearly not.

c This method works, but leads to some awkwardly constructed sentences.

4 **a** x is less than or equal to five.

b y is greater than or equal to seven.

c z is less than ten.

d b is greater than 19.

5 **a** Neither

b 'x is non-negative.'

6 **a** Courtney was at school on Friday.

b This chair is not broken.

c The hockey team either won or drew (tied) their match.

d The soccer team did not win the tournament.

e The hotel has running water.

7 **a** His signature is legible.

b James is my age or younger.

c The class contains at least eight boys.

d Her family name begins with a letter other than P.

e He has at most one sister.

8 **a** X is a male doctor.

b X is a female whose profession is something other than a doctor.

c X is a married woman.

d X is a single (unmarried) man.

e R is a positive rotation between 0° and 90° inclusive.

f R is a rotation of less than or equal to 90°.

Exercise 9D

1 **a** Susan speaks French and Susan speaks Spanish.

b Susan does not speak French and Susan speaks Spanish.

c Susan speaks French and Susan

does not speak Spanish.

d Susan does not speak French and Susan does not speak Spanish.

e It is not the case that Susan speaks French and Susan speaks Spanish.

2 **a** Jorge speaks Portuguese and Mei Ling speaks Malay.

b Jorge does not speak Portuguese and Mei Ling speaks Malay.

c Jorge speaks Portuguese and Mei Ling does not speak Malay.

d Jorge does not speak Portuguese and Mei Ling does not speak Malay.

e It is not the case that Jorge speaks Portuguese and Mei Ling speaks Malay.

3 **a** All dogs bark and All flowers are yellow.

b Not all dogs bark and All flowers are yellow.

c All dogs bark and Not all flowers are yellow.

d Not all dogs bark and Not all flowers are yellow.

e It is not the case that all dogs bark and all flowers are yellow.

4 **a** China is in Africa and Rwanda is in Asia.

b China is not in Africa and Rwanda is in Asia.

c China is in Africa and Rwanda is not in Asia.

d China is not in Africa and Rwanda is not in Asia.

e It is not the case that China is in Africa and Rwanda is in Asia.

5 **a** Chicago is the largest city in Canada and Jakarta is the largest city in Indonesia.

b Chicago is not the largest city in Canada and Jakarta is the largest city in Indonesia.

c Chicago is the largest city in Canada and Jakarta is not the largest city in Indonesia.

d Chicago is not the largest city in Canada and Jakarta is not the largest city in Indonesia.

e It is not the case that Chicago is the largest city in Canada and Jakarta is the largest city in Indonesia.

6 **a** $x \le 5$ and $x \ge 5$

b $x > 5$ and $x \ge 5$

c $x \le 5$ and $x < 5$

d $x > 5$ and $x < 5$

e It is not the case that $x \le 5$ and $x \ge 5$.

Yes, if $x = 5$. Hence **e** is better expressed as $x \ne 5$.

7 **a** ABCD is a parallelogram and ABCD is a rectangle.

b ABCD is a not a parallelogram and ABCD is a rectangle.

c ABCD is a parallelogram and ABCD is not a rectangle.

d ABCD is not a parallelogram and ABCD is not a rectangle.

e It is not the case that is a parallelogram and ABCD is a rectangle.

Statement **b** cannot possibly be true in this case.

8 **a** Triangle ABC is right-angled at C and $AB^2 = AC^2 + BC^2 + 1$

b Triangle ABC is not right-angled at C and $AB^2 = AC^2 + BC^2 + 1$

c Triangle ABC is right-angled at C and $AB^2 \ne AC^2 + BC^2 + 1$

d Triangle ABC is not right-angled at C and $AB^2 \ne AC^2 + BC^2 + 1$

e It is not the case that triangle ABC is (both) right-angled at C and $AB^2 = AC^2 + BC^2 + 1$

a, **b** and **c** cannot possibly be true in this case.

e, **d** must be true in this case.

9 **a** n is an odd integer and n is an even integer.

b n is not an odd integer and n is an even integer.

c n is an odd integer and n is not an even integer.

d n is not an odd integer and n is not an even integer.

e It is not the case that n is an odd integer and n is an even integer.

a cannot possibly be true. d cannot possibly be true, but only if the universal set is the set of integers.

b and c are necessarily true, but only if the universal set is the set of integers.

e must be true

10

p	¬p	p ∧ ¬p
T	F	**F**
F	T	**F**

$p \wedge \neg p$ is a logical contradiction because all the entries in its column are False.

11 $p \wedge \neg q$

p	q	¬q	p ∧ ¬q
T	T	F	**F**
T	F	T	**T**
F	T	F	**F**
F	F	T	**F**

12 $p \wedge q$

p	q	p ∧ q	n
T	T	**T**	20
T	F	**F**	18
F	T	**F**	15
F	F	**F**	7

Exercise 9E

1 a i x is less than or equal to 36.
  ii x is less than or equal to 36, but not both.
 b i
2 a i $p \vee r$ **ii** $p \veebar r$
  iii $q \vee r$ **iv** $(q \vee r) \wedge \neg p$
 b No
3 a i $p \vee q$ **ii** $p \veebar q$
  iii $p \vee r$ **iv** $q \veebar r$
  v $p \vee q \vee r$ **vi** $(p \vee q) \wedge \neg r$
  (Note that the brackets are required.)
 b i 1, 2, 3, 4, 6, 9, 12, 18, 24, 30, 36
  ii 1, 2, 3, 4, 9, 24, 30
  iii 1, 4, 6, 9, 12, 16, 18, 24, 25, 30, 36
  iv 2, 3, 6, 12, 16, 18, 25
  v 1, 2, 3, 4, 6, 9, 12, 16, 18, 24, 25, 30, 36
  vi 2, 3, 6, 12, 18, 24, 30
4 a $p \vee q$ **b** $q \veebar r$
 c $p \vee r$ **d** $r \wedge q$ or $q \wedge r$
5 a $p \vee \neg q$ **b** $\neg p \wedge \neg q$
6 a x ends in zero or x is not divisible by 5; 7
 b x ends in zero or x is not divisible by 5 but not both; 7
 c x ends in zero and x is not divisible by 5; FALSE
 d x ends in zero and x is divisible by 5; 10
 e x does not end in zero and x is divisible by 5; 15
7 a i $p \wedge q$ **ii** $p \veebar q$
  iii $p \vee q$ **iv** $\neg p \vee \neg q$
  v $\neg(p \vee q)$ **vi** $\neg(p \wedge q)$
  vii $\neg p \wedge \neg q$
 b i i **ii** iii
  iii v and vii
  iv iv and vi

Exercise 9F

1 a Final columns only given:

p	q	p∧q	p∨q	p⊻q	¬p∨¬q	¬(p∨q)	¬(p∧q)	¬p∧¬q
T	T	T	T	F	F	F	F	F
T	F	F	T	T	T	F	T	F
F	T	F	T	T	T	F	T	F
F	F	F	F	F	T	T	T	T

 b vi I am not studying both French and Chinese. (Equivalents in red in *a*.)

The columns in blue are the same; hence, the statements are equivalent.

2 a

p	¬p	¬(¬p)
T	F	T
F	T	F

b

p	p∧p
T	T
F	F

c

p	q	p∧q	p∨(p∧q)
T	T	T	T
T	F	F	T
F	T	F	F
F	F	F	F

d

p	q	¬p	¬p∧q	p∨(¬p∧q)	p∨q
T	T	F	F	T	T
T	F	F	F	T	T
F	T	T	T	T	T
F	F	T	F	F	F

3 $(p \wedge \neg q) \vee (\neg p \wedge q)$

p	q	p∧¬q	¬p∧q	(p∧¬q)∨(¬p∧q)	p⊻q
T	T	F	F	F	F
T	F	T	F	T	T
F	T	F	T	T	T
F	F	F	F	F	F

$(p \wedge \neg q) \vee (\neg p \wedge q) \Leftrightarrow p \veebar q$

4 a Tautology since:

p	¬p	p ∨ ¬p
T	F	T
F	T	T

b Contradiction since:

p	¬p	p ∧ ¬p
T	F	F
F	T	F

c Neither
d Tautology since:

p	q	p∨q	¬p∧¬q	(p∨q)∨(¬p∧q)
T	T	T	F	T
T	F	T	F	T
F	T	T	F	T
F	F	F	T	T

e Tautology
f Neither
g Neither
h Contradiction

Exercise 9G

1 $p \vee (q \wedge r)$: Neither

p	q	r	$(q \wedge r)$	$p \vee (q \wedge r)$
T	T	T	T	T
T	T	F	F	T
T	F	T	F	T
T	F	F	F	T
F	T	T	T	T
F	T	F	F	F
F	F	T	F	F
F	F	F	F	F

2 $(p \vee \neg q) \vee r$: Neither

p	q	r	$\neg q$	$(p \vee \neg q)$	$(p \vee \neg q) \vee r$
T	T	T	F	T	T
T	T	F	F	T	T
T	F	T	T	T	T
T	F	F	T	T	T
F	T	T	F	F	T
F	T	F	F	F	F
F	F	T	T	T	T
F	F	F	T	T	T

3 $(p \wedge q) \vee (p \wedge \neg r)$: Neither

p	q	r	$\neg r$	$p \wedge q$	$p \wedge \neg r$	$(p \wedge q) \vee (p \wedge \neg r)$
T	T	T	F	T	F	T
T	T	F	T	T	T	T
T	F	T	F	F	F	F
T	F	F	T	F	T	T
F	T	T	F	F	F	F
F	T	F	T	F	F	F
F	F	T	F	F	F	F
F	F	F	T	F	F	F

4 Neither
5 Contradiction
6 Neither
7 Neither
8 Neither; Question 1.

Exercise 9H

1 The two statements are equivalent. There is no need to use brackets.

2 Both Venn diagrams give:

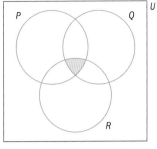

3 Both Venn diagrams give:

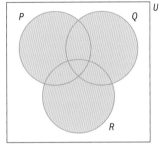

4 The final column of $p \wedge (q \vee r)$ is T T T F F F F F.
The final column of $(p \wedge q) \vee r$ is T T T F T F T F.
There is a need to use brackets.

5 $P \cap (Q \cup R)$ and $(P \cap Q) \cup R$ are **not** equivalent.

6 The final column of $p \vee (q \wedge r)$ is T T T T T F F.
The final column of $(p \vee q) \wedge r$ is T F T F T F F.
There is a need to use brackets.

7 $P \cup (Q \cap R)$ and $(P \cup Q) \cap R$ are **not** equivalent.

8 The final column of
$(\neg p \wedge q) \vee (\neg q \wedge r) \vee (\neg r \wedge p)$
is F T T T T T T F
The final column of
$(\neg p \vee q) \wedge (\neg q \vee r) \wedge (\neg r \vee p)$
is T F F F F F F T
They are not equivalent.

Exercise 9I

1

p	q	$p \wedge q$	$p \Rightarrow p \wedge q$	$p \vee q$	$p \Rightarrow p \vee q$
T	T	T	T	T	T
T	F	F	F	T	T
F	T	F	T	T	T
F	F	F	T	F	T

$p \Rightarrow p \wedge q$: invalid argument;
$p \Rightarrow p \vee q$: tautology

2

p	q	$p \wedge q$	$p \wedge q \Rightarrow p$	$p \vee q$	$p \vee q \Rightarrow p$
T	T	T	T	T	T
T	F	F	T	T	T
F	T	F	T	T	F
F	F	F	T	F	T

$p \wedge q \Rightarrow p$: tautology;
$p \vee q \Rightarrow p$: invalid argument

3

p	q	$p \wedge q$	$p \vee q$	$p \vee q \Rightarrow p$	$p \Rightarrow p \vee q$	$(p \vee q \Rightarrow p) \wedge (p \Rightarrow p \wedge q)$
T	T	T	T	T	T	T
T	F	F	T	T	F	F
F	T	F	T	F	T	F
F	F	F	F	T	T	T

Invalid argument

4

p	q	$p \wedge q$	$p \vee q$	$p \wedge q \Rightarrow p$	$p \Rightarrow p \wedge q$	$(p \wedge q \Rightarrow p) \wedge (p \Rightarrow p \wedge q)$
T	T	T	T	T	T	T
T	F	F	T	T	F	F
F	T	F	T	T	T	T
F	F	F	F	T	T	T

Invalid argument

5 $(p \wedge q \Rightarrow p) \vee (p \Rightarrow p \wedge q)$. Final column TTTT. Tautology.

6 $\neg(p \wedge q) \Rightarrow \neg p \vee \neg q$. Final column TTTT. Tautology.

7 $\neg(p \vee q) \Rightarrow \neg p \vee \neg q$. Final column TTTT. Tautology.

8 $\neg p \vee \neg q \Rightarrow \neg(p \wedge q)$. Final column TTTT. Tautology.

9 $\neg(p \vee q) \Rightarrow \neg p \wedge \neg q$. Final column TTTT. Tautology.

Exercise 9J

1 $[(p \Rightarrow q) \land \neg p] \Rightarrow \neg q$; invalid

p	q	$\neg p$	$p \Rightarrow q$	$(p \Rightarrow q) \land \neg p$	$\neg q$	$[(p \Rightarrow q) \land \neg p] \Rightarrow \neg q$
T	T	F	T	F	F	T
T	F	F	F	F	T	T
F	T	T	T	T	F	F
F	F	T	T	T	T	T

2 $[(p \Rightarrow q) \land q] \Rightarrow p$; invalid

p	q	$p \Rightarrow q$	$(p \Rightarrow q) \land q$	$[(p \Rightarrow q) \land q] \Rightarrow p$
T	T	T	T	T
T	F	F	F	T
F	T	T	T	F
F	F	T	F	T

3 $[(p \Rightarrow q) \land \neg q] \Rightarrow p$; valid

p	q	$\neg q$	$p \Rightarrow q$	$(p \Rightarrow q) \land \neg q$	$[(p \Rightarrow q) \land \neg q] \Rightarrow p$
T	T	F	T	F	T
T	F	T	F	F	T
F	T	F	T	F	T
F	F	T	T	T	T

4 $[(p \Rightarrow q) \land (q \Rightarrow r) \land \neg r] \Rightarrow \neg p$; valid

p	q	r	$\neg p$	$\neg r$	$p \Rightarrow q$	$q \Rightarrow r$	$[(p \Rightarrow q) \land (q \Rightarrow r) \land \neg r]$	$[(p \Rightarrow q) \land (q \Rightarrow r) \land \neg r] \Rightarrow \neg p$
T	T	T	F	F	T	T	F	T
T	T	F	F	T	T	F	F	T
T	F	T	F	F	F	T	F	T
T	F	F	F	T	F	T	F	T
F	T	T	T	F	T	T	F	T
F	T	F	T	T	T	F	F	T
F	F	T	T	F	T	T	F	T
F	F	F	T	T	T	T	T	T

5 $[(p \Rightarrow q) \land (q \Rightarrow r)] \Rightarrow (\neg p \Rightarrow \neg r)$; invalid

p	q	r	$\neg p$	$\neg r$	$\neg p \Rightarrow \neg r$	$p \Rightarrow q$	$q \Rightarrow r$	$[(p \Rightarrow q) \land (q \Rightarrow r)]$	$[(p \Rightarrow q) \land (q \Rightarrow r)] \Rightarrow (\neg p \Rightarrow \neg r)$
T	T	T	F	F	T	T	T	T	T
T	T	F	F	T	T	T	F	F	T
T	F	T	F	F	T	F	T	F	T
T	F	F	F	T	T	F	T	F	T
F	T	T	T	F	F	T	T	T	F
F	T	F	T	T	T	T	F	F	T
F	F	T	T	F	F	T	T	T	F
F	F	F	T	T	T	T	T	T	T

6 $[(p⇒q)∧(q⇒r)] ⇒ (¬p⇒¬r)$; *invalid*

p	q	r	$¬p$	$¬r$	$¬p⇒¬r$	$p⇒q$	$q⇒r$	$[(p⇒q)∧(q⇒r)]$	$[(p⇒q)∧(q⇒r)]⇒(¬p⇒¬r)$
T	T	T	F	F	T	T	T	T	T
T	T	F	F	T	T	T	F	F	T
T	F	T	F	F	T	F	T	F	T
T	F	F	F	T	T	F	T	F	T
F	T	T	T	F	F	T	T	T	F
F	T	F	T	T	T	T	F	F	T
F	F	T	T	F	F	T	T	T	F
F	F	F	T	T	T	T	T	T	T

7 $[(p⇒q)∧(q⇒r)∧(r⇒s)] ⇒ (¬s⇒¬p)$; *valid*

p	q	r	s	$¬s$	$¬p$	$p⇒q$	$q⇒r$	$r⇒s$	Triple conjunction	$¬s⇒¬p$	Final statement
T	T	T	T	F	F	T	T	T	T	T	T
T	T	T	F	T	F	T	T	F	F	F	T
T	T	F	T	F	F	T	F	T	F	T	T
T	T	F	F	T	F	T	F	T	F	F	T
T	F	T	T	F	F	F	T	T	F	T	T
T	F	T	F	T	F	F	T	F	F	F	T
T	F	F	T	F	F	F	T	T	F	T	T
T	F	F	F	T	F	F	T	T	F	F	T
F	T	T	T	F	T	T	T	T	T	T	T
F	T	T	F	T	T	T	T	F	F	T	T
F	T	F	T	F	T	T	F	T	F	T	T
F	T	F	F	T	T	T	F	T	F	T	T
F	F	T	T	F	T	T	T	T	T	T	T
F	F	T	F	T	T	T	T	F	F	T	T
F	F	F	T	F	T	T	T	T	T	T	T
F	F	F	F	T	T	T	T	T	T	T	T

Exercise 9K

1 $¬(p∧q) ⇔ (¬p∨¬q)$ is a tautology.

2 $¬(p∨q) ⇔ (¬p∧¬q)$ is a tautology.

3 $(p∧q) ⇔ p$ has final column TFTT, $(p∨q) ⇔ p$ has final column TTFT. The statements are not equivalent.

4 The statements $¬(p∧¬q)$ and $¬p∨q$ are equivalent.

5 The statements $¬(p∨¬q)$ and $¬p∧q$ are equivalent.

6 The statement $(p∨¬q) ⇔ (¬p∧q)$ is a contradiction.

7 The statement $¬(p∨q) ⇔ (p∧q)$ is neither a contradiction nor a tautology.

8 The statement $(p∧¬q) ⇔ (¬p∨q)$ is a contradiction.

Exercise 9L

p	q	Statement $p⇒q$	Converse $q⇒p$	Inverse $¬p⇒¬q$	Contrapositive $¬q⇒¬p$
T	T	T	T	T	T
T	F	F	T	T	F
F	T	T	F	F	T
F	F	T	T	T	T

Exercise 9M

1 **a** Valid   **b** Valid   **c** Valid
   d Invalid; counterexample 18
   e Valid   **f** Valid
   g Invalid; counterexample 12
   h Invalid; counterexample 3 and 7
   i Invalid; counterexample 2 and 5
   j Valid   **k** Valid
   l Valid   **m** Valid
   n Valid
   o Invalid; counterexample rhombus
   p Invalid; counterexample $x = -5$
   q Valid
   r Invalid; counterexample $x = -10$
   s Valid

2 & 3

a Converse: If ABCD is a quadrilateral, then ABCD is a square. Invalid: Counterexample rectangle.

Inverse: If ABCD is not a square, then ABCD is not a quadrilateral. Invalid: Counterexample rectangle.

Contrapositive: If ABCD is a not a quadrilateral, then ABCD is not a square. Valid.

b   Converse: If ABCD is a parallelogram, then ABCD is a rectangle. Invalid: Counterexample any parallelogram with internal angles not equal to 90°.

Inverse: If ABCD is not a rectangle, then ABCD is not a parallelogram. Invalid: Counterexample rhombus.

Contrapositive: If ABCD is not a parallelogram, then ABCD is a not rectangle. Valid.

c   Converse: If an integer is divisible by two then it is divisible by four. Invalid: Counterexample 10.

Inverse: If an integer is not divisible by four then it is not divisible by two. Invalid: Counterexample 10.

Contrapositive: If an integer is not divisible by two then it is not divisible by four. Valid.

d   Converse: If an integer is odd then it is divisible by three. Invalid: Counterexample 25.

Inverse: If an integer is not divisible by three then it is an even integer. Invalid: Counterexample 25.

Contrapositive: If an integer is an even integer then it is not divisible by three. Invalid: Counterexample 18.

e   Converse: If an integer is even then it is divisible by two. Valid.

Inverse: If an integer is not divisible by two then it is not an even integer. Valid.

Contrapositive: If an integer is not an even integer then it is not divisible by two. Valid.

f   Converse: If an integer is divisible by twelve then it is divisible by both four and by three. Valid.

Inverse: If an integer is not divisible by both four and by three then it is not divisible by twelve. Valid.

Contrapositive: If an integer is not divisible by twelve then it is not divisible by both four and by three. Valid.

g   Converse: If an integer is divisible by eight then it is divisible by both four and by two. Valid.

Inverse: If an integer is not divisible by both four and by two then it is not divisible by eight. Valid.

Contrapositive: If an integer is not divisible by eight then it is not divisible by both four and by two. Invalid: Counterexample 12.

h   Converse: If two integers are both even, then their sum is even. Valid.

Inverse: If the sum of two integers is not even, then the two integers are not both even. Valid.

Contrapositive: If two integers are not both even, then their sum is not even. Invalid: Counterexample 3 and 7.

i   Converse: If two integers are both even, then their product is even. Valid.

Inverse: If the product of two integers is not even, then the two integers are not both even. Valid.

Contrapositive: If two integers are not both even, then their product is not even. Invalid: Counterexample 2 and 5.

j   Converse: If one integer is odd and the other is even, then their sum is odd. Valid.

Inverse: If the sum of two integers is not odd, then either the two integers are both odd or the two integers are both even. Valid.

Contrapositive: If either the two integers are both odd or the two integers are both even, then their sum is not odd. Valid.

k   Converse: If two integers are both odd, then their product is odd. Valid.

Inverse: If the product of two integers is not odd, then the two integers are not both odd. Valid.

Contrapositive: If two integers are not both odd, then their product is not odd. Valid.

l   Converse: If $a^2 + b^2 = c^2$, then triangle ABC is right angled. Valid.

Inverse: If triangle ABC is not right angled, then $a^2 + b^2 \neq c^2$. Valid.

Contrapositive: If $a^2 + b^2 \neq c^2$, then triangle ABC is not right angled. Valid.

m   Converse: If the square of an integer is odd, then the integer is odd. Valid.

Inverse: The square of an even integer is even. Valid.

Contrapositive: If the square of an integer is even, then the integer is even. Valid.

n   Converse: If triangle ABC has three equal sides, then triangle ABC has three equal angles. Valid.

Inverse: If triangle ABC does not have three equal angles, then triangle ABC does not have three equal sides. Valid.

Contrapositive: If triangle ABC does not have three equal sides, then triangle ABC does not have three equal angles. Valid.

o   Converse: If quadrilateral ABCD has four equal angles, then ABCD has four equal sides. Invalid: Counterexample rectangle.

Inverse: If quadrilateral ABCD does not have four equal sides, then ABCD does not have four equal angles. Invalid: Counterexample rectangle.

Contrapositive: If quadrilateral ABCD does not have four equal angles, then ABCD does not have four equal sides. Invalid: Counterexample rhombus.

p   Converse: If $x = 5$, then $x^2 = 25$. Valid.

Inverse: If $x^2 \neq 25$, then $x \neq 5$. Valid.

Contrapositive: If $x \neq 5$, then $x^2 \neq 25$. Invalid: Counterexample $x = -5$

q   Converse: If $x = 3$, then $x^3 = 27$. Valid

Inverse: If $x^3 \neq 27$, then $x \neq 3$. Valid

Contrapositive: If $x \neq 3$, then $x^3 \neq 27$. Valid

r   Converse: If $x > 5$, then $x^2 > 25$. Valid

Inverse: If $x^2 \leq 25$, then $x \leq 5$. Valid

Contrapositive: If $x \le 5$, then $x^2 \le 25$. Invalid: Counterexample $x = -10$

s Converse: If $x < 3$, then $x^3 < 27$. Valid

Inverse: If $x^3 \ge 27$, then $x \ge 3$. Valid

Contrapositive: If $x \ge 3$, then $x^3 \ge 27$. Valid

ii Picasso did not paint picture A and van Gogh painted picture A.

b

p	q	$\neg p$	$\neg q$	$p \lor \neg q$	$\neg p \land q$
T	T	F	F	T	F
T	F	F	T	T	F
F	T	T	F	F	T
F	F	T	T	T	F

Review exercise

Paper 1 style questions

1 a

p	q	$p \lor q$	$\neg(p \lor q)$	$\neg p$	$\neg q$	$\neg p \land \neg q$	$\neg(p \lor q) \Rightarrow \neg p \land \neg q$
T	T	T	F	F	F	F	T
T	F	T	F	F	T	F	T
F	T	T	F	T	F	F	T
F	F	F	T	T	T	T	T

b She does not dance well and she does not sing beautifully.

2 a If the train leaves from gate z, then it leaves today and not from gate 8.

b $\neg r \Leftrightarrow (p \lor q)$

c

3 a

p	q	$p \Rightarrow q$	$\neg p$	$\neg q$	$\neg q \lor \neg p$	$\neg p \lor q$
T	T	T	F	F	T	T
T	F	F	F	T	T	F
F	T	T	T	F	F	T
F	F	T	T	T	T	T

b $(p \Rightarrow q) \Leftrightarrow (\neg p \lor q)$

4 a

p	q	$\neg p$	$\neg p \lor q$
T	T	F	T
T	F	F	F
F	T	T	T
F	F	T	T

b i False: p is T and q is F.

ii True: p is F and q is T.

5 a b

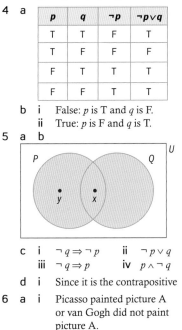

c i $\neg q \Rightarrow \neg p$ ii $\neg p \lor q$

iii $\neg q \Rightarrow p$ iv $p \land \neg q$

d i Since it is the contrapositive

6 a i Picasso painted picture A or van Gogh did not paint picture A.

c

p	q	r	q/r	$\neg p$	$(q/r) - \neg p$
T	T	T	T	F	F
T	F	F	T	F	F
T	F	T	T	F	F
T	F	F	F	F	F
F	T	T	T	T	T
F	T	F	T	T	T
F	F	T	T	T	T
F	F	F	F	T	F

p	q	r	$\neg q$	$p/\neg p$	$r \Rightarrow (p/\neg q)$
T	T	T	F	T	T
T	T	F	F	T	T
T	F	T	T	T	T
T	F	F	T	T	T
F	T	T	F	F	F
F	T	F	F	F	T
F	F	T	T	T	T
F	F	F	T	T	T

d

p	q	r	x
F	T	T	3
F	T	F	12
F	F	T	2

e

p	q	r	$(q/\neg r) - \neg p$	$r \Rightarrow (p/\neg q)$
T	T	T	F	T
T	T	F	F	T
T	F	T	F	T
T	F	F	F	T
F	T	T	T	F
F	T	F	T	T
F	F	T	T	T
F	F	F	F	T

x is not a multiple of 5 and is either a multiple of 3 or a factor of 90, but not both.

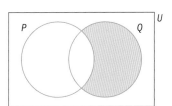

d i FFFF: the truth values of $p \lor \neg q$ and $\neg p \land q$ are never the same.

ii The regions $p \lor \neg q$ and $\neg p \land q$ do not overlap, hence the truth values of $(p \lor \neg q) \Leftrightarrow (\neg p \land q)$ are all false.

e logical contradiction.

7 a x is a multiple of 3 or a factor of 90 and is not a multiple of 5.

b $r \Rightarrow (p \lor \neg q)$

Chapter 10

Skills check

1 a $x = 5.85\,\text{m}$ b $y = 51.3°$

2 a $x = 51.2°$ b $2740\,\text{m}^2$

Investigation – drawing a prism

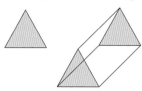

Relationships between volumes:

Square based pyramid $= \frac{1}{3}abh$

Cuboid $= abh$

The volume of a pyramid is $\frac{1}{3}$ the volume of a cuboid with the same base and height.

Cone $= \frac{1}{3}\pi r^2 h$

Cylinder $= \pi r^2 h$

The volume of a cone is $\frac{1}{3}$ the volume of a cylinder with the same base and height.

Exercise 10A

1 a i triangular prism
  ii 5 faces, 9 edges
     6 vertices
  iii 5 plane faces
  b i rectangular-based pyramid
  ii 5 faces, 8 edges
     5 vertices
  iii 5 plane faces
  c i hemisphere
  ii 2 faces, 1 edge, no vertices
  iii 1 plane face, 1 curved face

2 a

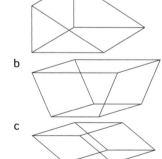

b

c

Exercise 10B

1 a

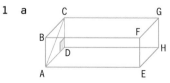

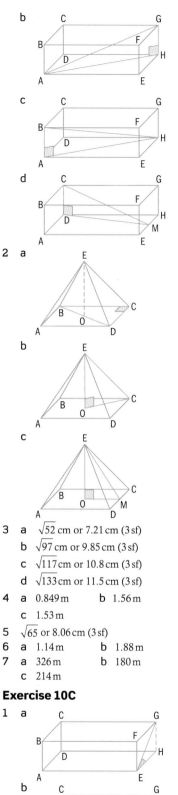

3 a $\sqrt{52}$ cm or 7.21 cm (3 sf)
  b $\sqrt{97}$ cm or 9.85 cm (3 sf)
  c $\sqrt{117}$ cm or 10.8 cm (3 sf)
  d $\sqrt{133}$ cm or 11.5 cm (3 sf)
4 a 0.849 m   b 1.56 m
  c 1.53 m
5 $\sqrt{65}$ or 8.06 cm (3 sf)
6 a 1.14 m   b 1.88 m
7 a 326 m   b 180 m
  c 214 m

Exercise 10C

1 a

b

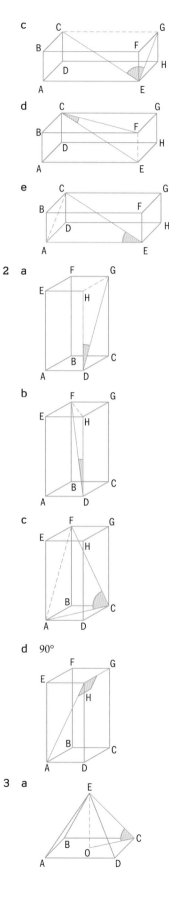

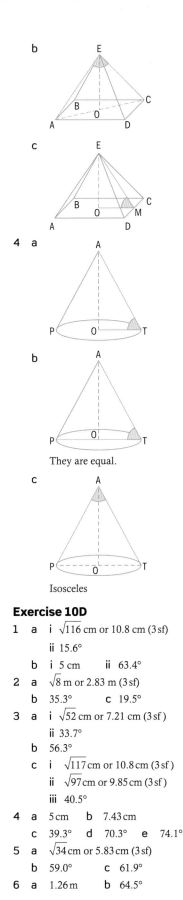

b

c

4 a

b

They are equal.

c

Isosceles

Exercise 10D

1 a i $\sqrt{116}$ cm or 10.8 cm (3 sf)
    ii 15.6°
  b i 5 cm   ii 63.4°
2 a $\sqrt{8}$ m or 2.83 m (3 sf)
  b 35.3°   c 19.5°
3 a i $\sqrt{52}$ cm or 7.21 cm (3 sf)
    ii 33.7°
  b 56.3°
  c i $\sqrt{117}$ cm or 10.8 cm (3 sf)
    ii $\sqrt{97}$ cm or 9.85 cm (3 sf)
    iii 40.5°
4 a 5 cm   b 7.43 cm
  c 39.3°   d 70.3°   e 74.1°
5 a $\sqrt{34}$ cm or 5.83 cm (3 sf)
  b 59.0°   c 61.9°
6 a 1.26 m   b 64.5°

Exercise 10E

1 a 24 cm²   b 23.5 m²
  c 73.9 cm²
2 a 3.90 cm²   b 5.20 cm
  c 52.6 cm²
3 a 6.5 cm   b 16.25 cm²
  c 90 cm²
4 1000 cm
5 a 175 m²   b 1.75 × 10² m
6 a 43.4 m²   b 53 litres
  c US$243.80 (2 dp)

Exercise 10F

1 a 30π cm² or 94.2 cm² (3 sf)
  b 4π cm² or 12.6 cm² (3 sf)
  c 6.75π cm² or 21.2 cm² (3 sf)
  d 4.125π m² or 13.0 m² (3 sf)
2 a 38π cm² or 119 cm² (3 sf)
  b 22.08π cm² or 69.4 cm² (3 sf)
3 8.92 cm (3 sf)
4 a 120π cm³ or 377 cm² (3 sf)
  b 12 cm (2 sf)

Exercise 10G

1 a 23.4 dm³   b 90 m³
  c 8000 cm³   d 160 cm³
  e 12 m³   f 210 cm³
2 a 5.03 m   b 15.1 m²
  c 151 m³
3 a 60°   b 10.8 cm²
  c 65.0 cm²   d 877 cm³
4 a x^3 b $3x^3$
  c $\dfrac{3x^3}{8}$ or 0.375 x^3 d 10x^2
5 a 25x^2 b 11025 = 25x^2
  c 21
6 a i 21 cm   b 2205 cm²

Exercise 10H

1 a 75 140π mm³ or
    236000 mm³ (3 sf)
  b $\dfrac{1}{6}$ π = 0.524 cm³ (3 sf)
  c 32.7 cm³   d 1130 cm³
  e 32.7 cm³   f 8 dm³
2 a 13.6 m³   b 13600 dm³
  c 13600 litres
3 a $V = \dfrac{x^2 h}{3}$ b $V = 2\pi x^3$
  c $V = 9\pi x^3$ d $V = 4.5\pi x^3$
4 a 36 cm²   b 6 cm²
  c 60°   d 3.72 cm

5 a 3.63 cm
  b 36 mm
6 a 6750π cm³ or 21200 cm³ (3 sf)
  b No. The second container has
    a volume (20400 cm³) smaller
    than the first
7 a i 1.2 cm
    ii 1.25 cm
  b i 28.8 cm²
    ii 4.89 cm³
  c number of pencils in one layer
    = 5.6 ÷ 0.7 = 8
    number of layers
    = 1.4 ÷ 0.7 = 2
    total number of pencils
    = 8 × 2 = 16
  d 27.6 cm³
  e 26%

Review exercise

Paper 1 style questions

1 a 5896 cm²
  b 28.56 dm³
2 a $\sqrt{116}$ cm or 10.8 cm (3 sf)
  b 24.9°
3 a $\sqrt{41}$ cm or 6.40 cm (3 sf)
  b 8.62 cm
  c 43.6°
4 a $\sqrt{90}$ cm or 9.49 cm (3 sf)
  b 28.5 cm²
  c 150 cm²
5 a 8 cm
  b 11.3 cm
  c Yes, as the greatest distance
    between two points in this
    cube is 13.9 cm (3 sf) which is
    bigger than 13.5 cm
6 a 71.8°
  b i 7.60 cm
    ii 49.7 cm³
7 a 2.71 m²   b 9.47 m³

Paper 2 style questions

1 a 27.0 m   b 93.7 m
  c 61.3°   d US$677502
2 b 9 cm   c 9.49 cm
  d 71.6°   e 1.53 kg
3 a 58.3 cm³   b 508 g

c 7.842 cm   d 63.2°

e 37.2°   f 99.3 cm²

4 a 8.58 cm

   b i 9.46 cm ii 45.8°

   c 215 cm²   d 183 cm³

Chapter 13

Exercise 1A

1 a 11   b 10   c 8

   d 4   e 5   f 3

   g 20   h 3

2 a 5   b 1.5

   c 1.25   d 24

3 a 12   b 540

   c 16   d 5

4 a 5   b 8

   c 8   d 2

5 a 2   b 4   c 34

Exercise 1B

1 a 1, 2, 3, 6, 9, 18

   b 1, 3, 9, 27

   c 1, 2, 3, 5, 6, 10, 15, 30

   d 1, 2, 4, 7, 14, 28

   e 1, 2, 3, 6, 13, 26, 39, 78

2 a $2^2 \times 3^2$ b $2^2 \times 3 \times 5$

   c 2×3^3 d 2^5 e $2^4 \times 7$

3 a 40   b 240

4 a 8   b 18

Exercise 1C

1 a $\dfrac{11}{12}$ b $1\dfrac{1}{15}$ c 1

   d $2\dfrac{49}{81}$

2 a $\dfrac{4}{9}$ b $\dfrac{7}{20}$

   c $\dfrac{2}{3}$ d $\dfrac{5}{8}$

3 a $\dfrac{18}{5}$ b $\dfrac{22}{7}$

   c $\dfrac{93}{4}$ d $\dfrac{167}{72}$

4 a $4\dfrac{4}{7}$ b $33\dfrac{1}{3}$

   c $4\dfrac{1}{4}$ d $14\dfrac{8}{11}$

5 a 0.32   b 0.714

   c 3.8   d 2.65

Exercise 1D

1 a 52%   b 70%

2 a 2.24 CHF   b 0.54 GBP

   c 187.57 EUR   d 10 400 JPY

Exercise 1E

1 576 GBP

2 14 875 JPY

3 7%

4 26.5%

5 26 500 000

6 32 USD

7 0.60 GBP

8 No. 1% decrease

Exercise 1F

1 5 : 4

2 95.1 : 100

3 21 : 160

4 11.2 m

5 200 000 : 1; 0.4 cm

6 45 USD, 27 USD

7 75, 45 and 30

Exercise 1G

1 7500 USD, 10 500 USD, 6 000 USD

2 18 min, 27 min, 30 min.

Exercise 2A

1 a $3x^2 - 6x$ b $x^3 - xy + \dfrac{x^2}{y}$

   c $b^2 + 3ab - 2ac$

2 a $3pq\left(1 - 2pq^2r\right)$

   b $3c\left(4ac + 5b - c\right)$

   c $abc\left(2a + 3b - 5c\right)$

Exercise 2B

1 $t = \dfrac{u - v}{g}$ 2 $c = \sqrt{a^2 - b^2}$

3 $r = \dfrac{C}{2\pi}$ 4 $b = \dfrac{a \sin B}{\sin A}$

5 $\cos A = \dfrac{b^2 + c^2 - a^2}{2bc}$

Exercise 2C

1 2.49   2 3.73   3 40.1

Exercise 2D

1 4   2 4   3 −3

4 3   5 5   6 9

7 2   8 −2   9 3

10 1.5   11 1   12 2

Exercise 2E

1 a $x = 1, y = 1$ b $x = 1, y = -2$

   c $x = -3, y = 4$

2 a $x = 6, y = -1$ b $x = 2, y = -1$

   c $x = -2, y = 2$

   d $x = 2, y = 1$

   e $x = 3, y = -1$

Exercise 2F

1 a 17   b 144   c 64

2 a 1   b $\dfrac{1}{9}$ c $\dfrac{1}{16}$

3 a 525.21875   b 4.08

   c 1.667

Exercise 2G

1 a $x \le 3$

   -5 -4 -3 -2 -1 0 1 2 3 4 5

   b $x > 8$

   -10 -8 -6 -4 -2 0 2 4 6 8 10

   c $x < 2$

   -5 -4 -3 -2 -1 0 1 2 3 4 5

2 a $x \le 5$ b $x > -1\dfrac{1}{2}$

   c $x \ge -1$

Exercise 2H

1 a 3.25   b 6.18   c 0

2 2, 3

3 a 2   b 2   c 2

Exercise 3A

1 27.6 cm   2 2.24 m

3 5.03 cm

Exercise 3B

1 a

b

c

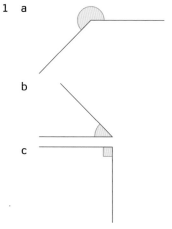

d

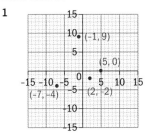

2 a reflex b obtuse c acute
3 a obtuse b acute
  c reflex d acute
  e reflex f reflex

Exercise 3C

1

Diagonals	Irregular	Rectangle	Parallelogram	Rhombus	Square	Trapezium	Kite
Perpendicular	✗	✗	✗	✓	✓	✗	✓
Equal	✗	✓	✗	✗	✓	✗	✗
Bisect	✗	✓	✓	✓	✓	✗	(one)
Bisect angles	✗	✗	✗	✓	✓	✗	(two)

2 a Kite, isosceles triangle, parallelogram, right angled triangles (2), scalene triangle, rhombus, arrowhead.
  b Square, isosceles triangle, right angled triangle, trapeziums (2), rhombus.

Exercise 3D

1 a 10.7 cm b 16.4 cm
  c 20.4 cm d 62.8 cm
  e 14.6 cm f 17.5 cm

Exercise 3E

1 63.6 cm² 2 23.0 cm²
3 37.7 cm² 4 10.3 cm²
5 6.48 m² 6 42.3 cm²

Exercise 3F

1

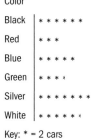

2 A $(4, 9)$, B $(-4, 2)$, C $(-8, -6)$ and D $(8, -8)$

Exercise 3G

1 $(5, 5)$ 2 $(-1, -1)$

3 $\left(1\frac{1}{2}, 2\frac{1}{2}\right)$

Exercise 3H

1 5 2 9.43 3 14.8

Exercise 4A

1

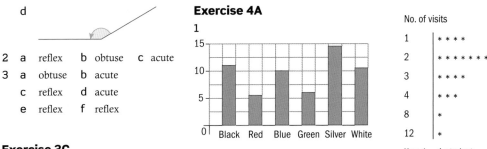

No. of visits

1	* * * *
2	* * * * * * *
3	* * * *
4	* * *
8	*
12	*

Key: * = 1 student

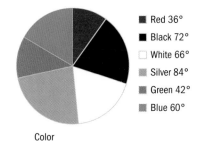

Red 36°
Black 72°
White 66°
Silver 84°
Green 42°
Blue 60°

Color

Black	* * * * * *
Red	* * *
Blue	* * * * *
Green	* * * ˙
Silver	* * * * * * * *
White	* * * * * ˙

Key: * = 2 cars

2

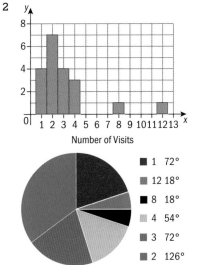

Number of Visits

1 72°
12 18°
8 18°
4 54°
3 72°
2 126°

Mark scheme

Practice paper 1

1 **a** $\dfrac{\sqrt{7^2-(6.4)(-5)}}{3125}$ M1

       $= 0.00288$ A1

   b **i**   0.003   A1

        ii   0.00029   A1

        iii 2.88×10^{-3} A1 A1

2 **a** 24   A1

   b $\dfrac{66}{24}$ M1

      $= 2.75$ A1

   c Standard deviation $= \sqrt{\dfrac{915}{30}} = 1.13$ A1

   d Median is the middle value   A1

      1, 1, 1, 2, 2, 2, 2, 2, 2, 2, 2, 3, 3, 3, 3, 3, 3, 3, 4, 4, 4, 4, 5, 5,

      Median = 3

3 **a** $\dfrac{0-4}{6-0}$ M1

      $= \dfrac{-2}{3}$ A1

   b $y = -\dfrac{2}{3}x + 4$ A1

   c $m(L_2) = \dfrac{3}{2}$

      $(y-2) = \dfrac{3}{2}(x-3)$ M1

      $c = -2.5$ A1

4 **a**

p	q	$\neg p$	$\neg p \Rightarrow q$	*Inverse*
T	T	F	T	F
T	F	F	T	T
F	T	T	T	T
F	F	T	F	T

                          A1 A1 A1 ft

   b $p \Rightarrow \neg q$ A1 A1

           Award A1 for correct negations, A1 for correct order

   c Above

   d The final two columns are not the same.   A1

5 **a** $u_1 r = 162,\ u_1 r^4 = -6$ A1 A1

      $\dfrac{u_1 r^4}{u_1 r} = \dfrac{-6}{162}$ M1

      $r = -\dfrac{1}{3}$ A1

   b $u_1\left(\dfrac{-1}{3}\right) = 162$ M1

      $u_1 = -486$ A1 ft

6 a $BD = \sqrt{3^2 + 4^2}$ M1

       $= 5\,\text{m}$ A1

b $BE = \sqrt{5^2 + 2.5^2}$ M1

       $= 5.59\,\text{m}$ A1 ft

c $\tan(\theta) = \dfrac{2.5}{5}$ M1

        Award M1 for correct trigonometric ratio

       $\theta = 26.6°$ A1 ft

7 a   When $x = 0$, $f(x) = 5$, $(0, 5)$ A1

b $f'(x) = 6 - 4x \cdot f'(x) = 0$ when $6 - 4x = 0$, so $x = \dfrac{3}{2}$.   M1

       When $x = \dfrac{3}{2}, f(x) = 5 + 9 - \dfrac{9}{2} = 9.5$. B $= (1.5, 9.5)$ A1

c   A1

d $5 = 5 + 6x - 2x^2$ A1 A1

       $0 = x(6 - 2x)$

       $x = 0$ or $x = 3$, so C $= (3, 5)$

8 a

                                        A1 A1 A1

            Award A1 for each correct pair

b $0.8 \times 0.3 + 0.2 \times 0.1$ A1 M1

          Award A1 for two correct products, M1 for adding
          their products

       $= 0.26$ A1

9 a $f(0) = 10 - (8)\,a^{-0}$ M1

       $y = 2$ A1

b $y = 10$ A1A1

          Award A1 for $y = a$ constant, A1 the constant
          being 10

c $10 - (8)\,a^{-1} = 8$ M1

       $a = 4$ A1

10 a 58 kg   A1

   b $66 - 52$ M1

                Award M1 for correct quartiles seen

        $= 14$ A1

   c $\dfrac{10}{40} \times \dfrac{9}{39}$ A1M1

                Award A1 for two correct fractions, M1 for multiplying their fractions

        $= \dfrac{3}{52}$ A1

11 a $FV = 4000\left(1 + \dfrac{3}{1200}\right)^{5 \times 12}$ M1A1

                Award M1 for substituted compound interest formula, A1 for correct substitutions

OR

       $N = 5$

       $I\% = 3$

       $PV = -4000$ M1 A1

       $P/Y = 1$

       $C/Y = 12$ Award (A1) for $C/Y = 12$ seen, M1 for other correct entries.

      OR

       $N = 60$

       $I\% = 3$

       $PV = -4000$ M1 A1

       $P/Y = 12$

       $C/Y = 12$ Award (A1) for $C/Y = 12$ seen, M1 for other correct entries.

       $= 4646.47$ A1 C3

   b $FV = 4000\left(1 + \dfrac{3}{400}\right)^{5 \times 4}$ M1

                Award M1 for correctly substituted compound interest formula,

      OR

       $N = 5$

       $I\% = 3$

       $PV = -400$ M1

       $P/Y = 1$

       $C/Y = 4$

                M1 for all correct entries seen.

      OR

       $N = 20$

       $I\% = 3$

       $PV = -4000$ M1

       $P/Y = 4$

       $C/Y = 4$

                M1 for all correct entries seen.

       $FV = 4644.74$ A1

       Difference $= €1.73$ A1 C3

> Illustrating use of GDC notation acceptable in this case only. However on P2 an answer given with no working would receive G2.

12 a −15   A1

   b $S_{50} = \dfrac{50}{2}\big(2(437) + 49(-15)\big)\ S_{50} = 3475$ M1

   c $437 - 15(k - 1) < 0$ M1

         Award M1 for correct substitution in correct formula

         $k > 30.13...$ A1

         $k = 31$ A1

13 a $(A \cap C) \cup B$ A1 A1

         Award A1 for $A \cap C$ seen

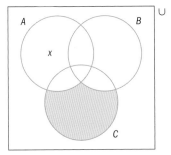

   b x in correct position on the Venn diagram   A2

   c $(A \cup B)' \cap C$ A2

         Award A1 for all of C shaded

14 a $f'(x) = 2x - k$ A1 A1

         $y = f(x)$ has a minimum point with coordinates $(3, p)$.

   b $2x - k = 0$ M1

         $k = 6$ A1

   c $f(3) = 3^2 - 6 \times 3$ M1

         $p = -9$ A1

15 a If the four sides of a quadrilateral are not equal
         then the quadrilateral is not a rhombus.   A1 A1

   b If the four sides of a quadrilateral are equal
         then the quadrilateral is a rhombus.   A1 A1

   c The converse of **this** statement p **is** always true.
         A square is also a rhombus.   A1 A1

Mark scheme

Practice paper 2

1 a

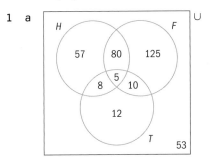

         (A1) (A1) (A1) (A1)   [4 marks]

b 12   (A1)   [1 mark]

c $350 - (57 + 80 + 125 + 8 + 5 + 10 + 12)$ (M1)

    $= 53$ (A1) (G2)   [2 marks]

d $\dfrac{200}{350}\left(= \dfrac{4}{7},\ 0.571,\ 57.1\% \right)$ (A1) (A1) (G2)   [2 marks]

e $\dfrac{13}{350}(= 0.371,\ 37.1\%)$ (A1) (A1) (G2)   [2 marks]

f $\dfrac{15}{350} \times \dfrac{14}{349}$ (A1) (M1)

    $= \dfrac{3}{1745}(0.00172,\ 0.172\%)$ (A1) (G2)   [3 marks]

Total [14 marks]

2 a i $r = 0.982$ (G2)

   ii   (very) strong, positive   (A1) (A1)   [4 marks]

b $y = 1.60x + 67.3$ (A1) (A1) (G2)   [2 marks]

c $y = 1.60(6) + 67.3$ (M1)

    $= 77$ (Accept 76.9 or 76)   (A1) (G2)   [2 marks]

d H_0: the time of the lesson and score (in the examination) are independent.   (A1)   [1 mark]

e 2   (A1)   [1 mark]

f $\dfrac{40 \times 71}{146}$ (M1) (A1)

    $= 19.4\ldots = 19$ (AG)   [2 marks]

g $\chi^2_{calc} = 3.42$ (G2)   [2 marks]

h $\chi^2_{calc} < \chi^2_{crit}$ (5.991)   (R1)   [1 mark]

   or

   $0.18122 > 0.05$

Total [15 marks]

3 a i $x(2x - t)$ (A1)(A1)

   ii $x = 0;\ x = 4$ (A1) (A1)

   iii $2 \times 4 - t = 0$ (M1)

        $t = 8$ (A1) (G2)   [6 marks]

b $a = 2$ (A1)   [1 mark]

c i -6 (G1)

   ii   10   (G1)   [2 marks]

d $5m + c = 10$ (A1)

    $m + c = -6$ (A1)   [2 marks]

e Subtracting the terms in the second equation from the terms in the first equation gives $4m = 16$ $m = 4$ (A1)

    $c = -10$ (A1)   [2 marks]

f $g(x) = 0$ (may be implied)   (M1)

    $4x - 10 = 0$

    $x = 2.5$ (A1) (G2)   [2 marks]

g $1 < x < 5$ (A1) (A1)   [2 marks]

Total [17 marks]

4 a   area PVR $= \dfrac{1}{2} \times 45 \times 60 \times \sin 75$ (M1) (A1)

         $= 1303.99\ldots$ (A1)

         $= 1304$ (A1) (G3)   [4 marks]

b $x^2 = 45^2 + 60^2 - 2 \times 45 \times 60 \times \cos 75$ (M1) (A1)

    $x = 65.0$ km   (A1) (G2)   [3 marks]

c $\dfrac{\sin 75}{65.01...} = \dfrac{\sin \alpha}{60}$ (M1) (A1)

    $\alpha = 63.0$ (allow 63.1 if 65 is used)   (A1) (G2)   [3 marks]

d $MR = \dfrac{60 \sin 75}{2}$ (M1) (A1) (M1)

      $= 28.97...$ (A1)

      $= 29$ km   (AG)   [4 marks]

e   volume $= 150^2 \times 2.85$ (M1)

      $= 64125$ m³ $(= 64100$m³$)$ (A1) (G2)   [2 marks]

> The answer includes the units.

f 64125×1.25 (M1)

    $= 80156.25$ (A1)(G1)   [2 marks]

g $\dfrac{64125 \times 1000}{3}$ (M1)

     $= 21375000 \ (= 21400000)$ (A1) (G2)   [2 marks]

> Answer should be consistent with part **g**.

h $2.1375 \times 10^7 \ (2.14 \times 10^7)$ (A1)(A1)   [2 marks]

                       Total [22 marks]

5 a

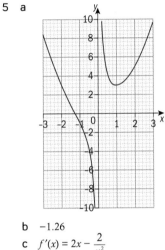

                  (A1) (A1) (A1) (A1)   [4 marks]

b -1.26 (G1)   [1 mark]

c $f'(x) = 2x - \dfrac{2}{x^2}$ (A1) (A1) (A1)   [3 marks]

d $f'(x) = 0$ (M1)

    $2x - \dfrac{2}{x^2} = 0$

    $2x = \dfrac{2}{x^2}$ (M1)

    $x^3 = 1$ (A1)

    $x = 1$ (AG)   [3 marks]

e   3   (G1)   [1 mark]

f   the curve is increasing (or equivalent)   (A2)   [2 marks]

g   **i** $2(-2) - \dfrac{2}{(-2)^2}$ (M1)

        $= -4.5$ (A1) (G2)

    ii $y = -4.5x - 6$ (A1) (A1)

        $4.5x + y + 6 = 0$ (or any multiple)   (A1) (G3)   [5 marks]

h   use of their $(0, -6)$ (M1)

    $\sqrt{(1-0)^2 + (3+6)^2}$ (M1)

    $= 9.06 \left(\text{or } \sqrt{82}\right)$ (A1) (G3)   [3 marks]

                       Total [22 marks]

Subject index

real numbers, 38
  set of, 9–11
real-life situations, graphs of, 189–92
record keeping, projects, 464
rectangles, 536
regression lines, 228–32, 250
regular polygons, properties, 115
regular tetrahedrons, 436
relationships, numerical, 93, 135
Rényi, Alfréd (1921–70), 452
representative samples, 45
results interpretation, in projects,
  459–60
rhombuses, 116–17, 536
  properties, 114
right cones, 424
right prisms, 422, 423, 450
right pyramids, 423, 450
right-angled triangles, 535
  finding angles of, 110–13
  finding in other shapes, 113–17
  finding sides of, 107–10
  trigonometric ratios, 103–19
roots, of equations, 157, 198
roulette, 379
rounding, 12–17
  decimal places, 13–15
  decimals, 13–15, 38
  GDCs and, limitations, 472
  rules, 12, 38
  significant figures, 15–18, 39
Russell, Bertrand (1872–1970), 200

sample space diagrams, 364–6
samples, 45–7
  biased, 45
  random, 45
  representative, 45
satisfying, equations, 95
scalene triangles, 535
scatter diagrams, 203
  on GDCs, 502–6
  using Data & Statistics pages, 502–4
  using Graphs pages, 505–6
Schrödinger, Erwin (1887–1961), 201
second sets, 134, 198
secondary sources, 457
selling rates, 310
sequences, 295
  Fibonacci, 295, 326–7
  of numbers, 296, 325
  see also arithmetic sequences; geometric
    sequences
set builder notation, 331
set theory, basic, 331–4
sets, 328–79
  empty, 332
  finite, 332
  first, 134, 198
  infinite, 332
  of integers, 5–6
  number, 3–11, 38
  of rational numbers, 6–9
  of real numbers, 9–11
  second, 134, 198
  three, 343–5
  universal, 334–7
  see also subsets
shapes, two-dimensional, 114–15,
  535–6

shortest distance, 118
SI *see* Système international d'unités (SI)
sides, finding, of right-angled triangles,
  107–10
significance, levels of, 234
significant figures, 38–9
  rounding, 15–18, 39
  GDCs and, 472
similar triangles, 103
simple discrete data, 47–8
simple statements, compound statements
  from, 397–401, 416
simultaneous equations, 530
  linear models involving, 151–2
  solving
    elimination method, 530
    on GDCs, 484–6
    graphical method, 529
    substitution method, 529–30
simultaneous linear equations, 529–30
  solving, GDCs and, 469–70
sine ratio, 103–19, 129
sine rule, 119–21, 129
slant height, 424
SOHCAHTOA, 104
solids
  distance between points in, 426–9
  surface area, 438–40
  see also three-dimensional solids
sources
  acknowledging, 464
  secondary, 457
specific values, 44
spheres, 424
  surface area, 439, 450
  volume, 444, 451
spirals, Fibonacci, 327
square metres, 26
square roots, rational numbers, 9
squares, 536
  rational numbers, 9
Sridhara (*c*.870–*c*.930), 157
standard deviation, 74–7
standard form, 22–5, 39
  GDCs and, 471–2
statements, 382–3
  brackets in, 399–401
  conditional, 407–12
  contrapositive of, 409–11
  converse of, 409–11
  inverse of, 409–11
  simple, compound statements from,
    397–401, 416
  structure, 396
  true or false, 382, 415
  see also compound statements; direct
    statements
stationary points *see* turning points
statistical graphs, 541–3
statistics, 541–3
  applications, 202–53
    GDCs and, 500–8
  calculating, on GDCs, 478–81
  morals and, 84–5
  summary, 478
  using, on GDCs, 481
  see also descriptive statistics
straight lines, 535
  equations of, 95–103
strong negative correlations, 225

strong positive correlations, 226
structure, in projects, 461–2
subsets, 335
  proper, 335
substitution method, 529–30
Sulba Sutras, 533
sum
  integers, 6
  natural numbers, 4
summary statistics, 478
surface area
  cones, 439, 451
  cylinders, 438, 450
  solids, 438–40
  spheres, 439, 450
  three-dimensional solids, 450–1
surfaces, 422
symbols, and compound statements,
  383–5, 415
symmetric difference, 391–2
symmetry, axes of, 153, 198
Système international d'unités (SI)
  prefixes, 26–7
  units, 25–34, 39
    base, 25–6, 39
    derived, 26, 39
    non-SI in, 31–2

tables
  on GDCs, 487–8, 490
  see also contingency tables; frequency
    tables; truth tables
tails, 352
tally charts, 47–8
tangent ratio, 103–19, 129
tangents, 255, 257–9
  equations of, 258
  gradients of, 273
  to curves, 257–8, 271–5, 291
    on GDCs, 509–10
tautologies, 396–7, 405, 416, 417
temperature, 33–4
terminology, in projects, 462
terms, 296, 325
  in arithmetic sequences, 299–302
  in geometric sequences, 304–8
testing validity, 402
tetrahedrons, regular, 436
thinking, logical, 382
three-dimensional solids
  geometry of, 422–5, 450
  surface area, 450–1
  volume, 441–7, 451
topics, choosing, 465–7
transformations, quadratic function
  modeling, 496–7
trapeziums, 536
  properties, 115
tree diagrams, 367–71
triangles, 535
  area, 124–6
  equilateral, 535
  isosceles, 535
  properties, 114
  scalene, 535
  similar, 103
  see also right-angled triangles
trigonometric ratios, 103–19, 129
trigonometry, 86–131, 420–53
  GDCs and, 482–6